Analytical Methods
for
Coal and Coal Products

Volume III

CONTRIBUTORS

A. ATTAR

ALFRED B. CAREL

G. L. FISHER

JONATHAN S. FRUCHTER

B. C. GERSTEIN

RON GOOLEY

JAMES A. GUIN

MYNARD C. HAMMING

RAY L. HANSON

G. P. HUFFMAN

F. E. HUGGINS

S. BRUCE KING

RICHARD G. LUTHY

ROBERT A. MAGEE

DAN P. MANKA

D. F. S. NATUSCH

MICHAEL R. PETERSEN

JOHN W. PRATHER

FREDERICK J. RADD

ROBERT RAYMOND, JR.

S. M. RIMMER

S. J. RUSSELL

E. STEINNES

ARTHUR R. TARRER

SAMUEL P. TUCKER

N. E. VANDERBORGH

ALEXIS VOLBORTH

KARL S. VORRES

RICHARD W. WALTERS

S. ST. J. WARNE

Analytical Methods
for
Coal and Coal Products

Edited by CLARENCE KARR, JR.
U.S. Department of Energy
Morgantown Energy Technology Center
Morgantown, West Virginia

Volume III

ACADEMIC PRESS
A Subsidiary of Harcourt Brace Jovanovich, Publishers

New York   London   Toronto   Sydney   San Francisco   1979

ACADEMIC PRESS, INC.
111 Fifth Avenue, New York, New York 10003

United Kingdom Edition published by
ACADEMIC PRESS, INC. (LONDON) LTD.
24/28 Oval Road, London NW1 7DX

Library of Congress Cataloging in Publication Data
Main entry under title:

Analytical methods for coal and coal products.

Includes bibliographies.
1. Coal--Analysis. I. Karr, Clarence.
TP325.A58   662'.622   78–4928
ISBN 0–12–399903–0

PRINTED IN THE UNITED STATES OF AMERICA

79 80 81 82   9 8 7 6 5 4 3 2 1

Contents

Part IX   ANALYSIS OF GASES

Chapter 38   Coke Oven Gas Analysis
Dan P. Manka

Chapter 39   Characterization of Recovered Volatiles from a High Methane Coal and the Significance of These Findings
Mynard C. Hamming, Frederick J. Radd, and Alfred B. Carel

Chapter 40   Characterization of Coals Using Laser Pyrolysis—Gas Chromatography
Ray L. Hanson and N. E. Vanderborgh

Chapter 45   Environmental Characterization of Products and Effluents from Coal Conversion Processes
Jonathan S. Fruchter and Michael R. Petersen

Part XI   SPECIAL INSTRUMENTAL TECHNIQUES FOR ANALYSIS OF COAL AND ITS PRODUCTS

Chapter 46   Instrumental Activation Analysis of Coal and Coal Ash with Thermal and Epithermal Neutrons
E. Steinnes

Chapter 47   Fast-Neutron Activation Analysis for Oxygen, Nitrogen, and Silicon in Coal, Coal Ash, and Related Products
Alexis Volborth

Chapter 56 **Sulfur Groups in Coal and Their Determinations**
A. Attar

List of Contributors

Numbers in parentheses indicate the pages on which the authors' contributions begin.

A. ATTAR (585), Chemical Engineering Department, University of Houston, Houston, Texas 77004

ALFRED B. CAREL (29), Research and Development Department, Continental Oil Company, Ponca City, Oklahoma 74601

G. L. FISHER (489), Radiobiology Laboratory, University of California, Davis, California 95616

JONATHAN S. FRUCHTER (247), Physical Sciences Department, Battelle Pacific Northwest Laboratory, Richland, Washington 99352

B. C. GERSTEIN (425), Ames Laboratory, U.S. Department of Energy and Department of Chemistry, Iowa State University, Ames, Iowa 50011

RON GOOLEY (337), Los Alamos Scientific Laboratory, Geological Research Group, Los Alamos, New Mexico 87545

JAMES A. GUIN (357), Department of Chemical Engineering, Auburn University, Auburn, Alabama 36830

MYNARD C. HAMMING (29), Research and Development Department, Continental Oil Company, Ponca City, Oklahoma 74601

RAY L. HANSON (73), Lovelace Biomedical and Environmental Research Institute, Albuquerque, New Mexico 87115

G. P. HUFFMAN (371), U.S. Steel Corporation, Research Laboratory, Monroeville, Pennsylvania 15146

F. E. HUGGINS (371), U.S. Steel Corporation, Research Laboratory, Monroeville, Pennsylvania 15146

S. BRUCE KING* (105), Department of Energy, Laramie, Wyoming 82071

RICHARD G. LUTHY (189), Department of Civil Engineering, Center for Energy and Environmental Studies, Carnegie–Mellon University, Pittsburgh, Pennslyvania 15213

ROBERT A. MAGEE (105), Radian Corporation, Austin, Texas 78766

* Present address: World Energy Incorporated, Laramie, Wyoming 82070.

DAN P. MANKA (3),   Consultant, 1109 Lancaster Avenue, Pittsburgh, Pennsylvania 15218

D. F. S. NATUSCH (489),   Department of Chemistry, Colorado State University, Fort Collins, Colorado 80523

MICHAEL R. PETERSEN (247),   Physical Sciences Department, Battelle Pacific Northwest Laboratory, Richland, Washington 99352

JOHN W. PRATHER (357),   CIBA-GEIGY Corporation, McIntosh, Alabama 36553

FREDERICK J. RADD (29),   Research and Development Department, Continental Oil Company, Ponca City, Oklahoma 74601

ROBERT RAYMOND, JR. (337),   Los Alamos Scientific Laboratory, Geological Research Group, Los Alamos, New Mexico 87545

S. M. RIMMER (133),   Illinois State Geological Survey, Urbana, Illinois 61801

S. J. RUSSELL (133),   Illinois State Geological Survey, Urbana, Illinois 61801

E. STEINNES (279),   Institutt for Atomenergi, Isotope Laboratories, 2007 Kjeller, Norway

ARTHUR R. TARRER (357),   Department of Chemical Engineering, Auburn University, Auburn, Alabama 36830

SAMUEL P. TUCKER (163),   National Institute for Occupational Safety and Health, Robert A. Taft Laboratories, Cincinnati, Ohio 45226

N. E. VANDERBORGH (73),   Los Alamos Scientific Laboratory, Los Almos, New Mexico 87545

ALEXIS VOLBORTH* (303, 543),   Nuclear Radiation Center and Department of Geology, Washington State University, Pullman, Washington 99163

KARL S. VORRES (481),   Institute of Gas Technology, Chicago, Illinois 60616

RICHARD W. WALTERS (189),   Department of Civil Engineering, Center for Energy and Environmental Studies, Carnegie–Mellon University, Pittsburgh, Pennsylvania 15213

S. ST. J. WARNE (447),   Department of Geology, The University of Newcastle, Shortland, New South Wales 2308, Australia

* Former address: Chemistry Department, University of California, Irvine; and Chemistry and Geology Departments, North Dakota State University, Fargo, North Dakota.

Preface

The world reserves of coal are larger than those of petroleum and natural gas, and the production capacity for coal currently exceeds that for oil shale, tar sands, peat, and the other less developed fossil energy reserves. As a result, coal is the projected resource alternative for energy and materials well into the next century. With this increasing importance of coal there is also an increasing need to meet improved environmental standards and to supply economic uses for by-products. This is seen in the current work on modifications of combustion and ancillary technologies for the generation of electric power without deterioration of the environment, and the carbonization of unfamiliar as well as traditional coals for the manufacture of metallurgical coke with controlled emissions. In addition, there are the emerging technologies for commercial production of synthetic liquid and gaseous fuels from coal that present special problems in their development, as well as the need to meet restrictions in air and water pollution and to utilize wastes.

A major key for solving these problems in the use of unfamiliar coals, the development of new coal conversion processes, the marketing of new solid, liquid, and gaseous fuels, the utilization of by-products, and the control of emissions is the use of appropriate analytical methods. The goal of these three volumes is to supply, insofar as feasible, a detailed presentation of what constitutes the first comprehensive reference work devoted exclusively to the subject of analytical methodology for coal and coal products. I have divided these volumes into a total of twelve parts, each part containing several chapters devoted to a particular subject. Some parts are based on major processes or products, such as liquefaction (Volume I), carbonization (coke, pitch) and combustion (Volume II), and gases (Volume III). However, the complex analytical problems involved are generally not limited to any specific process, and there are many problems that are held in common. Thus Volume III also includes discussion of waste products, by-products, environmental problems, and miscellaneous analytical problems, as well as special instrumental techniques for solving various problems.

Because different aspects of a particular subject are frequently scattered through various chapters in the volumes, cross-references between chapters have been entered. In addition, the subject indexes have been made as detailed as was practical, and the reader will benefit from examination of pertinent subjects in the indexes of all three volumes. Many individual subjects are located in the indexes of two or three volumes. The multiauthorship of the chapters has permitted a diversity of viewpoints and opinions on various analytical problems. A careful reading of these volumes will show that definitive solutions are not yet available in a number of instances. This is not an unexpected situation because coal and some of its products may well comprise the most difficult materials to analyze of all the world's major resources. There is therefore a clear need for continued research on the fundamentals of analysis of coal and coal products, and the development of reliable and accurate analytical instrumentation, including on-stream applications.

I am deeply indebted to all the many experts, both in the United States and abroad, who have made the publication of this multivolume reference work possible, and to the organizations involved for their generous cooperation. The interested reader can find more about the circumstances and details of the preparation of this work in the Prefaces to Volumes I and II.

Contents of Volumes I and II

VOLUME I

Analytical Methods
for
Coal and Coal Products

Volume III

Part IX

ANALYSIS OF GASES

Chapter 38

Coke Oven Gas Analysis

Dan P. Manka

PITTSBURGH, PENNSYLVANIA

I. INTRODUCTION

Major volumes of coke oven gas are generated by the carbonization of coal for the production of blast furnace coke. Typically, 80 million ft³ of gas is produced daily as a result of the coking of 7500 tons of coal. Approximately 11,000 ft³ of dry gas is generated per ton of typical coking coal. The volume of gas is dependent on the volatile matter of the coal mixture charged to the coke ovens. An additional volume is present from the volatilization of the moisture present in the charged

3

coal. Although a major portion of this moisture is condensed, coke oven gas is saturated with water at the gas temperature prevailing throughout the chemical products recovery system.

The gas is treated for the removal of tar, ammonia, naphthalene, and aromatics in the recovery system. A portion of the treated gas flows to the coke batteries for underfiring of the coke ovens to supply heat for the carbonization of the coal in the ovens. The major portion of the treated gas flows to the steel producing mills where it serves as fuel for heating many furnaces.

Methods of analysis for the major chemical products in the coke oven gas† are described in this chapter, beginning with the analysis of fixed gases flowing from the top of the coke oven to the analysis of sulfur compounds and hydrogen cyanide in the treated gas. The latter is important in the desulfurization of coke oven gas in order to meet EPA regulations for concentrations of inorganic and organic sulfur compounds in the treated gas prior to its use as a fuel.

Because space is limited in this chapter for a complete description of the analytical methods, it is assumed that the analyst is familiar with standard methods of analysis and with the operation of a gas chromatograph. It is of interest that all of the coke oven gas components discussed in this chapter are also included in the analysis of stack gas for environmental assessment (Volume II, Chapter 36).

II. COKE OVEN GAS FLOW DIAGRAM

The flow diagram of gas is practically the same in all coke plants. As more desulfurization processes are installed, these will vary in location depending on the type of system being installed.

In the flow diagram in Fig. 1, coke oven gas rises from the coke oven a, through standpipe b, to gooseneck c, where it is contacted with flushing liquor (ammonia liquor). Tar and moisture are condensed. Ammonia chloride, and a portion of the ammonia, fixed gases, hydrogen cyanide, and hydrogen sulfide are dissolved by the liquor. The gas, liquor, and tar enter the gas collecting main d, which is connected to all the ovens of a battery. In some cases there may be two gas collecting mains to a battery. The gas, liquor, and tar are separated in tar decanters. The tar f, separated from the liquor e, flows to tar storage g. A portion of the liquor e is pumped to the gooseneck c on the top of each oven. The remainder of the liquor is pumped to the ammonia liquor still h, where it is contacted with live steam to drive off free

† Analyses of coke oven effluents for polynuclear aromatics are given in Chapter 43.

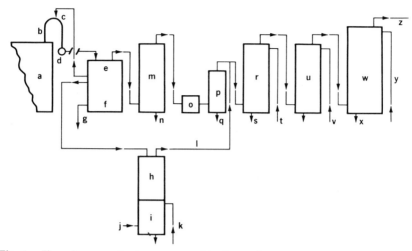

Fig. 1. Flow diagram of coke oven gas. The lettered items are identified in the text.

ammonia, fixed gases, hydrogen cyanide, and hydrogen sulfide. As the liquor flows from the free still h to the fixed still i, it is contacted with lime or sodium hydroxide k to liberate free ammonia from ammonium chloride. Live steam, admitted at j flows up through the fixed and free stills, and the ammonia, fixed gases, hydrogen cyanide, and hydrogen sulfide are added through l to the main coke oven gas stream ahead of the ammonia saturator r.

The coke oven gas, separated from liquor and tar in e, is cooled indirectly with water in the primary coolers m. The fine tar that separates from the gas is pumped through n to the tar storage tank. The cooled gas is pumped by exhausters o to the electrostatic precipitators p, where additional fine tar is condensed and pumped through q to the tar storage tanks. The gas is contacted with a dilute solution of sulfuric acid in the ammonia saturator r to remove free ammonia. The ammonium sulfate-laden acid flows through s to the ammonia crystallizer (not shown) where crystals of ammonium sulfate are separated and the remaining sulfuric acid is pumped back to the ammonia saturator through t.

The ammonia-free gas flows to the final coolers u, where it is further cooled by direct water contact. The water plus condensed naphthalene flows from the cooler through tar which absorbs the naphthalene. The water is cooled and recirculated into the final cooler through v.

The cooled gas enters the wash oil scrubbers w, also known as benzole scrubbers, where it contacts wash oil, a petroleum oil, pumped into the scrubbers through y. The aliphatic and aromatic compounds are extracted from the gas by the wash oil. The principal components are benzene, toluene, xylenes, indene, and solvent, also known collectively as light oil. The benzolized wash oil is pumped through x to the wash oil still (not shown) where live steam strips out the light oil compounds. The debenzolized wash oil is cooled and returned to the wash oil scrubber. In some plants the light oil is further processed and fractionated into benzene, toluene, and xylenes and into a high-boiling solvent fraction. Naphthalene is also present in the light oil. Plants with low volumes of light oil do not have facilities for refining, therefore, the oil is sold to large refineries.

The gas from the wash oil scrubbers flows through z to a gas holder which tends to equalize the pressure. Booster pumps distribute a portion of the gas for underfiring of the coke ovens, but the major portion flows to the steel plant where it is used as a fuel in the many furnaces.

III. DETERMINING END OF COKING CYCLE BY GAS ANALYSIS

The length of a coking cycle for carbonizing of coal in an oven is determined from experience and the range of the flue temperature. Most coke plants also have experimental coke ovens to determine coking cycles for variable coal mixes and at different flue temperatures. Another method to determine the end of this cycle is by analysis of the gas flowing from the oven into the standpipe. This section will describe this approach.

A. Sampling System

The gas sample from the standpipe must be cooled, separated from tar and water, and filtered before it is analyzed for H_2, O_2, N_2, CH_4, CO, CO_2, and illuminants.

The sampling train in Fig. 2 has been used successfully for the continuous pumping of gas to a sample bottle or to a gas chromatograph. It is designed to operate for several hours so that many samples can be analyzed before there is an accumulation of tar in probe c or in the separators d.

Standpipe b is located on the gas discharge end on top of coke oven a. Normally, there is an opening on the side of the standpipe where live steam is admitted to remove accumulated tar in the pipe. This opening is ideally located for gas sampling since it is generally about 2–3 ft above the oven.

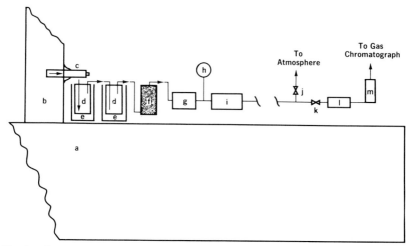

Fig. 2. Sampling gas from top of coke oven. The lettered items are identified in the text.

The probe c, extending to approximately the center of the standpipe, is a $\frac{3}{4}$- or 1-in. heavy wall stainless steel pipe. The end of the pipe in the gas stream is sealed, and the other end has a plug for removing accumulated tar. A $\frac{1}{4}$ in. by approximately 3 in. slot is cut along the length of the pipe in the gas stream extending from the closed end. When the probe is inserted into the standpipe, *it is important* that the slot face the direction of gas flow. This position decreases the amount of tar pumped in with the sample gas. The probe is inserted into a plug similar to the one used for the steam line and is pushed through the opening in the standpipe until the plug seals the opening. This insertion should be done rapidly to prevent excessive flow of coke oven gas through the opening. The probe extending outside of the pipe and the $\frac{1}{4}$ in. pipe from the probe to the first separator d should be insulated.

The two separators d for cooling the gas and for separation of tar and water are identical. These are made from 4 in. pipe, 10- to 12-in. long, one end sealed, and a flange with a gasket on the top end. The inlet pipe extends approximately half-way into the separator. Both separators are kept in a bucket of water e. The separators should be pressure tested before they are used. The remaining piping is $\frac{1}{4}$ in. copper tubing with Swaglock fittings. Although most of the tar is condensed in the separators, the gas is drawn through glass wool in a glass tube f to remove the lighter tar oil. This glass tube can be a glass bottle normally used in the laboratory to dry gases with Drierite. It has an inlet connection on the bottom, an outlet connection near the top, and a metal screw cap with a seal on the top. Tygon tubing is used for the metal tube to glass connection. The gas flows through a final filter g that removes submicron particles, such as the Pall Trinity "Junior Size" Epocel cartridge. The gas is drawn from the standpipe by a peristaltic pump i. Good results are obtained with this type of pump using a $\frac{1}{2}$- or $\frac{5}{8}$-in.-diameter plastic tube for conveying the gas. The tubing is easily replaced if it becomes broken or contaminated. A vacuum gauge h ahead of the pump is valuable to detect plugging of lines or separators. When vacuum reaches 20 in., it is most probable that tar is accumulating in the probe. A rod inserted into the probe through the plug opening of the probe will generally be sufficient to reopen the pipe.

Gas from the pump passes through valve k and is dried in tube l containing Drierite. The flow to the sample loop of the chromatograph is maintained at 50 cm³/min as indicated by rotometer m. Excess sample gas is vented to atmosphere through valve j.

B.   Orsat Analysis

If a portable gas chromatograph is not available, a gas sampling bottle can be filled with the dried gas after rotometer m. Analysis of hydrogen, oxygen, carbon dioxide, carbon monoxide, methane, and illuminants concentrations in the gas can be determined on an Orsat equipped with combustion apparatus to determine hydrogen and methane. However, the analysis time is lengthy so that the number of analyses made near the end of the coking cycle are too limited to obtain a true picture of the coking cycle end point.

C.   Gas Chromatographic Analysis

The use of a portable gas chromatograph located in a sheltered area near the sampling system is the best method to follow the course of the coking cycle.

The curves in Fig. 3 are based on the analytical results obtained on a gas chromatograph with a thermal conductivity detector. The separation column is a $\frac{1}{8}$ in. × 10 ft stainless steel tube containing No. 5A Molecular Sieve, and the reference column is a $\frac{1}{8}$ in. × 67 in. stainless steel tube containing Porapak Q. The operating conditions are argon flow 12 cm³/min to each column, 100 mA cell current, column and cell temperature is 40°C. The chromatograph is standardized with a gas containing 50% H_2, 30% CH_4, 10% CO, 5% CO_2, and N_2 the balance.

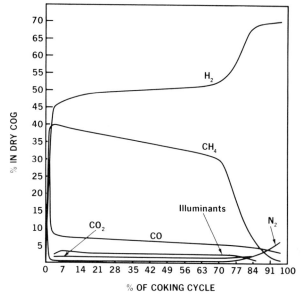

Fig. 3. Composition of coke oven gas during coking cycle.

The standard gas is admitted into the sample line ahead of the Drierite in l and the flow rate maintained through the sample loop the same as for coke oven gas. The sample loop capacity is 0.5 cm³ and is dependent on the sensitivity of the detector using argon as carrier gas.

If the detector has insufficient sensitivity for CO and CH_4, a carrier gas containing 8% hydrogen and 92% helium may be used at a cell current of 150 mA.

Concentrations of CO_2 and illuminants, as ethylene, can be obtained on the Porapak column. The chromatograph must also be equipped with a sample valve for this column and a polarity switch so that the peaks are positive on the chart paper of the recorder. However, CO_2 and illuminants are not necessary to determine the end point; therefore, these can be ignored or periodically determined by the Orsat method. A chromatogram of these same gas components, but in much different proportions from underground gasification of coal, is shown in Chapter 41, Fig. 3.

D. Results

The plot of the analytical results in Fig. 3 are those for coking dry coal for 14 hr at a flue temperature of 2300°F. Concentration of hydrogen approaches a maximum of over 70%, and the methane approaches the minimum of less than 1% near the end of the coking cycle. Generally, coking is continued for an additional 30 min after these analytical values are reached, and then the coke is pushed from the oven.

Similar curves are reported (Beckmann *et al.*, 1962) for the coking of wet coals. However, the coking cycle is 18–19 hr because a large amount of water must be evaporated from the wet coal charged to the ovens.

IV. AMMONIA

Ammonia is recovered from coke oven gas because it is marketable and it is deleterious to the heating system if it remains in the gas used as fuel in the steel plant.

This section describes analysis of ammonia in coke oven gas entering the ammonia saturators by wet chemical methods and by gas chromatography.†

In the flow diagram in Fig. 1, coke oven gas containing ammonia enters the saturator r, where it is contacted with a dilute solution of sulfuric acid which recovers the ammonia as ammonium sulfate.

A. Wet Chemical Methods

Most wet chemical methods are similar; therefore, one method will be described in more detail.

1. *Sampling System*

The $\frac{1}{4}$ in. sampling line, preferably stainless steel, in Fig. 4 is extended horizontally to the center of the gas main which carries the gas to the ammonia saturator. Either the end of the probe in the gas main is bent to a 90° angle, or an elbow is attached to the end. The probe a should be inserted so that the bent end, or the elbow, faces in the direction of gas flow. This method reduces the amount of naphthalene included with the gas sample. A $\frac{1}{4}$ in. valve b is attached to the end of the probe located outside of the gas main.

The pipe is extended from the valve to two traps, c and d in series, to collect entrained and condensed water. The traps are glass bottles about 10-in. long which have neck openings large enough for a neoprene rubber stopper fitted with two $\frac{1}{4}$ in. glass tubes. Two glass bottles

---

† Analysis of coke plant ammonia still effluent and waste liquor is given in Chapter 44, Table VII.

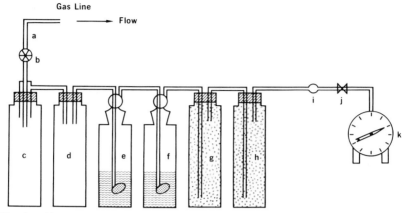

Fig. 4. Absorption apparatus for ammonia in coke oven gas. The lettered items are identified in the text.

with ground-glass heads also serve well as traps. Short connections between glass tubing and pipe can be made with Tygon tubing.

The traps are connected by means of glass tubing and short Tygon tubing to two gas washing bottles in series, e and f (Ace Glass No. 7162, 500 mliter capacity, porosity A) containing sulfuric acid solution. They connect to empty trap g for collecting any entrained sulfuric acid solution. Finally, they connect to trap h filled with iron oxide shavings to remove hydrogen sulfide. The inlet tube in this trap extends to within 2 in. of the bottom of the bottle. The outlet tube is connected to a tube i, loosely filled with glass wool, then to a needle valve j (Fisher No. 14-630-8A). The end of the valve is connected to a wet gas meter k with Tygon tubing.

2. Procedure

Add 140 mliter of 10% sulfuric acid and a drop of methyl orange to each of the two gas washing bottles. Assemble the apparatus, except the probe, according to the description. Open the valve on the probe to blow out accumulated water. Then connect the probe end to the first trap. The valve should be fully open. Make certain that the meter is level and that the indicating dials are set to zero.

Pass gas through the apparatus at the rate of 3-6 ft³/hr. Record the time, temperature of the meter thermometer, meter reading, and barometric pressure each hour. Pass 15 ft³ of gas through the apparatus depending on the expected ammonia content of the gas. Use the needle valve to control gas flow.

Combine the contents of the two gas washing bottles and the trap g into a 1 liter Florence flask. Rinse each bottle three times and add the washings to the flask. Neutralize the contents with about 125 mliter of 50% sodium hydroxide solution. Distill through a Kjeldahl trap and bubble the vapors into 50 mliter of N/1 sulfuric acid solution in a 500 mliter Erhlenmeyer flask in a water-cooled trough. Distill until there are about 400 mliter of liquid condensate. Transfer the liquid into a 600 mliter beaker, rinse the flask three times with distilled water, and add the washings to the beaker.

Back titrate the acid in the beaker with N/1 sodium hydroxide using methyl orange indicator.

Measure the amount of condensate in the first two traps, c and d. Wash out the traps several times with distilled water and add the washings to the condensate. Add 18% sodium hydroxide solution and determine the ammonia content as above.

3.   Calculations

Calculate the average meter temperature and barometric pressure during the absorption test. Correct the meter reading to standard conditions of 60°F and 30 in. Hg pressure.

a. NH_3 Calculations for Absorption Train

(1)   Milliliters of H_2SO_4 × normality − milliliters of NaOH × normality = net milliliters of N/1 H_2SO_4 combined with NH_3.

(2)   Net milliliters of H_2SO_4 × 0.017 = grams of NH_3.

(3) $\dfrac{\text{grams of } NH_3}{\text{corrected number of cubic feet used in test}}$ = grams of NH_3/cubic foot.

(4) $\dfrac{\text{grams of } NH_3/\text{cubic foot} \times 1,000,000}{453.6}$ = pounds of NH_3/1,000,000 ft³.

b. NH_3 Calculations for Condensate

(1)   Net milliliters of 1N H_2SO_4 combined with NH_3 × 0.017 = grams of ammonia in condensate.

(2) $\dfrac{\text{grams } NH_3 \text{ in condensate} \times 1,000,000}{\text{corrected cubic feet of gas in test} \times 453.6}$ = pounds of NH_3/1,000,000 ft³.

If only the total ammonia content is desired, the condensates in the first two traps are added to the sulfuric acid solutions from the two gas washing bottles and all the ammonia determined together. This procedure eliminates one analysis.

4. Other Methods

In other procedures, Kagasov and Zharkova (1969) titrate the acid washings with NaOH using methyl orange as indicator for the ammonia content and phenolphthalein as indicator for the pyridine base content. Ammonia is determined by difference.

B. Gas Chromatographic Analysis

Castello and Riccio (1975) analyze chromatographically for H_2S and NH_3 in a gas sampled ahead of the plant water scrubbing towers which remove hydrogen sulfide and ammonia. The sample is transferred to the laboratory in gas-tight glass bottles of about 250 mliter capacity, with stopcocks at both ends to purge the bottle with the sample gas. The analysis is made in a chromatograph with a thermal conductivity detector, a $\frac{1}{4} \times 80$ in. stainless steel column packed with Porapak Q 80/100 mesh, 60 mliter/min helium flow rate, and operated at 60°C. Because the retention times of the two compounds are very close, and the peak for water can obscure them, two samples are analyzed: the first after elimination of water and ammonia with anhydrone; the second after elimination of water and hydrogen sulfide with a filter containing solid sodium hydroxide. The two filters are alternately connected between the sample bottle and the sampling valve in the chromatographic system. The first is the analysis for hydrogen sulfide, and the second is the analysis for ammonia. The total time of the two determinations is about 13 min. Standard gases are used for calibration.

V. NAPHTHALENE

Naphthalene, which had not condensed with tar in the gas collecting main, d in Fig. 1, is carried with the coke oven gas throughout the by-product recovery system. A portion of the naphthalene is recovered in the final coolers where the gas is cooled by direct contact with water and in the benzole scrubbers where the gas is contacted directly with wash oil, u and w, respectively, in Fig. 1. Naphthalene recovery is determined by analysis of the gas stream entering the final cooler tower and the inlet and outlet of the benzole scrubbers.

A.   Wet Chemical Method

The probe for the sampling system is identical to that in Fig. 4. The gas is bubbled through two small wash bottles containing 150 cm^3 of 15% and 75 cm^3 of 75% sulfuric acid, respectively. Then the gas is bubbled through 75 cm^3 of 20% sodium hydroxide, passed into a trap, then into two wash bottles, each containing 75 cm^3 of 1.0% solution of picric acid. The two wash bottles are immersed in ice water. The gas flows to a trap and is then dried in a small tube containing Drierite before it is measured in a wet gas meter. A fine needle valve ahead of the meter is used to regulate the gas flow at 3 ft^3/hr. Experiments should be run on the individual gas stream to determine the total volume of gas required for the test. Generally, 3-6 ft^3 of gas is sufficient for high concentrations of naphthalene.

The crystals of the naphthalene picrate in the picric acid solution are filtered through a fine Gooch crucible, washed with distilled water followed by iodometric titration of the solution to the sharp color change from green to yellow.

Castello and Riccio (1975) measure the naphthalene content by bubbling the gas, purified from ammonia, into a solution of picric acid. The concentration of the picric acid–naphthalene complex is determined by titration with sodium hydroxide to the phenolphthalein end point. The concentration is calculated on the corrected cubic feet of gas passed through the apparatus.

In both methods, naphthalene may crystallize out ahead of the picric acid absorbers, particularly the gas sampled ahead of the final coolers. If this happens, the gas should be kept warm, at least to the temperature of the gas main in the range of 40°-45°C up to the picric acid washer.

B.   Gas Chromatographic Method

The probe for the gas sampling system is similar to that in Fig. 4. The gas is bubbled through three scrubbers in series with each scrubber containing 50-100 cm^3 of benzene, depending on the concentration of naphthalene in the gas. Following the last scrubber, the gas enters a trap which is immersed in an ice bath to condense any benzene remaining in the gas. The gas is measured by a gas meter at a flow rate of 3 ft^3/hr. Generally, 6 ft^3 of gas is sufficient for a test. The first benzene scrubber should be located as close as possible to the valve on the end of the probe so that naphthalene does not crystallize in this short section from the gas main to the scrubber.

The gas volume in the gas meter is corrected as in Section IV,A,3.

The benzene solutions are combined, separated from water, measured, and a portion dried with Drierite prior to analyses by gas chromatography. A 2 μliter sample is separated on a $\frac{1}{8}$ in. × 13 ft stainless steel column filled with 80/100 mesh Chromasorb PAW coated with 10% *tris* (cyanoethoxy) propane. The column is temperature programmed at 10°/min from 80° to 150°C and held at 150° through the elution of naphthalene. Cell and injector temperatures are 200°C with helium flow of 15 cm.^{3}/min. This column will also separate any forerunnings, such as aliphatic compounds, carbon disulfide, and cyclopentadiene, which are absorbed from the gas by the benzene solvent. These forerunnings elute before the benzene. Also there will be some toluene and xylenes.

The chromatograph is standardized with a synthetic solution of pure benzene and naphthalene with the naphthalene concentration in the range expected in the benzene extract sample. The peak area of naphthalene in the extract sample is compared to the peak area in the standard solution to determine the concentration in the sample. The volumes of the standard and sample solutions must be identical. Finally, the grams of naphthalene in the benzene extract is divided by the corrected cubic feet of gas extracted by the benzene to obtain grams of naphthalene per cubic foot of coke oven gas. The analyses can also be made using heptane as internal standard.

Pinchugov and Zharkova (1976) absorb naphthalene from coke oven gas with benzene. The benzene solution is chromatographed at 160°C on a column filled with Chromasorb W coated with 10% polyethylene glycol (molecular weight 40,000). Helium is the carrier gas; the detector is flame ionization; and acetophenone is the internal standard.

VI. AROMATICS

Coke oven gas contains a significant volume of aromatics, commonly called light oil, to make recovery economical. The main constituents are benzene, toluene, xylenes, naphthalene, coumarone, and indene. Recovery is made by direct contact of the gas with wash oil, a petroleum absorption oil, in scrubbers, commonly called wash oil or benzole scrubbers. The light oil-laden wash oil, or benzolized wash oil, is heated and the light oil recovered by steam distillation. The resulting debenzolized wash oil is cooled in heat exchangers by the incoming benzolized oil and is recirculated to the scrubbers.

A. Wet Chemical Method

Adsorption of aromatics or light oil from coke oven gas by activated carbon is the most common method to determine the concentration of the oil in the gas.

The schematic of the adsorption train is illustrated in Fig. 5. The probe a is similar to that used for ammonia in Section IV with a shut-off valve b on the line emerging from the duct. The gas is passed through 10% sulfuric acid solution in scrubber c and through 20% sodium hydroxide solution in scrubber d to remove ammonia, hydrogen sulfide, and phenols. Trap e collects any carry-over liquid. The activated carbon adsorption tubes can be installed in separate insulated boxes f and g for the short test or in one large insulated box which will accommodate the tubing for the short test as well as the glass tubes for the long test. The gas flows through iron oxide shavings in h to a needle valve i and finally to a wet test meter j, where it exits at k.

1. Short Test

This test is intended for plant control which, in most cases, is run once or twice per week.

The carbon tubes are $\frac{1}{4}$ in. × 10 ft stainless tubing filled with 35 × 45 mesh activated carbon, such as Barnabey–Cheney KI-1 carbon. Other mesh size that fits into the tube may be used. These either are inserted into separate insulated boxes or are connected together and inserted into one large insulated box.

The lines carrying the gas to the adsorption apparatus must be of metal or glass. Direct contact of the coke oven gas with rubber tubing should be avoided because rubber absorbs aromatics. Two pieces of glass tubing connected with rubber tubing should be butted against each other to minimize the contact of the gas with rubber tubing.

The length of tubing from the probe in the plant gas line to the first trap should be minimal to prevent condensation of naphthalene and

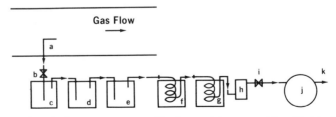

Fig. 5. Light oil adsorption system. The lettered items are identified in the text.

light oil. It is advised that this line be heated at least to the temperature of the gas in the plant line. A temperature 20°–30°F above the gas temperature is advisable.

Ice should be placed in the bottom of the box holding the carbon tubes to keep the temperature of the tubes below that of the outside atmosphere, but insufficient to freeze any water in the gas stream.

Gas is passed through the apparatus for 4–6 hr at the rate of 2.5 ft³/hr. Each hour record the time, temperature of meter, meter reading, meter pressure, atmospheric pressure, and atmospheric temperature as described in Section IV,A,2 under wet chemical methods for ammonia.

At the end of the adsorption period, the carbon tubes are disconnected and ends are plugged and brought to the laboratory for desorption. Each tube is connected separately to a source of low-pressure steam. The end of the tube is connected to a condenser through which flows ice-cold water. The discharge of the condenser is fitted with an adapter so that the water–oil condensate flows into a separatory funnel. Add sodium chloride to the funnel to facilitate separation of water from the oil.

The rate of steaming out the tubes should be regulated so that the outlet condensed steam and oil is not warm to the touch. The separatory funnel, which collects the condensed steam and oil, should be cooled by ice. The steaming should be continued until no additional oil is driven off. The separatory funnel should be replaced with a similar one to determine the end point of oil desorption. Combine all the oil and water. Add additional sodium chloride if necessary. After settling, discard the water layer. Add a very small amount of anhydrous sodium sulfate to dry the oil. Decant the oil and measure the volume. Measure the volume of oil from each tube, then combine them. Weigh the total dry oil.

Reduce the volume of gas passed through the adsorption apparatus to standard conditions. Calculate the grams of oil per cubic foot of gas and convert this to pounds per daily gas production. The daily production in gallons is determined from the pounds produced per day and the gravity of the oil.

After complete steaming of the carbon tubes, they should be thoroughly dried with dry air. Clean air should be pumped through a tube of molecular sieves prior to passing it through the carbon columns. *Never* use plant compressed air for drying because it always contains water and oil.

After the columns are dry, disconnect them from the drying apparatus and cap both ends. Mark the columns No. 1 and No. 2. They should be placed in the same order in the adsorption train. Eventually, the No. 1 column will lose adsorption efficiency because of accumulation of high molecular weight compounds and metallic compounds which cannot be removed by steam. Discard this column and replace it with No. 2 column, which will now become No. 1 column. Add a new carbon column, which will be No. 2 column.

It is very important that all new carbon columns be exposed at least once to coke oven gas followed by complete steaming before the columns are used for this analytical adsorption test.

To determine the efficiency of the benzole scrubbers, it is necessary to perform two tests simultaneously, one on the inlet gas to the scrubbers and the other on the outlet gas from the scrubbers. The inlet gas adsorption is conducted as described above. Adsorption of the scrubber outlet gas is operated in the same manner, but at a faster gas rate, so that at least 100 ft^3 of gas are passed through the adsorption apparatus. The difference in the volume of light oil recovered in both tests indicates the efficiency of the scrubbers.

Some plants may have a large amount of carryover of wash oil or absorption oil in the scrubber outlet gas. This oil will also be adsorbed by the activated carbon and on desorbtion will be included with the light oil. In this case, comparison of the light oil volumes from the two gases should be based on the volumes distilled under identical conditions to a predetermined temperature. This temperature should be less than the initial boiling point of the new dry absorbent oil used in the plant benzole scrubber system. A more exact method is by gas chromatography, which will be described in the next section.

2. Long Test

Tests by the author and by others have shown that the benzene concentration in coke oven gas is not uniform during the entire coking cycle. Much depends on the rate of coking in the various ovens. Therefore, it is important to run the long test at definite intervals to obtain the total volume of light oil in coke oven gas. The short test for a period of 4–6 hr is sufficient for a control test, but to obtain the complete volume of light oil the long test should be used, which requires adsorption by the carbon for a minimum period of 24 hr.

The adsorption apparatus is similar for the long test as for the short test except that the two stainless tubes are replaced by four glass tubes. Each glass tube is 1 in. in diameter and 3 ft in length. The best tubes are those either with ball-joint or standard-taper joints so that all connections can be made with glass rather than with rubber stoppers. Glass wool is inserted into the bottom of each tube, filled with the activated carbon, and sealed in with glass wool. The tubes are connected in series with glass U bends. These are placed into a large insulated box and surrounded by ice to keep the carbon cool during the adsorption cycle.

Coke oven gas is passed through the apparatus for 24 hr at the rate of 2.5 ft³/hr. After adsorption, each tube is desorbed with steam as in the short test. The total dry light oil from the four tubes is collected, measured, and weighed.

A fractionation of the oil can be made through a macrofractionation column. The chaser is Tetralin, which has previously been fractionated to a temperature of 200°C.

The oil may also be fractionated by following ASTM D 850-70 (1975).

All techniques described under short test apply to the procedure for the long test.

B.   Gas Chromatographic Method

Any form of fractionation of the light oil inherently does not show the many aromatic constituents of the oil. At best, it produces a reasonable boiling point curve for the oil. Significant concentrations and types of components are best determined by gas chromatographic separations.

The oil for chromatographic analysis is obtained by the short method or the long method described in the previous section on wet chemical method. The oil is dried, measured, and weighed as in the wet chemical method.

The following columns have been used for this separation and analysis of the light oil constituents:

(1)   The column is a ⅛ in. × 13 ft stainless steel tubing filled with 80/100 mesh Chromasorb PAW coated with 10% *tris* (cyanoethoxy) propane. Thermal conductivity cell temperature is 220°C, injector 200°C; helium flow 15 cm³/min. Temperature is programmed at 10°C/min from 80° to 170°C, then held at 170°C until indene and naphthalene are eluted. Ethyl benzene is the internal standard, and concentrations are determined from peak areas. This column has been used successfully by the author for evaluating light oil.

(2)   Castello and Riccio (1975) use ¼ × 80 in. stainless steel tubing filled with 80/100 mesh Varaport coated with 3% SE 30. The column is temperature programmed at 8°C/min from 40° to 160°C to the elution of naphthalene.

(3)   The author has developed a coating for the separation of light

oil which has been used for many years to analyze various types of
light oil samples, such as those in various stages of purification and
hydrogenation. The column is a ¼ in. × 15 ft stainless steel tubing filled
with 60/80 mesh Chromasorb coated with 10% glycerine carbonate. The
column is activated by gradually heating it to 160°C with a constant
helium flow rate of 30 cm³/min. The column is held at 160°C for approximately 5 hr. The gases eluting from the column during the activation should not be passed into the thermal conductivity cell. Following heat activation, the column is connected to the detector, and
mixtures of benzene, toluene, and xylenes are injected for further activation. The column is maintained at 35°–40°C until elution of benzene. It is then temperature programmed at 10°C/min to 150°C. The
helium flow rate is 30 mliter/min. Sample size for light oil ranges from
1 to 5 μliter, depending on the sample. The column is temperature
programmed in the same manner as above and is held at 150°C until
naphthalene has eluted. Excellent separations are obtained for the forerunnings, thiophene and its homologues, and other aromatics. However, *m*- and *p*-xylenes are not separated. If the aromatics do not separate completely, decrease the rate of temperature programming. Since
heptane is a minor constituent of light oil or almost insignificant, it
has been used as an internal standard for the analysis of light oil
desorbed from the carbon columns and for benzolized and debenzolized wash oils. In some analysis, a column 20 ft in length was used.

Response factors should be determined for the major constituents for
use in the calculation of peak areas. Concentrations are determined
from comparison of these to the area of heptane, the internal standard.
Proper calibration and standardization will give a reliable analytical
method.

C. Analyses of Wash Oils

Castello and Riccio (1975) analyze rich straw or benzolized oil and
lean straw or debenzolized oil for light oil content using the SE 30
column with temperature programming at 20°C/min from 40° to 200°C.
The light oil is separated through naphthalene, then the column is
backflushed. However, their straw or wash oil is a mixture of medium
fractions from the distillation of tar which contains phenols and naphthalene.

In the United States, the majority of the coke plants use a petroleum
oil as a wash oil to recover aromatics from coke oven gas. The wash oil
contacts coke oven gas in the benzole scrubbers where it extracts light
oil from the gas. This rich or benzolized oil is pumped to the wash oil

still where the light oil is stripped from the wash oil by steam distillation. The resultant lean or debenzolized wash oil is recycled to the benzole scrubbers for absorption of light oil from the gas.

Light oil content of both the dry benzolized and debenzolized wash oils is readily determined using the No. 3 glycerine carbonate column, adding heptane to the oil as the internal standard and determining concentration from the peak areas. A preliminary chromatographic separation should be made with the dry, fresh wash oil to determine if any constituent in the oil will interfere with elution of indene and naphthalene. Normally, wash oil does not elute at these temperatures. However, this chromatographic test can be a control to determine that the batch of new wash oil is not contaminated with light boiling compounds which would steam distill and cause problems with the recovered light oil, such as increasing the paraffin content.

The procedure for the chromatographic analyses of the benzolized and debenzolized wash oils is the same as described above for light oil. However, it is advisable that wash oil be flushed out of the column after the elution of naphthalene by means of a backflush valve in the chromatograph.

The difference in the light oil content of the two wash oils indicates the efficiency of the scrubbing system. A high percentage of light oil in the debenzolized wash oil indicates inefficient steam distillation of light oil from the wash oil. A low light oil content in the benzolized wash oil indicates either poor contact between the oil and the gas in the benzole scrubbers or channeling of the wash oil. Furthermore, the oil could be losing its absorption ability because of increased concentration of polymers or high-boiling compounds, such as those from tar.

VII.   H₂S and HCN

Government regulations state that the steel industry must reduce the sulfur content of coke oven gas to 50 grains per 100 ft^3 (800 ppm) so that emission of SO_2 formed in the combustion of the gas is drastically reduced. Furthermore, the regulation states that monitoring of the sulfur content must be on a 24-hr basis.

The monitoring regulation requires a dependable wet chemical method and a continuous operational system for analyzing H_2S.

A.   Wet Chemical Method

Because most full scale methods for desulfurizing coke oven gas also remove HCN, this wet chemical analytical method section will be divided into one method for H_2S and another method for HCN.

1. H_2S

The original method developed by Shaw (1940) for analyzing H_2S in coke oven gas was modified by the author and John L. Kieffer, Chief Chemist, Pittsburgh Coke Plant, Jones & Laughlin Steel Corporation. Considerable time and effort was put into the sampling and analytical techniques to make the method reliable and rapid. It has been tested repeatedly against various standard gases and chromatographic results. For more than 3 yr it has been used to verify new cylinders of standard gases and to verify results obtained on the plant continuous analyzer. It is equally accurate on the sour coke oven gas (high sulfur content, as much as 9000 ppm H_2S) and on the lean desulfurized gas (less than 800 ppm H_2S).

Reagents
   Concentrated hydrochloric acid
   Cadmium chloride solution—10% anhydrous salt
   Sodium carbonate $N/1$ solution
   Hydrochloric acid $N/1$ solution
   Dilute hydrochloric acid solution—80 mliter $N/1$ acid per liter
   Standard iodine $N/10$ solution—(2.3 mole KI: 1 mole I_2)
   Standard sodium thiosulfate $N/10$ solution
   Starch solution indicator
   Methyl orange indicator

The sampling probe is a $\frac{1}{4}$ in. stainless steel tubing inserted into the plant gas main with the end of the tubing bent in the direction of gas flow. This is similar to the probe used for ammonia and light oil. A shut-off valve is attached to the end of the probe immediately outside of the gas main. The shortest possible length of stainless steel tubing connects the valve with the absorption train. This line should be insulated and heated to prevent condensation of water, particularly in cold weather. All condensed water *must be added* to the $CdCl_2$ solution. The stainless steel tubing after the absorption train is fitted with a needle valve for flow control. The line extends to a gas meter.

A Shaw sulfide flask (Cat. No. J-2242 Scientific Glass Apparatus Co., Bloomfield, N. J. 07003) and an 8 in. test tube trap are each charged with 15 mliter of 10% $CdCl_2$ solution and 2 mliter of $N/1$ Na_2CO_3 solution. The groove in the funnel top of the flask is closed. The probe is flushed with the sour coke oven gas by opening the shut-off valve before connecting the apparatus. The flask, trap, and meter are connected to the probe in this order. If the gas contains a high concentration of NH_3, an auxiliary trap containing about 20 mliter of dilute HCl is added to the system ahead of the Shaw flask. Approximately 0.1 ft³ of gas are passed through the apparatus at a rate not to exceed 2 ft³/hr. At the end of the sampling period, the HCl solution in the NH_3 trap is made slightly alkaline to methyl orange and is added to the sulfide flask prior to evacuation.

During the test, log the meter temperature and pressure. After disconnecting the apparatus, log the meter reading and the barometric pressure.

The contents of the trap(s) are washed into the flask with a minimum of water. The flask is evacuated by connecting the outlet stopcock to a water aspirator for no more than 10 sec.

Add 10 mliter of concentrated hydrochloric acid through the grooved funnel top followed by a small amount of water. Mix by shaking the flask. A measured amount of an excess of standard iodine solution is added in the same manner and the residue washed in with two small portions of water. Shake thoroughly. Slowly remove the stopper to equalize the pressure gradually. Titrate the excess iodine with standard thiosulfate solution. Add 5 mliter of starch solution as the end point is approached. The thiosulfate is mixed with the sample by attaching an 18 in. length of Tygon tubing to the inlet tube of the flask and blowing gently through it during the titration.

In addition to H_2S, mercaptans and thiosulfates also react in this method. However, tests have shown that the concentration of the latter two constituents are less than 1 ppm, therefore, these are negligible.

Calculation:

grains H_2S/100 ft³ of gas
$$= \frac{(\text{net mliter } N/10 \text{ iodine}) \times 0.0014 \times 15.43 \times 100}{\text{corrected cubic feet of gas in sample}}$$

The test is similar for the sweet gas, but the volume of gas passed through the sulfide flask is 0.5 ft³.

2. HCN

In most desulfurization processes, HCN must be destroyed before H_2S is converted to sulfur or to sulfuric acid. Although continuous analysis of HCN is not required, the concentration should be determined at intervals.

Since description of the method is too extensive to include in this chapter, the analyst is directed to the original article by Shaw *et al.* (1944). Also see Chapter 41, Section V,B,4 for analysis of HCN in gas from coal.

The sampling system should be identical to the one described above for H₂S.

This method has been used extensively for checking standard gases and to verify the results of the chromatographic analyses.

H_2S concentration in coke oven gas is generally consistent depending on the sulfur content of the coal charged to the ovens. Actually, a change in coal sulfur is readily discernible in the H_2S concentration in the coke oven gas.

The HCN concentration in coke oven gas varies from day to day and does not follow a consistent pattern.

B.   Gas Chromatographic Method

The second government regulation states that the sulfur content of the sour and sweet gas must be monitored on a 24-hr basis. Manka (1975, 1977) developed a continuous sampling and analysis system based on chromatography using a thermal conductivity detector for high concentrations of H_2S and HCN in the sour gas and the flame photometric detector (FPD) for low concentrations of H_2S, COS, and CS_2 in the sweet gas.

The probe is a Teflon tube, 1¼-in. o.d. and ¾-in. i.d., which extends across the full width of the plant gas duct. The tube is inserted inside a 2-in. diameter stainless steel pipe with one-half of the pipe cut along the length of the probe, except at the top and bottom, so that the six ⅜ in. probe openings in the Teflon tube are exposed to the gas. The probe is positioned so that the openings are on the same course as the gas flow.

The coke oven gas flows to a knockout drum to remove entrained oil and tar, then to a filter to remove plus 1.0 μ solids.

The gas flows or is pumped through 1 in. stainless steel tubing which is heated at 65°C. Although a major portion of the clean gas is returned to the plant gas duct downstream from the sampling probe, a portion of the gas flows to two rotometers located in the gas chromatograph, then to separate sample valves.

The flow schematic for separation and analysis of the sulfur compounds in the sweet gas is shown in Fig. 6. In the flame photometric chromatograph (FPD), a 600 μliter sample of coke oven gas is injected by the sample valve onto the ⅛ × 12 in. Teflon column, containing 100/120 mesh Teflon-6 coated with 5% SF-96, a methyl silicone stationary phase. The sulfur compounds pass through the column and backflush valve onto the second column, leaving behind oils, naphthalene, tar, and water. On a signal from the control system, the sample valve and

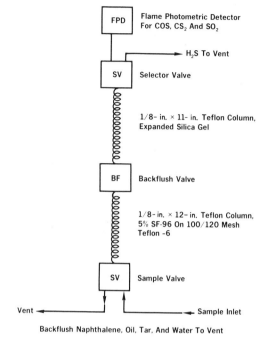

Fig. 6. Schematic for analysis of low concentration of sulfur compounds.

backflush valve are switched to permit one source of nitrogen to back-flush the tar, oils, and water to atmosphere and a second source to separate the sulfur compounds on the $\frac{1}{8} \times 11$ in. Teflon column filled with a special silica gel. The separated COS, CS_2, and H_2S flow to the FPD. The columns and valves are maintained at 63°C, and nitrogen is the carrier gas at a flow rate of 16 cm³/min.

These sulfur compounds, except H_2S, can also be analyzed in the sour gas. The high concentration of H_2S saturates the detector. Therefore, the signal from the control system opens the selector valve as H_2S is being eluted from the column and H_2S is eliminated to the atmosphere through this vent.

The flow schematic for the separation of high concentration of H_2S and HCN in the sour gas is shown in Fig. 7. The thermal conductivity unit is in a separate compartment with its own valves and columns. A 1000-μliter sample of coke oven gas is injected by the sample valve onto the precolumns which consist of a $\frac{1}{16} \times 3$ in. stainless steel tube filled with 100/120 mesh Porapak Q plus a $\frac{1}{16} \times 12$ in. stainless steel tube filled with 100/120 mesh Chromasorb P coated with 15% Carbowax

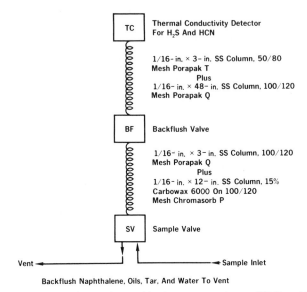

Fig. 7. Schematic for analysis of high concentrations of H₂S and HCN.

6000. HCN and the unresolved sulfur compounds, except CS_2, elute through the backflush valve onto the second column, leaving behind the tar, oils, water, and CS_2. On a signal from the control, the sample valve and backflush valves are switched. One source of helium back-flushes the tar, oils, water, and CS_2, and another source of helium flows to the separation column. The H_2S and HCN are separated on two columns in series, a $\frac{1}{16} \times 3$ in. stainless steel tubing containing 50/80 mesh Porapak T plus $\frac{1}{16} \times 48$ in. stainless steel tubing containing 100/120 mesh Porapak Q. All components flow through the thermal conductivity detector, but only the response to H_2S and HCN is re-corded.

The columns and detector are maintained at 60°C. Helium is the carrier gas at a flow rate of 13 cm³/min.

Concentrations of the components are determined by peak height. The voltage at the maximum height of the peak is compared to the voltage of the same component in the standard gas. The results for each component are printed out in parts per million.

REFERENCES

Beckmann, R., Simons, W., and Weskamp, W. (1962). *Brennst.-Chem.* **43**(8), 241.
Castello, G., and Riccio, M. (1975). *Fuel* **54**(3), 187–192.

Kagasov, V. M., and Zharkova, N. A. (1969). *Koks Khim* **9,** 29–33.

Manka, D. P. (1975). *Instrum. Technol.* **22**(2), 45–49.

Manka, D. P. (1977). *Int. Symp. Anal. Chem., Birmingham Univ., Birmingham, Eng.*

Pinchugov, V. N., and Zharkova, N. A. (1976). *Koks Khim* **26,** 41–42.

Shaw, J. A. (1940). *Ind. Eng. Chem., Anal. Ed.* **12,** 668–671.

Shaw, J. A., Hartigan, R. H., and Coleman, A. M. (1944). *Ind. Eng. Chem., Anal. Chem.* **16,** 550–552.

Chapter 39

Characterization of Recovered Volatiles from a High Methane Coal and the Significance of These Findings

Mynard C. Hamming   *Frederick J. Radd*   *Alfred B. Carel*
RESEARCH AND DEVELOPMENT DEPARTMENT
CONTINENTAL OIL COMPANY
PONCA CITY, OKLAHOMA

I. INTRODUCTION

For many years chemists have conducted an arduous search for an understanding of the complex, transformed, primarily fossil plant residue, coal. In the work reported here, we were concerned with an understanding of coal surfaces in deep mines and how deep mine

29

safety relates to gas and films produced on fresh coal surfaces. Accordingly, answers to the following questions were pursued.

(1) What happens, chemically and physically, at the surface of a newly exposed high-methane coal surface, such as the Pittsburgh No. 8 seam? (The Pittsburgh seam is a gigantic coal seam located in West Virginia, Pennsylvania, and Ohio. See Fig. 1.)

(2) Chemically speaking, what happens as a newly exposed coal surface is air oxidized and how are these surface changes effected by time?

(3) What kinds of chemical products are inherently and/or geochemically present in high-methane coals, and how do they relate to hydrocarbons found in that same freshly mined and air-exposed coal?

(4) How do these coal surface chemistries vary with the vertical position of the coal in the seam? [There is a time-deposition stratigraphy visible even in the numerous layers (40–60+) present in a single seam.]

(5) What sort of handling, storage, and shipping problems are related to these surface effects and internal coal variations?

Some of the answers to the above questions depend on a knowledge

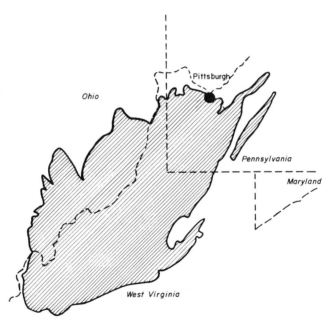

Fig. 1 Tristate outline of the Pittsburgh seam bituminous coal.

of the coal surface compositional properties. Therefore, it was our purpose to characterize the coal surfaces of the high-methane-content Pittsburgh No. 8 seam coal as carefully as possible. While other coal seams in the West Virginia–Pennsylvania–Ohio area could have been chosen, this specific seam is of the highest economic concern. Furthermore, it is considered representative of Pennsylvanian Period bituminous coals throughout the world.

In these studies, it was desired to document the silent, unseen oxidation of fresh coal surfaces. These oxygen contacts may come from: (a) newly mined, methane-charged coal contacted with air, (b) near-surface stripping of coal seams that may have been air contacted for long periods, or (c) deep or anaerobic coal seams that contact aerobic groundwaters and are surface reacted.

To find answers to the above coal chemistry questions, these specific analytical objectives were undertaken.

(1) To construct a pressure-vacuum vessel for volumetrically collecting gases produced from the fine grinding of coal in a Bleuler mill.

(2) To measure, analyze, classify, and compare coal samples taken 24 hr earlier at different seam levels from the Blacksville No. 2 mine, Wana, West Virginia, by studying the gases evolved during grinding of the samples in a Bleuler mill. These gases would be collected over various time periods from various atmospheres.

(3) To study the effluent and residue fractions obtained from thermal gravimetric analyses (TGA) of raw roof and rib coal samples. To collect the volatile effluent to 500°C from TGA and to analyze the effluent and residue fractions by mass spectrometry.

(4) To relate the data obtained to operational usefulness concerning precautions as to release of combustible gases from coal surfaces in mining, in transport, and in storage.

(5) To study coal surface oxidation processes and to provide information and/or techniques for future coal gasification or liquefaction studies.

The above objectives are especially germane to the safe mining of coal, to increased safety in the silo storage and ship transport of methane-containing coals, and to safety matters concerning mine fires and the proper recovery of operations after a mine fire.

By way of literature background, devolatilization of coal by rapid heating has been reported by Menster *et al.* (1974) from a study at the U.S. Bureau of Mines in Pittsburgh. That work, for the most part, covered a higher temperature range and involved more rapid heating than the slower rates used in this study (8250°C/min versus 0.08°C/

min). A previous publication by Girling (1963) reported the identification of products in the range of C_4 to C_{10}. The work reported here covers organic products well beyond C_{10}. The application of low-resolution mass spectrometry used to identify the compounds in coal-carbonization products has been reported by Shultz *et al.* (1972). Both high- and low-resolution mass spectrometry were used in this study.

II. EXPERIMENTAL

The ideal place to have done the analytical work would have been in the mine at a fresh coal surface. While the sampling was so done, the in-mine analysis, of course, was not possible. Samples of roof and rib coal collected (see Fig. 2) from a new operation section at the Blacksville No. 2 mine in West Virginia were airline carried in special containers in less than 24 hr to the Bleuler volumetric milling in Ponca City,

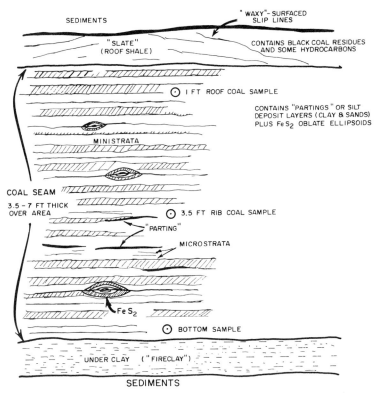

Fig. 2 Sampling of Pittsburgh coal seam. (Sample taken in West Virginia.)

Oklahoma. We had shown from earlier data that this analytical delay time was not significant.

A. Grinding the Coal Samples with a Bleuler Mill and Collection of the Released Gases

A Bleuler mill vessel was constructed to pulverize the coal and to contain the released gases. The vessel was fabricated from AISI-4340 steel and then heat treated by quenching and tempering to a Rockwell "C" hardness of 45 for the case and 33 for the rings. The lid was equipped with an O-ring seal for both vacuum and pressure operation. The overall size of the modified Bleuler mill vessel was the same as commercially supplied Bleuler vessels which are not constructed to contain released gases. The vessel was equipped with vacuum-pressure bellows valves on the inlet and outlet lines, and these lines were fitted with fritted steel filtering plugs to contain the coal dust. An exploded view of the grinding vessel is shown in Fig. 3. Figure 4 shows a cutaway view of the grinding vessel with the glass 50-mliter gas bomb attached through a standard taper joint and Kovar glass-to-metal seal. A complete drawing of the gas-collecting and measuring apparatus is shown in Fig. 5. The assembly includes the source gas, the vessel, gas-collecting bomb, Wallace and Tiernan 0–800-mm manometer, Zimmerli gauge 0–80 mm, cold trap, and vacuum pump.

The freshly mined coal sample was placed in the Bleuler vessel and sealed; the gas-collecting vessel was attached; and the gas-collecting and measuring apparatus was connected. Appropriate vacuum was applied with three-way valve (e) open to (g) and both vacuum bellows valves closed. When high vacuum was obtained as observed on the Zimmerli gauge (d), the bellows valve leading to (f) was opened until high vacuum was again obtained. The bellows valve was then closed, the three-way valve (e) was opened to (i), and the Bleuler vessel (f) was removed. The sample was ground for the specified time and the vessel reattached to (g). Three-way valve (e) was then opened to (a), (i), and (g) until a high vacuum was observed on (d) and (i). Valve (e) was then closed to (a), the readings of manometer (i) and the leveling bulb were recorded, the bellows valve below (g) was opened, and the gas was allowed to equilibrate. The leveling bulb was returned to its original position and the final reading of manometer (i) taken. The difference in the two readings of the manometer was the pressure of the system because of the evolved gases. The volume of the system (f), (g), (h), and (i) was 850 mliter. The gas calculations for weight percent CH_4 were then made from this volume, from the percent CH_4 in the gas as

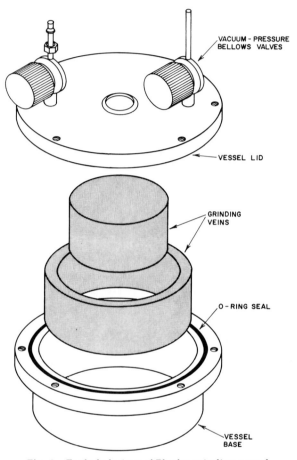

VACUUM - PRESSURE
BELLOWS   VALVES

VESSEL LID

GRINDING
VEINS

O - RING SEAL

VESSEL
BASE

Fig. 3   Exploded view of Bleuler grinding vessel.

obtained by mass spectrometry and from the 50 g of coal (in all cases) that was charged to the grinding vessel.

This same basic procedure was used to obtain gas samples in the air and oxygen studies, except the system was not evacuated. In the case of the oxygen atmosphere, the vessel was flushed with oxygen for 10 min prior to grinding. A similar procedure was followed for air.

B.   Thermogravimetric Analyzer Operations†

The instrumentation used for this study consisted of a DuPont Model 950 Thermogravimetric Analyzer (TGA) powered by a DuPont Model

† Thermogravimetric analysis of coal is detailed in Volume II, Chapter 37.

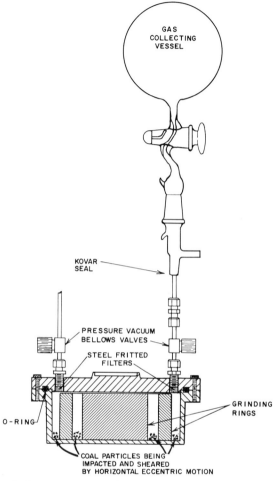

Fig. 4   Bleuler grinding apparatus with gas-collecting vessel. (From Radd *et al.*, 1976.)

990 Thermal Analyzer unit, both of which combine to form the operating TGA system. Figure 6 shows a cutaway view of the TGA sample-collecting apparatus.

The sample was loaded into an aluminum sample pan (a) connected to the balance assembly (b) containing a thermocouple for monitoring the sample temperature. The furnace tube (c) was then connected to the balance assembly and inserted into the furnace (d). Connector (e), heater (f), sample trap (g), and water diffusion trap (h) were then connected. Dewar (i) containing liquid nitrogen was used to condense portions of the effluent, and pyrometer (j) was used to monitor the

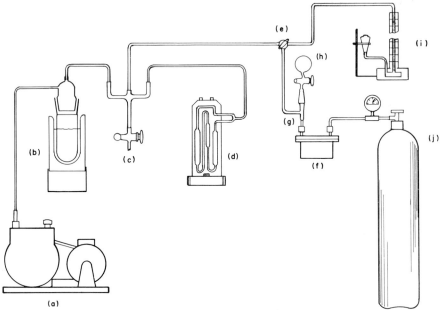

Fig. 5 Complete gas-collecting and measuring apparatus. (a) vacuum pump, (b) cold trap (liquid N_2), (c) vent valve, (d) Zimmerli gauge 0–80 mm, (e) three-way valve, (f) Bleuler vessel, (g) Kovar glass to metal, (h) gas collection bomb, (i) Wallace and Tiernan 0–800 mm manometer, (j) source gas for Bleuler mill. (From Radd *et al.*, 1976.)

temperature of heater (f). Nitrogen sweep gas was passed over the sample at 50 mliter/min carrying the evolved coal components into trap (g). The entrance point for the nitrogen sweep gas is at valve (k). This nitrogen gas sweeps through the furnace tube (c) and over the sample pan (a) and sample thermocouple (l). The samples were weighed into the sample pan, the weight being simultaneously recorded on the chart paper of the DuPont Model 990 Thermal Analyzer module. The temperature range of ambient to 500°C was used for collecting the volatile materials from the coal. Programmed heating rates of 1° and 5°C/min were used. The first fraction was collected from ambient to 100°C. After each fraction was collected for low-resolution mass spectrometry, the coal residue sample was removed for analysis by high-resolution mass spectrometry. The remaining fractions were collected in 50°C increments from ambient to 500°C. A fresh coal sample was then loaded into the TGA unit, and the next temperature increment of the coal volatiles was collected (volatiles evolving prior to the initial temperature of this increment were not collected); the residue was again removed; and the

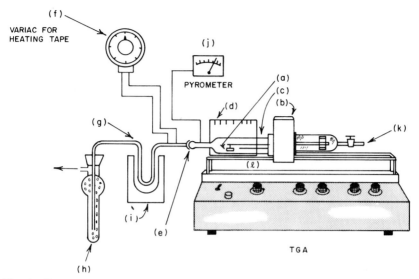

Fig. 6 Cutaway view of sample collecting apparatus. (a) aluminum sampling pan, (b) balance assembly, (c) furnace tube, (d) furnace, (e) ball joint connector, (f) heater, (g) sample trap, (h) water diffusion trap, (i) dewar, (j) pyrometer, (k) nitrogen inlet, (l) sample thermocouple.

thermal splitting process continued until the desired 500°C temperature was obtained.

It was not possible to condense all of the TGA effluent as it was swept through the condensation tube by the nitrogen sweep gas. By weighing the recovered fraction and comparing its weight to the weight loss from the TGA curve, we found that about 75% recovery was obtained over the full temperature range of 100° to 500°C. No attempt was made to weigh each 50°C fraction that was collected from the TGA since the amount of volatile material was extremely small.

C. Mass Spectrometry Techniques†

The low-resolution mass spectra were obtained on a DuPont (CEC) Model 21-103C instrument. High-resolution data were obtained on a DuPont (CEC)-21-110B mass spectrometer using photoplate recording. Fractions collected from the TGA were loaded into the 103C mass spectrometer from the U tubes that were used for collecting the TGA fractions) at an inlet temperature of 300°C by the normal procedures.

† See Volume I, Chapter 16, Section IV, and Chapter 17, Section II,B for more on mass spectrometric quantitative analysis.

The original coal sample, ground to ~10 mμ size in a Bleuler mill, was introduced via a probe distillation into the high-resolution mass spectrometer. Residue samples from TGA were also probe loaded. In probe loading, the material is admitted directly into the ion source of the mass spectrometer (not from an inlet system) whereby different fractions are successively vaporized by increasing the probe temperature (from 150° to 350°C). Spectra were obtained from the "fraction" evolved at each temperature interval taken (usually 50°C).

Precise mass measurements were determined from 70 eV spectra. Only molecular formulas whose measured masses deviated less than ±0.004 (unified) amu from the theoretical values were considered. In most cases, more than one spectrum was used to support the identification of molecular species. Thus the qualitative data from the high-resolution mass spectrometer was excellent. The high-resolution quantitative data, however, is not as precise as the low-resolution quantitative data. This is because only approximately 70% of the total sample was vaporized into the high-resolution instrument at the operating conditions of 300°C and 10^{-6} torr. The difference in the quantitative data from the low-resolution and high-resolution instruments is reflected in the tables of data. The high-resolution quantitative data are used only to show relative differences among samples.

The high-resolution data on photoplates were reduced via the photodensitometer and the Supernova computer with final processing on the IBM 370/155 computer system. Peak intensities and line positions were processed by a modified version of a program by Tunnicliff and Wadsworth (1968). Accuracy and repeatability of mass measurements are shown in Table I. These precision data were obtained from a standard sample used for mass spectrometer calibration purposes. This standard sample (obtained from the U.S. Bureau of Mines, Bartlesville, Oklahoma) is a 370°–535°C boiling range distillate of Gach Saran (Iranian) crude oil.

Five separate analytical runs of this distillate were made. The average masses of selected formulas, the standard deviation, deviation from calculated values, the 95% confidence interval, and the percent relative error at 95% confidence interval were obtained. The data show that the agreement between experimental and theoretical mass is excellent with a standard deviation ranging from ±0.00017 to ±0.00923 mass values and an absolute deviation from theoretical ranging from +0.00410 to −0.00851 mass values. The 95% confidence interval ranges from ±0.01339, and the percent relative error at the 95% confidence interval ranges from ±0.00031 to ±0.00895.†

† The resolving powers required for various levels of agreement between experimental and theoretical mass are shown in Volume I, Chapter 16, Tables I and II.

TABLE I *Repeatability of Selected Measured Masses from Five Spectra of a Gach Saran Distillate*

Formula	Theoretical mass	Measured mass	Deviation from theoretical	95% confidence interval	Percent relative error
C_5H_8	68.06260	68.06161	−0.00099	±0.00064	±0.00094
C_5H_{10}	70.07825	70.07788	−0.00037	±0.00022	±0.00031
C_5H_{12}	72.09390	72.08989	−0.00401	±0.00047	±0.00065
C_6H_6	78.04695	78.04535	−0.00160	±0.00096	±0.00122
C_7H_8	92.06260	92.06264	+0.00004	±0.00095	±0.00103
C_8H_{12}	108.09390	108.09335	−0.00055	±0.00053	±0.00049
$C_{10}H_8$	128.06260	128.06670	+0.00410	±0.01146	±0.00895
$C_{10}H_{10}$	130.07825	130.07788	−0.00037	±0.00134	±0.00103
$C_{10}H_{12}$	132.09390	132.09247	−0.00143	±0.00267	±0.00202
$C_{10}H_{14}$	134.10954	134.10918	−0.00036	±0.00257	±0.00192
$C_{11}H_{10}$	142.07825	142.07812	−0.00013	±0.00685	±0.00482
$C_{11}H_{12}$	144.09390	144.09286	−0.00104	±0.00084	±0.00059
$C_{14}H_{14}$	182.10954	182.10820	−0.00134	±0.00346	±0.00190
$C_{15}H_{12}$	192.09390	192.08539	−0.00851	±0.01080	±0.00562
$C_{17}H_{14}$	218.10954	218.10608	−0.00346	±0.01339	±0.00614
$C_{18}H_{20}$	236.15649	236.15518	−0.00131	±0.00257	±0.00109
$C_{19}H_{26}$	254.20344	254.20195	−0.00149	±0.00261	±0.00103
$C_{20}H_{28}$	268.21909	268.21509	−0.00400	±0.00312	±0.00116

The low-resolution mass spectrometer was operated under the conditions shown in Table II. The high-resolution mass spectrometer was operated as shown in Table III.

III. DATA PROCESSING

A. Interpretations of Bleuler Mill Data

The volume of the gas-collecting system was determined to be 850 mliter. The samples were ground in duplicate and duplicate analyses of the collected gases were made by low-resolution mass spectrometry. The weight percent methane was obtained by using the ideal gas law and converting the gas to milliliters at standard conditions, then to liters, moles, grams, and finally to weight percent. It should be noted that some increase in temperature of the Bleuler vessel occurred as a result of the grinding. This was found to be <5°C for a 5-min grinding period. Most of this temperature increase was dissipated during gas collections, and no correction was made in the calculations for random temperature fluctuations.

B. TGA Weight-Loss Thermograms

The TGA was simply used to obtain effluent and residue fractions from roof and rib coal for mass spectrometry analyses. Since a fresh

TABLE II *Conditions for Operating the Low-Resolution Mass Spectrometer*

Ionizing voltage	70 eV	Sensitivity	High
Inlet temperature	100°C	Accelerating voltage	3,200–500 V
Collector slit width	30 mil at 2,398 G (m/e 12 to 90)	226/57 intensity ratio of nC$_{16}$	Adjusted to 3.5 ± 0.1
	7 mil at 4,138 G (m/e to end of spectrum)	Source temperature	230°C
Ionizing current	100 μA		
Ion beam	Focused mode		

TABLE III *Conditions for Operating the High-Resolution Mass Spectrometer*

Ionization voltage	70 eV	Source pressure	1×10^{-6} torr
Source temperature	150°C	Analyzer pressure	1×10^{-8} torr
Inlet system	100°C	Recorder	Photoplate
Ionizing current	100 μA		

sample was loaded into the unit for each 50°C increment studied, complete thermograms of weight loss versus temperature data were not obtained during the fraction collecting. However, complete TGA thermograms were obtained on the roof and side coal samples at the same heating rates of 1° and 5°C/min and temperature range of 50°–500°C that were used for the fractions. These thermograms are shown in Figs. 7 and 8.

The weight loss for both roof and rib coal samples was greater at 1°C/min than at 5°C/min in the temperature range of 50°–450°C. The comparative weight loss figures are 16.6 and 17.2% loss for 1°C/min heating rates and 17.3 and 18.6% weight loss at 5°C/min heating rate for the roof and rib coal samples. The weight loss was greater for both coal samples at 500°C at the 5°C/min heating rate. Significant differences in the percent volatiles from bituminous coals were also reported by Menster *et al.* (1974) when rapid pulse heating (8250°C/min) was used as compared to slower (0.08°C/min) heating techniques. Menster postulated a reason for this difference was because of the bond-breaking reactions that occur in the coal structure which give rise to initial decomposition fragments. Another possible explanation of this observation is that at the slower heating rate, polymerization reactions are increased and/or pyrolysis reactions are reduced.

C. Mass Spectral Data

To handle such large volumes of mass spectrometer data properly computer-processing procedures are imperative. In the case of high-resolution mass spectral data, the computer processing should be well understood to avoid serious errors in interpretation.

The high mass spectra data presented the greatest problems in data processing. Line positions on photoplates are the source for assigning accurate masses which then are transformed into formulas. While the spectral line positions are well defined by programs such as those of Tunnicliff and Wadsworth (1968), these positions have to be given accurate mass assignments. Often a calibration compound, such as

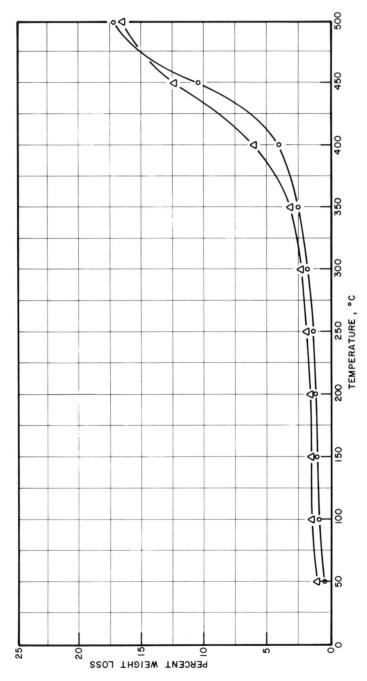

Fig. 7 Thermogram scan of roof coal. Δ, 1°C/min heating rate; O, 5°C/min heating rate.

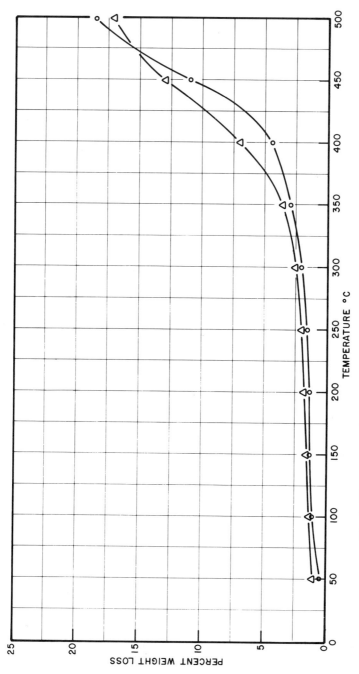

Fig. 8 Thermogram of rib coal. Δ, 1°C/min heating rate; O, 5°C/min heating rate.

perfluorokerosene, is used to relate voltage parameters to mass to aid in making mass assignments. However, this was not carried out in this work because of problems of overloading the perfluorokerosene with the coal samples being analyzed. A computer program developed in the Analytical Research Laboratory at Continental Oil Company was used to make mass assignments of the observed line positions. In place of the internal calibration compound, the common fragment ions of hydrocarbon were used as input candidate calibration peaks to aid the computer in its calculation of accurate masses. The assignment of masses to line positions has with it an uncertainty when a complete spectrum is processed (over 500 peaks per spectrum). In the course of this work a unique additional data check was observed. The cluster of five major isotopic peaks of cadmium found in these coal spectra were one of the several useful checks as to the accuracy of mass measurements for any given spectrum. These cadmium isotopes have values of 109.0903, 110.9042, 111.9028, and 113.9036 and are thus clearly separated from other possible elemental combinations of hydrocarbons and heteroatoms. The intensities of these cadmium isotopes also provide a measurement of the relative intensities of the peaks. A typical example is given in Table IV.

The next data-processing step is to take into account all the possible atomic compositions (referred to here as empirical formulas) for the assignment of accurate masses. This data-processing step has been a source of errors in interpretation. A computer program (Robertson and Hamming, 1977) can be used which calculates all possible empirical formulas for each accurately measured peak. This program has a feature

TABLE IV　*Intensities of Cadium Isotopes*

Accurate masses (AMU)		Relative intensities of natural abundances	
Theoretical mass[a]	Measured versus theoretical	Reported[a]	Measured
109.0930	0.003	12.39	13.02
110.9042	0.004	12.75	12.90
111.9028	0.002	24.07	23.96
112.9046	0.001	12.26	12.93
113.9036	0.002	28.86	27.50

[a] *Handbook of Chemistry and Physics* (1971–1972).

of using a library of formulas which allows the user to select a certain series of peaks, such as even numbered hydrocarbon molecular ions or odd numbered hydrocarbon fragment ions to give additional confidence to identifications. After the correct formulas are determined, the compound types can then be rapidly assigned from a series of interpretation maps presented by Hamming and Foster (1972a).

Several additional computerized subroutines can be included in the data processing. One is the normalized weight percent of selected elements present in a sample as determined by its mass spectrum. This was done by a Conoco computer program that uses the concentration of molecular ions to compute the elemental composition of the samples and the concentration distribution of elements within each species. In a complex sample, such as the volatiles from coal, this method offers a new and simple means of sample-to-sample comparison. Other techniques, such as elemental analysis, may then be used to verify the mass spectrometer method. The capability of high-resolution mass spectrometry for determining elemental composition of single compounds has been recognized for some time (Chait *et al.*, 1968). This cumulative approach provides a new means of comparing results in a much more simplified form which differs from the listing of compound types. Typical results are shown in the accompanying tabulation.

Element	Normalized weight percent	Element	Normalized weight percent
Carbon	87.98	Nitrogen	0.32
Hydrogen	8.80	Sulfur	0.88
Oxygen	2.03		

Various matrix calculations using computer techniques have been discussed by Hamming and Foster (1972b) to provide semiquantitative results.

IV. DISCUSSION OF RESULTS

The experimental outline given in Diagram I is intended to aid in the discussion of results.

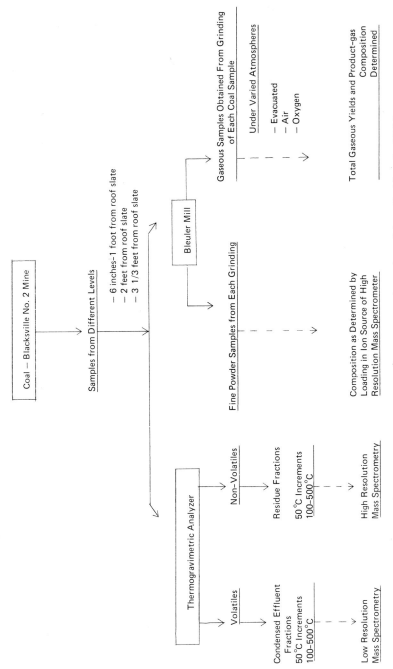

Diagram I. Experimental outline. (Experimental outline adapted from Radd *et al.*, 1976.)

A. Composition of Gases and Residue Volatiles from Bleuler Mill Grinding of Coal

1. Composition of Gases

First to be discussed are data on the gaseous samples obtained from grinding coal under three different atmospheres in the Bleuler mill, these being an evacuated system, air, and pure oxygen. Table V gives a summary on the methane data obtained on 22 separate samples. The percentages of methane found (excluding air, hydrogen, and oxygen) range from 68 to 80%. The higher yield of methane in the evacuated atmosphere agrees with the work of Kekin *et al.* (1973) in which they found a higher yield of hydrocarbon products in the thermal breakdown of coals in vacuum.

In Table VI, the total composition of the gases collected from the three different levels in the mine is given. Methane has been converted to percentage of the original sample. The amount of methane released ranged from 0.30 to 0.43% based on a 50 g coal sample used in the grinding. The weight percent CH_4 was obtained by using the ideal gas law and converting the gas to milliliters at standard conditions, then to liters, moles, grams, and finally to weight percent. The percent methane released by this Bleuler grinding is in agreement with that predicted from adsorption data in a study by Radd and Hassell carried on at Continental Oil Company, Ponca City, Oklahoma. That study showed that the methane–coal system behaves as a Freundlich adsorption system (the methane having a van der Waals-type attraction) following this basic equation:

$$nP = K' \ln V + K''V + K''',$$ (1)

where $\ln P$ = natural log of P, $\ln V$ = natural log of V, and the terms K', K'', K''' are separate constants.

TABLE V *Methane in Evolved Gases Obtained by Bleuler Grinding Pittsburgh No. 8 Seam Coal Samples[a]*

Atmosphere	Minutes grinding time	CH_4 (Vol %) in evolved gases collected from samples taken at these depths in seam		
		1 ft	2 ft	3.5 ft
Evacuated	5	79.0—79.5 [a]	76.2—76.8	74.8—75.3
Air	5	74.1—74.3	72.5—73.6	73.7—73.8
Oxygen	5	75.4—77.8	68.2—71.9	71.9—72.3
Oxygen	10	—	72.0—74.0	—

[a] From Radd *et al.* (1976).
[b] All results obtained from duplicate Bleuler ground samples.

TABLE VI   *Total Composition of Bleuler Mill Gases (Excluding Gases Added before Grinding) and CH$_4$ Converted to wt. % of Original Sample[a]*

Level in mine	Gases evolved by grinding under vacuum for 5 min (vol. %)				CH$_4$ converted to wt. % of original sample
	Methane	Unidentified	CO and/or N$_2$	CO$_2$	
1 ft	79.5[a]	2.7[b]	6.4	11.3	0.39	Average 0.36
	79.0	2.5	6.6	11.9	0.32	
2 ft	76.8	2.0	5.8	15.3	0.40	0.40
	76.2	2.5	5.5	15.8	0.41	
3.5 ft	75.3	2.0 c	7.2	15.5	0.41	0.41
	74.8	2.6	7.3	15.2	0.41	

Gases evolved by grinding with air for 5 min (vol%).

1 ft	74.1	4.9	11.3	9.7	0.30	0.32
	74.3	4.7	11.0	10.1	0.34	
2 ft	73.6	3.2	9.7	13.5	0.39	0.38
	72.5	3.4	10.1	13.9	0.37	
3.5 ft	73.8	3.4	8.9	13.9	0.42	0.42
	73.7	3.6	8.8	14.0	0.43	

Gases evolved by grinding with oxygen for 5 min (vol%).

1 ft	75.4	3.3	11.5	9.8	0.35	0.33
	77.8	4.0	9.9	8.2	0.31	
2 ft	68.2	3.7	14.7	13.3	0.35	0.36
	71.9	4.0	12.4	11.7	0.38	
3.5 ft	72.3	4.0	11.8	11.9	0.39	0.40
	71.9	3.6	12.4	12.0	0.41	

Gases evolved by grinding with oxygen for 2 min (vol%).

2 ft	70.6	3.7	13.5	12.2	0.31	0.34
	71.8	3.2	12.6	12.5	0.36	

Gases evolved by grinding with oxygen for 10 min (vol%).

2 ft	72.0	3.5	12.9	11.7	0.40	0.40
	74.0	3.3	10.3	12.3	0.40	
					Average of 22 =	0.38

[a] From Radd *et al.* (1976).

[b] Results from duplicate samples.

[c] For compositional data by high-resolution mass spectrometric analysis, see Table VII.

[d] For compositional data by high-resolution mass spectrometric analysis, see Table VIII.

By using Eq. (1) and considering the local hydrostatic water head, it is possible to define the methane solubility in any coal measure. The detailed units and constants are given in Fig. 9. The values of the appropriate constants will depend on the internal specific surface of the particular coal.

In considering the data in Table VI, we believe the small increase in

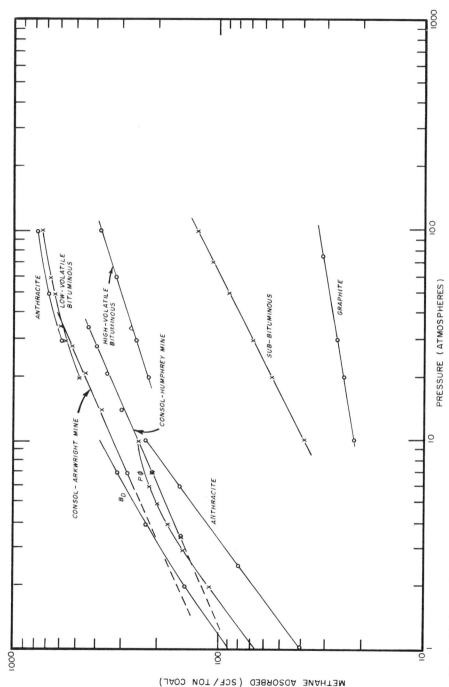

Fig. 9 Methane adsorption on coals and carbon versus pressure. Arkwright equation: $\log_{10} P = 3.323/\log_{10} V - 0.001189V - 7.0342$. Humphrey No. 7 equation $\log_{10} P = 2.159 \log_{10} V - 4.2446$. P = atmospheres, V = SCF methane (70°F) per ton coal. (1-ft active hydrostatic head = 0.4331 psi, 1 atm = 14.696 psi.)

the weight percent of methane from the roof coal (1 ft) to the rib coal (3.5 ft) is due to the fact that the roof coal was the oldest sample. When methane/carbon dioxide ratios are computed from Table VI a consistent pattern emerges. There is an increase in the ratio moving progressively from the Bleuler mill vacuum collection system through the air atmosphere and then to the oxygen atmosphere.

The Bleuler mill itself may impart some active metallic, catalytic surfaces (by very slight erosion effects) for gas phase reactions, as with H_2, CO_2, and/or H_2O. These metal surfaces were present and might be oxygen active and/or hydrogen active in addition to the coal's pyrite content. While a small surface activation energy activity would probably have been present, even for the vacuum milling case, the principal chemistry effects expected were found.

2. Composition of Volatiles Evolved from Coal Dust Remaining in the Bleuler Mill after Grinding

The ground portion of the coal samples remaining in the Bleuler mill were examined by high-resolution mass spectrometry. The data are shown in Tables VII and VIII. The components listed in these tables are those which could not be identified by low-resolution mass spectrometry. The volume percents shown are normalized to the total amount of unidentified components listed in Table VI designated by footnotes b and c. For example, the 1 ft level sample had 2.7% unidentified components (Table VI), and the compositional data for that amount is listed in the column head "Vacuum" in Table VII.

The data in Tables VII and VIII show what appears as an increase in the percent of oxygenated components with increases in the amounts of oxygen in the Bleuler mill atmosphere (going from vacuum, to air, to oxygen). The oxygen compound types found include ketones, aldehydes, alcohols, ethers, and oxygenated aromatics. The percentages shown are intended more as indicators of chemical change rather than of absolute values. The difference in the concentration of oxygenated compounds with increasing amounts of oxygen in the controlled atmosphere of the Bleuler mill can be revealed more dramatically by analysis of the CHO fragment ion peaks in the high-resolution mass spectral data. This is because oxygenated compounds usually give more intense fragment peaks than their molecular ions (Hamming and Foster, 1972c). Also the fragment ions give clues as to the various structures present. This is shown in Table VIII which lists the diagnostic significance of the characteristic fragment ions. The relative *increase of oxygenated compounds* as shown by the fragment ions in Table IX can be expressed as shown in the accompanying tabulation. These percentages

TABLE VII   *Composition of Bleuler Mill Gases other Than* CH_4, CO *and/or* N_2, *and* CO_2 *Evolved by Grinding Coal Taken from 1 Ft Below Roof Slate[a,b]*

Compound or compound type	Volume percent		
	Vacuum[c]	Air[c]	Oxygen[c]
Ethane	0.22	0.64	0.34
Propane	0.21	0.30	0.18
Butane	0.19	0.41	0.22
Pentanes	—	0.26	0.10
Ethylene	0.25	0.05	0.11
Propylene	0.28	0.28	0.18
Butenes	0.11	0.31	0.19
Pentanes and/or cyclopentanes	0.21	0.37	0.22
Hexenes and/or cyclohexanes	0.08	0.17	0.11
Heptenes and/or cycloheptane	—	0.06	0.05
Octenes and/or cyclooctane	0.08	0.14	0.10
Acetylene	0.08	—	—
Cyclopentene	0.07	0.06	0.10
Cyclohexene	0.10	0.21	0.17
Cycloheptene	—	0.08	0.08
Cyclooctene	—	0.13	0.11
Cyclopentadiene	0.10	0.17	0.11
Cyclohexadiene	—	—	0.10
Benzene	0.04	0.20	0.10
Toluene	0.27	0.43	0.26
Naphthalenes	—	—	0.10
Polyaromatics	0.17	—	0.09
Oxygenated aromatics	0.05 ⎫	0.06 ⎫	0.05 ⎫
Mixed Ketones	0.09 ⎪	— ⎪	0.08 ⎪
Ketones or aldehydes	0.11 ⎬ 0.25	0.24 ⎬ 0.4	0.25 ⎬ 0.4
Alcohols or ethers	— ⎪	0.08 ⎪	— ⎪
CHNO compounds	— ⎭	0.04 ⎭	0.02 ⎭

[a] From Radd *et al.* (1976).

[b] 'Slate' is the mining term for the roof shale which looks slatelike from the many slickensides present therein, and miners generally refer to the shale bottom as "fireclay." The roof shales herein are heavily deformed in the plastic flow sense.

[c] For 5 min grinding time.

of total ion yield are not to be considered as percentages of oxygenated compounds in the total sample but are intended to serve instead as additional spectral evidence for an increase of CHO compound types present in the various gases.

Grinding atmosphere	Relative quantity of oxygenated species expressed as percent of total ion yield
Vacuum—5 min	3.8
Air—5 min	5.4
Oxygen—2 min	5.7
Oxygen—5 min	6.2
Oxygen—10 min	7.5

TABLE VIII *Composition of Bleuler Mill Gases other Than CH_4, CO and/or N_2, and CO_2 Evolved by Grinding Coal Taken from $3\frac{1}{2}$ Ft Below Roof Slate*

Compound or compound type	Vacuum[a]	Volume percent Air[a]	Oxygen[a]
Ethane	0.23	0.36	0.49
Butane	0.15	0.28	0.28
Ethylene	0.15	0.04	0.17
Propylene	0.22	0.21	0.28
Butenes	0.17	0.23	0.21
Pentenes and/or cyclopentanes	0.06	0.26	0.33
Hexenes and/or cyclohexanes	0.06	0.14	0.14
Heptenes and/or cycloheptane	—	0.07	0.06
Octenes and/or cyclooctane	0.07	0.13	0.12
Acetylene	0.08	0.08	0.11
Cyclopentene	0.07	0.12	0.13
Cyclohexene	0.09	0.21	0.21
Cycloheptene	—	0.11	0.11
Cyclopentadiene	0.08	0.12	0.14
Cyclohexadiene	—	0.10	—
Benzene	0.07	0.12	0.13
Toluene	0.23	0.35	0.39
Naphthalenes	—	0.08	0.15
Polyaromatics	0.12	0.11	0.12
Oxygenated aromatics ⎫	— ⎫	0.05 ⎫	0.05 ⎫
Mixed Ketones ⎬	0.07 ⎬ 0.15	0.10 ⎬ 0.35	0.13 ⎬ 0.4
Ketones or aldehydes ⎬	0.08 ⎭	0.17 ⎭	0.23 ⎬
CHNO compounds ⎭	—	0.03 ⎭	0.02 ⎭

[a] For 5 min grinding time.

These volatiles are the same as those which would be released from the inner surfaces of the coal upon increased heating. Among these components are those which eventually would find their way into the atmosphere and surface waters.

The coal industry has long known that stored coals oxidize and slowly change (decrease) in BTU values, but the specific chemistry of such coal surface-produced natural changes has not been analytically defined heretofore. While we have not shown that long-term aging is the same as short-term Bleuler milling in mixed atmospheres, the basic oxidative chemistries are expected to be parallel.

B. Data Obtained on the TGA Volatile† and Residue Fractions

The TGA volatile fractions gave rise to all possible aromatic compound types from C_nH_{2n-6} to C_nH_{2n-10}. The carbon numbers for these series ranged from C_6 to C_{20}.

Fractions for the roof coal in general showed a higher concentration of lower molecular weight aromatics than did the rib coal fractions. A

† The trapped volatiles vacuum distilled from bituminous coal at 150°C are shown in Volume II, Chapter 21, Fig. 7.

TABLE IX *Increase of Oxygenated Compound Types with Atmospheres of Increasing Concentrations of Oxygen[a]*

Atmosphere	Accurate mass	Formula	Compound or compound type
Vacuum 5 min [a]	43.0183	C_2H_3O	Diagnostic of CH_3CO—R compounds
	44.9976	CHO_2	Diagnostic of R—COOH compounds (acids)
Air 5 min [a] (20% oxygen)	39.9949	C_2O	C=C=O+. Ion of $C_nH_{2n-4}O$ compounds
	43.0183	C_2H_3O	Diagnostic of CH_3CO—R compounds
	105.0340	C_7H_5O	Diagnostic of aromatic CHO compounds
	44.9976	CHO^2	Diagnostic of R—COOH compounds (acids)
	67.9898	C_3O^2	O=C=C=C=O+. Ion of pyrones
Oxygen 5 min [a]	39.9949	C_2O	C=C=O+. Ion of $C_nH_{2n-4}O$ compounds
	43.0183	C_2H_3O	Diagnostic of CH_3CO—R compounds
	45.0340	C_2H_5O	Diagnostic of $HOCH_2CH_2$—R compounds
	57.0340	C_3H_5O	Diagnostic of C_2H_5—CO—R compounds
	105.0340	C_7H_5O	Diagnostic of aromatic CHO compounds
	44.9976	CHO_2	Diagnostic of R—COOH compounds (acids)
	67.9898	C_3O_2	O=C=C=C=O+. Ion of pyrones
Oxygen 10 min [a]	39.9949	C_2O	C=C—O+. Ion of $C_nH_{2n-4}O$ compounds
	43.0183	C_2H_3O	Diagnostic of CH_3CO—R compounds
	45.0340	C_2H_5O	Diagnostic of $HOCH_2CH_2$—R compounds
	53.0027	C_2HO	Diagnostic of furans
	55.0183	C_3H_3O	Diagnostic of H_2C=CHCO—R compounds
	57.0340	C_3H_5O	Diagnostic of C_2H_5—CO—R compounds
	59.0496	C_3H_7O	Diagnostic of C_2H_5—O—CH_2—R compounds
	71.0496	C_4H_7O	Diagnostic of C_3H_5—CO—R compounds
	93.0340	C_6H_5O	Diagnostic of aromatic CHO compounds
	105.0340	C_7H_5O	Diagnostic of aromatic CHO compounds
	44.9976	CHO_2	Diagnostic of R—COOH compounds (acids)
	67.9898	C_3O_2	O=C=C=CO+. Ion of pyrones

[a] From Radd *et al.* (1976).

[b] Grinding time.

typical comparison for selected fractions from four temperature ranges is illustrated in Table X. The aromatic compounds included here are alkylbenzenes, naphthalenes, and anthracenes.

The roof coal TGA fractions also showed a higher concentration of monoaromatics than were found in the rib coal fractions (data are on volatiles from a collection rate of 1°C/min). These data are listed in Table X and include the TGA fractions from 100° to 400°C.

Monoaromatics are defined here as C_nH_{2n-6}, C_nH_{2n-8}, C_nH_{2n-10} compounds. Examples of these structures are shown below.

C_nH_{2n-6}	C_nH_{2n-8}		C_nH_{2n-10}

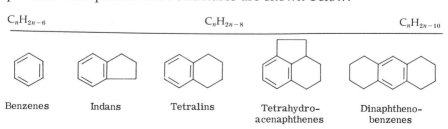

Benzenes	Indans	Tetralins	Tetrahydro- acenaphthenes	Dinaphtheno- benzenes

TABLE X   *Concentration Comparison of Selected Lower Molecular Weight Aromatics in TGA Volatiles Expelled from Roof and Rib Coal*

Fraction temperature, °C	Compound types	Concentration comparison		
		Roof coal	Rib coal	Ratio
100-150	Alkylbenzenes, C_6-C_9	10.0[a]	5.2	1.9
150-200	Alkylbenzenes, C_6-C_9	2.3	1.1	2.0
200-250	Naphthalenes, C_{10}-C_{13}	7.5	5.7	1.3
250-300	Anthracenes, C_{14}-C_{15}	2.0	1.6	1.3

[a] The highest concentration value was set at 10.0 for a frame of reference. All other values are related to 10.0.

The ratio of the roof coal to rib coal monoaromatics in the TGA fractions is shown in the last column of Table XI. The ratio decreases from 1.9 in the 100°-150°C fraction to 1.0 in the 350°-400°C fraction.

The ratio for the polyaromatic compound types determined (C_nH_{2n-12}, C_nH_{2n-14}, C_nH_{2n-16}, C_nH_{2n-18}, and C_nH_{2n-20}) showed no important variation for the roof and rib coal TGA-collected fractions as is shown in Table XII. This is true for both 1° and 5°C/min collection rates. Since only about 75% of the TGA volatiles are collected, some preferential collection could occur which would mask true differences in the original samples. Analysis of the original roof and rib coal samples by high-resolution mass spectrometry does show a preferential concentration of the polyaromatics in the roof coal. This is discussed in Section IV,B.

It will be noted that the roof and rib coal species do not involve a different spectrum of molecular types but only a selective accenting of lighter species in the roof coal samples. This shifting of the concentra-

TABLE XI   *Concentration Comparison of Selected Monoaromatics in TGA Volatiles Expelled from Roof and Rib Coal*

Fraction temperature, °C	Roof coal	Rib coal	Ratio
100-150	10.0[a]	5.2	1.9
150-200	4.8	2.5	1.9
200-250	3.4	2.0	1.7
250-300	2.5	2.3	1.1
300-350	2.3	2.0	1.2
350-400	3.6	3.6	1.0

[a] The highest concentration value was set at 10.0 for a frame of reference. All other values are related to 10.0.

TABLE XII   *Concentration Comparison of Selected Major Polyaromatic Compound Types Found in TGA Volatiles Expelled from Roof and Rib Coal*

Fraction temperature, °C	Compound types	Roof coal TGA heating rate @ 1°/min	@ 5°/min	Rib coal TGA heating rate @ 1°/min	@ 5°/min
100-150	Naphthalenes	6.9	5.6	10.0[a]	5.8
150-200	Naphthalenes	8.5	8.3	9.2	7.7
200-250	Naphthalenes	6.9	4.8	5.8	3.7
250-300	Perinaphthalenes	4.0	3.3	3.6	3.3
300-350	Perinaphthalenes	3.1	2.7	3.5	3.1
350-400	Hexahydropyrenes	3.7	2.7	2.7	2.7
400-450	Hexahydropyrenes	3.1	2.7	3.3	2.1
450-500	Hexahydropyrenes	3.3	2.9	NA	NA

[a] The highest concentration value was set at 10.0 for a frame of reference. All other values are related to 10.0.

tion of aromatic components is because of an increase in the concentration of paraffinic hydrocarbons. The aromatic compounds (often over a narrow carbon range) are of interest due to the possible value of expelling such compounds from coal at rather low temperatures.

Tables XIII, XIV, and Fig. 10 deal with the polyaromatic compound types found in the TGA 5°C/min roof coal fractions. In Table XIII the comparative concentrations (see footnote at bottom of Table XIII) of C_nH_{2n-12} through C_nH_{2n-24} compounds are listed for each of the eight TGA fractions. The highest concentration of total ion yield for the most abundant C_nH_{2n-x} species was set at 10 with all other C_nH_{2n-x} species related to it. The tabulated data in Table XIII show that the highest concentration of C_nH_{2n-12} species was found in the 200°–250°C TGA fraction, the highest concentration of C_nH_{2n-14}, C_nH_{2n-16}, and C_nH_{2n-18} was found in the 300°–350°C fraction, and the highest concentration of C_nH_{2n-20}, C_nH_{2n-22}, and C_nH_{2n-24} was found in 350°–400°C fraction.

In Table XIV the carbon range and the most abundant polyaromatics in this range are given for each of the eight TGA fractions covering C_nH_{2n-12} through C_nH_{2n-24}. The most abundant polyaromatics range from C_{10} to C_{20} with compounds as high as C_{28} being detected. The highest carbon number reported in the most abundant peak series of the outlined carbon range is C_{20}. The carbon range of the TGA fractions was therefore considered to end with C_{20}.

Chemical structures for the possible polyaromatic compound types for $n = 10$ through $n = 26$ are given in Fig. 10. The more abundant species found are between $n = 10$ and $n = 20$. The structures for the compounds where $n = 24$ and $n = 26$ are included since lesser concentrations of these compounds were found as reported in Table XIV. A

TABLE XIII *Mass Spectrometric Examination of TGA (5°C/Min) Roof Coal Fractions*

Fraction number	Temperature range of fraction, °C	Concentration comparison of polyaromatic compound types						
		C_nH_{2n-12}	C_nH_{2n-14}	C_nH_{2n-16}	C_nH_{2n-18}	C_nH_{2n-20}	C_nH_{2n-22}	C_nH_{2n-24}
1	100-150	3.3x	1.3x	1.0x				
2	150-200	9.2x	2.4x	1.9x	0.6x			
3	200-250	10.0x[a]	7.3x	4.5x	3.7x	2.2x	0.5x	0.6x
4	250-300	7.9x	7.7x	7.6x	7.3x	5.5x	2.5x	0.7x
5	300-350	6.5x	10.0x	10.0x	10.0x	9.7x	9.7x	6.9x
6	350-400	5.5x	8.6x	9.1x	7.3x	10.0x	10.0x	10.0x
7	400-450	4.6x	5.8x	6.5x	5.1x	7.1x	5.6x	5.2x
8	450-500	4.2x	5.3x	6.2x	5.4x	5.4x	4.6x	4.2x

[a] For a given compound type, the highest concentration of total ion yield for the different runs was set at 10.0x. The other values for the same compound type are relative to 10.0x.

TABLE XIV *Most Abundant Carbon Species and Carbon Range of Polyaromatics in the 5°C/Min TGA Roof Coal Fraction*

Fraction number	Temperature range of fraction, °C	C_nH_{2n-12}	C_nH_{2n-14}	C_nH_{2n-16}	C_nH_{2n-18}	C_nH_{2n-20}	C_nH_{2n-22}	C_nH_{2n-24}
1	100-150	C_{10}[a] C_{10}-C_{13}[b]	Small[c]	Small	—	—	—	—
2	150-200	C_{10} C_{10}-C_{14}	C_{12} C_{12}-C_{14}	C_{12} C_{12}-C_{12}	Small	—	—	—
3	200-250	C_{12} C_{10}-C_{16}	C_{13} C_{12}-C_{19}	C_{14} C_{12}-C_{16}	C_{14} C_{14}-C_{17}	C_{14}-C_{15} C_{14}-C_{17}	Small	Small
4	250-300	C_{13} C_{10}-C_{19}	C_{15} C_{12}-C_{19}	C_{15} C_{12}-C_{19}	C_{14} C_{14}-C_{18}	C_{15}-C_{16} C_{14}-C_{19}	C_{16} C_{16}-C_{19}	Small
5	300-350	C_{13}-C_{14} C_{10}-C_{22}	C_{16} C_{12}-C_{24}	C_{16}-C_{17} C_{12}-C_{22}	C_{15} C_{14}-C_{22}	C_{16} C_{14}-C_{22}	C_{16} C_{14}-C_{22}	C_{18} C_{18}-C_{23}
6	350-400	C_{16}-C_{17} C_{10}-C_{24}	C_{16} C_{12}-C_{25}	C_{17} C_{12}-C_{23}	C_{16}-C_{17} C_{14}-C_{23}	C_{16}-C_{17} C_{14}-C_{24}	C_{18} C_{16}-C_{26}	C_{18} C_{18}-C_{25}
7	400-450	C_{17} C_{10}-C_{27}	C_{16} C_{12}-C_{27}	C_{17} C_{12}-C_{27}	C_{17} C_{14}-C_{28}	C_{19} C_{14}-C_{28}	C_{19} C_{15}-C_{28}	C_{19} C_{18}-C_{28}
8	450-500	C_{17} C_{10}-C_{23}	C_{17} C_{12}-C_{23}	C_{18} C_{12}-C_{24}	C_{18} C_{14}-C_{24}	C_{19} C_{14}-C_{24}	C_{19} C_{16}-C_{24}	C_{20} C_{18}-C_{24}

[a] Most abundant species.
[b] Carbon range.
[c] Carbon range is less than three carbons.

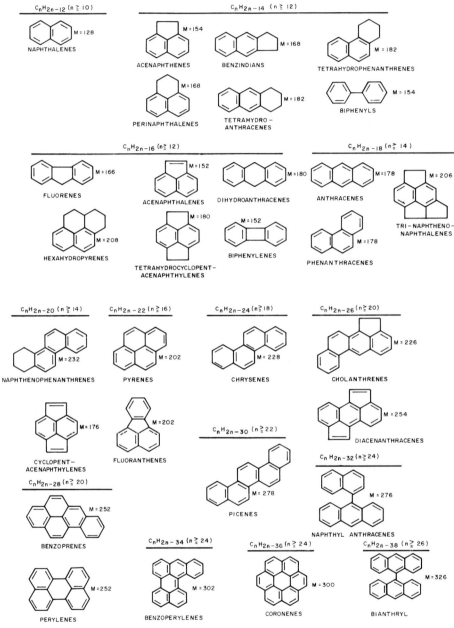

Fig. 10 Possible compound types for the polyaromatics.

greater concentration of these compounds would probably be found if larger TGA samples were used and/or the collection of fractions was continued to higher temperatures. Many of these same structures and their homologs were identified by Dark *et al.* (1977) in coal liquids using liquid chromatography–mass spectrometry (LC–MS).

Since the analysis of the TGA fractions was done, no attempt was made to classify the heteroatom compounds contained in the TGA fraction. However, the heteroatom compound types were classified by high-resolution mass spectrometry on analysis of the original roof and rib coal samples, and the classification is discussed in Section IV,D.

The data presented in Table XV were obtained from the 300°–350°C TGA roof and rib coal residues and illustrate the expected decrease of the aromatic compound types in the TGA residue samples as compared to the original roof and rib coal samples. The decrease was expected since the volatiles removed by TGA heating were largely aromatic compound types. Consequently, the alkanes and cycloalkanes (paraffinic compound types) increased.

Data from the lower-temperature TGA residues showed higher concentrations of aromatics, i.e., the concentration of aromatics in the residue was proportional to the temperature to which the residue was taken. The composition of the 300°–350°C roof and rib coal residues was selected to illustrate the change in the coal composition as the samples were heated.

Table XV gives the comparative data for the roof and rib coal samples, respectively. The value of 10 was assigned to all compound types listed for both the original roof and rib coal samples. The ratio of paraffin types remaining is higher in the roof coal than rib coal (25.1 versus 15.8). Since the volatiles from the roof coal were richer in aromatics, it appears that the reduced paraffinic content of the roof coal samples remains in the residue. The aromatic compound types present are the

TABLE XV   *TGA Residue Samples Compared to the Original Coal Sample*

Compound types	Original coal[a]	300°–350°C TGA residue from:	
		Roof coal	Rib coal
Alkanes and cycloalkanes	10.0	25.1	15.8
Monoaromatics	10.0	5.0	8.5
Naphthalenes (−12)	10.0	2.4	1.8
Tetrahydroanthracenes (−14)	10.0	2.0	2.2
Acenaphthalenes (−16)	10.0	4.4	3.7
Anthracenes (−18)	10.0	2.2	1.1
Other polyaromatics	10.0	3.7	4.3

[a] The values of 10.0 is assigned for comparison between compound types.

same for both original and residue samples of the roof and rib coal samples.

C. Analysis of Raw Coal Samples Taken from Different Levels in the Mine

Several differences were noted between samples from the roof coal (1 ft or less below the slate) and the rib coal (taken at about 3½ ft below the roof coal). (See Fig. 2.) Some of these differences have been referred to in data so far presented. Additional differences will be presented with data on the compounds with heteroatoms.

Differences are shown in Table XVI for five classes of polyaromatics. The concentration of these polyaromatics was much greater in the original roof coal than in the rib coal (ratio was 10.0:5.6). These data resulted from a summation of data increments obtained from the high-resolution mass spectrometer, i.e., the analytical data were accumulated from each fraction as the temperature of the sample probe was raised from 50° to 350°C.

Tabulated below is a comparison summary of three general classes of hydrocarbon compounds showing the relative concentration of these hydrocarbons in the roof and rib coal samples.

	Roof coal	Rib coal
Alkanes	9.6	18.0
Mono, bi, and tricycloalkanes	19.7	18.2
Total aromatics C_nH_{2n-6} through C_nH_{2n-26}	50.9	43.4

The above data were obtained from many fractions produced by high-resolution mass spectrometry.

An additional and useful study using steam distillation was made by Professor E. J. Eisenbraum of Oklahoma State University on these same samples; this provided additional proof of these seam position chemistry differences. In this study the coal samples were placed in flasks which had previously been fired in an annealing furnace. Other glassware, e.g., sintered funnels, was also fired. Hot deionized water (150 mliter) was added, and the flask was connected to the steam distillation apparatus. Steam was introduced, and the flask containing the coal sample was heated with a heating mantle. Two 300 mliter fractions of condensate were collected. These fractions were invariably contaminated with coal dust. Consequently, they were filtered through a bed of Dicalite contained in a sintered glass funnel. A fresh funnel

TABLE XVI *A Mass Spectrometric Relative Comparison of Selected Hydrocarbon Compound Types Found in Roof and Rib Coal*

Examples	Relative concentration of polyaromatics detected		
	Roof coal	Rib coal	Ratio
C_nH_{2n-12} compounds — Naphthalenes	5.3	2.8	1.8
C_nH_{2n-14} compounds — Perinaphthenes	2.6	1.5	1.6
C_nH_{2n-16} compounds — Hexahydropyrenes	1.0	0.6	1.6
C_nH_{2n-18} compounds — Anthracenes	0.8	0.5	1.6
C_nH_{2n-20} compounds — Naphthenophenanthrenes	0.3	0.2	1.5
TOTAL	10.0[a]	5.6	

[a] The sum of the polyaromatics for the roof coal was set at 10.0 for a frame of reference for comparison.

and Dicalite bed was used with each coal sample. The condensate fractions were each treated with 100 g NaCl and then extracted with two 20 mliter portions of ethyl ether. Each ether extract was dried with 1.5 g MgSO$_4$ and filtered. A blank determination was also carried out on a single 600 mliter sample of condensate. When these steam distillates from roof and rib coal samples were examined at OSU, they showed the same general characteristics as we had previously found in our work. The roof coal showed an increased concentration of lower homologs of the aromatic compound types as previously shown in Table X.

D.   Compounds with Heteroatoms Found in Coal Samples

1.   CHO Components Evolved at 300° and 400°C in the Ion Source

High-resolution mass spectrometry is particularly useful in the detection of heteroatomic species present in a complex mixture. This is because in high-resolution mass spectrometry each integral mass is determined precisely. For example, a $C_{13}H_{10}O$ dibenzofuran has a nominal or integral mass of 182, just as does a $C_{14}H_{14}$ tetrahydroanthracene. It is the precise mass measurements of 182.0731 and 182.1095, respectively, which allows the identification of the two different compounds.

The concentrations of the CHO compounds varied greatly for the samples taken at the three different levels in the coal seam as shown in Table XVII. The concentrations of CHO components were found to be greater (twice) at levels nearer the roof slate. This is where the aerobic ground percolation waters are most likely to be present and first contacting.

Examples of compounds or compound types found from the precise mass measurements include the following:

Mixed ketones	Phenylfurans	Naphthalic	Dibenzalacetone
Mixed acids	Chromones or coumarins	anhydrides	Benzonaphthofurans
Mixed esters	Phthalic anhydrides	Gallic acids	Benzanthrones
Furans	Dibenzofurans	Indanones	Naphthoxanthenes
Phenols	9-Fluroenones	Tetralones	Dinaphthofurans
Phenyl alkanones	Benzoic anhydrides	Cinnamic acids	

The carboxyl and hydroxyl groups (among the more dominant) shown above have also been reported by Van Krevelen (1961) from coal oxidation. In his discussion, it is pointed out that evidently part of the

TABLE XVII *Comparison of Heteroatom Compound Types in Bleuler Coal Dust Samples Taken from Different Levels in Consol Blacksville No. 2 Mine*

Compound types and carbon range	Examples of structure	Relative percentages from different seam levels					
		300°C probe fraction		400°C probe fraction		
		1 ft	2 ft	3½ ft	1 ft	2 ft	3½ ft
CHO Compounds C_4 to C_{19}		10.0 [a]	6.4	3.6	8.3	7.4	5.5
Other Heteroatomic Compounds C_4 to C_{14}		3.4	1.3	1.8	4.0	3.1	0.8

[a] The value of 10.0 is assigned for comparison between compound types.

heterocyclic oxygen is opened and converted into carboxyl and hydroxyl groups by the following mechanism:†

The depth of ground cover over a coal seam and the type of cover (lithology) and the presence or absence of aerobic groundwater all help to determine the methane content of a coal seam. Therefore, these data may be interpreted along the following lines: namely, that the shallow-covered coals will have more of the oxygenated species listed here, while the deeper, high-methane coals will have fewer of these oxygenated species.

2. CHN and CHNO Components Evolved at 300° and 400°C in the Ion Source

The preferential concentration of CHN and CHNO compounds was at the upper level in the coal seam as was true for the CHO compounds (Table XVII).

Percent of Ion Yield of CHN
and CHNO Components at
Indicated Probe Temperature

Seam level	300°C	400°C
1 ft or less	3.1	2.1
$3\frac{1}{2}$ ft	1.4	0.4

A partial listing of the identical heteroatomic compounds includes the following:

Pyridines	Benzotriazoles	Bicycloamines	Hydroxyquinolines
Quinolines	Phenanthrolines	Pyrroles	Pyrazinoic acids
Amines	2-Pyridones	Primaclones	Nitroanilines
		Benzo(def)carbazoles	

Of the above compounds, pyridines and quinolines have been reported as typical nitrogen containing compounds isolated by solvent

---

† Similar mechanisms for oxidative degradation of coal are detailed in Chapter 55, Section III.

extraction (Kirmer, 1945). The early date of this reference and the only such reference discussed in a 1974 paper by Fine *et al.* (1974) on the total nitrogen in coal indicate the limited amount of data reported in the literature on the determination of CHN and CHNO compounds in coal. The limited information concerning the nitrogen groups in coal was also noted by Van Krevelen (1961). Francis and Wheeler (1925) had concluded several years ago that nitrogen occurs mainly in cyclic structures. Recently, Schiller (1977) used gas chromatography–mass spectrometry and found quinolines, benzoquinolines, and carbazoles in coal derived liquids.

3. CHS Components Evolved at 300° and 400°C in the Ion Source

The data in Table XVII show the characteristic preferential concentration of heteroatom compounds in the roof layer of coal in a coal seam. Comparative data are shown below:

Percent of Ion Yield of CHS
Components at Indicated Probe
Temperature

Seam level	300°C	400°C
1 ft or less	2.2	3.8
3½ ft	1.3	1.2

The selection of compound types was based upon the work of Akhtar *et al.* (1974). The small mass difference of 0.0034 amu between nominal ion peaks of hydrocarbons and of CHS compound types makes positive identification of sulfur compounds difficult. Akhtar made his selection of sulfur compounds by additional mass-matching techniques. Compounds thus identified by Akhtar† were those also found in this study and include the following:

Thiacycloalkanes	Dimethylbenzothiophenes
Bicycloalkylthiols	Benzylthiophenes
Methylthiophenes	Dibenzothiophenes
Aromatic thioethers	Methyldibenzothiophenes
Tetrahydrobenzothiophenes	Alkyldibenzothiophenes
Octahydrothiafluorenes	Dithiaalkylbenzenes
Benzothiophenes	

† For additional details on the aromatic sulfur compounds in Synthoil products see Volume I, Chapter 17, Table IV.

4. Comparisons Possible between Compound Types Determined

With an improved knowledge of the hydrocarbon and heteroatomic compounds present in coal, it then is possible to make certain comparisons between the various components. Such comparisons could provide a detailed means of categorizing different coals.

It has been pointed out by Karr (1963) that the same fundamental structures are found throughout the various classes of coal-derived aromatics with analogous structures containing oxygen, sulfur, or nitrogen. An example is that of fluorenes (I) and dibenzofurans (II).

and

I   II

In a comparison of the concentration of the above two compound types which included the methyl-substituted compounds, it was found that the dibenzofurans (II) were only about one-third the concentration of the analogous hydrocarbon structures (I). Such comparisons could be easily computerized from the analytical data.

E. Potential Operational Usefulness of Compositional Data

These data may be found operationally useful in explaining how newly mined coals produce heat and/or combustible or explosive gases in silo storage, in rail transport, or in ship transport and why freshly mined coal surfaces are so combustible.

We may also postulate from these data a very interesting thing about coal storage and/or transport and oxidizing coal surfaces that are emitting gaseous hydrocarbons. If the coal pile is partially wetted with liquid water, as from a rain, then the packed coal will have many capillary channels and pores filled with water. These water-plugged envelopes and capillarity produced little "tanks" will have increased amounts of hydrocarbons available as reactants per unit pore volume and per unit catalytic reaction site. This means that the reaction rates will be speeded up in at least three ways. First, from the kinetic rate effects according to the following equation,

$$dx/dt = K(pC_xH_y)^m(pO_2)^n, \tag{2}$$

the increased partial pressure of (C_nH_m) results in an important increased rate effect. Second, the water vapor will allow even more hydrocarbons to be released for reaction by a water vapor displacement effect or stripping effect on the coal surfaces.† And third, the increased number of chemical events per unit area mean even more heat effects, hence more outdiffusion of hydrocarbon gases per unit time. So, seemingly, a rather odd event may occur. The addition of some liquid water to a coal pile may result in a much higher heat release and faster oxidation of the coal pile.

It will also be noted that there apparently is a vertical gradient effect in the amount and nature of the evolved gases over the height of the coal seam. This "chromatographic column" effect is a very striking point and, we believe, a quite useful concept.

F. Additional Comments

Figure 10 shows the basic governing equation (Freundlich's adsorption system) of the amount of methane and other gases being held by the complex of van der Waals' bonds and also that there are many, many different physical adsorption sites operative within the coal structure. With coal being a complex, transformed, primarily fossil plant residue, this Freundlich equation adherency is not at all surprising.

It has been reported elsewhere (Morrie *et al.*, 1976) that the local topographic hydrostatic pressure determines the local methane gas pressure (generally, this amounts to 0.8–0.9 of the prevailing hydrostatic pressure). Therefore, it is not surprising that for a given seam, the amount of gas adsorbed within and upon the coal is a linear function of the depth of overburden. The fracture and joint (cleat) system within the coal helps to govern the actual rates of coal seam gas efflux (Popp and McCulloch, 1976; McCulloch *et al.*, 1974), as might be expected. With a coal seam at or near the surface, as with surface minable coals, there could be no "excess" ΔP methane gas available to measure.

As already noted, the coal seam gases in the Pittsburgh seam are not uniformly distributed on the vertical seam face but have some interesting geochemical chromatographic aspects. While we have not pursued a full microstratigraphic, intraseam exploration of these hydrocarbon differences, from our partial analytical explorations we must

† Dr. Kang Yang at Continental Oil Company suggested this water-vapor-film displacement of hydrocarbons from observations on how charcoal columns are hydrocarbon cleaned using steam-stripping methods. (See also Chapter 38, Section VI,A, for steam stripping of hydrocarbons from activated carbon.)

consider that such microseam stratigraphic differences very likely do exist on a layer-by-layer basis. Our samplings were on a seam basis, not a microseam basis. The near-roof top seam section evidently has the higher hydrocarbon gas content. (Furthermore, the P, the S, the ash, and the N contents are similarly higher in the top Pittsburgh No. 8 zone, but these variables are also higher in the bottommost section of the seam. The oxygen contents are inversed, being lowest in the topmost and the bottommost zones of the seam.) Perhaps these near-roof hydrocarbon accumulations happen just because their adsorption-limited mobilities are least. It may be true that these near-roof coal sections have a somewhat different rate and set of coalification reactions. And it may be that the chemical pyrolysis transforms of hydrocarbon gases have most occurred there, and that this coal has normal fossil plant residues plus hydrocarbon gas transforms formed over geological periods. What this means to us is that the roof coal is not coal alone. Instead, it is a coal stratum that has had ongoing gaseous hydrocarbon reactions for geological epochs and that it now represents coalified plant tissue residues plus hydrocarbon pyrolysis products plus current hydrocarbon gases per the Freundlich adsorption. In other words, it is a labile, solid hydrocarbon chromatographic column that has been continuously changed by reactions of and from the geologic time passage of numerous hydrocarbons.

These coal-adsorbed methane-rich gases are capable of early surface reactions with the mine or the storage area (silo) air. A large number of oxygen related organic compounds are formed. The presence of the alkenes and even acetylene within these coals might come as somewhat of a geochemical surprise, but the USBM-Maurice Deul team also amply and reliably confirms this to be the case (Kim, 1973). While these hydrocarbon types are present at the 0.1–0.3 vol % level individually, they offer a total volume of 2.5–3.0%. Further, as the methane preferentially flushes off from the new coal surfaces, these heavier hydrocarbons are left to react upon the surface. We must consider the newly mined, high-methane, fresh coal surface to be a most active, chemically changing, hydrocarbon-rich reaction zone.

Finally, it is clear that helium and hydrogen must also be considered as characteristic gases in many coal seams. Hydrogen sulfide also occurs in a few coal seams and areas, even to the point of limiting or eliminating coal production. This is similar to the H_2S/CH_4 mixtures that are found elsewhere, as in West Texas, Louisiana, Wyoming, or in Alberta. In those coal seams now having pyritic "sulfur" balls, it seems likely that similar H_2S/CH_4 mixtures met over long geochemical periods with

Fe(HCO$_3$)$_2$ from groundwaters that had percolated down through the overlying Permian redbeds, etc.

However strange it may at first appear to some, we must conclude that the Pittsburgh seam has had for eons a hydrocarbon trapping and reaction or transformation role, as evidence from the roof coal hydrocarbon analyses. While part of these hydrocarbon "shadows" so to speak, may come from normal coalification reactions, we believe that another part may come from migrating petroleum fractions, including natural gases and crude oils, which have been partly coalified. The roof coal contains these hydrocarbon shadows or traces, as evidenced by our measured higher methane and light hydrocarbons and aromatics contents. In fact, sometimes, in the mining of deep anerobic sections of the Pittsburgh coal seam, crude oil residues are still encountered. We have on occasions found small bits of the brilliant black solid hydrocarbon transform mineral called "grahamite" in this coal seam.

V. SUMMARY

The work described utilizes analytical techniques and research capabilities in mass spectrometry and thermogravimetry. A specially constructed volumetric Bleuler mill apparatus was used for grinding fresh coal samples from three positions of the Pittsburgh No. 8 seam at Blacksville, West Virginia. Grinding was carried out in vacuum, air, and oxygen atmospheres, and the released gases were collected and analyzed. Original coal samples and the Bleuler ground samples were then subjected to TGA analysis. Volatile fractions were collected to 500°C, and both the volatile and the TGA residues were further analyzed. The following conclusions are presented:

(1) A Bleuler grinding and gas-collecting apparatus was developed and shown to be applicable.

(2) The average calculated weight percent of methane released by grinding coal in a Bleuler mill was 0.38 wt%. Data are based upon 22 separate samples (50 g each) ground under vacuum, in air, and in oxygen atmospheres. This amount is in agreement with predicted values from adsorption data.

(3) An unusually comprehensive array of both stable and reactive hydrocarbons other than methane was found in the collected gases. These included alkanes, alkenes, and light aromatics ranging from C$_2$ to C$_{10}$ plus oxygenated compounds (ketones, aldehydes, alcohols, ethers, oxygenated aromatics).

(4)   The oxygen atmosphere in the Bleuler mill (as compared to vacuum and air) gave product gases showing greater concentrations of oxygenated compounds, thus defining the tendency of coal surfaces to take up large amounts of oxygen. This oxygen reactivity with fresh coal also implies a net BTU decrease for the coal.

(5)   The coal dust samples produced by Bleuler grinding were semiquantitatively analyzed for comparison of components in the roof coal and coal specimens taken at lower levels in the seam. The roof coal is characterized by (1) lower concentrations of aliphatic and alicyclic hydrocarbons, (2) higher concentrations of total aromatics and containing a higher ratio of polyaromatics to monoaromatics, and (3) higher concentration of heteroatom hydrocarbons (CHO, CHN, CHNO, CHS, etc.). These conclusions are similar to those found for the TGA coal residues.

(6)   These data are useful in the understanding of high-methane, fresh bituminous coal surface chemistry including various reactions of coal surfaces with air (oxygen) versus time. This information could lead to increased safety in underground coal mines. A further contribution of this research is the development of hardware and procedures for the analysis of gas phase reactions on or at a solid surface.

ACKNOWLEDGMENTS

We found Adler Spotte, President of the Blacksville Division, Consolidation Coal Company, most helpful in arranging for our coalface samplings and for useful discussions of the local area mining problems as herein related. Appreciation is due several individuals at Continental Oil Company, Ponca City, Oklahoma, including J. R. Harmon, C. L. Hassell, W. D. Leslie, A. S. Rosenberg, E. L. Sones, C. M. Starks, and R. M. Tillman. Acknowledgment is also due to the late W. K. Moore and the late Gerald Perkins, Jr., for their involvement in this work.

REFERENCES

Akhtar, S., Sharkey, A. G., Jr., Shultz, J. L., and Yavorsky, P. M. (1974). *Am. Chem. Soc., Natl. Meet., 167th, Los Angeles, Calif.*

Chait, E. M., Shannon, T. W., Amy, T. W., and McLafferty, F. W. (1968). *Anal. Chem.* **40,** 835–837.

Dark, W. A., McFadden, W. H., and Bradford, D. L. (1977). *J. Chromatogr. Sci.* **15,** 454–460.

Fine, D. H., Slater, S. M., Sarafim, A. F., and Williams, G. C. (1974). *Fuel* **53,** 120–125.

Francis, W., and Wheeler, R. V. (1925). *J. Chem. Soc.* **127,** 2236.

Girling, G. W. (1963). *J. Appl. Chem.* **13,** 24–29.

Hamming, M. C., and Foster, N. G. (1972a). "Interpretation of Mass Spectra of Organic Compounds," pp. 602–642. Academic Press, New York.

Hamming, M. C., and Foster, N. G. (1972b). "Interpretation of Mass Spectra of Organic Compounds," pp. 491–504. Academic Press, New York.

Hamming, M. C., and Foster, N. G. (1972c). "Interpretation of Mass Spectra of Organic Compounds," pp. 477–480. Academic Press, New York.

"Handbook of Chemistry and Physics" (1971–1972). 52nd Ed. Chem. Rubber Publ. Co., Cleveland, Ohio.

Karr, C., Jr. (1963). *In* "Chemistry of Coal Utilization" (H. H. Lowry, ed.), p. 555. Wiley, New York.

Kekin, N. A., Sklzar, M. G., Palaquta, T. S., and Nikitina, T. E. (1973). *Khim. Tverd. Topl.* **3,** 71.

Kim, A. G. (1973). *U.S. Bur. Mines, Rep. Invest.* No. 7762.

Kirmer, W. R. (1945). *In* "Chemistry of Coal Utilization" (H. H. Lowry, ed.), Chap. 13. Wiley, New York.

McCulloch, C. M., Deul, M., and Jeran, P. W. (1974). *U.S. Bur. Mines, Rep. Invest.* No. 7910.

Menster, M., O'Donnel, H. J., Erqun, S., and Friedel, R. A. (1974). *Adv. Chem. Ser.* No. 131, 1–8.

Morrie, G. W., Diamond, W. P., and Lambert, S. W. (1976). *U.S. Bur. Mines, Rep. Invest.* No. 8189.

Popp, J. T., and McCulloch, C. M. (1976). *U.S. Bur. Mines, Rep. Invest.* No. 0000.

Radd, F. J., Carel, A. B., and Hamming, M. C. (1976). *Fuel* **55,** 323–328.

Robertson, A. L., and Hamming, M. C. (1977). *Biomed. Mass Spectrom.* **4,** 203–208.

Schiller, J. E. (1977). *Anal. Chem.* **49,** 2292–2294.

Shultz, J. L., Kessler, T., Friedel, R. A., and Sharkey, A. G., Jr. (1972). *Fuel* **51,** 242.

Tunnicliff, D. D., and Wadsworth, P. A. (1968). *Anal. Chem.* **40,** 1826–1833.

Van Krevelen, D. W. (1961). "Coal," p. 170. Elsevier, Amsterdam.

Chapter 40

CHARACTERIZATION OF COALS USING LASER PYROLYSIS–GAS CHROMATOGRAPHY

Ray L. Hanson

LOVELACE BIOMEDICAL AND
ENVIRONMENTAL
RESEARCH INSTITUTE
ALBUQUERQUE, NEW MEXICO

N. E. Vanderborgh

LOS ALAMOS SCIENTIFIC LABORATORY
LOS ALAMOS, NEW MEXICO

I. INTRODUCTION

A. Pyrolysis of Carbonaceous Sediments

Organic geochemistry seeks to learn both of the nature of source rocks and of processes that have resulted in the formation of carbonaceous products. Dynamic conditions result in the transformation of organic sedimentary debris into a variety of different product types. The study of these processes, maturation, has been active for decades.

Simulation of maturation in the laboratory has proven difficult. Although one can readily expose natural immature sediments to suspected pressure and temperature conditions, unless unrealistically high

73

temperatures are reached (near 200°C), degradation and condensation of lipids and lignin are remarkably slow. It is now clear that a variety of factors, including chemical and biological environments, must also influence maturation in addition to the better understood effects of pressure and temperature.

Breger's (1977) model for sediment maturation, Fig. 1, is based on a series of experimental observations and agrees well with such data. Two possible, different depositional environments are assumed, aquatic and terrestrial. (Actually, combinations of these may also occur.) Terrestrial plants deposit organic debris rich in cellulose and lignin; aquatic debris consists largely of lipids, condensed isoprenes, and fatty acids, for example. This model depicts lignin degrading to coaly material and kerogen. Lipids degrade to petroleum and boghead coal and into kerogen. (Kerogen is a complicated polymeric carbonaceous substance which is sparingly soluble.) The mixtures of liquid and solid products are cogenerated along with the gases methane and carbon dioxide.

Depending upon geological depositional conditions, methane and other cogenerated gaseous products can either remain trapped within the carbonaceous residue or can drain into adjacent geological traps. Breger's model represents well the many experimental observations that show thermal maturation is a disproportionation process—sediments are converted simultaneously into lower and higher molecular weight compounds. Such an admixture of starting material and products is the nature of coal.

Pyrolysis of such complex composite materials leads to a variety of product types. Heating accelerates maturation-type reactions simulta-

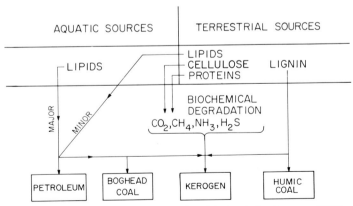

Fig. 1 Interrelationships among fossil fuels. (After Breger, 1977.)

neously releasing dissolved and entrapped gaseous species. (Actually moisture removal, accelerated by increased temperature, is an important first step in low molecular weight gas drainage.) Consequently, after heating the solid graphitic residue will be a "matured" sediment, quite unlike the original material. Organic characterization needs to recognize that naturally occuring coal contains low molecular weight gases and high molecular weight polymers, both products from the maturation processes, mixed together with starting materials.

Sampling this sort of mixture is a nontrivial problem. Although various statistically sound sampling procedures are well established to yield a well-blended, "average" material, such sample handling may mask the necessary subtleties that suggest the details of coal chemistry. Although such details have been of long interest to the geochemist, the coal technologist remains committed to well-blended samples. That situation may change when increased scientific information is needed for desulfurization, pyrolysis, and gasification.

B. Grinding of Coal Samples

Grinding coal samples is a well known way to remove volatiles partially,† to cause localized heating and accompanying bond rupture, and to open reactive surfaces to oxidative reactions. Yet, without such grinding and blending to homogenize a coal sample, the heterogeneity of coaly sediments dominate analytical results. It has been possible to obtain reproducibility by using well blended samples for coal characterization. This has been done, however, at the expense of changing the natural material to a characteristic "fuel." Working with the known heterogeneities of coal seams will only be possible if rapid analytical techniques are developed. For, either one seam will require numerous and sophisticated analysis or one sample will require similar analyses during dynamic processing of coal. New, rapid methods of coal characterization must serve as the data base for developing new coal technologies.

II. THERMAL TREATMENT AND LASER INTERACTIONS WITH COALS

The analysis of coal is complicated by the fact that the solid material is difficult to dissolve and by the fact that the organic compounds within coal exist in a complicated interdependent grouping of gases,

---

† Grinding coal samples with a Bleuler mill and collection of the released gases are described in Chapter 39, Section II, A.

liquids, and solids. Pyrolysis on this material is quite different from that on polymeric material, polystyrene, for example. For in other polymeric solids, commercial polymers, for example, thermal treatment can result predominantly in decoupling reactions. Coal, like certain vinyl polymers, clearly gives products from pyrolytic reactions which are both lower (product gases) and higher (product condensed polymers) in molecular weight than the starting material. This fact requires that attention be paid to the way in which the pyrolysis reactions are completed.

Modern analytical pyrolysis requires this sort of control. Polymeric materials typically degrade under several quite dissimilar mechanisms. As temperatures are varied, the kinetics of particular processes begin to dominate product formation mechanisms. As other temperatures are reached, different mechanisms may be the preferential product formation reactions. The results are that a series of product types will be produced during the thermal excursion. Product distributions depend on thermal treatment as well as on sample chemistry. All modern pyrolysis techniques, especially those coupled with gas chromatography, are designed to avoid this problem by rapidly heating a sample—heating rapidly enough to minimize the effects of low-temperature pyrolysis routes.

Two different approaches to analytical pyrolysis have been developed. The first, the utilization of a pulsed laser, is the subject of this chapter. The second, filament pyrolysis, has certain unique advantages and disadvantages for coal analysis. Filament pyrolyzers operate either as current controlled resistance heaters or as Curie point devices. Although the heating caused by each of these devices is rather similar (thermal rise time, thermal control, etc.), the Curie point devices utilize a far simpler sample handling design. The use of a new, clean ferromagnetic wire for each analysis avoids possible sample contamination. Curie point systems are also more easily employed within mass spectrometer systems.

The concept behind contemporary filament pyrolyzers is to utilize a thin sample evaporated or coated on the surface of a wire so that heat transfer through the sample is sufficiently fast to avoid large thermal gradients. If the wire is rapidly heated, the entire sample will be brought to the desired pyrolysis temperature. If the sample is sufficiently thin, rapid heat transfer occurs. In practice, this sampling procedure is rather restrictive. First, the sample must be in a solution or finely dispersed. Coals cannot be reliably dissolved. Grinding and blending to form a mull is feasible, although it is recognized that grinding releases uncontrolled fractions of volatiles and gases. The sample may only faintly resemble the actual material of interest. Should a thin, even layer be obtained, heating the back side of this film generates gases (thermal maturation) at the metal–film interface. These gases then drive the remaining sample away from the heated surface. Degradation then can occur at a temperature well removed from the predicted temperature of the filament. Last, it is well known that pyrolysis reactions are easily influenced by catalytic agents. Metals such as nickel or platinum are the frequent choices for filaments. These metals are also used to direct polymerization reactions along desired pathways. It seems apparent that such catalysis would also occur on heated wires in these systems and that product distributions emanating from hot filaments may well be influenced by the catalytic properties of the filament metals.

This does not mean that filament pyrolyzers have limited utility. In many systems, especially for the analysis of trace quantities of high molecular weight organic compounds, the utility of these devices has been firmly demonstrated (Gough and Jones, 1975). We feel that laser-induced pyrolysis may indeed offer equal utility for coal characterization.

Laser-induced pyrolysis represents a different type of thermal excursion. It is ideally suited for rapid pyrolysis of solid materials with a minimum of sample pretreatment. However, during this thermal event a series of different temperatures is reached. This is rather different from the basic design of filament pyrolysis, Curie point, for instance.

Laser energy deposition into solid materials has been widely studied since this technology has been adapted in cutting and welding operations (Charschan, 1972). Although chemical studies with lasers typically utilize the high spectral purity of these instruments, for pyrolysis we are concerned with the possibilities of the deposition of precise, controlled quantities of energy into coals. Moreover, this is done in a short time period so that secondary degradation pathways are minimized.

Even though considerable work has been reported on laser interactions with matter, it would be misleading to indicate that the resulting processes, especially with high-power devices, are well understood (Ready, 1965). Clearly, the usual models for spectrochemical excitation are only applicable during initial absorption, if at all, for laser energy deposition results in ionization. Following ionization, the system preferentially absorbs energy by electron absorption of the incoming laser photons.

A typical experiment is shown schematically in Fig. 2. A pulse of laser energy containing 0.5–10 J in a pulse width of 10^{-3} s is deposited onto a sample. The amount of energy absorbed by the sample is a

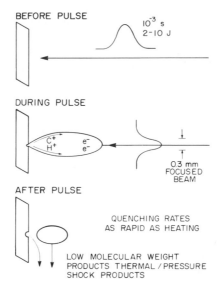

Fig. 2 Laser pyrolysis product mechanisms.

complex function of several variables. Likewise, the temperature generated during the initial portion of the pulse depends upon reflectivity,
heat capacity, and thermal diffusivity. If the sample has high reflectivity, for instance, little or no energy may be coupled into the molecular
system. Likewise, if heat is rapidly transferred away from the site of
energy deposition, the temperature excursion may be small. However,
with coals the absorption occurs with high probability at the focal
point of the focused beam. There are two well established mechanisms
for depositing photon energy of 694.3 nm (ruby) or 1060 nm (Nd).
Classically, neither of these lasers produces photons of sufficient energy
to cause ionization. Yet there is clear evidence that such ionization
occurs through either electron tunneling or multiphoton absorption.
Data suggest that many classes of compounds which should show only
limited absorption at laser frequencies strongly absorb laser radiation.
Through such absorption processes, part of the sample is pumped into
a plasma consisting of energetic ions, radicals of unusual stability, and
electrons. Velocities of plume growth in vacuum are measured at 10^5
cm/s (Pirri *et al.*, 1972). Data show that the absorptivity of laser light
by the plasma is far more probable than by the sample (Afanasyev *et
al.*, 1967). Consequently, during all but the initial stages of the pyrolysis, the sample is shielded by an opaque plasma generated between
the sample and the laser source. During the continued laser radiation,

hot electrons move most rapidly into the path of the laser beam; laser energy is converted into electron thermal energy. The electron-rich plasma reaches temperatures, as indicated by spectroscopic measurements, in excess of 10^4 K (Howe, 1963). Radiation from this plasma heats a region of the sample in the immediate vicinity of the original laser pulse.

At the termination of the 0.001 s laser pulse, the entire system rapidly collapses to ambient conditions. Energy leaves the system by electron deceleration (inverse bremsstellung), by collision with cold gaseous molecules, and by interactions with the sample chamber wall and solid materials. Because of the selective accelerations of electrons into the laser beam, recombination reactions of electrons and positively charged atoms generate a shock (laser spark). The interaction of this shock wave with the heated region of the sample leads to additional molecular fragmentation and formation of pyrolysis products. This is one product distribution. The second comes from the recombination of atomic fragments in the laser-induced plasma. Both of these product ensembles, low molecular weight products and thermal/pressure shock products, are useful for characterization of coals.

Laser pyrolysis is a very rapid process. Figure 3 illustrates the meas-

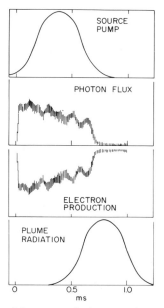

Fig. 3 Time relationship of the source flash pump, laser photon flux, plasma electron production, and black-body radiation from the plasma.

ured time scale for lasers in the normal burst mode. (It is possible to also use a Q-switched mode. Details are not given here.) The laser is pumped with a flash lamp. The controlling L-C time constant of the laser power supply is set to produce a light flash with approximately 1 ms lifetime. Shortly after the onset of the flash lamp, the laser fires. Actually, as shown in Fig. 3, fast recording shows that that the single normal burst is composed of a series of closely spaced short laser pulses. The laser pulse follows the decay of the flash lamp. Electron production from the sample closely follows the laser intensity. Plume radiation, however, shows a time delay, that period needed to move and heat matter in the plasma. Likewise, the radiation from the plasma decays after the termination of the pulse. However, the entire thermal event, thermal rise and fall, is of the same time duration as the laser pulse, 10^{-3} s. This rapid energy deposition is unlike other pyrolysis techniques.

Such rapid kinetics lead to simpler degradation patterns from complicated polymeric samples. Free radicals, generated in the solid material, have no opportunity to couple and produce alternate pyrolysis mechanistic paths. It is probably somewhat incorrect to consider that the higher molecular weight fragments result only from "pyrolysis." However, this short-lived thermal excursion expels a series of molecular fragments. Coupled free radicals in solid coal, formed well after the termination of the pulse, do not contribute to product formation during the usual single pulse interrogation.

The fact that the plasma products quench out from a hot ensemble suggests that laser pyrolysis gives sample information about the composition of this plasma. Figure 4 shows this in a schematic fashion. During process A, part of the sample, especially the surface components, is pumped into a plasma. Because of the selective absorption of photons by free electrons, this plasma rapidly reaches an elevated temperature. Pressures in the plasma probably remain at at least 1000 torr in the experiments described here. Although the elevated temperature does result in increased gas velocity, the results suggest that the plasma expands into the direction of the laser beam and that absorbing centers, unbound electrons, remain within the plasma system. Because of these elevated temperatures and because of the rapid motion of free electrons, heat transfer within the plasma should be highly efficient. Consequently, one can conceive of this plasma as an "isothermal" condition, for any imbalance in temperature would be efficiently removed. (This does not account for the steady-state heat losses at the plasma–atmosphere boundary.) At the termination of the laser pulse, the system quenches back to ambient conditions.

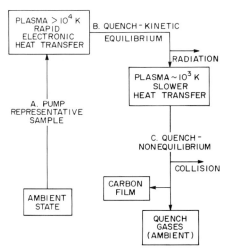

Fig. 4 Mechanistic representation of product formation during plasma quenching reactions.

During the quenching, process B, initially the system remains in kinetic equilibrium until a temperature is realized that first permits the recombination of free electrons into the atomic manifolds. Without the possibility of efficient heat transfer processes caused by the unbound electrons the plasma cools in a nonequilibrium manner. Consequently then a third process, C, cools the plasma by either collision or additional radiative heat losses. Because of the mechanisms of this sort of cooling process, the plasma-quenching products reflect the equilibrium distribution that is frozen out of the plasma at that point where electron recombination occurs, at a temperature significantly cooler than the peak excursion of the thermal pyrolysis. This point has been called the "quenching temperature" and represents that condition during the quenching process where rapid heat transfer caused by unbound electrons is first not possible.

Because of the nature of this process it is possible to compare the quantitative composition of the low molecular weight products (plasma products) with theoretical results calculated from Gibbs free energy values at the quenching temperature and know the atomic stoichiometry of the plasma. Since the stoichiometry of the plasma should be identical to the sample, such results indicate that laser pyrolysis permits a measurement of the atomic composition of a sample by the analysis of low molecular weight gases produced during analyses. This technique has been successful in hydrocarbon analyses.

III.   CONTROL OF LASER THERMAL PYROLYSIS

A schematic representation of a solid-state lasing device is shown in Fig. 5. The heart of the system is a ruby or glass rod surrounded by a high-intensity flash lamp. (Typically the rod and helically coiled lamp are positioned within a water cooled reflective cavity.) The rod is carefully aligned between two mirrors, the fully reflecting back mirror and the partially reflecting front mirror. An intense xenon flash lamp pulse pumps electrons within atomic manifolds into a long-lived ex-cited state. Spontaneous photon emission from this higher state trig-gers a cascade of additional photons which pass back and forth within the laser cavity. A small fraction of the total photon energy is removed during each pass through the front mirror. That fraction forms the output beam which can be used for pyrolysis experiments. In order to achieve high-power densities it is necessary to focus the beam. This is easily done by using a single element lens after the front partially reflecting mirror.

Table I lists commercially available lasing devices. Either pulsed gas (CO_2) or solid state (ruby, Nd/glass, or Nd/YAG) are available. As can be seen the output wavelength varies from the infrared (CO_2) into the visible (ruby). Maximum output power (per pulse) on modest com-mercially available devices runs from 100 to 10^6 W over the beam diameter that is set by the laser optics. It is possible to obtain power densities in excess of 10^7 W/cm^2 with these sources by use of a focusing lens.

The wide flexibility for using lasers as pyrolysis sources has increased the possibilities for pyrolysis experiments over conventional devices.

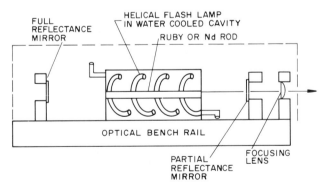

Fig. 5   Schematic representation of laser for coal pyrolysis. Solid laser rod surrounded by water cooled flash lamps positioned between back and front mirrors. Focusing lens concentrates beam for sample pyrolysis.

TABLE I *Characteristics of Pulsed Lasers as Pyrolysis Sources*

Laser Type	Output wavelength (nm)	Beam divergence (mrad)	Possible rep. rate (Hz)	Output power (W)	Efficiency[a] (%)
CO_2 (gas)	1.06×10^4	6	1000	600	20
Cr^{3+}/Al_2O_3 (ruby)	6.94×10^2	10	1	10^5	0.7
Nd^{3+}/glass	1.06×10^3	10	1	10^6	3
Nd^{3+}/YAG	1.06×10^3	8	20	10^2	3

[a] Efficiency, calculated as output energy as percent of input (flash) energy or electrical current.

However, that same flexibility has complicated the control problem. Basically, control of laser pyrolysis begins with a reproducible laser and then involves a careful manipulation of experimental geometry to ascertain that the energy density deposited into the sample is similar on replicate shots. Unfocused beams, of a diameter, dictated by the diameter of the laser rod, seldom will cause thermal pyrolysis unless large lasing systems are used. Highly focused beams can cause the majority of products to emanate from plasma quenching. Usually one designs laser pyrolysis experiments to set the sample just beyond the spot of intense focus, i.e., at 10.5 cm from a lens of 10.0 cm focal length.

It is not possible to focus laser beams precisely. Rather, as can be seen in Fig. 6, a lens or lens system will form a minimum focused beam diameter d_2 that is equal to the product of the focal length times the wavelength divided by π times the diameter of the beam entering the lens system. Thus the focused spot diameter is directly proportional to the lens focal length divided by the entrance aperture. This ratio, f_1/d_1, has been called the effective f/number of the lenses combination.

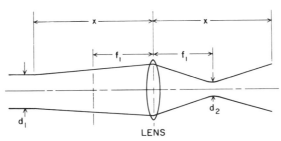

Fig. 6 Focusing of collimated laser beam. Beam diameter d_1 is lead into lens of focal length f_1 to form high intensity zone of diameter d_2. $d_2 \sim f_1 \lambda / \pi d_1$.

Quite obviously, large diameter, short focal length lens offer the highest degree of focus and the highest power densities. However, such high-power densities lead to enhanced plasma products and control difficulties. [If large entrance apertures are required, an expanding telescope can be first utilized (before the focusing lens focal length) to expand the laser beam to fill the lens aperture completely. This involves increasing the value of d_1 (Fig. 6).]

In practice a partially focused beam is simpler to use than a highly focused one. Ten cm lenses with 2.54-cm diameter permit precision of analyses without the necessity of extremely careful spatial control of sample-laser geometry. In practice, too, because of the multimode nature of most pulsed laser radiation it is impossible to focus as sharply (10^{-4} cm) as the equation shown in Fig. 6 indicates. Rather the beam diameter approaches 10^{-2} cm as its minimum. In the experiments reported here evidence suggests beam diameters on the surface were of this magnitude.

The frequency of laser radiation is selected when one chooses the lasing system. Any of the three frequencies listed in Table I is suitable. One would first assume that a higher frequency device (shorter wavelength) is more suitable for pyrolysis. However, laser photons can be degraded to heat by the presence of free electrons, lattice vibrations, and other absorbing centers so that photon energy is less important. Experience shows that almost all materials absorb intense CO_2 radiation (10,400 nm) even though normal infrared transitions are not active at that wavelength. Likewise, limited experience shows that not much difference in pyrolysis products results when one uses ruby (694.3 nm) or neodymium (1060 nm) (Hanson, 1975). Although energy may be absorbed into one particular region of the molecular manifold, rapid lattice vibrations in solid species efficiently transfer that energy. Reflectance generally increases with higher wavelength; however, reflectance decreases with increased temperature. The end result is that most samples, especially coal, will efficiently absorb laser radiation.

IV. LASER PYROLYSIS METHODOLOGY

Two different sample types for coal pyrolysis are possible. Either one can use small chips removed from run-of-mine or core samples or one can use blended samples. It is important that approximately uniform density be obtained in each sample. If this is not the case, different quantities of material will be irradiated in the laser beam. Sieved and blended samples are most easily handled as pellets. Coal is readily pelletized. It is best to obtain a reasonably constant density by pressing a fixed weight into a preset volume (the volume of the die–punch combination). Small variations in sample density are compensated somewhat by deeper laser penetration into less-dense samples.

The sample chamber used for these pyrolysis studies is shown in Fig. 7. A short section of quartz tubing [0.25-in. o.d. (6 mm) × 2-in. long] is connected between two glass-to-metal seals. Quartz tubing gives superior performance since it does not fracture as easily as glass. Although various other sample chambers are possible, none has proven as simple and reliable (Vanderborgh, 1977). Tubing must be rigorously cleaned to remove traces of organics, and clean quartz should be handled with gloves prior to analyses. An important part of the cleaning operation is a high-temperature flaming to remove traces of organic compounds. (Such contaminants can degrade the laser intensity by absorbing light before the beam impinges on the sample as well as lead to spurious product formation.) It is, therefore, possible to utilize these tubes repeatedly, although the low cost would permit single analysis per tube. Thorough cleaning is necessary in either case.

The curvature of the quartz tubing wall causes some additional focusing. However, the narrowness of the beam at the point the laser beam meets the tubing wall suggests that this is a small effect. There is consequently little advantage in going to a planar cell window, and such a modification would remove the possibility of using readily available and inexpensive tubing.

Laser pyrolysis requires flow modification of normal gas chromatographic instrumentation in the same way required with other thermal analysis–gas chromatography techniques. This is shown in Fig. 8. The coal chip or pellet is inserted into the quartz sampling tube connected on line to a gas chromatograph. It is efficient to isolate the pyrolysis chamber so that by closing toggle valves the pyrolysis chamber can be

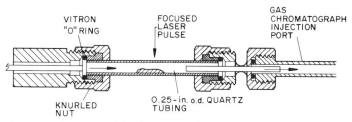

Fig. 7 Laser pyrolysis assembly. Commercially available glass-to-metal seals are modified (hard solder) to attach directly to the inlet port of a gas chromatograph. Sample tube is 6 mm o.d. quartz tubing connected to 0.25 in. tubing fittings.

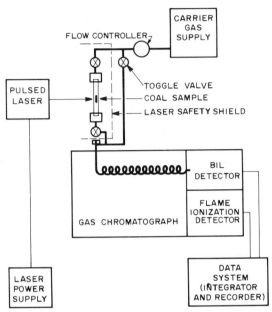

Fig. 8 Pulsed laser-gas chromatograph instrumentation for laser pyrolysis-GC. Sampling chamber connected directly to GC.

made ready for another shot while the previous analysis is being completed. The sample is inserted into the quartz tube and carefully positioned so that the required section of the sample will be in the laser beam. This positioning is simplified by the use of a second, low-power laser (HeNe, 1–3 mW) that is positioned to run continuously and coaxially with the high-energy system. In this way a laser "pointer" marks the exact analysis location.

Pyrolysis products consist of a complex mixture of gaseous compounds. These compounds range from hydrogen to high molecular weight species and include both inorganic (H_2S, for example) and organic species (C_2H_2, for example). Complete analysis would also include graphitic carbon and high-temperature compounds formed by heating of clay particles. Rather obviously these low vapor pressure solid materials are ignored by gas chromatography. Consequently, one does not analyze all pyrolysis products. Yet a wide range of compound types is of interest. Principally, one would need to know the composition of low molecular weight gases emanating from the plasma-quenching process and intermediate molecular compounds that arise from thermal shock into coals.

Of interest in the low molecular weight gas ensemble are carbon oxides (CO and CO_2), H_2O, H_2S, etc. Many of these low molecular weight gases are not detected with a flame ionization detector. The commercially available thermal conductivity detector lacks sufficient sensitivity to detect low concentrations of these gases. For these analyses previous work has utilized a beta-induced luminescence (BIL) detector. This device shown schematically in Fig. 9 is built around a small section (1×0.5 cm) of the titanium foil that has been impregnated with tritium. This is cemented into a $\frac{1}{8}$ in. branch tee. A quartz rod is then cemented into the NPT. The end of this rod is positioned against the window of a standard photomultiplier tube. Gas connections are made and then this detector is run in series before the flame ionization detector.

This device, the BIL detector, operates on the principle of lumines-

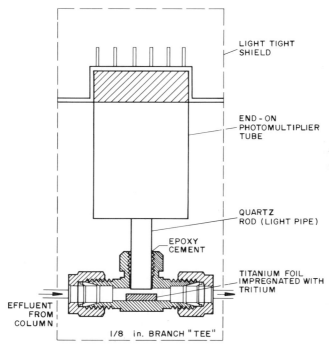

Fig. 9 Schematic of Beta induced luminescence detector. Generalized high-sensitivity detector for quantification of pyrolysis gases. Detector is constructed of a tritiated foil cemented into the bottom of a standard $\frac{1}{8}$ in. branch tee. Measurement is by following photo current continuously from the photomultiplier tube. Necessary high voltage supply (1.5 kV), microammeter, and recorder are not shown.

cence quenching. Trace impurities of nitrogen gas, approximately 20 ppm, in the helium carrier are irradiated with beta activity from tritium. Collisional deactivation results in photon emission which is registered as a steady-state luminescence signal. The presence of another compound in the effluent such as one of the low molecular weight gases of interest transfers energy away from· the system with a decreased amount of photon emission, i.e., the photomultiplier signal is decreased. The quenching response of this device is logarithmic with concentration.

It is important to protect personnel against the hazards of focused coherent radiation. The instrumentation schematic (Fig. 8) shows a "laser safety shield" completely enclosing the experimental area. It is recommended that such an opaque shield be constructed completely enclosing the laser pyrolysis area. A hinged door to give access to the pyrolysis chamber should be arranged with an electrical interlock switch so that it is impossible to fire the laser unless the safety shield is in place. Eye protection is also required during these experiments. The focusing lens serves as another safety device. Although an extremely high-radiation field exists shortly after the beam emerges from the lens, past the exiting focal length, radiation is divergent and should present little hazard to personnel or to the safety shield.

V. LASER PYROLYSIS OF COAL

A measure of the flexibility of pyrolysis–gas chromatography involves the possibilities for carrying out pyrolysis in different environments. Reactions can be carried out in inert gases, the carrier gas He, for instance. Reactions can also be carried out in vacuum, and the carrier gas can be utilized to sweep pyrolysis products into the gas chromatograph. Or, reactions can be conducted in a reactive environment. Pyrolysis products can react with reactive gases. For example, when oxygen is present hydrocarbon products will oxidize to carbon monoxide and carbon dioxide. Primary gaseous products from all coal pyrolysis are carbon monoxide, hydrogen, methane, acetylene, ethylene, and carbon dioxide. Less-significant quantities of larger hydrocarbon compounds, hydrogen sulfide, nitrogen, and hydrogen cyanide also are formed.

Several alternative methods for coal pyrolysis have been reported (Romovacek and Kubat, 1968; Groom, 1969). None of these early techniques utilized rapid-heating methods. Because of the large difference in heating rates and strategies, it is difficult to compare results even in a qualitative sense. Samples have been sealed in tin ampoules and then dropped into molten tin baths at various temperatures (Romovacek and Kubat, 1968). Slow pyrolysis was completed in 50 s. Various types of quartz pyrolysis furnaces have also been utilized (Banerjee *et al.*, 1973; Ignasiak *et al.*, 1970). Typically, such studies involve moving a container of coal sample into a heated region. It is well known that pyrolysis products are significantly altered by heating rate. Typically, high heating rates lead to more significant quantities of low molecular weight hydrocarbons and graphitic char. The significance of controlled heating rates has been appreciated, and resistance heating devices are available in a ribbon configuration. However, even though the heating element can be heated at a controlled rate, sample temperatures are not easily controlled. Other workers have utilized high-intensity xenon flash lamps to cause pyrolysis (Granger and Ladner, 1970). An intense flash of 0.001 s is sufficient to effect pyrolytic reactions.

There have been several reports where pyrolysis data was compared to known properties of coals (Giam *et al.*, 1977). For instance, pyrolysis using coals of different ranks have been reported (Romovacek and Kubat, 1968). The production of benzene and toluene, relative to pentane, increased with higher coal rank. Nonaromatic products compared to aromatic products increased in the opposite order with rank. Other workers have noted that pyrolysis products total yield decreases with higher rank (Suggate, 1972). In these experiments the gas yield was correlated with the volatile matter and hydrogen content of the coal samples. Evidence suggests that the quantity of nitrogen produced during pyrolysis increases with temperature (Groom, 1969). Likewise, water is a product of coal pyrolysis formed both by drying and as a pyrogenic product. Pyrolysis has been used to measure the extent of oxidation of coals (Ignasiak *et al.*, 1970). Data suggest that the volume of carbon monoxide evolved per gram at 350°C gives a measure of the extent of prior oxidation. Other workers showed that higher rank coals require increased temperature to reach a maximum rate in product formation (Barker, 1974). Ultraviolet photolysis experiments have shown higher yields of CO, H_2, and C_2H_2 are formed using the complete spectrum of the flash lamp compared to experiments which absorbed part of the UV radiation (Granger and Ladner, 1970).

It is well known that high heating rates yield higher amounts of volatile products (Kimber and Gray, 1967; Gray *et al.*, 1974). Rates typically approach 10^4-K/s, and under these conditions significant quantities of acetylene and ethylene are formed in the 900–1500 K range (Coates *et al.*, 1974). However, the yield of graphitic char increased during high heating rate experiments, as well.

First experiments using laser pyrolysis on coal were done in sealed ampoules using a series of different inert gases (Sharkey *et al.*, 1964). Multiple laser pulses were deposited into a coal sample, and then the composition of gaseous products in the ampoule was sampled and analyzed by mass spectrometry. Ruby radiation (694.3 nm) was used as the pyrolysis source. Mass spectrometry showed products with mass as high as 130. Calculations suggested that approximately 60 wt % of the volume of the crater formed in the coal sample could be accounted for by the mass of the gaseous products. Hydrogen, acetylene, and carbon monoxide were the predominant products.

Laser pyrolysis gases have a significantly higher H/C ratio than the original coal (Karn and Sharkey, 1966). This fact again shows that pyrolysis reactions disproportionate—products include both hydrogen enriched hydrocarbons and hydrogen deficient (compared to the original coal) char. As would be expected, a fourfold increase in the total gas yield occurred as coals were varied from high rank anthracite to low rank lignite. Likewise, the oxygen content of the coal samples increases in lower rank coals, and pyrolysis products showed a fivefold increase in the CO/CO_2 ratio as the rank was lowered. Again, acetylene yields increased using rapid laser heating.

Coal macerals have been investigated with laser heating (Karn and Sharkey, 1968). Rather similar results were found with data from bulk coal samples. Total gas yields increased with volatile content. Total hydrogen yield increased with hydrogen content.

Several investigations have explored the effect of changing the gaseous atmospheres within the pyrolysis chamber upon pyrolysis distributions (Shultz and Sharkey, 1967). Although one might assume that similar products would be obtained for "inert" pyrolysis conditions, data show appreciable differences as various inert atmospheres are used. Relative amounts of acetylene yield decrease in He atmosphere compared to vacuum. These data and other show that inert laser pyrolysis is actually not approached. The thermal properties of the quenching gas (thermal diffusivity, heat capacity, and ionization potential) dictate the rates of heat loss from the laser plasma. These interrelated properties then influence the quenching reactions and can lead to changes in the product distributions.

Other laser pyrolysis studies have utilized laser heating on a coal sample contained in the source section of a time-of-flight mass spectrometer (Knox and Vastola, 1967; Joy *et al.*, 1968, 1970; Vastola and Pirone, 1966).† Analyses showed that benzene, toluene, xylenes, ethyl-

---

† For experimentals with coals heated at 360°C on a metal ribbon in the source of a time-of-flight mass spectrometer see Volume II, Chapter 21, Sections II, A, 1, and III, A, 1, and Fig. 4.

benzene, naphthalene, methyl-naphthalenes and, ethyl-naphthalenes result from laser pyrolysis (Vastola and Pirone, 1966). With zero ionization voltage, charged fragments of C_2, C_2H_2, and mass 28 (CO or N_2) were detected (Knox and Vastola, 1967). At the reduced voltage of 25 eV additional peaks for C_2H_3, C_2H_5, C_2H_6, HS, H_2S, and mass 32 fragments were measured. Other workers showed that acetylene is again the major pyrolysis product (Joy *et al.*, 1968, 1970). The coal sample produced products to yield detectable spectra for 10 ms using a 1 ms laser pulse.

VI. PLASMA STOICHIOMETRIC ANALYSIS ON COAL SAMPLES

Analyses of the low molecular weight gases resulting from the rapid quenching of a laser-induced plasma from coal samples permit a direct and rapid method for determining the stoichiometry of the sample (Hanson *et al.*, 1977). Plume quenching occurs at some elevated temperature. By comparison with data obtained with calibration samples, it is obvious that sensible results are obtained if one assumes that the quenching temperature is 3000 K. Also, at this elevated temperature thermodynamic predictions show that free radicals species exist in significant concentration.

Calculations are made using standard thermodynamic functions assuming that an isothermal distribution occurs in the plume. Then calculations are completed using increments of 100 K from 2000 to 3500 K. This calculation requires a specified pressure; 2280 torr, the pressure of the inlet side of the gas chromatograph, was used. Equilibrium concentrations were then calculated for the species CH_4, C_2H_4, C_2H_2, C_2H, CH_3, CH_2O, HCO, CO, CO_2, H_2, H, H_2O, C_2N_2, HCN, CS_2, N_2, NH_3, SH, S_2, H_2S, SN, SO, NO, and $C_{(s)}$. These calculations also require specified atomic stoichiometries. Then using these, *and* pressure and temperature, a list of the most stable molecular distribution for these conditions can be obtained.

Input data included the analytical results for five coal samples shown in Table II. These samples included lignites, western sub-bituminous coal, and Illinois Basin bituminous coal. Calculated values for equilibrium molecular distributions resulting from these input parameters are shown in Tables III–VII.

Data in Tables III–VII do well predict many of the observed data resulting from laser pyrolysis on coal. Data is only shown for the 11 most significant gaseous species. The most prominent component of the equilibrium distribution is solid carbon. These data are shown in Table VIII, which shows results at both 2000 and 3500 K. Higher tem-

TABLE II   *Coals Used for Laser Pyrolysis Investigations*

Sample no. Location Type	1 Texas Lignite	2 New Mexico Sub-bituminous	3 Louisiana Lignite	4 N. Illinois Bituminous	5 Central Illinois Bituminous
Carbon, %	43.13	65.26	60.10	72.21	65.40
Hydrogen, %	3.39	4.73	4.23	5.03	4.59
Nitrogen, %	0.73	1.07	1.26	1.29	1.08
Chlorine, %	0.08	0.07	0.14	0.04	0.06
Sulfur, %	2.45	1.47	0.83	2.99	4.41
Oxygen, %	11.96	12.78	14.60	9.60	8.65
Ash, %	38.26	14.62	18.84	8.84	15.81

peratures (3500 K) show a decreased graphite yield and lower rank coals, not surprisingly, show decreased solid carbon concentration in the distribution.

Laser pyrolysis of various hydrocarbon compounds produces hydrogen and acetylene as the primary gaseous products along with significant concentrations of carbonaceous residue. Laser pyrolysis on hydrogen deficient coal yields hydrogen, acetylene, and carbon

TABLE III   *Calculated Equilibrium Distributions from Coal No. 1: Texas Lignite*

Gaseous species	Equilibrium temperature			
	2000 K	2500 K	3000 K	35000 K
C_2H_2	0.00147	0.0189	0.0706	0.0819
C_2H	0.00000196	0.000735	0.0208	0.112
CO	0.301	0.297	0.282	0.248
H_2	0.661	0.632	0.501	0.295
H	0.00111	0.0165	0.0879	0.229
HCN	0.00185	0.00719	0.0122	0.0121
CS_2	0.00612	0.00764	0.00651	0.00461
N_2	0.00951	0.00671	0.00371	0.00252
SH	0.00147	0.00687	0.0109	0.0125
S_2	0.000911	0.00133	0.00139	0.00141
H_2S	0.0143	0.00572	0.00239	0.000949

Stoichiometric mole fractions of original coal (as received basis).

C = 0.456
H = 0.431
O = 0.095
N = 0.007
S = 0.010

TABLE IV   *Calculated Equilibrium Distributions from Coal No. 2: New Mexico Sub-Bituminous*

Gaseous species	Equilibrium temperature			
	2000 K	2500 K	3000 K	3500 K
C_2H_2	0.00198	0.0253	0.0925	0.102
C_2H	0.00000408	0.00101	0.0282	0.147
CO	0.249	0.246	0.233	0.201
H_2	0.722	0.682	0.523	0.289
H	0.00124	0.0183	0.0949	0.235
HCN	0.00228	0.00870	0.0143	0.0138
CS_2	0.00209	0.00287	0.00239	0.00163
N_2	0.0107	0.00734	0.00390	0.00261
SH	0.00147	0.00415	0.00635	0.00691
S_2	0.000280	0.000450	0.000455	0.00439
H_2S	0.00810	0.00336	0.00135	0.000501

Stoichiometric mole fractions of original coal (as received basis).

C = 0.490
H = 0.426
O = 0.072
N = 0.007
S = 0.004

TABLE V   *Calculated Equilibrium Distributions from Coal No. 3: Louisiana Lignite*

Gaseous species	Equilibrium temperature			
	2000 K	2500 K	3000 K	3500 K
C_2H_2	0.00172	0.0222	0.0818	0.0924
C_2H	0.00000358	0.000897	0.0252	0.134
CO	0.297	0.294	0.279	0.242
H_2	0.676	0.640	0.495	0.278
H	0.00117	0.0174	0.0907	0.228
HCN	0.00238	0.00940	0.0162	0.0160
CS_2	0.000946	0.00139	0.0113	0.000746
N_2	0.0134	0.00976	0.00564	0.00390
SH	0.000974	0.00284	0.00432	0.00466
S_2	0.000132	0.000226	0.000223	0.000208
H_2S	0.00531	0.00228	0.000911	0.000334

Stoichiometric mole fractions of original coal (as received basis).

C = 0.488
H = 0.412
O = 0.089
N = 0.009
S = 0.003

TABLE VI   *Calculated Equilibrium Distributions from Coal No. 4: Northern Illinois Bituminous*

Gaseous species	Equilibrium temperature			
	2000 K	2500 K	3000 K	3500 K
C_2H_2	0.00239	0.0305	0.109	0.117
C_2H	0.00000502	0.00124	0.0342	0.174
CO	0.188	0.186	0.175	0.149
H_2	0.769	0.723	0.540	0.285
H	0.00134	0.0198	0.100	0.239
HCN	0.00277	0.0105	0.0172	0.0163
CS_2	0.00592	0.00736	0.00630	0.00450
N_2	0.0130	0.00895	0.00477	0.00319
SH	0.00247	0.00662	0.0101	0.0110
S_2	0.000743	0.00108	0.00112	0.00113
H_2S	0.0134	0.00527	0.00210	0.000770

Stoichiometric mole fractions of original coal (as received basis).

C $=$ 0.509
H $=$ 0.425
O $=$ 0.052
N $=$ 0.007
S $=$ 0.008

TABLE VII   *Calculated Equilibrium Distribution from Coal No. 5: Central Illinois Bituminous*

Gaseous species	Equilibrium temperature			
	2000 K	2500 K	3000 K	3500 K
C_2H_2	0.00234	0.0299	0.107	0.115
C_2H	0.00000492	0.00122	0.0335	0.171
CO	0.186	0.183	0.172	0.146
H_2	0.761	0.717	0.537	0.284
H	0.00133	0.0196	0.0997	0.238
HCN	0.00264	0.00997	0.0162	0.0153
CS_2	0.0113	0.0133	0.0116	0.00842
N_2	0.0120	0.00817	0.00429	0.00285
SH	0.00340	0.00889	0.0137	0.0151
S_2	0.00143	0.00197	0.00207	0.00217
H_2S	0.0184	0.00708	0.00285	0.00107

Stoichiometric mole fractions of original coal (as received basis).

C $=$ 0.505
H $=$ 0.425
O $=$ 0.050
N $=$ 0.007
S $=$ 0.013

TABLE VIII *Calculated Mole Fractions of $C_{(s)}$ for Equilibrium Distributions*

Sample	2000 K	3500 K
Coal No. 1	0.5304	0.3490
Coal No. 2	0.5894	0.3943
Coal No. 3	0.5694	0.3822
Coal No. 4	0.6278	0.4251
Coal No. 5	0.6245	0.4223

monoxide. This should be compared to low-temperature pyrolysis when methane, ethylene, and carbon dioxide predominate. For studies on plasma stoichiometry it is assumed that hydrogen, acetylene, and carbon monoxide emanate from plume quenching while methane, ethylene, and carbon dioxide result as thermal blow-off products from lower temperature pyrolytic zones in the vicinity of the crater formed in the coal sample. Little methane yield is predicted as part of the high-temperature distribution.

Analytical pyrolysis data on coal must be obtained under closely duplicated conditions. Coal samples are formed from compacted powders of blended standards. Compaction is easily accomplished in a standard stainless steel laboratory punch–die combination operated at 20,000 psi. Pellets are transferred to clean pyrolysis tubes and pyrolyzed with a single normal-pulse laser. Both a ruby laser with an output of 2.6 J and a neodymium–glass device with 1.8 J (nominal)/pulse have been used for coal pyrolysis. Resulting data are shown in Figs. 10 and 11.

Figure 10 shows a dual trace of a pyrogram from a coal sample. The upper, downward going trace is the response from the BIL detector. Data on the bottom is the more normal FID response. These data were obtained with a Porapak Q column and show initially, at the highest attenuation the ensemble of low molecular weight gases and then substituted benzenes and higher molecular weight compounds. Notice that several peaks on the BIL trace do not appear in FID data.

Figure 11 shows BIL detector response from another laser pyrogram. The response characteristics quenching of the beta-induced luminescence lead to broader peak shapes than found for the linear FID response. These data were obtained using a II Carbosieve B column run isothermally at 373 K.

The results from a typical thermodynamic calculation are shown in Fig. 12. This data only considers gaseous compounds and shows these

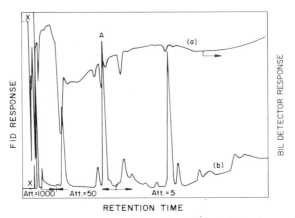

Fig. 10 Laser pyrogram from coal No. 4–Northern Illinois bituminous. (a) Upper trace is from BIL detector, (b) lower from FID. Pyrolysis caused by neodymium laser pulse of 1.8 J. Column is 1 M Porapak Q programmed for 2 min at 323 473 K. 8 K/min to 473 K, hold at 473. 323 K Peak A represents retention time for benzene under these conditions.

distributions as a function of temperature. As can be seen, the predominate gaseous species throughout the temperature range of 2000 to 3400 K is H_2 although at higher temperatures both acetylene and atomic hydrogen H·, are also significant. Methane yields decrease with temperature. This data is quite sensitive to beginning input atomic ratios, i.e., to the chemical composition of sample coal. These data show that several free radical species exist at significant concentration. Naturally, these will quench to more stable species as the temperature is lowered

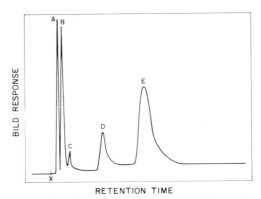

Fig. 11 Laser pyrogram showing BIL detector response. X represents laser pulse time. Peak A is hydrogen; peak B is CO; peak C is methane; peak D is water and H_2S, peak E is C_2H_2. Analysis determined on II Carbosieve B column run isothermally at 373 K.

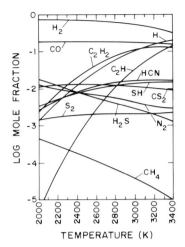

Fig. 12 Equilibrium molecular and atomic distributions for gaseous products of stoichiometry of coal no. 5. Data show log mole fraction as a function of temperature. Compounds not shown have concentrations below 10^{-5} mole fraction.

to the temperature of the gas chromatographic separation. Thus, the total acetylene yield measured on the gas chromatograph corresponds to the sum of C_2H_2 and $C_2H\cdot$ actually produced in the plasma. Likewise, for this model the hydrogen concentration corresponds to the summaration of H_2 and $H\cdot$.

These two assumptions were tested by comparing the experimentally produced hydrogen and acetylene produced with a single laser pulse fired into standard compounds, anthracene and phenanthrene. Calculated thermodynamic equilibrium values for a compound with a H/C ratio of 0.70 at a temperature of 2900 K and pressure of 2280 torr show that the sum for acetylene from combined C_2H_2 and $C_2H\cdot$ should be 19.98% while the calculated value for hydrogen yield should be 80.08%. (This value is the summation of both H_2 and $\frac{1}{2} H\cdot$.) Experimentally, it was found that anthracene produced 20.2% acetylene and 77.0% hydrogen. Phenanthrene data showed 16.4% acetylene and 80.8% hydrogen. Both of these compounds have identical chemical formulas, $C_{14}H_{10}$, and should give such similar results. Pyrolysis data from these two standard compounds show methane at concentrations 10^2 times greater than one would predict from thermodynamic considerations. Compounds such as these aromatic hydrocarbons have no methyl groups to account for high methane concentrations by methyl cleavage. Rather, methane must result from pyrolysis mechanisms that involve molecular cleavage of aromatic linkages in these molecules.

Using these two assumptions, data for these five coal samples were compared to calculated data, at 3300 K, for H_2, CH_4, CO, and C_2H_2. Data is shown in Table IX. There is excellent agreement between the experimental yields and the yields predicted by calculation except for methane. Experimental methane values are at least a factor of 10^3 higher than those suggested for equilibrium distributions. Certainly in these coal samples there is good probability that methyl groups could be cleaved in low-temperature reactions.† The anthracene–phenanthrene data suggest that carbon aromatic cleavage would also contribute to this methane yield. (One would think that relatively little adsorbed methane would be incorporated in these ground samples.)

These data show that predictive calculations are reliable estimates for gaseous yields. There is evidence that oxygen content can be deduced from these data. Figure 13 shows the stoichiometric O/H ratio of the solid coal samples compared to the CO/H_2 ratio found in the pyrolysis products. Increasing oxygen content, as would be expected, leads to greater quantities of CO compared to H_2. This is clearly shown in Fig. 13. Data such as these are rather precise. Although actual quantities of pyrolysis products are often not highly reproducible (±15%), ratios of particular product types show far better precision (±3%). This improvement in precision of data arises from the fact that total quantity

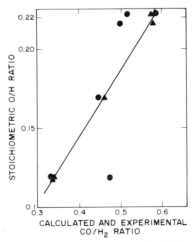

Fig. 13   Comparison of calculated and experimental CO/H_2 ratio with the atomic O/H ratio in a series of American coals. Circles are experimental data; triangles are theoretical.

† An example of this would be the composition at about 2000 K for coke oven gas, Chapter 38, Fig. 3

TABLE IX Comparison of Experimental Low Molecular Weight Gas Products with Products Predicted at 3300 K

Sample	Experimental composition mole percentages from BIL detector				Calculated gaseous percentage composition at 3300 K			
	H_2	CH_4	CO	C_2H_2	$H_2 + H$	CH_4	CO	$C_2H_2 + C_2H$
Coal 1	44.5	2.05	22.9	15.3	46.08	1.36×10^{-3}	26.35	15.6
Coal 2	46.6	3.70	20.9	20.5	46.69	1.42×10^{-3}	21.50	20.3
Coal 3	50.2	1.93	25.1	17.4	47.73	1.30×10^{-3}	25.90	18.2
Coal 4	51.4	3.05	17.7	22.1	47.19	1.46×10^{-3}	16.20	23.7
Coal 5	45.4	3.83	21.7	22.0	47.04	1.45×10^{-3}	15.77	23.4

is determined by the reproducibility of the laser absorption in such samples. Quantities of gaseous products such as CO and H_2 are effected equally by variations in laser absorption. Consequently, the *ratio* of product concentration is less influenced by irreproducible laser energy deposition.

VII. LASER PYROLYSIS OF COALS IN REACTIVE ATMOSPHERES

Several studies have been reported that show that the reactive atmosphere alters pyrolysis distribution (Shultz and Sharkey, 1967; Banerjee *et al.*, 1973, 1977; Anthony *et al.*, 1976; Krukonis *et al.*, 1974). This results from two types of interaction. Free radical coupling can be retarded by "inert" atmospheres so that lower molecular weight products result than would result in the absence of the inert atmosphere. As mentioned previously, specific heats and thermal conductivities are also important, for these dictate the time duration for heat concentration in the pyrolysis experiments. The second type of interaction involves similar free radical quenching in the solid molecular backbone. If gaseous reactants can prevent the coupling of two carbon-free radicals, then lower molecular weight products, including greater acetylene yields, will occur.

These effects are clearly visible as one considers the variations in product yield as one changes pyrolysis atmosphere from helium to hydrogen. Experiments on coals have been done using these two gases as carrier gas for chromatographic separation. The changes are truely dramatic. Although thermodynamic predictions have been made over a range of C/H stoichiometries, the use of hydrogen gas moves the experimental system into hydrogen-rich regimes. For instance, calculations predict minimal solid carbon formation and large yields of hydrocarbon gases.

Experimental results are shown in Table X. Upper values list results found under helium atmospheres. The predominant gaseous product is acetylene. Although pyrolysis under hydrogen does lead to increased methane, ethylene, and ethane concentration, the most dramatic effect is a factor of ten increase in total low molecular weight gas quantity. Not shown here is the experimental fact that hydrogen atmosphere leave a clean interior to the pyrolysis chamber unlike the carbon film found in helium. Secondary free radical condensation is retarded by quenching in this way to increase greatly hydrocarbon yields, including the acetylene yield.

TABLE X *Laser Pyrolysis Results on Coal From the San Juan Basin (Fruitland Seam) Showing Pyrolysis Products in Helium and Hydrogen Atmospheres*

Carrier gas		Pyrolysis product			Average corrected response
	CH_4	C_2H_2	C_2H_4	C_2H_6	
Helium	3.43	92.15	4.14	0.26	
Helium	3.63	92.55	3.63	0.20	
Average	$\overline{3.53} + 0.10$	$\overline{92.35} \pm 0.20$	$\overline{3.88} \pm 0.25$	$\overline{0.23} \pm 0.03$	3,660,000
Hydrogen	4.50	90.62	4.63	0.24	
Hydrogen	5.55	88.29	5.85	0.31	
Hydrogen	5.79	87.66	6.24	0.31	
Hydrogen	3.97	91.98	3.97	0.17	
Hydrogen	4.69	90.07	5.03	0.20	
Hydrogen	4.44	90.83	4.53	0.20	
Average	$\overline{4.82} \pm 0.69$	$\overline{89.89} \pm 1.55$	$\overline{5.04} \pm 0.87$	$\overline{0.24} \pm 0.06$	37,100,000

VIII. STATUS OF LASER PYROLYSIS AND AREAS FOR FUTURE WORK

Sufficient data exists to show clearly the utility of laser pyrolysis for the characterization of coal samples. The main attribute of laser pyrolysis compared to other more usual pyrolysis–gas chromatographic techniques is the ease in sample handling and preparation. It is not reasonable to assume that the actual sample utilized following extensive grinding and blending well represents the sample of interest. Laser pyrolysis permits characterization without preliminary grinding. Other pyrolysis tools used for analytical characterization require the same rapid heating that is effected in laser pyrolysis. Yet these other methods demand sample preparation that may well change the chemical state of the sample.

The other significant advantage of laser pyrolysis over conventional pyrolysis–gas chromatography techniques is that lasers can be utilized in difficult sampling situations. This is clearly shown by the experiments that have used laser heating inside the high vacuum chamber of a mass spectrometer. It is also easy to conceive of devices that permit direct coal analyses either on open mining faces or in bore-hole geometry for direct analysis of coal samples (Schott, 1978).

All pyrolysis techniques offer the possibility of rapid analyses. Data shown for the quantification of low molecular weight gases can be easily obtained in less than 1 min. This short analysis period greatly speeds up through-put for analysis and, consequently, reduces the expense for both instrumentation and personnel.

Laser pyrolysis–gas chromatography instrumentation is currently not commercially available. However, rugged and reliable pulsed lasers are available, and it is a simple task to connect these devices for pyrolysis operations. Once that has been done, operation is inexpensive since these devices give many thousands of shots with good reliability.

Experimental data on coal has shown that this technique offers promise. Yet many possibilities for rapid and definitive coal measurements have not been fully explored. By the nature of this technique, surfaces are selectively sampled. It would seem that laser pyrolysis should be uniquely suited to study surface oxidation phenomena. The fate of nitrogen and sulfur in these coals should be determined. Clearly, N_2 and H_2S yields from pyrolysis may yield new and exceedingly rapid methods for the characterization of coals for these constituents. These tasks remain to be done.

REFERENCES

Afanasyev, Y. V., and Krokhin, O. N. (1967). *Sov. Phys.–JETP* **25**(4), 639.
Anthony, D. B., Howard, J. B., Hottel, H. C., and Meissner, H. P. (1976). *Fuel* **55**(2), 121.
Bannerjee, N. N., Murty, G. S., Rao, H. S., and Lahiri, A. (1973). *Fuel* **52**, 168.
Banerjee, N. N., Ghosh, B., and Nair, C. S. B. (1977). *Fuel* **56**, 192.
Barker, C. (1974). *Fuel* **53**, 176.
Breger, I. A. (1977). "Geochemical Interrelationships Among Fossil Fuels, Marsh, and Landfill Gas, in Future Supply of Nature-Made Petroleum and Gas." p. 913. Pergamon, New York.
Charschan, S. S. (1972). "Lasers in Industry." Van Nostrand-Reinhold, New York.
Coates, R. L., Chen, C. L., and Pope, B. J. (1974). *In* "Coal Gasification" (L. G. Massey, ed.), Advances in Chemistry Series, No. 131, pp. 92–107. Am. Chem. Soc., Washington, D.C.
Giam, C. S., Goodwin, T. E., Giam, P. Y., Rion, K. F., and Smith, S. G. (1977). *Anal. Chem.* **49**, 1540.
Gough, T. A., and Jones, C. E. R. (1975). *Chromatographia* **8**, 696.
Granger, A. F., and Ladner, W. R. (1970). *Fuel* **49**, 17.
Gray, D., Cogoli, J. G., and Essenhigh, R. H. (1974). *In* "Coal Gasification" (L. G. Massey, ed.), Advances in Chemistry Series, No. 131, pp. 72–91. Am. Chem. Soc., Washington, D.C.
Groom, P. S. (1969). *Fuel* **48**, 161.
Hanson, R. L. (1975). Ph.D. Thesis, Univ. of New Mexico, Albuquerque.
Hanson, R. L., Vanderborgh, N. E., and Brookins, D. G. (1977). *Anal. Chem.* **49**, 390.
Howe, J. A. (1963). *J. Chem. Phys.* **39**, 1362.
Ignasiak, B. S., Nandi, B. N., Montgomery, D. S. (1970). *Fuel* **49**, 214.
Joy, W. K., Ladner, W. R., and Pritchard, E. (1968). *Nature (London)* **217**, 640.
Joy, W. K., Ladner, W. R., and Pritchard, E. (1970). *Fuel* **49**, 26.
Karn, F. S., and Sharkey, A. G., Jr. (1966). *Am. Chem. Soc., ISI Nat. Meet., Pittsburgh, Pa.* p. C-44.
Karn, F. S., and Sharkey, A. G., Jr. (1968). *Fuel* **47**, 193.
Kimber, G. M., and Gray, D. (1967). *Combust. Flame* **11**, 360.

Knox, B. E., and Vastola, F. J. (1967). *Laser Focus* **3**(1), 15.

Krukonis, V. J., Cannon, R. E., and Model, M. (1974). *In* "Coal Gasification" (L. G. Massey, ed.), Advances in Chemistry, Series No. 131, pp. 29–41. Am. Chem. Soc., Washington, D.C.

Pirri, A. N., Schier, R., and Northan, D. (1972). *Appl. Phys. Lett.* **21,** 79.

Ready, J. F. (1965). *J. Appl. Phys.* **36,** 462.

Romovacek, J., and Kubat, J. (1968). *Anal. Chem.* **40,** 1119.

Schott, G., Schott, G. L., ed. (1978). Stimulation and Characterization of Eastern Gas Shales," LA-7320-PR. Los Alamos Sci. Lab., Los Alamos, New Mexico.

Sharkey, A. G., Jr., Shultz, J. L., and Friedel, R. A. (1964). *Nature (London)* **202,** 988.

Shultz, J. L., and Sharkey, A. G., Jr. (1967). *Carbon* **5,** 57.

Suggate, R. P. (1972). *N. Z. J. Sci.* **15,** 601.

Vanderborgh, N. E. (1977). *In* "Analytical Pyrolysis" (C. E. R. Jones, ed.), p. 235. Elsevier, Amsterdam.

Vastola, F. J., and Pirone, A. J. (1966). *Am. Chem. Soc., ISI Nat. Meet., Pittsburgh, Pa.* p. C-53.

Chapter 41

PRODUCT STREAM ANALYSIS FROM AN UNDERGROUND COAL GASIFICATION TEST

S. Bruce King†   *Robert A. Magee*

DEPARTMENT OF ENERGY   RADIAN CORPORATION
LARAMIE, WYOMING   AUSTIN, TEXAS

I. INTRODUCTION

A. The Coal Resource

The current national energy policy is that during the 1980s and 1990s there will be a greater emphasis on coal conversion technology in an effort to increase the nation's utilization of its domestic coal reserves and decrease its dependence on foreign fossil energy supplies. Recent

† Present address: World Energy Incorporated, Laramie, Wyoming.

estimates (Averitt, 1975) place the United States coal resources at approximately four trillion tons, yet only 10–25% of this resource is
considered recoverable with present day mining techniques. The majority of coal conversion processes currently under investigation deal
with affecting this 10–25% portion of the national resource. They are
not intended to expand the total amount of coal recoverable, only its
final usable form. A major area of development in conversion technology is coal gasification, within which resides the process called underground coal gasification or UCG. Underground coal gasification is
the single process concerned not only with converting coal to a combustible gas but also with providing an alternate method to remove
that coal from beneath the earth without mining. The portion of the
domestic coal resource to be affected by this process is the 75–90%
unsuitable for mining.

While not receiving extensive notoriety, UCG has nevertheless been
demonstrated in field tests that it is not only technically feasible
(Fischer *et al.*, 1976, 1977a) but economically viable (Moll, 1976). As
with any new process, there are certain parameters, both operational
and environmental, that must be determined in the developmental
stage prior to construction of commercial plants. This chapter shall deal
with the techniques and procedures used to measure some of these
parameters. Before discussion of the methods, the authors feel that,
because of the uniqueness of UCG, a description of the process would
benefit those not familiar with the current technology.

B.   Description of an Underground Coal Gasification
Test

There are various approaches to gasifying a coal seam, but only one
has been demonstrated in the United States by way of a field test. The
technique is called the linked vertical well (LVW) process and is currently being investigated at the Hanna, Wyoming, field site operated
by DOE's Laramie Energy Technology Center.

The LVW process is a two-step operation which first involves preparing the coal seam for gasification. This preparation is to provide the
necessary permeability that a virgin coal seam lacks. Once sufficient
permeability exists, the large volumes of air at low pressure, required
for the second part of the operation, gasification, can be pumped into
the seam. To give a quantitative feel for the process, reference will be
made to the operational parameters used in some of the Hanna, Wyoming, tests.

The coal used was a sub-bituminous coal seam referred to as the

Hanna No. 1 seam. It averaged about 30-ft thick and, because of a 7° dip, tests were conducted at depths varying from 165 to 375 ft. The analysis of the coal, determined from core samples taken from the Hanna IV test pattern, is shown in Table I and, as typical of most Western coal seams, it was water saturated. Figure 1 depicts the process for a two-well system, where well separation may be from 60 to 150 ft. Holes are drilled from the surface through the coal seam, then cased and cemented from ~7 ft above the bottom of the seam to the surface.

Without any preparatory work, the injection of air into one well at 200–400 psig (depending on lithostatic pressure) will result in ~100 scfm (standard cubic feet per minute) of air acceptance and only 10–50% recovery of that air at the adjacent well. To prepare the seam or increase its permeability, a linkage path (hence linked vertical well) between the two wells must be created. This is done by starting a fire (Fig. 1b) in the coal at the bottom of a well (well No. 2) and injecting air to substain that fire. After a period of several hours, the air injection is switched to well No. 1 (Fig. 1c); the majority of the air is "lost" to the seam, but the remainder intercepts well No. 2. and the fire. The fire moves in the direction of the oxygen source (reverse combustion) creating a carbonized path between the two wells. This carbonized path, or char matrix, provides the high permeability required for gasification. Once the fire reaches well No. 1 (Fig. 1d) air injection continues at low pressures, ~50 psig, and at higher rates, >1500scfm. The combustion zone grows around well No. 1 radially (Fig. 1e) until it reaches the top of the seam, then moves toward well No. 2 (Fig. 1f) with production gases moving through the linkage path and up the production well.

During the gasification stage air injection rates range from 1500 to 7500 scfm with typical production rates ~1.7 times greater than injection rates (2500–12,500 scfm). As in all air blown gasifiers, the gas

TABLE I *Hanna No. 1 Coal Analysis*

Proximate analysis		Ultimate analysis	
% Moisture	8.55	% Moisture	8.55
% Ash	27.48	% Carbon	47.88
% Volatiles	33.21	% Hydrogen	3.74
% Fixed carbon	30.76	% Nitrogen	1.15
Heating values (Btu/lb)	8286	% Sulfur	0.75
		% Ash	27.48
		% Oxygen	
		(by difference)	10.84

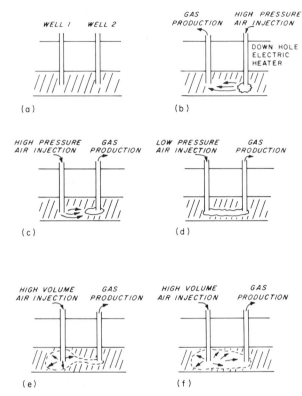

Fig. 1 Linked vertical well processes. (a) Virgin coal, (b) ignition of coal, (c) combustion linking front proceeds to source of air, (d) linkage complete when combustion zone reaches injection well (system ready for gasification), (e) combustion front proceeds in the same direction as injected air, (f) combustion front eventually reaches production well.

produced is a low-Btu gas ranging from 125 to 175 Btu/scf. Material balance calculations show that ~2800 tons of coal can be removed with use of a two-well system and 60 ft spacings (Brandenburg *et al.*, 1976). Comparison of energy input to energy production results in ~4.5 Btu produced for each Btu put into the system and in calculated cold gas efficiencies of >80% (Fischer *et al.*, 1977b).

C.   Production Stream Components

The major gas components of UCG product gas and their maximum ranges of concentration are listed in Table II.

In addition to the major gas components of the production stream,

TABLE II *Underground Coal Gasification Major Gas Components*

Components	Concentration range[a]
Hydrogen	0.0–20.0
Nitrogen	45.0–79.0
Oxygen	0.0–21.0
Argon	0.5–1.1
Carbon monoxide	0.0–18.0
Methane	0.0–6.0
Carbon dioxide	0.0–20.0
Ethylene	0.0–0.5
Ethane	0.0–1.0
Propylene	0.0–0.5
Propane	0.0–0.5
iso-Butane	0.0–0.3
n-Butane	0.0–0.3

[a] Concentration in mole %.

other materials need to be considered. First among those is the condensible liquids, comprised of an aqueous phase and an organic phase. The condensibles are normally maintained in the vapor phase since the production stream temperatures range from 600° to 900°F. The water vapor forms a major part of the product stream (10–20% by volume), and its source is primarily water influx from the coal seam. The results from past tests indicate that the quality of the produced gas is directly related to the volume percent water vapor in the produced gas (Fischer *et al.*, 1976). This should be expected since the water from the coal seam acts as a crucial source of hydrogen for the gasification reaction. In addition, any excess water, above what is needed for gasification, tends to lower the efficiency of the gasification process since heat is lost raising the water to steam. Monitoring this parameter is particularly important both from an operational standpoint and as a design factor in gas cleanup for a commercial plant. The organic phase, or coal tar, is produced through devolatilization of the coal prior to gasification and constitutes ~1% (by weight) of the product stream. The eventual use for this by-product is not certain at present, but its presence in the product stream must be considered with respect to material balance calculations and again as a design parameter in cleanup for a commercial operation.

The remainder of the production stream is made up of some particulate matter and gaseous components appearing in trace amounts.

Both the qualitative and quantitative nature of these materials are of prime interest from operational and environmental viewpoints. Concentrations of trace elements (such as sodium, lithium, and potassium) and particulate matter (total loading and size distribution) are required to formulate the ultimate design of commercial clean-up facilities. They are especially important when considering the possibility of using the produced gas in a gas turbine to generate electrical power. Monitoring of sulfur species and trace elements, such as mercury, lead, and arsenic, along with the components listed previously, are extremely critical from an environmental standpoint.†

The techniques and procedures to be described were used during the Hanna IV test, which at the time of this writing had not been completed and, therefore, the raw data will not be presented until the Department of Energy has finished interpreting the data. The determination of coal tar, water, and major gas component concentrations in an automated manner was performed by Department of Energy personnel. The monitoring of selected trace components and particulate sampling in a manual manner was performed by Radian Corporation personnel under contract to the Department of Energy.

II. COAL TAR AND WATER DETERMINATION

As alluded to earlier (Section I,C), the amount of water vapor in the product stream is an important operational parameter. Since it is such an important factor, the need for an automated measurement system is evident. There are many commercially available instruments, using various techniques, to monitor water vapor in a product stream in a continuous manner. These products and their mode of operation have been examined, but to date none has been found applicable to the product stream from an underground coal gasification process. The primary problem with most instruments is the coal tar in the process stream and its ability to either foul or interfere with the operation of the equipment. At the same time, no commercially available instruments are capable of measuring coal tar concentration on an independent basis. For these reasons, a system was designed that would provide not only moisture information but also coal tar concentrations. The system is automated to a large extent, requiring minimal intervention from operational personnel, but instead of a true continuous monitoring system it is actually a "batch mode" operation.

---

† A detailed sampling and analysis program for emissions from fluidized-bed combustion of coal is given in Volume II, Chapter 36.

The final apparatus constructed is relatively simple in concept and allows the accomplishment of three objectives with a single piece of equipment:

(1) to provide a measurement of the amount of water and coal tar in the product stream on a repetitive basis,

(2) to supply a clean, relatively dry gas stream for analysis by gas chromatography, which required removal of most of the water and coal tars from the analysis stream, and

(3) to make available a convenient location for collection of liquid samples (coal tar and water) for laboratory analysis (King *et al.*, 1975, 1977; King, 1977).

The system (Fig. 2) basically has three parts: a condenser, a sump for liquid collection, and a flow meter. The condenser is maintained at ~40°F and condenses ~98% of the water from the stream and ~99% of the coal tars. These liquids collect in the sump while the relatively dry gas moves through the flow meter. Prior to the flow meter a packed column provides sufficient surface area and residence time for elimination of the aerosol normally produced when cooling the gas. The filter provides an extra measure of safety in the event any liquid droplets are carried beyond the packed column. The control valve (Teledyne

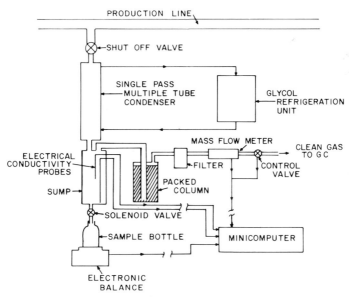

Fig. 2 Tar and water measurement system.

Hastings–Raydist control valve) is connected to the flow meter (Teledyne Hastings–Raydist mass flow meter†) that maintains a specified
flow rate through feedback from the flow meter. Beyond the control
valve a portion of the gas is continuously bled through the sample loop
of the gas chromatograph. The determination of water and coal tar
concentrations is controlled by the interfacing of the field site's minicomputer (Hewlett–Packard 21MX) to the described system and allowing the minicomputer to control the system in a "batch mode." This
is possible because the water and coal tar in the sump separate into
aqueous (lower) and organic (upper) layers in a relatively short period
of time, and the electrical conductivity of water is several orders of
magnitude greater than that of the coal tar.

The two probes in the top of the sump sense the electrical conductivity between the walls of the sump and the probe. Every 5 min the
minicomputer checks the conductivity reading between the sump and
the upper probe. When the aqueous (lower) layer reaches the upper
probe, resulting in a dramatic change in conductivity, the minicomputer executes a specific program that functions as follows:

(1)   The program first reads the electrical output of the balance (Met-
tler P1200) to obtain a tare weight on the empty bottle.

(2)   The program then opens the solenoid valve at the bottom of the
sump allowing the liquid level to drop. Simultaneously, the minicomputer reads the conductivity of the lower probe (~20 Hz).

(3)   When the lower probe is out of the aqueous and into the organic
(upper) layer, the solenoid valve is closed.

(4)   From electrical output of the balance, a second weight is determined, and the difference from the first weight is the weight of the
water collected.

(5)   The solenoid valve is again opened (~10 sec) to allow the coal
tar (upper) layer to drain into the bottle and then closed.

(6)   The third reading from the balance allows the calculation of the
weight of tar collected.

(7)   During the period the sump is collecting liquids, the minicomputer monitors the electrical output of the flow meter. When the sump
is emptied, the total volume of dry gas that has flowed through the
system is calculated. From this volume is computed total weight of gas,
and, thus, for a specific time period, the ratio of coal tar to water to
dry gas on a weight percent basis is determined.

---

† Mention of a specific manufacturer does not constitute an endorsement of that
specific product by the authors.

(8)   In addition, a very convenient sample of coal tar and water has been collected for laboratory analysis.

The only intervention by personnel is the replacing of bottles on the balance periodically. Cycle time for the system is dependent on flow rate through the system and water content of the product stream. Typical cycle times range from 45 to 80 min with a dry gas flow rate through the system of 2 scfm.

III.   MAJOR GAS COMPONENT ANALYSIS

With the primary UCG product being a low-Btu gas, whose composition varies with changes in operational conditions, a reliable gas analysis system was required. Gas chromatography was selected since past experience had shown this to be the most satisfactory analysis method for the components with the desired accuracy. In addition to the columns, detectors, and carrier gas several other factors particular to the Hanna IV field site were considered. The environment, primarily dust and ac power line failures, and its effect on any instrument was a prime factor in equipment selection. Another point considered was the operating personnel, though not trained in the theory or operation of laboratory equipment were nevertheless required to operate and/or maintain the chromatograph. A final consideration was interfacing to other pieces of equipment, particularly the minicomputer used for field site data acquisition. This was a very important factor in light of the number of anlayses anticipated during the life of the test and the eventual insertion, either by hand or automatically, of gas analyses into the data base. The final selections of equipment and operating parameters for the gas analysis were not in themselves unique, rather they were part of an analysis system that included sample collection, analysis, and data storage, which has been proven reliable in a field site operation.

Specifically, the gas chromatograph is a Hewlett–Packard model 5840A, which is built around a microprocessor and includes a thermal conductivity detector. Options include a heated gas sampling valve (0.5 cm^3 vol), cryogenic oven capability, and an ASCII interface card. The column is 20 ft × $\frac{1}{8}$-in. stainless steel tubing packed with Johns–Manville Chromosorb 102 (80/100 mesh). A mixed carrier gas of 8% hydrogen in helium at a flow rate of 30 cm^3/min is used and results in a linear response for hydrogen from 0 to >50 mole %. By temperature programming the column oven at −60°C for 6.5 min, heating at 30°C/min to 225°C, and remaining there till the end of the analysis, a gas chromatographic trace as shown in Fig. 3 is obtained. The hydrogen peak is made to appear positive, allowing proper integration, by starting the analysis with the detector polarity reversed. After elution of the hydrogen peak, the polarity is returned to normal for the elution of all remaining components.

With the use of prepared gas mixtures, response factors and retention times are determined and entered into the microprocessor portion of the chromatograph. This allows one of several reports to be computed by the microprocessor and automatically or on demand written by the chromatograph. The typical report is a normalized analysis for the components listed in Table II. A complete cycle for the chromatograph requires ~32 min, including analysis, cooling down of the oven, and stabilization, which allows for a possible 45 analyses per day. These analyses are stored in the total data file during the life of the UCG test to allow access through the data acquisition system as needed.

The ASCII interface and its operation will not be discussed in detail but basically it allows remote operation of the chromatograph through the use of programming instructions in the minicomputer. The complete control of the chromatograph, from an injection of the sample to transfer of the normalized analysis into data files, may be relinquished to the minicomputer.

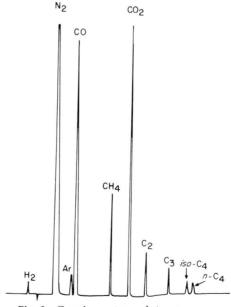

Fig. 3   Gas chromatograph trace.

This provides for rapid and convenient access to the gas analyses by various methods, such as digital formats, computation of material balances, or graphic displays.

IV.   REMOTE SAMPLING OF GAS COMPONENTS

One of the disadvantages of monitoring an underground gasification test is the inherent remoteness of the actual gasification process. The result is a very expensive and to some degree limited instrumentation program. To gain as much information as possible, especially in the early development of the process, instrumentation packages were placed in separate wells within the coal seam to be gasified (Northrop, 1977). One area of particular interest is the gas compositional variations prior to, within, and after the gasification zone. In light of this interest, gas samples are collected and analyzed from various locations within the coal seam.

The most difficult portion of the undertaking is the collection of a gas sample or, more specifically, provision of a gas sampling port within the seam. The ultimate design (Fig. 4) was supplied and installed by Sandia Laboratories of Albuquerque, New Mexico, which also supplied other remote sensing instrumentation for the Hanna field tests. The sampling ports, which are termed "sniffer tubes," actually serve a dual purpose. The system consists of a continuous length of $\frac{1}{4}$

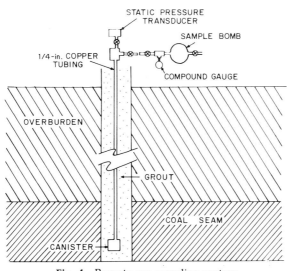

Fig. 4   Remote gas sampling system.

in. copper tubing extending from the surface to a canister located at a specific location in the coal seam. The wells in which these tubes are located contain other instrumentation packages, all of which necessitate grouting the entire package in place. To open the canister to the coal seam, once it has been grouted, involves detonating two separate charges previously attached to the canister at installation. This ruptures the walls of the canister and fractures the grout allowing communication between the coal seam and the sniffer tube. At the surface, the tubing is split to allow attachment of a pressure transducer to measure static pressure within the tube and a valve to allow purging of the gas in the tube and collection of a gas sample. Gas sampling is accomplished by evacuating a stainless steel bomb and connecting it to the valve, then opening the valve and allowing the bomb to pressurize to ~5 psig. The bomb is then transferred to a gas chromatograph and attached to the sample loop. At this point, an identification number is entered into the chromatograph using a standard option available on the Hewlett–Packard 5840A chromatograph. This identification number indicates the location of the sniffer tube and the time of collection. After purging the sample loop for ~2 min, the analysis sequence is started by the operator pressing a single button which starts the automated process of sample injection, analysis, and normalized data transfer to the minicomputer. The identification number, also transmitted to the computer, allows insertion of the analysis data in the proper location of the total data base.

The results from the early stages of the Hanna IV test indicate the system of remote sampling has worked very satisfactorily. The variations in gas composition at selected locations downstream of the gasification zone have shown very good agreement with mathematical model predictions. The use of microprocessor controlled chromatographs has been very successful not only from reliability but also with respect to ease of operation for untrained personnel and interfacing with existing equipment.

V.   MANUAL MONITORING TECHNIQUES

The sampling and analytical techniques described in this section are designed to characterize particulates and trace components in a pressurized, hot (up to 1000°F) gas stream. All of the techniques described have, at a minimum, been applied to the monitoring of the product gas from a UCG test (Magee *et al.*, 1976) and process gas from the CO_2 Acceptor Process operated by Conoco Coal Development Company

(Radian Corporation, 1978). The analytical parameters are those which are being determined during monitoring of the Hanna IV UCG test.

The overall approach is divided into two distinct operations: sample acquisition and analysis. Sample acquisition includes the isokinetic removal of a portion of the gas from the process, collection of particulate material from the sampled gas, and collection of individual or groups of gaseous components in a form suitable for analysis. Collection of gaseous components is either by sorption in impinger solutions or by preconditioning and containment of a gas sample in a sampling bomb. Analysis is then performed on the samples by a combination of wet chemical and instrumental techniques including titrimetry, colorimetry, gas chromatography, atomic absorption spectroscopy, or ion selective electrodes.

A. Sample Acquisition

The sample collection system described here for sampling a pressurized gas stream is designed to

(1)   provide controlled access to the stream without interruption of process operation,
(2)   isokinetically remove a portion of the gas,
(3)   collect particulates from the sampled gas for particulate loading and size distribution and to provide particulate samples for analysis,
(4)   provide a particulate-free gas stream for gaseous component determination,
(5)   measure the volume of gas sampled, and
(6)   monitor the flow rate and temperature of the process gas.

Figure 5 provides a schematic presentation of the sampling system. The system as shown consists of four main segments:

(1)   probe assembly,
(2)   particulate collection,
(3)   gas collection, and
(4)   metering.

1. Probe Assembly

The probe assembly performs three main functions:

(1)   interface with the gas stream,
(2)   transport of sampled gas, and
(3)   process gas temperature and flow measurement.

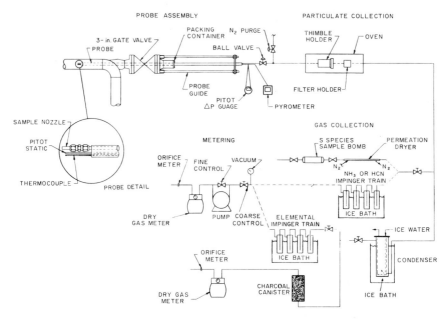

Fig. 5 Sampling system for monitoring UCG product gas.

a. Interface with the Gas Stream   The probe consists of a 1½-in. diam, 6-ft length of 316-stainless steel tubing. It enters the gas stream through a packing gland (Fig. 6) containing graphite–asbestos packing which forms a lubricated seal through which the probe may slide while maintaining a pressure-tight integrity. Final entry to the process gas is through a 3-in. full open gate or ball valve. The length of the probe is adequate to allow insertion to a point at least two pipe diameters upstream of the process line elbow to minimize turbulence. Two all-thread rods mounted on the packing gland/valve flange on either side of the probe, and passing through a yoke welded to the probe, provide mechanical assistance for insertion and withdrawal. The ends of the probe are sealed with welded caps to contain the process pressure.

b. Sample Transport   A sample transport tube (½ in. 316-S.S. tubing) passes through the probe and is seal welded to the caps on each end. As shown in the probe detail in Fig. 5, the sample transport tube is tipped with interchangeable nozzles. Nozzles are provided in diameters from ⅛ to ½ in. to allow isokinetic sampling over a range of product gas velocities while maintaining sampled gas flow within the capacity of the remainder of the sampling system. A ½ in. ball valve at the

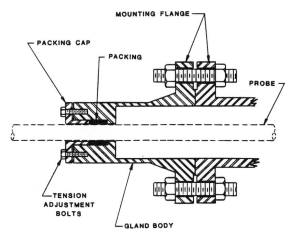

Fig. 6 Probe packing gland assembly.

sample transport tube exit provides on/off control of flow. The capability to purge the probe with nitrogen through a tee behind the valve provides a means to avoid introduction of oxygen to the combustive process gas and to blow out the probe should it plug.

c. Temperature and Flow Rate Measurement Two guide tubes are mounted in the probe with tube fittings at the rear of the probe. A $\frac{1}{4}$-in. diam standard pitot is in one guide tube while the other contains a $\frac{1}{4}$-in. sheathed Type-K thermocouple. The differential pressure across the pitot is measured on a differential pressure gauge (Midwest Instrument Company) with range of 0–10 in. H_2O and maximum static pressure rating of 300 psig. A digital pyrometer provides temperature measurement.

2. Particulate Collection

The particulate collection portion of the sampling system removes particulates from the sampled gas at process pressure and temperature. This approach avoids the collection of condensable materials and minimizes particulate losses which would occur in any pressure reduction device. Process temperature is maintained by a cylindrical oven containing the various collection devices described below. The selection of particulate collection temperature provides a means to define the distinction between particulates and condensables. In this application, collection at process temperature defines this separation to be the same as in the process gas line. Other applications may dictate the use of a preselected collection temperature. Three modes of particulate collection are provided:

(1) filtration with a glass-fiber filter for particulate loading,
(2) collection with a cascade impactor for particulate size distribution, and
(3) filtration with an alundum thimble followed by a glass-fiber filter to provide samples for analysis.

a. Particulate Loading The particulate loading in the process gas is determined by collection on a preweighed (±0.1 mg on a Cahn Electrobalance) glass-fiber filter. Prior to weighing, the filters are preconditioned by heating at 300°F overnight. Following collection, filters are reweighed and particulate loading calculated from the weight gain and sampled gas volume.

b. Particulate Size Distribution An Andersen Mark III Cascade Impactor (Andersen 2000, Inc.) is used to aerodynamically separate the particulates into nine size fractions. The cascade impactor is shown schematically in Fig. 7. Particulates in each fraction are collected by impaction on glass-fiber substrates and a backup filter. The substrates and backup filter are preconditioned by heating at 300°F overnight and preweighed on a Cahn Electrobalance.

Following collection, the substrates and backup filter are reweighed and the distribution among the various size ranges calculated as fractions of the total weight gain. The 50% cutoff diameter (D_p in centimeters) for each plate is calculated from

$$D_p{}^2 = 2.43\mu \, WC/P_p V_o \tag{1}$$

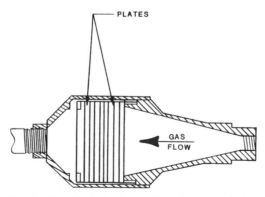

Fig. 7 Cascade impactor particulate sizing device.

where W is the diameter of acceleration jets (holes in plate) in centimeters, μ the gas viscosity (poise), P_p is 1 g/cm^3 (equivalent aerodynamic diameter), V_o the gas velocity for that plate, and C is a correction factor for gas resistance, defined by

$$C = 1 + (2.514\lambda/D_p) \tag{2}$$

where D_p is the particle diameter (cm), and λ the mean-free path of gas (cm).

Since C is a function of D_p, the calculation requires an iterative solution beginning with $C = 1$. A typical set of cutoff diameters (microns) for the *in situ* gasification product gas at 890°F and 58 psig are

$$
\begin{array}{cc}
19.8\mu & 3.6\mu \\
12.9\mu & 2.1\mu \\
8.5\mu & 1.3\mu \\
6.0\mu & 0.83\mu \\
\end{array}
$$

c. Particulate Collection for Analysis   An alundum thimble mounted in a modified thimble holder is used to provide the additional filtration surface area required to collect an adequate quantity of particulate material for analysis. A glass-fiber filter is used for a backup to avoid contamination of the gas to the vapor collection systems in the event of thimble leakage. The collected material is removed from the thimble by sonification to achieve maximum recovery.

3.   Gas Collection

Gaseous components are collected from the sampled gas by two general approaches:

(1)   impinger collection (NH_3, HCN, vapor phase elements) and
(2)   grab samples in gas bombs (sulfur species).

a. Impinger Collection   Impinger collection of gaseous components removes the species of interest from the sampled gas by sorption in a reagent solution. This approach simultaneously collects the species in a form compatable with most analytical techniques and averages their concentration over the time necessary to collect adequate quantities for analysis. Smith–Greenburg impingers (500 mliter capacity) are used for all collection. Each collection impinger set is followed by a dry impinger and a modified Smith–Greenburg containing silica gel to collect moisture, providing dry gas for metering.

(1) Ammonia (NH_3)   For collection of ammonia, a portion of the sample gas exiting the particulate collection oven is passed through

a pair of impingers in series containing 200 mliter of 5% sulfuric acid. During sampling of *in situ* gasification product gas, it has been found that the collection efficiency of the first impinger exceeds 99%. This indicates that the second impinger may be eliminated with negligible loss of efficiency.

(2) Hydrogen Cyanide (HCN), Chloride (Cl⁻), and Fluoride (F⁻)   Two impingers in series, each containing 200 mliter 10% sodium hydroxide solution, are used to collect hydrogen cyanide from a portion of the sample gas exiting the particulate collection oven. Collection efficiency of the first impinger has been found in field tests to be ~90% if the collection solution pH remains above 11. For sampling of gases containing high levels of acid gases, either the sample volume must be limited or the sodium hydroxide concentration must be increased to avoid neutralization of the collection solution. Losses of cyanide may occur if the pH falls below 10.5. The same impinger solutions are used to determine chloride and fluoride.

(3) Vapor Phase Elements   A series of five impingers is used to collect samples for analysis of nine elements in the vapor state:

arsenic	sodium
lead	lithium
cadmium	potassium
selenium	calcium
vanadium	

The main sample gas stream, as shown previously in Fig. 5, exits the particulate collection oven to a condenser. The condenser lowers the gas temperature to 35°F, condensing moisture and organics. A portion of the exit gas from the condenser enters an acid/base impinger train consisting of two impingers containing 10% nitric acid, a dry impinger to collect carryover, and two impingers containing 10% sodium hydroxide. Determination of the elements then requires analysis of five samples: the condensate and four impinger solutions.

Mercury in the vapor phase is collected by a separate technique by amalgamation on a plug of gold wire. Sample gas exiting the particulate oven is passed through a fritted bubbler containing 3% hydrogen peroxide and a quartz tube containing 8 g of 0.010 in. gold wire molded into a plug. The flow rate is measured at ~500 mliter/min on a rotometer and total flow determined from flow rate and sampling time (~15 min).

b. Gas Bomb Collection   Gas samples for determination of sulfur species by gas chromatography are collected in flow-through glass sample bombs. A low flow (<100 cm³/min) of sample gas is passed through a permeation dryer (Perma Pure Products, Inc.) to remove moisture by permeation distillation across an extrudable desiccant. The dried gas purges a 125-cm³ glass bomb fitted with Teflon stopcocks on inlet and outlet. Prior to use, the glass bomb is silinized with a silyl donor such as DMDCS (dimethyldichlorosilizane, 3% in xylene) to neutralize active sites. One silinization treatment should last indefinitely. Following ~10 min of purging the exit stopcock is closed and the bomb pressured to ~5 psig.

4. Metering

The sample train configuration shown in Fig. 5 requires two sample gas metering systems: one for the impinger trains and one for the additional gas flow required to achieve isokinetic rates. Total gas volume for each is measured with dry gas meters with inlet and outlet temperature monitoring. Immediately downstream of the dry gas meters, each system includes an orifice meter to monitor flow rate conveniently. These are calibrated against the dry gas meter with actual sample gas to eliminate the effects of gas composition.

The use of dry gas meters requires that the gas be dry prior to metering. The impinger trains accomplish this by a combination of condensation and silica gel desiccant. The remaining gas is partially dried by the condenser and any residual moisture or condensable organics removed by a canister of activated charcoal.

B. Analysis

1. Elemental Analysis of Particulates by Atomic Absorption

Particulates collected from the sample gas are characterized for the following elements:

sodium	aluminum
calcium	magnesium
iron	manganese
potassium	silicon

The particulates are mixed with lithium metaborate (Perkin–Elmer Corporation, 1973) for fusion at 1000°C in a muffle furnace. The melt is then dissolved with dilute hydrochloric acid and is diluted to volume. Analysis of the digested sample is performed by atomic absorption (Perkin–Elmer Corporation, 1974) for each of the elements listed above. Samples are compared with a standard calibration curve for each element. In case of matrix interferences, the method of standard additions is used. Quality control is monitored and maintained by analysis of NBS coal ash (SRM 1633).† Table III lists the detection limit, optimum concentration range, and sensitivity for each element.

2. Elemental Analysis—Vapor Phase

Fluoride and chloride are determined on aliquots of the sodium hydroxide impinger solution used to collect hydrogen cyanide. The gold amalgamation plug and its preceding hydrogen peroxide solution are analyzed for mercury. The remaining nine elements are determined in the condensate and acid/base impinger solutions.

Pretreatment of impinger samples consists of extraction of oils and tars with methylene chloride. The organic phase is subjected to perchloric acid reflux digestion (Smith, 1965), and the resulting solution is recombined with the aqueous phase.

Detection limits, optimum concentration range, and sensitivity for each element are listed in Table IV.

a. Chloride An aliquot of the impinger solution is boiled with sulfuric acid, oxidized with hydrogen peroxide, and boiled with sodium

TABLE III *Analytical Information for Particulates*

Element	Detection limit[a]	Optimum concentration range[b]	Sensitivity[b]
Sodium	1	0.01–1	0.015/1% abs
Calcium	5	0.1–4	0.08/1% abs
Iron	25	0.2–5	0.12/1% abs
Potassium	10	0.04–2	0.04/1% abs
Aluminum	200	1–40	1/1% abs
Magnesium	2	0.01–0.4	0.007/1% abs
Manganese	30	0.1–3	0.055/1% abs
Silicon	900	5–125	1.8/1% abs

[a] For solid sample microgram per gram assuming digestion of 0.2 g of particulates in a 100 mliter solution.
[b] Concentration listed as microgram per milliliter for the aqueous digested sample.

† Details on atomic absorption spectrophotometry and SRM 1633 are given in Volume I, Chapter 14, Sections II, B and III, B.

TABLE IV *Analytical Informaton for Gaseous Elemental Analyses*

Element	Detection limit[a]	Optimum concentration range[a]	Sensitivity
Mercury[b]	0.001	0.002–0.025	0.005 μg/1% abs
Arsenic[c]	0.005	0.02–0.5	20 pg/1% abs
Lead[c]	0.001	0.01–1	10 pg/1% abs
Cadmium[c]	0.0005	0.001–0.1	1 pg/1% abs
Selenium[d]	0.0002	0.0008–0.03	0.0002 ppm/1% F
Sodium[e]	0.002	0.01–1	0.015 ppm/1% abs
Lithium[e]	0.003	0.01–2	0.035 ppm/1% abs
Potassium[e]	0.02	0.04–2	0.04 ppm/1% abs
Calcium[e]	0.01	0.1–4	0.08 ppm/1% abs
Vanadium[c]	0.005	0.01–1	35 pg/1% abs
Fluoride[f]	0.01	0.05–25	—
Chloride[g]	0.01	0.05–10	0.04 ppm/1% abs

[a] Detection limit and concentration range expressed on a microgram per milliliter basis in collection solution, except for mercury.
[b] Flameless atomic absorption spectrophotometry.
[c] Graphite furnace atomic absorption spectrophotometry.
[d] Fluorimetry.
[e] Flame atomic absorption spectrophotometry.
[f] Specific ion electrode.
[g] Colorimetry.

hydroxide as preliminary treatment. The sample is then mixed with ferric ammonium sulfate and mercuric thiocyanate according to Method D512C from the *1977 Annual Book of ASTM Standards, Part 31* (ASTM, 1977). Chloride reacts with mercuric thiocyanate to produce thiocyanate ion which combines with the ferric ion to form red ferric thiocyanate. The color intensity, proportional to chloride concentration, is measured photometrically at 463 nm and compared with a set of standard chloride solutions in the range 0.05–10 μg/mliter.

b. Fluoride Fluoride in the impinger solution is determined by the standard addition technique utilizing a specific ion electrode. A citrate buffer is added to release fluoride complexed by uranium, thorium, aluminum, and iron and to cancel out variances in pH and ionic strength. The observed change in potential is directly related to fluoride concentration in the range 0.05–25 μg/mliter.

c. Mercury Mercury sorbed in the H_2O_2 impinger solution is reduced to the elemental state with potassium permanganate, hydroxylamine hydrochloride, and stannous chloride. Elemental mercury is then swept through the absorption cell of the atomic absorption spectrophotometer for determination. Mercury amalgamated on the gold plugs is thermally desorbed and reamalgamated on a second gold plug to minimize interferences. Upon thermal desorption from the second gold plug, the mercury is passed through the optical cold-vapor cell of the atomic absorption spectrophotometer for determination. Vapor phase mercury standards are injected into the system, and samples are compared with the standard calibration curve.

d. Arsenic   Arsenic is determined by a flameless atomic absorption procedure based on the method of Ramakrishna *et al.* (1969) incorporating the techniques of Ediger (1975). The usually volatile arsenic is complexed as the molybdenum heteropoly acid with ammonium molybdate in a nitric acid medium. The complex formation allows analysis on the graphite furnace by reducing the volatility of the arsenic so that high-temperature charring can be used to rid the sample of matrix interferences.

e. Lead and Cadmium   Lead and cadmium are determined by injection of impinger solutions into the graphite furnace. Samples are compared either to a standard calibration curve or determined by use of standard additions to cancel matrix interferences if necessary.

In the case of low concentrations of lead or cadmium, these elements are separated and concentrated by chelation and extraction. Kinrade and VanLoon's (1974) method utilizing two chelating agents, ammonium pyrrolidinedithiocarbamate (APDC) and diethyl ammonium diethyldithiocarbamate (DDDC), to complex lead and cadmium has been highly successful. A citrate buffer is used because of its stability, strong buffering capacity, and nonparticipation in any reaction. The sample solution is extracted into methyl isobutyl ketone and the organic phase injected into the graphite furnace of the atomic absorption spectrophotometer. Concentration is determined by use of a standard calibration curve.

f. Selenium   Selenium in the form of selenite reacts with aromatic orthodiamines to form piazselenols which fluoresce in various organic solvents. This method of determination of selenium is a Radian modification of a procedure published by Levesque and Vendette (1971). Aqueous samples are heated with nitric and perchloric acids to oxidize selenium to selenate which is reduced to selenite by addition of hydrochloric acid. Isolation of selenite from interfering substances is accomplished by adding hydroxylamine, EDTA, and formic acid. The selenite is then reacted with 2, 3-diaminonaphthalene to form naphtho-

(2, 3-*d*)-2 selena-1,3-diazole, which fluoresces upon extraction into cy-clohexane. This method is free of interferences and is capable of de-tecting nanogram quantities of selenium.

g. Sodium, Lithium, Potassium, and Calcium Sodium, lithium, potas-sium, and calcium are determined by direct aspiration into an air/acetylene flame of the atomic absorption spectrophotometer. Sodium, potassium, and calcium are diluted with lanthanum chloride to mask interferences as well as to prevent self-ionization. All elements are compared with a standard calibration curve.

h. Vanadium Vanadium is determined by flameless atomic absorp-tion using the heated graphite atomizer. The method is an adaption of the work of Cioni *et al.* (1972). Standard additions are performed on the impinger solutions. National Bureau of Standards samples of coal (SRM 1632) and coal ash (SRM 1633) yield results within 10% of the certified values.

3. Ammonia

Ammonia in the sulfuric acid impinger solution is determined by a distillation–acid titration procedure from *Standard Methods for the Ex-amination of Water and Wastewater* (American Public Health Associa-tion, 1975). A portion of the impinger solution is extracted with meth-ylene chloride to remove condensed organics, and an aliquot of the aqueous phase is retained for analysis. Free ammonia is distilled at pH 9.5 into a boric acid solution containing methyl red indicator. Ammonia in the distillate is then titrated to the indicator end point with standard sulfuric acid. (Compare with the procedure for ammonia in coke oven gas, Chapter 38, Section IV.)

4. Hydrogen Cyanide

Condensed organics are extracted from a portion of the sodium hy-droxide impinger solution with methylene chloride and cyanide deter-mined in an aliquot of the aqueous phase. To remove interference from sulfide in the sample, lead nitrate is added, and the precipitated lead sulfide is removed by filtration. The cyanide content of the sample is then determined by a distillation–colorimetric procedure from *Standard Methods for the Examination of Water and Wastewater* (American Public Health Association, 1975). Hydrogen cyanide is distilled from the sam-ple during acidification and is sorbed in a Fischer–Milligan gas wash-ing bottle containing dilute sodium hydroxide solution. An aliquot of the sorbing solution is buffered, reacted with chloramine-T, and complexed with the pyridine–barbituric acid colorimetric reagent. The absorbance of samples is compared with standard solutions containing 0.2–10 μg cyanide at a wavelength of 578 nm.

5. *Sulfur Species (H_2S, COS, CS_2, SO_2, Methyl Mercaptan)*

Gaseous sulfur components in the gas bomb sample are analyzed by gas chromatography with a flame photometric detector specific for sulfur. Separation is achieved on a 4.7-m, 1.5-mm-i.d. Teflon column packed with 1% TCEP [*tris*-(cyanoethyl) propane] and 0.5% phosphoric acid on 60/80 mesh Carbopack B (Supelco, Inc.). The analysis is temperature programmed with an initial 7-min hold at 30°C followed by 16°C/min heating to 100°C and a 16 min hold at 100°C. A ten port sample valve is mounted in the GC oven providing two Teflon sample loops for sample injection. Retention times for the five sulfur species are

H_2S	2.2 min
COS	4.4 min
SO_2	6.8 min
CH_3SH	8.5 min
CS_2	13.8 min

The analytical approach described above has been used to analyze the product gas from the Hanna IV UCG test with a Hewlett–Packard 5730 GC and a 3380A Integrator/Plotter. With this system, a linear relationship exists between logarithm of peak area and logarithm of sulfur mass up to 60 ng sulfur to the detector for each compound. An example chromatogram from a 150 μliter injection of gas containing four of the five components is shown in Fig. 8. An attenuation change was made from ×512 to ×4 as indicated to give enhanced visual response for the lower concentrations of methyl mercaptan and carbon disulfide.

VI. CONCLUSION

Defining the qualitative and quantitative nature of the materials found in the product stream from an underground coal gasification test plays an important role in developing the technology. As seen, the gathering of this information is far from straightforward and in some instances requires the use of new techniques and equipment.

In achieving this, the original objective of developing the potential for utilization of deep and dirty coal seams, unsuitable for mining, will

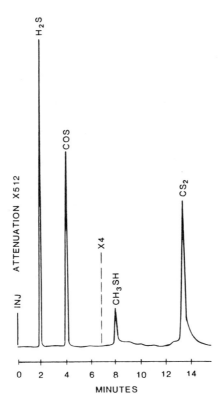

Component	Retention time (min)	Response (counts)	Sulfur (ng)	Concentration (ppm)
H_2S	2.2	2.6×10	3.1	21
COS	4.4	2.8×10	3.9	26
SO_2	6.8	0	<0.015	< 0.1
CH_3SH	8.5	5.8×10	0.15	1.0
CS_2	13.8	3.8×10	0.25	0.8

Fig. 8   Sulfur gas example chromatogram. Conditions: sample size: 150 μliter; column: 1% TCEP, 0.5% H_3PO_4 on 60/80 caropack B 10 ft, $\frac{1}{8}$-in.-o.d. FEP Teflon; oven—initial: 7 min at 30°C, program: 16°C/min at 100°C, final: 16 min at 100°C; barometric pressure: 595 mm Hg; attenuation: ×512, ×4.

help the attainment of the national goal of shifting our dependence from foreign to domestic fuel supplies.

REFERENCES

American Public Health Association, American Water Works Association, and Water Pollution Control Federation (1975). "Standard Methods for the Examination of Water and Wastewater," 14th Ed. Washington, D.C.

ASTM (1977). "1977 Annual Book of ASTM Standards. Part 31: ."Am. Soc. Test. Mater., Philadelphia, Pennsylvania.

Averitt, P. (1975). U.S. Geol. Surv. Bull. No. 1412.

Brandenburg, C. F., Fischer, D. D., Northrop, D. A., and Schrider, L. A. (1976). Proc. Underground Coal Gasif. Symp., 2nd, Morgantown, W.Va. pp. 36–52.

Cioni, R., Innocenti, F., and Mazzouli, R. (1972). At. Absorpt. Newsl. 11, 102.

Ediger, R. (1975). "Atomic Absorption Analysis with the Graphite Furnace Using Matrix Modification," Atomic Absorption Application Study No. 584. Perkin–Elmer Corporation, Norwalk, Connecticut.

Fischer, D. D., Brandenburg, C. F., King, S. B., Boyd, R. M., and Hutchinson, H. L. (1976). Proc. Underground Coal Gasif. Symp., 2nd, Morgantown, W.Va. pp. 354–371.

Fischer, D. D., King, S. B., and Humphrey, A. E. (1977a). Am. Chem. Soc., Div. Fuel Chem., Prepr. 22, No. 4, 49 (1977), Montreal

Fischer, D. D., Boysen, J. E., and Gunn, R. D. (1977b). Soc. Min. Eng. Am. Inst. Min. Eng., Atlanta, Ga.

King, S. B. (1977). Am. Chem. Soc., Div. Fuel Chem., Prepr. 22, No. 2, 169 (1977), New Orleans, La.

King, S. B., Brandenburg, C. F., and Lanum, W. J. (1975). Am. Chem. Soc., Div. Fuel Chem., Prepr. 20, No. 2, 131 (1975), Philadelphia, Pa.

King, S. B., Guffey, F. D., and Gardner, G. W. (1977). Am. Chem. Soc., Div. Fuel Chem., Prepr. 22, No. 4, 113 (1977), Montreal

Kinrade, J. D., and VanLoon, J. C. (1974). Anal. Chem. 46, 1894–1898.

Levesque, M., and Vendette, E. D. (1971). Can. J. Soil Sci. 51, 85–93.

Magee, R. A., Mann, R. M., and Collins, R. V. (1976). "Monitoring of an In-Situ Gasification Test of the Linked Vertical Well Concept, Final Report," ERDA Contract No. E(29-2)-3689, Radian Project No. 200-138, TID-27215. Radian Corporation, Austin, Texas.

Moll, A. J. (1976). Proc. Underground Coal Gasif. Symp., 2nd, Morgantown, W. Va. pp. 169–178.

Northrop, D. A. (1977). "Instrumentation and Process Control Development for In-Situ Coal Gasification," 10th and 11th Quarterly Report. SAND 77-1821. Sandia Lab., Albuquerque, New Mexico.

Perkin–Elmer Corporation (1973). "Analytical Methods for Atomic Absorption Spectrophotometry." Norwalk, Connecticut.

Perkin–Elmer Corporation (1974). "Instructions, Model 503, Atomic Absorption Spectrophotometer." Norwalk, Connecticut.

Radian Corporation (1978). "Environmental Characterization of the CO_2 Acceptor Process," Conoco Contract No. 1734-1598. Austin, Texas.

Ramakrishna, T. V., Robinson, J. W., and West, P. W. (1969). Anal. Chim. Acta 45, 43–49.

Smith, G. (1965). "The Wet Chemical Oxidation of Organic Compositions Employing Perchloric Acid," Bull. No. 213. G. Frederick Smith Co., Columbus, Ohio.

Part X

WASTE PRODUCT STREAMS AND ENVIRONMENTAL PROBLEMS

Chapter 42

Analysis of Mineral Matter in Coal, Coal Gasification Ash, and Coal Liquefaction Residues by Scanning Electron Microscopy and X-Ray Diffraction

S. J. Russell   *S. M. Rimmer* †
ILLINOIS STATE GEOLOGICAL SURVEY
URBANA, ILLINOIS

I. INTRODUCTION

Research on coal conversion wastes at the Illinois State Geological Survey has involved the development of techniques for the characterization of mineral matter in the wastes. Techniques for two analytical methods, x-ray diffraction and scanning electron microscopy, are discussed here.

The nature and occurrence of mineral matter in coal and the methods used for its analysis are discussed in Volume II, Chapter 26. Some of the methods reported in that chapter are also applicable to the analysis of mineral matter in coal liquefaction residues and gasification ash. The mineral assemblages present in these waste materials differ from

† Present address: Department of Geosciences, The Pennsylvania State University, University Park, Pennsylvania.

those present in coal and depend upon the temperature, pressure, and atmosphere to which the coal has been subjected.

A. Importance of Mineral Matter in Coal Conversion Processes and Resulting Waste Materials

The characterization of minerals in coal liquefaction residues and in coal gasification ash is necessary to assess the beneficial and detrimental effects which the mineral matter may have, both on the process in which it is involved and on its ultimate disposal. Beneficial effects of mineral matter to coal liquefaction processes include the possible catalysis of hydrogenation reactions and sulfur removal. Tarrer *et al.* (1977) ranked various types of mineral matter according to their performances as hydrogenation and desulfurization catalysts in experiments on liquefaction systems. In studying lignite in liquefaction systems, Given *et al.* (1975) reported that the highest liquefaction yields were obtained from a lignite with the greatest amount of mineral matter. Granoff *et al.* (1978), Granoff and Thomas (1978), Gray (1978), Mukherjee and Chowdhury (1976), and Henley (1975), among others, performed experiments to determine the individual catalytic effects of different minerals present in coal and coal liquefaction residues on the liquefaction processes. Some researchers have reported that the presence of pyrite and other iron-bearing minerals aids in the hydrogenation of coal during liquefaction (Tarrer *et al.*, 1977; Mukherjee and Chowdhury, 1976; Granoff *et al.*, 1978; Granoff and Thomas, 1978). Minerals may also act as catalysts in coal gasification; this behavior is detailed in Volume II, Chapter 32, Section III, B, 5.

Possible detrimental effects of mineral matter associated with coal conversion processes include the poisoning of catalysts in liquefaction and gasification processes and the leaching of harmful elements from residues and ashes after disposal (Griffin *et al.*, 1978). Certain minerals may cause undue abrasive or chemical wear in coal conversion plants as well as clogging (Harris and Yust, 1978) and buildup in reactor vessels (Walker *et al.*, 1977). Therefore it is important to characterize the mineral matter in liquefaction residues and gasification ashes in order to assess its positive and negative effects in specific conversion processes.

B. Mineral Composition of Coal Liquefaction Residues and Gasification Ash

The minerals found in coal and their nature and occurrence are reported in Volume II, Chapter 26, Table I, and Section I, A. Among those commonly found are pyrite, calcite, quartz, illite, kaolinite, and expandable clays. Dolomite, siderite, marcasite, feldspars, gypsum, barite, and sphalerite occur in lesser amounts.

Most coal liquefaction processes operate at pressures between 1500 and 4000 psig and temperatures of 400°–500°C in a hydrogen atmosphere. The major change in coal mineralogy which occurs under these conditions is the transformation of nearly all pyrite (FeS_2) to pyrrhotite ($Fe_{1-x}S$, where $X \doteq 0$ to 0.2). Wollastonite, a calcium silicate ($CaSiO_3$), has been identified in very small quantities in the residue of one liquefaction process and is thought to form from the calcite and quartz in the coal (Russell, 1977). The majority of the calcite and quartz, however, is unchanged. Walker *et al.* (1977) reported some dehydration of minerals during liquefaction. The minerals other than those mentioned essentially do not change with liquefaction.

The morphology of mineral matter in liquefaction residues collected after filtration of the product oil is very similar from process to process but varies within each process. Pyrrhotite occurs typically as aggregates of small rounded grains 2 μm and less in size. These aggregates may consist purely of pyrrhotite or may include other mineral matter such as quartz and clays. Pyrrhotite also forms fine granular coatings on pyrrhotite particles and on particles of other mineral matter. Pyrrhotite can occur, though less commonly, in network structures which resemble pyrite framboid aggregates (Russell, 1977). These structures may contain some unconverted pyrite.

Walker *et al.* (1977) described deposits of solids containing pyrrhotite which accumulated in the reactor vessel of a coal liquefaction process. Pyrrhotite occurred in these solids as agglomerates of pyrrhotite alone, as agglomerates of pyrrhotite and other minerals, as dispersed particles, and in radial arrangement in a "shell" surrounding a central mineral particle. Calcium carbonate is also present in the shells of these particles. The differences in particle morphology between the residues and the reactor solids are in part caused by the longer residence time under process conditions of the reactor solids.

Wollastonite was found in anhedral grains which, when etched in HCl, assumed a fibrous appearance. Other minerals in the residues appear essentially unchanged from their occurrence in coal. Euhedral forms of calcite and kaolinite have been seen. In some cases quartz and clay minerals are incorporated into aggregates along with pyrrhotite. Reactor solids described by Walker *et al.* (1977) also contain calcite, vaterite ($CaCO_3$ polymorph), quartz, anhydrite, clays, calcium salts, and some pyrite. Precipitates containing calcium were predominant in reactor solids derived from coal from the Wyodak seam in Wyoming. Walker *et al.* (1977) attributed this to an abundance of exchangeable calcium in the coal.

Characterization of the organic components of liquefaction residues is described in Mitchell *et al.* (1977) and Walker *et al.* (1977).

Coal gasification processes cover a wide range of temperature and pressure conditions, but most are conducted in an oxidizing atmosphere. Temperatures range from 600° to 1800°C and pressures from atmospheric to 1500 psig. The mineralogy of gasification ash is much different from that of the coal or of liquefaction residues.

Several studies have characterized the minerals in gasification ash (outside of fly ash). Yerushalmi *et al.* (1975), Harvey *et al.* (1975), and Kolodney *et al.* (1976) examined the mineralogy of ash agglomerates from gasification beds of Ignifluid boilers. Temperatures of the process were between 1200° and 1400°C. Only enough oxygen is present in this process to fluidize the moving bed of the gasifier. Ash agglomerates formed were composed of a matrix of amorphous silica with inclusions of quartz, mullite ($3Al_2O_3 \cdot 2SiO_2$), hercynite ($FeAl_2O_4$), and iron sulfides. Some work has been done on ash from the Lurgi process which operates at temperatures between 600° and 850°C and pressures of 350–450 psig (Griffin *et al.*, 1978). The Lurgi ash contained quartz, mullite, hematite, magnetite, plagioclase feldspar, and amorphous silica. Distinct crystals of iron oxide developed in a matrix of the other minerals and amorphous silica. O'Gorman and Walker (1973) and Mitchell and Gluskoter (1976) studied phase transformations of coal minerals at elevated temperatures. Among those phases forming from coal minerals under conditions similar to those of coal gasification are anhydrite, anorthite, clinopyroxene, hematite, mullite, quartz, and amorphous silica.

II. SCANNING ELECTRON MICROSCOPY IN THE ANALYSIS OF COAL AND COAL CONVERSION WASTES

In recent years electron microscopy has proven useful for characterizing coal and coal conversion wastes. Sutherland (1975) used the electron microprobe to determine organic sulfur in coal. Boateng and Phillips (1976) used the microprobe and the scanning electron microscope (SEM) to examine surfaces of coal for iron and sulfur distribution. Other studies using the electron microprobe on the determination of organic and inorganic sulfur in coal were conducted by Harris *et al.* (1977), Solomon and Manzione (1977), and Raymond and Gooley (1978). Using microprobe analysis, Cobb and Russell (1976) determined iron and cadmium contents in sphalerite present in coals of west central

Illinois. For an extensive discussion of the use of the electron microprobe on coal, see Chapter 48, and on lignite, see Volume II, Chapter
27.

The transmission electron microscope (TEM) was employed by Harris
and Yust (1976) to observe micropores in coal and by Strehlow *et al.*
(1978) to explore relationships between coal minerals and certain maceral types. With the SEM, Augustyn *et al.* (1976) examined fractured
and polished surfaces of coal for coal structure and composition of the
mineral constituents. Gluskoter (1977; Gluskoter and Lindahl, 1973)
routinely used the SEM for the characterization of mineral matter in
coal. Other studies involving the SEM and coal are pyrite occurrence
and distribution in coal—Greer (1977, 1978) and Scheihing *et al.* (1978);
in situ analysis of inorganic trace element sites in coal—Finkelman
(1978); correlated Mössbauer–SEM studies of the mineralogy of coal
and coal conversion products using an automated SEM—Lee *et al.*
(1978); and microstructural studies of chemically desulfurized coals—
Rebagay and Shou (1978). Only a few studies have been devoted to the
examination of coal conversion wastes with the SEM. Characterization
of coal ash with the SEM was instrumental in understanding the Godel
phenomenon of the Ignifluid process (Yerushalmi *et al.*, 1975; Harvey
et al., 1975; Kolodney *et al.*, 1976). Russell (1977) described the mineralogy of liquefaction residues from several coal liquefaction processes.
Harris and Yust (1978) identified the constituents of a carbonaceous
plug from a Solvent Refined Coal liquefaction plant.

The scanning electron microscope is an instrument which can be
used to study the surface features of specimens with a resolution of
100–200 Å on the average and 50 Å or better in some cases. With SEM,
the morphology and interrelationships of minerals in coal and coal
conversion residues and associated organic matter can be studied. An
energy dispersive x-ray analyzer ancillary to the SEM provides an
elemental analysis for elements with atomic numbers greater than ten
for specimens analyzed. A comprehensive treatment of the physics of
the SEM and energy dispersive x-ray analysis is given in Wells *et al.*
(1974) and Goldstein and Yakowitz (1975).

A. Preparation of Specimens for SEM

1. *Pretreatment of Samples*

Pretreatment of coal and coal ash from gasification processes consists
of low temperature ashing (Gluskoter, 1965). A discussion of this and
other methods of separating minerals from coal are reported in Volume

II, Chapter 26, Section II. During low temperature ashing organic matter is oxidized at less than 150°C in an atmosphere of electronically excited oxygen. This produces a comparatively unaltered mineral matter residue. Some minor dehydration effects are observed in bituminous coals, but these changes are thought to be reversible (Rao and Gluskoter, 1973). Miller and Given (1978) reported that significant changes in the mineral matter of lignite and low rank coals can occur with low temperature ashing.† They proposed a method of low temperature ashing whereby these effects can be minimized. The morphology of the mineral matter can most clearly be observed by examining the low temperature ash. The orientation, distribution, and association of minerals with respect to the organic constituents (macerals) in coals can be seen by examining an unashed specimen.

Some coal liquefaction residues containing a high percentage of organic material must be treated even before they are ashed. Residues can be extracted with tetrahydrofuran (THF) to remove the THF soluble organics leaving the mineral matter and THF insolubles, which are then ashed. The extraction process also homogenizes the residue. The authors attempted to ash a coal liquefaction residue without THF extraction and noted a precipitate of unknown composition which collected on the walls of the specimen container. The following procedure can be used for THF extraction (Dreher, 1978): The solid liquefaction residue is broken up and slurried in THF in a covered container for several days with occasional stirring. When the residue is completely dissolved, the solution is filtered with a Whatman No. 1 filter. The filter cake is air dried, ground with a mortar and pestle, and then low temperature ashed. The THF soluble portion is poured into Teflon cookie sheets and allowed to evaporate under a hood at ambient temperature. The THF soluble residue is then chipped off and stored. This extraction has no apparent effect on the mineral matter.

Preparation of gasification ash for low temperature ashing requires grinding the sample to −60 mesh.

2. Preparation of Low Temperature Ash (LTA)

a. Separation of Heavy and Light Minerals   A high percentage by weight of the mineral matter in coal consists of clay minerals [in Illinois coals clay minerals comprise approximately 50% of the mineral matter (Rao and Gluskoter, 1973)]. The clay minerals cling to the surfaces of particles of other mineral matter after low temperature ashing and

---

† The formation of nitrates is discussed in Volume II, Chapter 20, Section IV, B, and of sulfates in Chapter 26, Section II.

cause difficulty in observing the minerals by SEM. This is a problem with coals and liquefaction residues but not with the LTA of gasification ashes because the clay minerals are converted to higher temperature minerals during the gasification. To alleviate this problem, a heavy mineral separation with bromoform is recommended. Bromoform has a specific gravity of 2.88. Clay minerals and quartz float with the light fraction while sulfides and most carbonates (including calcite, specific gravity 2.72–2.94) settle with the heavy fraction. Separation of the LTA of coal is relatively simple and can be accomplished by gravity with a separatory funnel according to the method of Müller (1967) or Carver (1971). Enough heavy minerals can usually be obtained with this method to get a representative sample of those present. Low temperature ash of liquefaction residues is more difficult to separate. The heavy minerals in the liquefaction residues are intimately associated with the clay minerals, and during gravity separation very few heavies settle out. Some success in concentrating the heavy minerals of the liquefaction residue has been attained using a method modified slightly from Barsdate (1962). Low temperature ash of liquefaction residue is added (about 0.05–0.1 g) to a 15 mliter glass centrifuge tube containing bromoform. The tubes are centrifuged for 15 to 20 min at 3000–4000 rpm. Heavy material segregated at the bottom of the tubes after centrifuging is removed with a hypodermic needle, taking care to expel air slowly from the hypodermic as it is being lowered through the light mineral layer. The heavy minerals are then emptied from the hypodermic onto filter paper and washed in acetone to remove all traces of bromoform. The heavy minerals can then be mounted for SEM work using the techniques described below. Bromoform separated fractions should not be used for analytical purposes other than SEM examination. Clay minerals, especially, could be altered by the adsorption of some bromoform. After removal of the heavies, the light mineral fraction consisting largely of clays and quartz can be poured off, filtered, and washed with acetone. Float–sink separation of coal samples is described in Volume I, Chapter 15, Section II, B and C.

b. Sample Mounting Techniques of LTA for SEM Analysis One way to study the morphology of the light fraction of the LTA of coals and liquefaction residues is by dispersing this fraction in water and sedimenting it on an SEM specimen stub. Approximately 0.02 g of material is added to 40 mliter of distilled water. The suspension is dispersed in an ultrasonic bath for about 2 min. Using an eyedropper, 1–3 drops of the suspension are placed on an SEM stub and allowed to dry in a desiccator or dust-free environment. A method similar to this, which has been used for clays, is described by Walker (1978). If the above method does not successfully disperse the clays, add 0.002 g or less of sodium hexametaphosphate to the distilled water before the ultrasonic bath and proceed as in the above. However, if sodium hexametaphosphate is added, Na and P may be detected in the clays by energy dispersive x-ray analysis. Colleagues at the Illinois State Geological Survey have etched the sedimented clay minerals in HF fumes to bring out the boundaries between clay flakes. Copper SEM stubs rather than aluminum are used for sedimenting clays to allow detection of the aluminum in clays without interference from the stub.

Specimens of elements with low atomic numbers such as alumino-silicates have a larger specimen volume from which x rays are generated than do specimens composed of higher atomic number elements. The Al in the stub may be detected through a thin layer of clays more easily than through a heavy mineral particle.

Examination by electron microscope of some expandable clay minerals with high water sorption capacity may necessitate special preparation to preserve their original three-dimensional structure. The heat of the electron beam at high magnifications can drive out some adsorbed water, causing movement and alteration of the sample, disruption of the specimen's conductive coating, and contamination of the electron microscope. Bohor and Hughes (1971) used a method of spray drying described in Hughes and Bohor (1970) to prepare clay samples with high water content for SEM work. They reported that this procedure greatly reduced the adverse effects noted previously. Another method of preparing samples with high moisture content is freeze-drying, described in Wells et al. (1974) and in Goldstein and Yakowitz (1975). In freeze-drying, the water in a specimen is rapidly frozen and then sublimed under high vacuum. This technique preserves the original structure of the specimen and prevents damage caused by dehydration under the electron beam. Gillot (1969), Chen et al. (1976), and Erol et al. (1976) used freeze-drying to prepare clay minerals for the SEM. Erol et al. (1976) explained in detail the method of freeze-drying they used.

Heavy minerals from the low temperature ash of coal and liquefaction residues may be mounted on SEM stubs by dusting the particles on an adhesive applied to the stub. Glues are unsuitable because the particles tend to sink in and become covered over by the glue. Koda-flat®, a print flattener manufactured by Kodak, has been used successfully by the authors as an adhesive. A very thin coating of Kodaflat is applied full strength to the surface of a stub and allowed to dry in a dust-free environment. After a few minutes, the surface will dry to a tacky consistency and particles may be dusted on. Doublestick tape may also be used as an adhesive. However, Walker (1978) notes that when working with doublestick tape, care must be taken when coating the specimen to avoid letting the evaporator or sputterer become hot enough to crack the tape. This could cause charging of the specimen. If very fine particles of heavy minerals must be examined, such as those which may adhere to the filter paper in bromoform separation, polycarbonate filters manufactured by Nuclepore are suggested for use by Walker (1978). Sections of the filters may be mounted on SEM stubs. These filters have smooth rather than fibrous surfaces, and particles can easily be studied. Walker reports that polycarbonate filters do not have detectable x-ray spectra and therefore would not interfere with energy dispersive x-ray analysis.

After mounting, the heavy minerals should be cleaned in ethyl alcohol in a mild sonic bath for several seconds to remove surface debris. Acetone is unsuitable as it dissolves the adhesive material. Inevitably, some particles will be lost from the adhesive in cleaning. If energy dispersive x-ray analysis is to be performed, the adhesive employed should be analyzed to determine the composition in case of contamination. The LTA of gasification ash is prepared in the same manner as the heavy minerals of coal and liquefaction residue LTA. Other techniques for mounting particles for SEM work are reported in Brown and Teetsov (1976).

Preparation of LTA discussed so far has been for examining the morphology of minerals. Investigating their interrelationships requires a different approach. One method is to embed the low temperature ash of a sample (either the whole ash or just the heavy minerals) in epoxy (the authors have used Castolite®) and to polish the surface to expose sections through the particles. A small amount of LTA is placed in the bottom of an empty specimen mold of a diameter similar to that of the SEM stub on which it will be mounted. Epoxy is poured into the mold and the mixture stirred slightly to insure binding of the LTA particles with the epoxy. It is desirable to have the final chip as thin as possible to reduce the length of the working distance in the microscope; hence, one can either pour the chip thinner or polish it down to the desired thickness (2 mm thickness is recommended). The epoxy chip is glued to the stub with Duco® cement, allowed to dry, and then ground and polished, either manually or with an automatic device. The specimen is ground with 600 grit first (starting with coarser grit tends to remove too many of the specimen particles) until the surface is flat. After a thorough ultrasonic cleaning the sample is polished with 1 μm chrome oxide suspended abrasive for 3–5 min on a silk or nylon lap, cleaned

again, and finished for 1-2 min with 0.05 μm Buehler Finish-pol®.
After ultrasonic cleaning, the specimen is ready for application of a
conductive coating. If ultrasonic cleaning does not adequately remove
all the abrasive, gently wipe the polished surface with a cotton ball or
a fingertip (this will not significantly harm the polish) and then clean
it ultrasonically. Figure 1 shows a sample of Lurgi ash which has been
prepared in this way. Iron oxide crystals are visible in a matrix of
plagioclase feldspar, mullite, quartz, and amorphous silica.

3. *Preparation of Samples Which Have Not Been Low Temperature Ashed*

Coal gasification ash and small blocks of coal that have not been low
temperature ashed have been examined by SEM. No adverse effects on
the instrument because of volatility of coals and ash have been reported
by researchers studying these materials. Coal gasification ash should
present no contamination problems; however, some coals may. Most
coals examined by the authors have been high volatile B and C bituminous. Sub-bituminous coals and lignites are more volatile (much of
the volatility is moisture) and may cause some contamination of the

Fig. 1 Iron oxide crystals (probably magnetite) from Lurgi ash in a matrix of other
minerals. Scale bar = 5 μm.

instrument, although the volatility of the coal should decrease if a conductive coating is applied by evaporation or sputtering.

It is recommended that only THF-extracted coal liquefaction residues be examined with the SEM. The THF-soluble portion contains unidentified volatile organics which could contaminate the instrument.

4. Etching Techniques

Structures and relationships visible in polished specimens can sometimes be greatly enhanced by etching. The iron oxide crystals (probably magnetite) in Fig. 1 stand out dramatically against the matrix after the sample has been etched in concentrated HF for 1–2 min (Fig. 2). Etching can aid in the identification of minerals in coal materials. A calcium silicate, wollastonite, was identified in very small quantities in liquefaction residue of the H-Coal process (Russell, 1977). Its solubility in HCl aided in the identification of this mineral. Scheihing *et al.* (1978) investigated pyrite framboids in coal from the Meigs Creek seam and etched polished coal chips with nitric acid. The pyrite in the framboids dissolved, leaving interstitial networks which were determined to be kaolinite. Colleagues of the authors at the Illinois State Geological Survey have etched polished specimens of coal in a low temperature

Fig. 2 Iron oxide crystals in Lurgi ash after etching with HF. Scale bar = 5 μm.

asher to enhance the microscopic contrast between minerals and macerals in coal (Fig. 3). A period of about a half-hour in the asher is
sufficient to give a well-etched surface on bituminous coals. A similar
unetched coal specimen is shown in Chapter 48, Fig. 2.

5. Specimen Coating

A comprehensive treatment of specimen coating techniques is given
in Echlin (1978), but some aspects which are especially useful in coating
liquefaction residues and gasification ashes are discussed here.

The purpose of coating materials for SEM examination is to provide
a conductive surface which will minimize thermal damage to the specimen and provide an undistorted image for viewing. One must choose
a conductive coating of ultrafine structure which has good secondary
electron emission to provide good resolution and one that does not
interfere with the elements of interest during x-ray analysis. Several
elements may provide satisfactory coatings under different circumstances for the materials dealt with here.

In cases where x-ray analysis is most important, carbon provides the
best coating. However, its secondary electron yield is low, and reso-

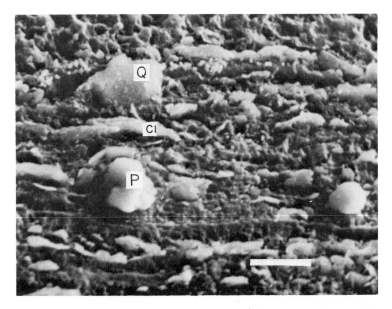

Fig. 3 Quartz (Q), pyrite (P), and clay (thin and elongate particles, as Cl) as they
occur in vitrinite (dark) in Illinois coal etched in a low-temperature asher. Scale bar =
5 μm.

lution over 5000× may be poor. A Au/Pd alloy commonly used, less grainy than Au alone, has a high secondary emission coefficient resulting in good resolution, but the gold M_α line interferes with the sulfur K_α line in x-ray analysis. This becomes important in the detection of sulfide minerals in liquefaction residues. Chromium is found to be a good compromise between resolution and x-ray interference. Chromium has a finer grain size than the Au/Pd alloy and provides good resolution. In x-ray analysis, Cr interferes with the detection of Na, Ti, and Ba, which have all been found in coal mineral matter, but to a minor extent. Chromium itself is a rarely occurring element in coals.

Fine mineral particles or very irregular particle surfaces can be difficult to coat and require a thicker layer of conductive coating than do flat or polished surfaces. Coating such specimens with carbon first helps to minimize charging effects because of its multiple scattering and surface diffusion properties. The use of a sputter coater on fine particles may be advantageous because of its ability to deposit a continuous coating (more so than vacuum evaporation) on all specimen surfaces without rotation or tilting of the specimen (Echlin, 1978).

B.   Energy Dispersive X-Ray Analysis

When an electron beam strikes a specimen in the SEM, several forms of radiation are released from the specimen; among them are x rays characteristic of the atom from which they are released. The energy dispersive x-ray analyzer available for use with the SEM rapidly collects x rays one at a time, sorts them by energy, and counts them. The resulting elemental analysis is displayed on a cathode ray tube (CRT), and the data is recorded in various ways. This technique is only qualitative unless correction factors are applied to the x-ray data similar to the processing of electron microprobe data (Chapter 48, Section II, B). A detailed discussion of the theory and electronics of energy dispersive x-ray analysis can be found in Goldstein and Yakowitz (1975) and Wells *et al.* (1974). Elemental analysis can be performed for elements of atomic number greater than that of fluorine. Depending upon the elements involved, it is possible to detect as little as a few tenths of a percent of an element present. Analysis is best on polished specimens as the topography of the specimen can interfere with x-ray generation and collection. Coupled with x-ray diffraction data on the mineral matter in liquefaction residue and gasification ash samples, energy dispersive x-ray analysis can be used to identify the mineral composition of particles observed in the SEM. There are, however, stipulations to the analysis which must be remembered. The beam energy must exceed the excitation potential of the elements to be analyzed. Twenty kilovolts is sufficient to excite at least the L lines of most elements. Analysis of submicrometer-sized particles in a matrix is difficult because of uncertainty in the shape and dimensions of the x-ray excited volume in the sample. This volume† depends upon the depth of penetration of the beam which is influenced by specimen composition. X rays are generated from a smaller sample volume in samples composed of heavier elements than those composed of lighter elements. Hence, submicrometer-sized particle analyses may contain x-ray data produced from surrounding matrix material. Wells *et al.* (1974) gives examples of calculations used to determine the x-ray excited volume for some elements.

Even though results of energy dispersive x-ray analysis cannot be quantitative without corrections for atomic number, absorption and secondary fluorescence effects, the ratios of the x-ray counts of each element from a sample can reveal its chemical composition. For example, in a polished sample it is not uncommon to be able to distinguish pyrite (FeS_2) from pyrrhotite ($Fe_{1-x}S$) by the sulfur counts approximately double those of iron in pyrite as opposed to pyrrhotite in which the sulfur counts are approximately equal to the iron counts.

In addition to a general chemical analysis, one can obtain x-ray distribution images‡ and line scans of an element of interest. Figures 4 and 5 show an aggregated particle of hematite grains from Lurgi ash and its corresponding x-ray distribution image for iron, respectively. The x-ray distribution image is obtained by setting the discriminators of the multichannel analyzer to the iron area of the spectra and adjusting the CRT image. Output from a ratemeter can produce single line scans of an element in a similar manner. These images are particularly useful for detecting compositional changes within a particle.

III. X-RAY DIFFRACTION IN THE ANALYSIS OF COAL AND COAL CONVERSION WASTES

A number of studies of coal and coal conversion wastes have used x-ray diffraction to identify the mineral matter. Rao and Gluskoter (1973) and Ward (1977) obtained quantitative data on quartz, pyrite, and calcite in low temperature ashes of coals. These techniques, like that described in Volume II, Chapter 26, Section III, A, may also be applied to the analysis of coal conversion wastes. Walker *et al.* (1977) performed quantitative x-ray diffraction analysis on mineral matter in

† See Chapter 48, Fig. 3.
‡ See Chapter 48, Figs. 4 and 5.

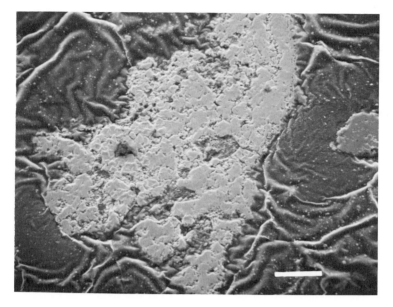

Fig. 4   Hematite from Lurgi ash. Scale bar = 50 μm.

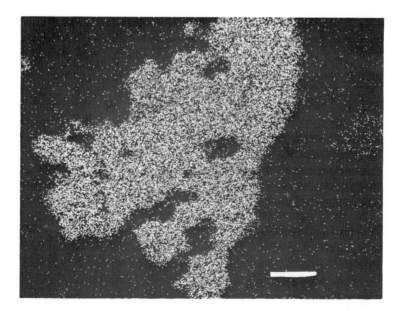

Fig. 5   Iron x-ray distribution image for hematite in Fig. 4. Scale bar = 50 μm.

coal and coal liquefaction residues. Granoff *et al.* (1978) used quantitative x-ray diffraction in the analysis of the low temperature ash of
feed coals in a study of the catalytic effects of mineral matter on coal
liquefaction. These researchers used the technique of Rao and Gluskoter with nickel oxide as an internal standard. Qualitative x-ray diffraction of mineral matter from gasification processes was determined
by Harvey *et al.* (1975) for Ignifluid boiler ash agglomerates and by
Griffin *et al.* (1978) for solid coal wastes including liquefaction residue
and gasification ash. O'Gorman and Walker (1973) and Mitchell and
Gluskoter (1976) identified by x-ray diffraction the changes in mineral
matter from coal experimentally heated to high temperatures.

For details on x-ray procedures and apparatus used we refer the
reader to Klug and Alexander (1974).

A.   Quantitative Analysis of Nonclay Minerals

1.   Sample Preparation

Prior to x-ray analysis of coals and coal conversion wastes, the mineral matter must be isolated. Organic material may be removed by
using an oxidizing agent (such as hydrogen peroxide) (Ward, 1974) or
by using the more widely accepted method of low temperature ashing
as described previously.

2.   Preparation of Standards

The method of quantitative x-ray diffraction using internal standards
involves preparing standard mixtures of varying composition, including each mineral to be analyzed, and constructing standard graphs
from diffraction data for each mineral. For liquefaction residues, the
major minerals present might be quartz, pyrrhotite, and calcite depending upon the composition of the feed coal. The internal standard
method is not applicable to quantitative determination of clay minerals
because of the numerous problems involved with selection of standard
clay minerals. For gasification ash, the minerals would be silicates and
oxides such as quartz, mullite, feldspar, and hematite. Minerals present
would depend upon the parent coal and also upon the process temperature and pressure conditions.

Minerals chosen for standards ideally should be crystallographically
identical to the minerals found in the unknown mixtures, diffracting
x rays at the same angles and intensities. Fulfilling this requirement
and securing monomineralic materials for standards can be difficult. A

synthesized pyrrhotite (Shiley *et al.*, 1979) is used as a standard for pyrrhotite from liquefaction residues. This pyrrhotite shows Mössbauer spectroscopic parameters and x-ray diffraction patterns very similar to those from the pyrrhotite in the liquefaction residues. Natural pyrrhotite is unsuitable for a standard because of its unavailability in quantity in monomineralic form.

Standard graphs are constructed only for those minerals present in an unknown which are to be determined quantitatively; the standards should approximate the composition of the unknowns as closely as possible. The remaining minerals should comprise a matrix which also closely approximates the composition of these minerals in the unknown. For example, in analysis of coal mineral matter from Illinois Basin coals, the major minerals present are quartz, pyrite, calcite, kaolinite, illite, and expandable clays (expand with ethylene glycol solvation†). Since this method of quantitative analysis is not suitable for clay minerals, the clay minerals will form the matrix for the standards, and the minerals quartz, pyrite, and calcite will be quantified. Minor minerals may also be quantified if desired. Kaolinite, illite, and montmorillonite fractionated by sedimentation techniques are used for the clay matrix. Montmorillonite is not ordinarily found in Illinois coals. The expandable clays that occur are a degraded mixed-layer material. Montmorillonite is used because it was not possible to find a similar mixed-layer material suitable for a standard. Clay minerals comprise approximately 50% of the low temperature ash of Illinois Basin coals, and it was determined from analyses by Rao and Gluskoter (1973) in what proportions the clays should be added: kaolinite 12%, illite 20%, and expandables 18% by weight. The proportions of constituents in a typical set of standards are shown in Table I. The minerals used as standards should be ground to −325 mesh (U.S. Standard Sieve Series) before being added to the standard mixtures.

Twenty percent by weight of 0.3 μm Linde A alumina (Al_2O_3) is added as an internal standard to each of the mixtures. Linde A alumina was chosen as an internal standard over other minerals, such as fluorite (as used by Rao and Gluskoter, 1973), for several reasons: (1) the smaller Al_2O_3 particle size minimizes orientation effects, (2) extremely pure Al_2O_3 is available, and (3) alumina produces a minimal amount of x-ray diffraction peak interference. The main objection to the use of CaF_2 is the interference of the (111) fluorite peak with that of the (111) of sphalerite and (111) of pyrite. The standard mixtures (including Al_2O_3) are ground together by hand in absolute alcohol with an agate or

† See Volume II, Chapter 26, Section III, A.

TABLE I *Typical Proportions of Mineral Constituents in Standards for Quantitative X-Ray Diffraction of Low Temperature Ash for Coals of the Illinois Basin*

| Standard no. | Percentages | | | | | |
	Calcite	Pyrite	Quartz	Illite	Kaolinite	Montmoril-lonite
1	20	0	30	20	12	18
2	25	5	20	20	12	18
3	30	10	10	20	12	18
4	10	15	25	20	12	18
5	15	20	15	20	12	18
6	5	25	20	20	12	18
7	15	30	5	20	12	18
8	10	35	5	20	12	18
9	0	40	10	20	12	18
10	5	45	0	20	12	18

mullite mortar and pestle (in our work for x-ray diffraction analysis no contamination has been noted from the use of a mullite mortar and pestle). Grinding should continue until satisfactory reproducibility of x-ray peaks is attained when the mixture is run by x-ray diffraction. Standard and unknown mixtures are ground by hand because unmixed material was routinely found around the top of sample containers from the mechanical mixers previously used.

For liquefaction residues analyzed, similar standards are established. The major minerals to be quantified include quartz, calcite, and pyrrhotite, and the matrix is kaolinite, illite, and expandable clays. For lack of contradictory data, composition of the clay matrix used in residue analysis is the same as that used for coal mineral matter standards. However, the amount of clay matrix in relation to the other minerals is greater, approximately 70% from analyses in progress. The reduction in nonclay minerals is due to the loss of sulfur from pyrite when pyrite is transformed to pyrrhotite during liquefaction.

Standards for gasification ash should be appropriate to the minerals present in the ash as determined by a qualitative x-ray diffraction analysis. For example, Lurgi ash from Illinois Herrin (No. 6) Coal contains hematitie, magnetite, mullite, plagioclase feldspar, quartz, and amorphous silica. The standards should contain these minerals. Where no information is available on the quantities in which these components are present, it is best to make up standards containing a wide compositional range of the components. All the methods for standard mixtures discussed above are applicable to gasification ashes.

Standard and unknown samples are run on a Phillips Norelco x-ray diffractometer using copper K_α radiation and a graphite monochrometer. A cavity powder mount consisting of a rectangular aluminum plate about 2-mm thick containing a rectangular window and a glass slide backing is used for the specimen. The mount is back loaded with powder while the aluminum plate and glass slide are either taped or held together by a clip. Powder is dusted from a spatula into the rectangular window with a minimum of packing. Excess powder is removed from the mount, and a second glass slide is lowered onto the aluminum plate so that it is sandwiched between the two slides. The mount is then inverted and the first glass slide carefully removed by lifting it directly off the aluminum plate without sliding it across the powdered surface. The powder should appear firmly packed, and the surface should be level. This procedure is designed to minimize orientation effects and is basically the same as that described in Rao and Gluskoter (1973). If a rotating sample mount or spinner is not available, it is advisable to run a sample for quantitative analysis three times, with at least one of the runs being $2°\ 2\theta$ to $60°\ 2\theta$ to include all the major diffraction peaks of the minerals. The sample is repacked each time it is run.

Slow scanning speeds ($\frac{1}{4}$–$\frac{1}{2}°$/min) are generally recommended for quantitative work. Reproducibility studies are in progress for scanning speeds of $\frac{1}{4}°$, $\frac{1}{2}°$, 1°, and 2°/min. Preliminary results show the best reproducibility for $\frac{1}{4}°$ and 1°/min runs. A 1°/min scanning speed is recommended when large numbers of samples must be analyzed.

3. Construction of Standard Curves

In constructing standard curves for calcite, quartz, pyrite, and pyrrhotite from coal and coal liquefaction residues, the following peaks were used for measurements: (104) calcite, (101) quartz, (200) pyrite, (200) pyrrhotite (hexagonal) [the pyrrhotite observed in liquefaction residues has a single (102) peak characteristic of hexagonal pyrrhotite rather than a doublet characteristic of monoclinic pyrrhotite], and (104) Al_2O_3. In choosing peaks for intensity measurements of gasification ash mineral matter, the peaks should occur relatively close to each other on the diffraction pattern because of accentuated intensities at lower angles of 2θ (Lorentz factor) and should be of high intensity. Feldspar, hematite, and mullite peaks occur too close to (104) Al_2O_3 making Al_2O_3 unsuitable as a standard for gasification ash. Fluorite, CaF_2, also used as an internal standard, may be used for gasification ashes with the same phases present as in Lurgi ash. The (220) peak of fluorite is suitable for intensity measurements in the presence of Lurgi ash mineral matter.

Construction of the standard curves will be explained using the standards in Table I as an example. Each standard is run three times, and the peak areas of the (104) calcite, (101) quartz, (200) pyrite, and (104) Al_2O_3 are determined. Peak areas, rather than peak heights, are preferable because of the effects of particle size. Peak heights are unreliable because height decreases and width increases with decreased particle size (<0.1 μm) (Klug and Alexander, 1974). Peak areas are calculated by measuring the height of the peak from a suitable baseline, which must be drawn in the same manner for each sample; measuring the width of the peak at half the total height of the peak; and multiplying these two numbers. The ratios of these peaks of calcite, quartz, and pyrite to the standard peak, (104) Al_2O_3, are calculated. The ratios of the three replicates for each mineral in each standard are averaged. There are then ten average values for each mineral—calcite, quartz, and pyrite—corresponding to ten percentage compositions. Three separate standard graphs are constructed, one each for calcite, pyrite, and quartz, by plotting the percentages of each mineral on the abscissa and the ratio of its peak area to the standard peak area on the ordinate. A least-squares line is then drawn through the points, and the composition of an unknown can be read directly from the curve by knowing the ratio of the mineral peak areas to the standard peak area.

The (104) Al_2O_3 peak was chosen over the (012) and (113) peaks because it produced the most nearly linear relationship with composition.

4. Preparation of Unknowns for Analysis

The preparation of unknowns (the low temperature ash of coals, coal liquefaction residue, or gasification ash) follows the same procedure as the preparation of standards. Twenty percent by weight of the standard material is added to a sample of low temperature ash; this mixture is ground by hand in absolute alcohol (for approximately 20 min). The sample is oven dried at about 35°C, reground slightly, and mounted in a cavity powder mount. The unknown is run three times in the same manner as the standard. The (104) calcite, (101) quartz, (200) pyrite, and (104) Al_2O_3 peaks (and any additional peaks for which standards have been prepared) are measured for coal LTA. Peak area ratios are then calculated, and the percentage of each mineral (calcite, quartz, pyrite, etc.) is read from the standard curve.

This method of x-ray diffraction analysis has an inherent error of

±5–7% absolute which is due to machine error and human error in sample preparation.

B. Qualitative and Semiquantitative Analysis of Clay Minerals

Clay minerals may account for approximately 50% of the mineral matter in low temperature ashes of coals (Rao and Gluskoter, 1973; Ward, 1977) and an even greater weight percentage in liquefaction residues (as an indirect result of the loss of sulfur during the conversion of pyrite to pyrrhotite). Clay minerals have not been observed in gasification ashes because of alteration of the aluminosilicates to other minerals under the high temperatures involved in this process. Despite this abundance of clay minerals, there have been relatively few attempts to quantify clay minerals in coals and liquefaction residues.

1. Pretreatment of Samples

With the use of low temperature ashing, mineral matter can easily be extracted from coal (Gluskoter, 1965). The effects of low temperature ashing on the mineral matter in coal have been discussed elsewhere (Rao and Gluskoter, 1973; Miller and Given, 1978). The only effect on the clay minerals is the dehydration of expandable clays, which is a reversible process because of the low temperatures involved (<150°C). However, analysis of the clay minerals remains a problem, since the <2 μm fraction must first be isolated from the remaining mineral matter. This separation is difficult because only a small amount of material is available and the clays are in a highly flocculated state. This tendency to flocculate may be attributed to the mode of formation of the clays (i.e., deposition in acid swamp waters) and also to the abundance of iron and other ions (Ward, 1977). Therefore, prior to analysis of the clay fraction, considerable pretreatment is necessary.

Gluskoter (1967) achieved dispersion of the clay fraction through repeated washings with distilled water and the addition of sodium hexametaphosphate. The disadvantage of this method is that residues of this dispersant can produce extraneous x-ray diffraction peaks. Of particular importance is the sodium hexametaphosphate peak at 7.03 Å which interferes with the (001) of kaolinite at 7.15 Å.

Ward (1977) adapted the sodium dithionite–citrate–bicarbonate method of iron removal (Jackson, 1975), to obtain dispersion of the clay fraction. Ward's method (1977) is basically a three-step process involving the removal of soluble ions and the removal of carbonates and acid soluble iron compounds, followed by the removal of soluble ferrous iron after reduction from the ferric state. The method we recommend is basically the same, with a few minor changes, and is summarized in Fig. 6.

A small amount of ash (0.2–0.5 g) is placed in a 100 mliter beaker with 80 mliter distilled water. The suspensions are thoroughly agitated and allowed to settle, usually overnight. After the ashes have flocculated, the supernatant is decanted. This liquid contains water soluble ions and may be discarded. Note that throughout this method, samples may either be left to flocculate at their own rate or they may be centrifuged prior to removing the supernatant. Approximately 50 mliter of 5% acetic acid is then added to each sample. After the samples are warmed for 2 hr, the beakers are allowed to cool, and after the clays have flocculated, the clear liquid containing dissolved carbonates and other acid soluble compounds is decanted. Each ash sample is then heated in a solution containing 40 mliter 0.3M sodium citrate and 5 mliter 1M sodium bicarbonate. When a temperature of 70° to 80°C has

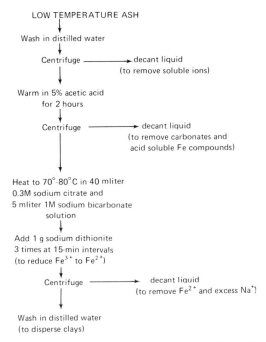

Fig. 6   Flow diagram for clay mineral dispersion. [From Ward (1977).]

been reached, 1 g of sodium dithionite is added at three 15-min intervals, with thorough stirring with a glass rod after each addition. At
this stage, ferric iron (Fe^{3+}) is reduced by the sodium dithionite
($Na_2S_2O_4$) to ferrous iron (Fe^{2+}). Sodium dithionite has a high reducing
power in a basic or neutral system but not in an acid system. Hence,
a buffer (sodium bicarbonate—$NaHCO_3$) is used. The use of HCO_3^- to
raise the pH also prevents sulfur precipitation. Precipitation of iron
sulfide is prevented by sodium citrate ($Na_3C_6H_5O_7 \cdot H_2O$) which is a
strong complexing agent for iron (Jackson, 1975). After the samples
have cooled and the suspensions have flocculated, the supernatant is
discarded. At this stage all that is needed to disperse the clays are two
or three washings in distilled water. The amount of water used in the
final wash is determined by the type of slide preparation technique
used as described below.

Advantages of clay dispersion by iron removal are that the clay
structures are not attacked and the method is fairly rapid and relatively
free of analytical difficulties. Jackson (1975) provides further discussion
of this method.

2. Sample Preparation

a. Isolation of the <2 μm Fraction   Once the clays have been dispersed, the <2 μm fraction can be isolated. The clay fraction may be
separated by centrifuge or by gravity settling. Tanner and Jackson
(1947) describe details of sedimentation times. Sedimentation by gravity settling has proven adequate in previous studies of clay minerals
in low temperature ashes (Ward, 1977). After the suspensions are thoroughly agitated, each is allowed to settle for 21 min, after which time
the top 0.5 cm of suspension is removed with a pipette. According to
Stokes' law, this will contain the <2 μm fraction.

Use of the <2 μm fraction for qualitative and semiquantitative clay
studies is widely accepted by clay mineralogists and soil scientists.
Clay minerals tend to concentrate in this size fraction, thus facilitating
the preparation of well-oriented mounts. This size fraction also tends
to be free of most nonclay minerals (Grim, 1968).

b. Slide Preparation   There are many different techniques of slide
preparation (Brindley, 1961; Gibbs, 1971) each with varying degrees of
accuracy and reproducibility (Gibbs, 1965). Several authors (Gibbs,
1965; Schultz, 1955) have discussed the advantages and disadvantages
of these methods with emphasis on errors introduced by segregation

of the clay particles during sample preparation. Stokke and Carson (1973) have discussed errors caused by the use of too much or too little material. Three methods which are routinely used in semiquantitative analysis of clays are: (a) smearing the <2 μm fraction across a slide, (b) centrifuging the <2 μm fraction onto a ceramic tile, and (c) pipetting the <2 μm fraction onto a glass slide.

According to Gibbs (1965), the reproducibility of these three methods is comparable ($\pm 10\%$). However, experience has shown that reproducibility can be greatly improved when human error is minimized. The smear technique is often favored over the centrifuge and pipette techniques because no particle settling takes place; thus the technique provides more accurate data (Gibbs, 1965). However, good orientation is important to increase the detection limits of the analysis. Thus, the method of slide preparation used must be rapid, internally consistent, and capable of providing good orientation [to emphasize the basal (00l) reflections]. Oriented <2 μm slides (pipette method) have been used in previous studies of clay minerals in coals (Gluskoter, 1967; Ward, 1977) and are currently used by the authors to attain high levels of reproducibility.

Centrifuging the <2 μm fraction onto a ceramic tile was suggested by Kinter and Diamond (1956). This method provides excellent orientation and good reproducibility and has been used successfully in semiquantitative analysis of clays in liquefaction residues (Russell, 1977). However, problems associated with this method include excessive sample preparation time, cracking ceramic tiles, and production of extraneous peaks on the diffraction patterns (from the mullite in the tiles). Of particular importance is the mullite peak which occurs at 5.44 Å, which may interfere with the $(002)_{10}/(003)_{17}$ and $(005)_{27}/(002)_{10}$ reflections of mixed-layer clays.

3.   X-Ray Diffraction of Clays

X-ray analyses are carried out on a Phillips Norelco diffraction unit using copper K_α radiation with a graphite monochrometer and employing a linear ratemeter. All slides are scanned at 2° 2θ/min over the region 2° 2θ–35° 2θ. A scanning speed of 2° 2θ/min is considered optimum in terms of information obtained and economy of time. A slower speed of 1° 2θ/min does not provide sufficiently better resolution to warrant the longer analysis time (an extremely important consideration when analyzing large numbers of samples).

An air-dried run is obtained to compare with the ethylene glycol run to check for the presence of small amounts of mixed-layer material.

Minor amounts of interstratified material are indicated when the glycol run shows an increase of intensity in the third order and decrease in the second order of illite. After solvation with ethylene glycol for 48 hr, the <2 μm slides are rerun. Two additional glycol runs of the 14° 2θ to 21° 2θ region are obtained to determine the exact position of the $(002)_{10}/(003)_{17}$ and $(005)_{27}/(002)_{10}$ reflections of mixed-layer illite–montmorillonite if they are present.

After heating to 375°C for 1 hr, each slide is then x rayed while still hot, as rehydration may occur as soon as the slide cools. The optimum temperature for reproducibility of semiquantitative data is 375°C. This temperature produces complete collapse of expandable layers (Austin and Leininger, 1976) without affecting the intensity of the chlorite reflection or cracking theslides, as might occur at higher temperatures.

4. Qualitative Clay Mineralogy

Clay minerals observed in ashes of coals and liquefaction residues include kaolinite, illite, a highly degraded mixed-layer material and, rarely, chlorite. Random and ordered mixed-layered illite–montmorillonite, although seen in associated strata such as underclays (Stepusin, 1978) and shale partings, have not been detected in coal ashes to date.

Clay minerals in the <2 μm fraction are identified by the $(00l)$ reflections determined from x-ray diffraction patterns of ethylene glycol treated and heat treated samples. The main peaks used to identify kaolinite are the (001) reflection at 7.15 Å (12.4° 2θ) and the (002) reflection at 3.57 Å (24.9° 2θ). These peak positions are changed by neither ethylene glycol treatment nor heating to 375°C. Because the (002) reflection of chlorite coincides with the (001) of kaolinite, the second order of kaolinite reflection (3.57 Å) is necessary for identification. Resolution between the (004) of chlorite at 3.54 Å (25.1° 2θ) and the (002) of kaolinite is observed at this position.

Chlorite, when present, is identified by the (003) peak at 4.68 Å (18.7° 2θ) and the (004) peak at 3.54 Å (25.1° 2θ). The (002) reflection of chlorite coincides with kaolinite (at 7.15 Å) and the (001) coincides with vermiculite (14 Å, 6.3° 2θ). Vermiculite and chlorite may be differentiated after heating the sample to 375°C. Upon heating, the chlorite spacing remains unaltered whereas vermiculite collapses to 10 Å (Walker, 1961). Vermiculite has not been reported in low temperature ashes.

Illite is identified by the (001) reflection at 10 Å (8.8° 2θ) and the (002) reflection at 5 Å (17.75° 2θ) in both the glycol and heated runs.

Discrete phases of mixed-layer materials may be identified using the techniques described by Reynolds and Hower (1970). However, in low temperature ashes studied so far, expandable materials exhibit a broad diffraction band between 10 and 17 Å in the glycol runs. Upon heating to 375°C, this degraded mixed-layer material collapses to 10 Å.

5. Semiquantitative Analysis of Clay Minerals

Various methods exist for calculating the relative percentages of clays in the <2 μm fraction. Each of these methods was devised for a particular study and was based on the method described in Johns et al. (1954). Pierce and Siegel (1969) have compared results obtained by several of these methods.

In recent studies of low temperature ashes (Ward, 1977; Russell, 1977), the percentages of kaolinite plus chlorite (K + C), illite (I), and expandables (Ex) in the <2 μm fraction are determined semiquantitatively by a procedure modified from Griffin (1971). Calculations are based on the relative intensities of the 7- and 10-Å peaks obtained from x-ray diffraction patterns of ethylene glycol and heat treated samples (Fig. 7). Although peak areas were used in these studies, the use of peak heights is suggested here since experience has shown that this method provides better reproducibility of data. In x-ray patterns of ethylene glycol treated samples, the broad diffraction band between 10 and 17 Å causes considerable error in estimating the peak widths and, ultimately, error in peak area calculations for the 10 Å illite peak. The modified formulas (Ward, 1977) used to determine the relative amounts of clays present are

$$\%(K + C) = \frac{7 \text{ Å h}/2.5}{(7 \text{ Å h}/2.5 + 10 \text{ Å h})} \times 100,$$

$$\%(I + Ex) = 100 - \%(K + C),$$

$$\%I = \frac{10 \text{ Å g}}{10 \text{ Å h}} \times \frac{7 \text{ Å h}}{7 \text{ Å g}} \times \%(I + Ex),$$

$$\%Ex = \%(I + Ex) - \%I,$$

where %(K + C) is the percentage kaolinite + chlorite, %I the percentage illite, and %Ex the percentage expandable clay minerals. The notations 7Å h and 10 Å h refer to the intensities of the 7 Å and 10 Å peaks, respectively, after heating the slide to 375° C for 1 hr, while 7 Å g and 10 Å g refer to these intensities following exposure to ethylene glycol vapor for two days.

The intensity of the 7 Å peak is affected by neither glycol nor heat treatment. The inclusion of the 7 Å h/7 Å g factor in the calculation provides an internal check on fluctuations caused by operating condi-

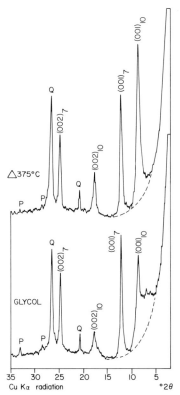

Fig. 7 X-ray diffraction patterns of ethylene glycol treated and heat treated <2 μm fraction of a low temperature ash. %(K + C) = 23, %I = 42, %Ex = 35. Q = quartz, P = pyrite.

tions and contributes to the high level of reproducibility achieved by this method.

ACKNOWLEDGMENTS

Development of the methods described here was supported in part by U.S. EPA Grant R804403 and U.S. Department of Energy contract EY-76-C-21-8004 which we gratefully acknowledge. The authors appreciated the suggestions and critical review of their colleagues G. B. Dreher, H. D. Glass, R. D. Harvey, and P. R. Johnson.

REFERENCES

Augustyn, D., Illey, M., and Marsh, H. (1976). *Fuel* **55,** 25–38.
Austin, G. S., and Leininger, R. K. (1976). *J. Sediment. Petrol.* **46,** 206–215.
Barsdate, R. J. (1962). *J. Sediment. Petrol.* **32,** 608–620.
Boateng, D. A. D., and Phillips, C. R. (1976). *Fuel* **55,** 318–322.
Bohor, B. F., and Hughes, R. E. (1971). *Clays Clay Miner.* **19,** 49–54.

Brindley, G. W. (1961). *In* "The X-Ray Identification and Crystal Structures of Clay Minerals" (G. Brown, ed.), pp. 1–50. Mineral. Soc. (Clay Miner. Group), London.

Brown, J. A., and Teetsov, A. (1976). *Scanning Electron Microsc./1976, IITRI, Chicago, Ill.* pp. 385–392.

Carver, R. E. (1971). *In* "Procedures in Sedimentary Petrology" (R. E. Carver, ed.), pp. 427–452. Wiley (Interscience), New York.

Chen, Y., Banin, A., and Schnitzer, M. (1976). *Scanning Electron Microsc./1976, IITRI, Chicago, Ill.* pp. 425–432.

Cobb, J. C., and Russell, S. J. (1976). *Geol. Soc. Am. Abstr. Programs* **8,** 816.

Dreher, G. B. (1978). Personal communication.

Echlin, P. (1978). *Scanning Electron Microsc./1978, SEM, AMF O'Hare, Ill.* pp. 109–132.

Erol, O., Lohnes, R. A., and Demirel, T. (1976). *Scanning Electron Microsc./1976, IITRI, Chicago, Ill.* pp. 769–776.

Finkelman, R. B. (1978). *Scanning Electron Microsc./1978, SEM, AMF O'Hare, Ill.* pp. 143–148.

Gibbs, R. J. (1965). *Am. Mineral.* **50,** 741–751.

Gibbs, R. J. (1971). *In* "Procedures in Sedimentary Petrology" (R. E. Carver, ed.), pp. 531–539. Wiley (Interscience), New York.

Gillot, J. E. (1969). *J. Sediment. Petrol.* **39,** 90–105.

Given, P. H., Cronauer, D. C., Spackman, W., Lovell, H. L., Davis, A., and Biswas, B. (1975). *Fuel* **54,** 34–39.

Gluskoter, H. J. (1965). *Fuel* **44,** 285–291.

Gluskoter, H. J. (1967) *J. Sediment. Petrol.* **37,** 205–214.

Gluskoter, H. J. (1977). *Energy Sources* **3,** 125–131.

Gluskoter, H. J., and Lindahl, P. C. (1973). *Science* **181,** 264–266.

Goldstein, J. I., and Yakowitz, H. (1975). "Practical Scanning Electron Microscopy." Plenum, New York.

Granoff, B., and Thomas, M. G. (1978). "Mineral Matter Effects and Catalyst Characterization in Coal Liquefaction." Sandia Lab., Albuquerque, New Mexico.

Granoff, B., Baca, P. M., Thomas, M. G., and Noles, G. T. (1978). "Chemical Studies on the Synthoil Process: Mineral Matter Effects." Sandia Lab., Albuquerque, New Mexico.

Gray, D. (1978). *Fuel* **57,** 213–216.

Greer, R. T. (1977). *Scanning Electron Microsc./1977, IITRI, Chicago, Ill.* pp. 79–93.

Greer, R. T. (1978). *Scanning Electron Microsc./1978, SEM, AMF O'Hare, Ill.* pp. 621–626.

Griffin, G. M. (1971). *In* "Procedures in Sedimentary Petrology" (R. E. Carver, ed.), pp. 541–569. Wiley (Interscience), New York.

Griffin, R. A., Schuller, R. M., Suloway, J. J., Russell, S. J., Childers, W. F., and Shimp, N. F. (1978). "Solubility and Toxicity of Potential Pollutants in Solid Coal Wastes, Environmental Aspects of Fuel Conversion Technology III." U.S. EPA, Research Triangle Park, North Carolina.

Grim, R. E. (1968). "Clay Mineralogy." McGraw-Hill, New York.

Harris, L. A., and Yust, C. S. (1976). *Fuel* **55,** 233–236.

Harris, L. A., and Yust, C. S. (1978). *Scanning Electron Microsc./1978, SEM, AMF O'Hare, Ill.* pp. 537–542.

Harris, L. A., Yust, C. S., and Crouse, R. S. (1977). *Fuel* **56,** 456–457.

Harvey, R. D., Masters, J. M., and Yerushalmi, J. (1975). *Ill. State Geol. Surv. Ill. Miner. Notes* **61.**

Henley, J. P. (1975). M.S. Thesis, Auburn Univ., Auburn, Alabama.

Hughes, R. E., and Bohor, B. F. (1970). *Am. Mineral.* **55,** 1780–1786.

Jackson, M. L. (1975). "Soil Chemical Analysis—Advanced Course," published by author, Madison, Wisconsin.

Johns, W. D., Grim, R. E., and Bradley, W. F. (1954). *J. Sediment. Petrol.* **24**, 242–251.

Kinter, E. B., and Diamond, S. (1956). *Soil Sci.* **81**, 111–120.

Klug, H. P., and Alexander, L. E. (1974). "X-Ray Diffraction Procedures for Polycrystalline and Amorphous Materials." Wiley (Interscience), New York.

Kolodney, M., Yerushalmi, J., Squires, A. M., and Harvey, R. D. (1976). *Trans. Br. Ceram. Soc.* **75**, 85–91.

Lee, R. J., Huggins, F. E., and Huffman, G. P. (1978). *Scanning Electron Microsc./1978, SEM, AMF O'Hare, Ill.* pp. 561–568.

Miller, R. N., and Given, P. H. (1978). "A Geochemical Study of the Inorganic Constituents in Some Low-Rank Coals," Tech. Rep. No. 1. Pennsylvania State Univ., University Park, Pennsylvania.

Mitchell, G. D., Davis, A., and Spackman, W. (1977). *In* "Liquid Fuels from Coal" (R. T. Ellington, ed.), pp. 255–270. Academic Press, New York.

Mitchell, R. S., and Gluskoter, H. J. (1976). *Fuel* **55**, 90–96.

Müller, G. (1967). "Sedimentary Petrology, Part I, Methods in Sedimentary Petrology" (H.-U. Schmincke, transl.). Hafner, New York.

Mukherjee, D., and Chowdhury, P. B. (1976). *Fuel* **55**, 4–8.

O'Gorman, J. V., and Walker, P. L., Jr. (1973). *Fuel* **52**, 72–79.

Pierce, J. W., and Siegel, F. R. (1969). *J. Sediment. Petrol.* **39**, 187–193.

Rao, C. P., and Gluskoter, H. J. (1973). *Ill. State Geol. Surv. Circ.* No. 476.

Raymond, R., Jr., and Gooley, R. (1978). *Scanning Electron Microsc./1978, SEM, AMF O'Hare, Ill.* pp. 93–108

Rebagay, T. V., and Shou, J. K. (1978), *Scanning Electron Microsc./1978, SEM, AMF O'Hare, Ill.* pp. 669–676.

Reynolds, R. C., and Hower, J. (1970). *Clays Clay Miner.* **18**, 25–36.

Russell, S. J. (1977). *Scanning Electron Microsc./1977, IITRI, Chicago, Ill.* pp. 95–100.

Scheihing, M. H., Gluskoter, H. J., and Finkelman, R. B. (1978). *J. Sediment. Petrol.* **48**, 723–732.

Schultz, L. C. (1955). *J. Sediment. Petrol.* **25**, 124–125.

Shiley, R. H., Russell, S. J., Dickerson, D. R., Hinckley, C. C., Smith, G. V., Saporoschenko, M., and Twardowska, H. (1979). *Fuel.* In press.

Solomon, P. R., and Manzione, A. V. (1977). *Fuel* **56**, 393–396.

Stepusin, S. M. (Rimmer) (1978). M.S. Thesis, Univ. of Illinois, Urbana.

Stokke, P. R., and Carson, B. (1973). *J. Sediment. Petrol.* **43**, 957–964.

Strehlow, R. A., Harris, L. A., and Yust, C. S. (1978). *Fuel* **57**, 185–186.

Sutherland, J. K. (1975). *Fuel* **54**, 132.

Tanner, C. B., and Jackson, M. L. (1947). *Soil Sci. Soc. Am., Proc.* **12**, 60–65.

Tarrer, A. R., Guin, J. A., Pitts, W. S., Henley, J. P., Prather, J. W., and Styles, G. A. (1977). *In* "Liquid Fuels from Coal" (R. T. Ellington, ed.), pp. 45–61. Academic Press, New York.

Walker, D. A. (1978). *Scanning Electron Microsc./1978, SEM, AMF O'Hare, Ill.* pp. 185–192.

Walker, G. F. (1961). *In* "The X-Ray Identification and Crystal Structures of Clay Minerals" (G. Brown, ed.), pp. 297–324. Mineral. Soc. (Clay Miner. Group), London.

Walker, P. L., Jr., Spackman, W., Given, P. H., White, E. W., Davis, A., and Jenkins, R. G. (1977). "Characterization of Mineral Matter in Coals and Coal Liquefaction Residues," 2nd Annu. Rep. to EPRI, Palo Alto, Project RP 366-1.

Ward, C. R. (1974). *Fuel* **53,** 220–221.
Ward, C. R. (1977). *Ill. State Geol. Surv. Circ.* No. 498.
Wells, O. C., Royde, A., Lifshin, E., and Rezanowich, A. (1974). "Scanning Electron Microscopy". McGraw-Hill, New York.
Yerushalmi, J., Kolodney, M., Graff, R. A., Squires, A. M., and Harvey, R. D. (1975). *Science* **187,** 646–648.

Chapter 43

Analyses of Coke Oven Effluents for Polynuclear Aromatic Compounds

Samuel P. Tucker
NATIONAL INSTITUTE FOR OCCUPATIONAL SAFETY AND HEALTH
ROBERT A. TAFT LABORATORIES
CINCINNATI, OHIO

I. INTRODUCTION

Polynuclear aromatic compounds (PNA's) are of particular interest in the analyses of coke oven effluents because many of these compounds are carcinogens or suspected carcinogens (Epstein *et al.*, 1964; Huberman and Sachs, 1976). One of the most potent carcinogens in the atmosphere is benzo[a]pyrene (Epstein, 1965). Exposure to the carcinogenic PNA's may pose a serious occupational hazard. There are several

163

ways in which exposure to polynuclear aromatic compounds can take place; namely, by skin contact, ingestion, and inhalation. Although the first two means can involve direct contact with liquids and solids which contain PNA's, all three means of exposure can result from vapors and aerosols in the air. Thus, the concentrations of various PNA's in the air, especially in the breathing zone of the worker, are important. This chapter emphasizes the analysis of air samples for PNA's.

The coke oven process† involves the baking of coal to produce coke, a highly carbonaceous residue. Depending on the type of process, temperatures near 1500°C may be attained (Schulte *et al.*, 1975; Keeling and Jung, 1949). High temperatures can promote the formation of polynuclear aromatic compounds from simpler compounds that evolve during the coke oven process. For example, the treatment of benzene at 700°C has produced a tar contntaining fluoranthene, benzo-[b]fluoranthene, chrysene, and other compounds (Badger and Novotny, 1961). Since coals contain various amounts of nitrogen, sulfur, and oxygen, it is expected that polynuclear aromatic compounds bearing these elements are produced during the coke oven process. However, most of the polynuclear aromatic compounds in coke oven emissions reported by Schulte *et al.* (1974) and Lao *et al.* (1975) were hydrocarbons.

A.   Structures of Selected Polynuclear Aromatic Compounds

Polynuclear aromatic compounds make up a broad class of compounds which generally includes those structures bearing two or more fused aromatic rings; i.e., rings which share a common border. Within this class are the polynuclear aromatic hydrocarbons (PAH's) and also heterocyclic compounds. (For the purpose of this chapter, PAH's are considered to be those PNA's which consist of only carbon and hydrogen.) Naphthalene is among the simplest of the PNA's. The PNA's which are present in coke oven emissions may number in the hundreds and perhaps in the thousands. The PNA's which have been identified in air samples from coke oven effluents generally range up to six- or seven-ring compounds in size and most of these PNA's have been hydrocarbons (Broddin *et al.*, 1977; Lao *et al.*, 1975; Schulte *et al.*, 1974). In the analysis of air samples for PNA's, at coke oven plants and elsewhere, attention has been focused on benzo[a]pyrene because of the possible health hazards posed by this compound (Jackson *et al.*,

† See Chapter 38, Section II, for the coke oven gas flow diagram.

1974; Mašek, 1971; Sawicki *et al.*, 1960). It would be difficult or impossible to identify all the PNA's in the coke oven environment on account of the following limiting factors: (1) the sampling techniques available, (2) the analytical techniques available, (3) the concentrations of the PNA's present, (4) the stability of the PNA's, and (5) the standards available for reference.

Figure 1 presents the structures of selected PNA's that may be found in coke oven emissions. Structures of all of these compounds may be found in *The Ring Index* by Patterson and Capell (1960) except indeno[1,2,3-cd]pyrene, which may be found in supplement I to this reference (Capell and Walker, 1963).

B. Physical Properties of Polynuclear Aromatic Compounds

Many polynuclear aromatic hydrocarbons are planar, nonpolar compounds which melt at temperatures above room temperature. The melting points and a few boiling points of selected PAH's and their corresponding molecular weights are listed in Table I. An important factor which tends to accompany an increase in melting point is the increase in the number of vertical planes of symmetry. For example, anthracene has a higher melting point than any of the following: benz[a]anthracene, benzo[a]pyrene, and benzo[e]pyrene, all three of which have higher molecular weights and contain the anthracene nucleus. Anthracene bears two planes of symmetry vertical to the plane of the molecule, while each of the other three compounds bears fewer than two vertical planes of symmetry. Perylene, benzo[a]pyrene, benzo[e]pyrene, and benzo[k]fluoranthene have the same molecular weight; yet, perylene has the highest melting point and the greatest number of vertical planes of symmetry of the four PAH's. Coronene has six vertical planes of symmetry in addition to its relatively high molecular weight and has a very high melting point. However, this type of observation is not made in the comparison of the melting points of pyrene, benzo[a]pyrene, and benzo[e]pyrene.

Many PAH's have significant vapor pressures near room temperature which may give rise to losses during air sampling and laboratory handling. The vapor pressures and equilibrium vapor concentrations at 30°C listed in Table II are approximate values based on the extrapolation of vapor pressures at higher temperatures. Collection efficiencies of particulate filters may be low for those compounds with equilibrium vapor concentrations near to or above the air concentrations. Also, some of the compounds once collected on the particulate filters

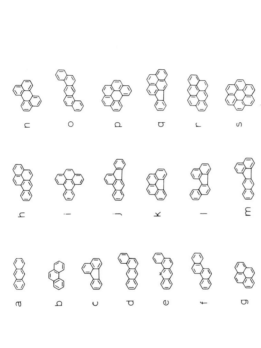

Fig. 1 Structures of selected polynuclear aromatic compounds. (For additional structures of volatile PNA's possible in coal see Chapter 39, Fig. 9.)

a. anthracene
b. phenanthrene
c. fluoranthene
d. benz[a]anthracene
e. benz[c]acridine
f. chrysene

g. pyrene
h. benzo[a]pyrene
i. benzo[e]pyrene
j. benzo[b]fluoranthene
   (benz[e]acephenanthrylene)

k. benzo[ghi]fluoranthene
l. benzo[j]fluoranthene
m. benzo[k]fluoranthene
n. perylene
o. dibenz[a,h]anthracene

p. benzo[ghi]perylene
q. indeno[1,2,3-cd]pyrene
r. anthanthrene
s. coronene

TABLE I *Melting Points and a Few Boiling Points of Selected PAH's*

Compound	Molecular Weight	Melting Point (°C)	Boiling Point (°C)[a]	Reference
Fluorene	166.22	116–117	295	b
Anthracene	178.24	216.5–217.2	339.9	c,d
Fluoranthene	202.26	110.6–111.0		e
Pyrene	202.26	152.2–152.9		c
Benzo[ghi]fluoranthene	226.28	147–149		f
Benz[a]anthracene	228.30	159.5–160.5	435	g,h
Chrysene	228.30	250	448	i
Benzo[a]pyrene	252.32	176.5–177.5	(310–312)	j
Benzo[e]pyrene	252.32	178–179		j
Benzo[k]fluoranthene	252.32	215.5–216		k
Perylene	252.32	273–274		l
Coronene	300.36	438–440		m

[a] Each boiling point is at a pressure of 1 atm except the boiling point of benzo[a]pyrene is at a pressure of 10 torr.

[b] Meyer and Hofmann (1916).

[c] Jones and Neuworth (1944).

[d] Timmermans and Burriel (1931).

[e] Orchin and Reggel (1947).

[f] Campbell and Reid (1952).

[g] Fieser and Hershberg (1937).

[h] Kruber (1941).

[i] Krafft and Weilandt (1896).

[j] Cook and Hewett (1933).

[k] Moureu et al. (1946).

[l] Morgan and Mitchell (1934).

[m] Melting point determined with sample in sealed tube (Newman, 1940).

may sublime during additional sampling and during storage. Sublimation losses may be lower when the compounds are adsorbed on particulate matter (Pupp *et al.*, 1974). Losses occur when solutions of the volatile compounds are taken to dryness, particularly under substantially reduced pressures. These losses are smaller if removal of the solvent from the solutions is halted prior to dryness (Grimmer and Hildebrandt, 1965).

C. Chemical Properties of Polynuclear Aromatic Compounds

The chemistry of PNA's is quite complex and may differ from one PNA to another. Generally, the PNA's are more reactive than benzene. The reactivities of PNA's toward methyl radicals tend to increase with

TABLE II *Vapor Pressures, Equilibrium Vapor Concentrations, and Sublimation Rates of Selected PAH's at 30°C*

Compound	Vapor pressure[a] (torr)	Equilibrium vapor concentration[b] (ng/m³)	Sublimation rate at 14 torr[c] (ng/0.5 hr)
Anthracene	1.2×10^{-5}	1.1×10^5	6370
Pyrene	1.2×10^{-5}	1.3×10^5	5130
Benz[a]anthracene	2.4×10^{-7}	2.9×10^3	810
Benzo[a]pyrene	1.2×10^{-8}	1.6×10^2	60
Benzo[e]pyrene	1.2×10^{-8}	1.6×10^2	0
Benzo[ghi]perylene	2.4×10^{-10}	3.5×10^0	70
Coronene	3.6×10^{-12}	5.8×10^{-2}	0

[a] The vapor pressures are approximate values based on extrapolations to 30°C using the equation, $\log P = -(A/T) + B$, where A and B are constants (Bradley and Cleasby, 1953; Murray *et al.*, 1974).

[b] The equilibrium vapor concentrations are approximate values based on the equation, $C = 1 \times 10^{-12} (PM)/(RT)$, where C is the equilibrium vapor concentration in ng/m³, P is the vapor pressure, M is the molecular weight, R is the gas constant, and T is the absolute temperature (Pupp *et al.*, 1974).

[c] The sublimation rates are crude values based on losses in taking cyclohexane solutions of 5.56-13.26 μg quantities of the individual compounds to dryness in a 100 mliter round-bottomed flask with a bath temperature of 30°C and a pressure of 14 torr (Grimmer and Hildebrandt, 1965).

greater conjugation. In comparison to benzene, naphthalene and benz[a]anthracene react with methyl radicals 22 and 468 times faster, respectively (Levy and Szwarc, 1955). Rates of electrophilic, nucleophilic, and free-radical substitution reactions may be greater for naphthacene and anthracene than for benzene. Diels–Alder reactions can take place between maleic anhydride and many PAH's (Dewar, 1952). Polymerization of reactive PAH's may be induced by an acid catalyst in attempts to perform a Friedel–Crafts alkylation of the PAH's (Schiller, 1977). Many PNA's form charge-transfer complexes with certain electron-acceptor compounds such as 2,4,7-trinitro-9-fluorenone (Heacock and Hutzinger, 1975). [For the formation of the complex between naphthalene and 2,4,6-trinitrophenol (picric acid) see Chapter 38, Section V,A.]

Some of the chemical reactions of interest would be those that affect the concentrations of PNA's in the coke oven environment and those that affect the concentrations in air samples during collection, storage, and analysis. Environmental factors which may influence the reactivity of selected PNA's include temperature, light, oxygen, ozone, other chemical agents, and the surface areas of particulate matter on which the PNA's may be adsorbed.

Small quantities (~100 mg or less) of numerous PAH's can be vaporized completely without apparent reaction in an inert atmosphere (argon) during the course of heating the container at 10°C/min toward 750°C (Lewis and Edstrom, 1963). Many of such PAH's have ionization potentials above 7.10 eV and consist of seven or fewer rings. These relatively stable PAH's include naphthalene, benz[a]anthracene, chrysene, benzo[a]pyrene, and coronene. Under the same heating conditions, PAH's with ionization potentials less than 7.10 eV generally are more reactive, and the noncatalytic thermal reactions may involve intermolecular transfer of hydrogen. The heating of naphthacene and pentacene, for example, has produced dihydro derivatives. Since naphthalene is relatively quite volatile, 100 mg quantities of this compound can vaporize from the container before high temperatures promote apparent reaction. However, treatment of carbon-14 labeled naphthalene at 700°C has produced a tar from which benzo[j]fluoranthene, benzo[k]fluoranthene, and other compounds were isolated (Badger *et al.*, 1964).

Photochemical change of benzo[a]pyrene in sunlight can take place when this compound is adsorbed on thin layers of silica gel. At initial loadings of 1–100 μg benzo[a]pyrene per square centimeter of the silica gel, much of this PAH may react during a 5-hr exposure to sunlight (Mailath *et al.*, 1974). A major decrease of benzo[a]pyrene on soot has been noted after exposure to light for 40 min (Thomas *et al.*, 1968). Ultraviolet irradiation of benzo[a]pyrene in methanol solution at a wavelength of 365.5 nm has given benzo[a]pyrene-1,6-dione, benzo[a]pyrene-3,6-dione, and benzo[a]pyrene-6,12-dione as photooxidation products. These same diones may be produced by oxidation of benzo[a]pyrene with chromic acid (Antonello and Carlassare, 1964). Treatment of a solution of benzo[a]pyrene in dichloromethane at 20°C with ozone in a 1:1 molar ratio has given a mixture containing benzo[a]pyrene-1,6-dione and benzo[a]pyrene-3,6-dione (Moriconi *et al.*, 1961).

II. TECHNIQUES OF COLLECTION OF AIR SAMPLES

Efficient collection of many polynuclear aromatic compounds from an air sample may be accomplished with a two-stage collector; the first stage consists of a filter for trapping particulate matter, and the second stage consists of a medium for trapping the vapors. However, most reports of collection techniques for PNA's in the literature have been limited mainly to the collection of particulate matter. The collection and analysis of dust samples in the air in coal mines for PNA's is described in Volume II, Chapter 28, Section V.

The following items should be recorded during sampling: sampling rate, length of period of sampling, sample volume, and the temperature and pressure of the atmosphere.

After sampling, the collection media should be stored in the dark in an airtight container under refrigeration.

A. Glass Fiber Filters

Glass fiber filters have been very popular in the collection of particulate matter bearing PNA's.

1.   The glass fiber filters should be nonhygroscopic and free of both organic and inorganic binders. The organic binders may cause interferences in analysis. The use of inorganic binders may result in some catalytic decomposition of various PNA's (Sollenberg, 1976). Interferences by trace organic material on the filters may be reduced substantially by Soxhlet extraction and vacuum drying at elevated temperatures.

2.   The sample volumes desired depend on: (a) the air concentrations of the PNA's of interest; (b) sample losses during air sampling, storage, and workup; and (c) the sensitivity of the analytical method. The desired volumes may range from less than one to a few thousand cubic meters (1000 liter = 1 m³). The sampling periods may exceed 24 hr (Sollenberg, 1976; Jackson *et al.*, 1974; Dong *et al.*, 1976; Sawicki *et al.*, 1962).

3.   Practical sampling rates depend upon the size of the filter, the capacity of the pump, and the pressure drop in the system. A rate of 170 liter/min has been employed with filters 10.16 cm in diameter (Krstulovic *et al.*, 1976). Rates ranging from 566 to 1770 liter/min have been employed with 20.32 × 25.40 cm filters (Schulte *et al.*, 1975; Jackson *et al.*, 1974). These rates span a range of 1.1–3.4 liter/min/cm².

4.   If the total weight of the particulate matter is desired, the filter should be weighed at equilibrium at constant humidity and temperature before and after sampling. However, it should be noted that treatment of a glass fiber filter in a Soxhlet extraction apparatus or in an ultrasonic bath for the purpose of extracting PNA's may dislodge some

of the glass fibers and result in a weight change of the filter (Richards *et al.*, 1967; Sawicki *et al.*, 1975).

B. Silver Membrane Filters

Silver membrane filters† have been used in air sampling near coke ovens (Searl *et al.*, 1970; Lao *et al.*, 1975). However, the resistance to airflow by such filters is considerable and limits the sampling rate. A 47-mm-diam filter with an 0.8 μm pore size creates a pressure drop of 8.1 cm of mercury at a flow rate of 22 liter/min (Richards *et al.*, 1967). A cubic meter of air could be sampled in ~45 min at this flow rate. Results of a coke oven study conducted by the National Institute for Occupational Safety and Health indicate that some particulate matter adheres poorly to the silver membrane filters (Schulte *et al.*, 1974). It is of interest to note that the *NIOSH Manual of Analytical Methods* includes a few methods which involve sampling at 2.0 liter/min with a combination of a glass fiber filter followed by a silver membrane filter (Taylor, 1977). These methods specify filters with a diameter of 37 mm. Results reported by Jackson and Cupps (1978) show that benzene-soluble material may be found on the silver membrane filter located behind the glass fiber filter.

C. Filter Papers

A series of filter papers in a single sampler may be used to collect airborne PNA's (Malý, 1971; Hlucháň *et al.*, 1974). According to these workers, three filter papers in series permit quantitative collection of the pentacyclic and hexacyclic hydrocarbons: 3,4-benzofluoranthene (benzo[b]fluoranthene), benzo[a]pyrene, 1,2:3,4-dibenzopyrene (dibenzo[def,p]chrysene), 1,12-benzoperylene (benzo[ghi]perylene), and anthanthrene. Four filter papers mounted in separate frames may be arranged in series in a sample holder. Flow rates of 100 and less than 33.3 liter/min may be achieved with series of filters 50 and 30 mm in diameter, respectively. Individual filter papers which have been pretreated with a 3:1 mixture of methanol and glycerine have been used in collecting air samples of benzo[a]pyrene at coke oven plants (Mašek, 1971). Sampling rates of 25–30 liter/min have been achieved with such filters 110 mm in diameter.

† The use of silver membrane procedures for analysis of respirable coal dust is described in Volume II, Chapter 28, Section III,B,1.

D. Solid Sorbents

The porous polymer, Tenax GC† (poly 2,6-diphenyl-*p*-phenylene oxide) is efficient in collecting vapors of PNA's (Jones *et al.*, 1976). Anthracene, which is relatively quite volatile, is collected efficiently by this sorbent. Other compounds collected with greater than 90% efficiency include chrysene, perylene, pyrene, and benzo[ghi]perylene. A 7 × 3 cm tube may be packed with 12 g of Tenax GC and enclosed with a glass frit in the inlet and a glass wool plug at the outlet. Such a collector permits a sampling rate of 14 liter/min.

Chromosorb 102 (a styrene-divinylbenzene polymer) has been used as a solid sorbent behind a glass fiber–silver membrane filter combination (Jackson and Cupps, 1978).

E. Cascade Impactors

Cascade impactors (Andersen samplers) have been used to collect airborne PNA's on particulate matter (Broddin *et al.*, 1977; Pierce and Katz, 1975a). These multistage samplers fractionate particles according to their aerodynamic sizes. Such samplers are valuable in determining concentrations of respirable particulates. The stages may be equipped with glass fiber filters (Burton *et al.*, 1973).

III. TECHNIQUES OF ANALYSIS

A. Extraction of Polynuclear Aromatic Compounds from Collection Media

1. *Glass Fiber Filters*

a. Soxhlet Extraction A proper choice of solvent is necessary in the efficient extraction of selected polynuclear aromatic compounds from particulate matter by means of Soxhlet extraction. Benzene has been used substantially in Soxhlet extraction and may be efficient in extracting many PNA's. Recoveries of benzo[a]pyrene from enriched particulate matter and glass fiber filter material were 94% or better after six or more hours of Soxhlet extraction with benzene (Pierce and Katz, 1975a; Stanley *et al.*, 1967; Fox and Staley, 1976). In addition, eight hours of Soxhlet extraction with benzene permitted 95–100% recoveries

---

† Mention of commercial products does not constitute endorsement by the National Institute for Occupational Safety and Health.

of seven other PAH's including chrysene, benz[a]anthracene, benzo[k]-fluoranthene, benzo[ghi]perylene, and coronene (Pierce and Katz, 1975a). These workers reported the efficient extraction (93 ± 4%) of two oxygenated species, 7-H-benz[de]anthracen-7-one and phenalen-1-one, from particulate matter with benzene. However, Stanley *et al.* (1967) reported the mean recovery of 7-H-benz[de]anthracen-7-one from enriched particulate matter was only 42% after six hours of Soxhlet extraction with benzene; a mean recovery of 80% was achieved for this compound with dichloromethane. The recovery of benz[c]acridine from the same enriched particulate matter was 63% with benzene but was 100% with chloroform and a 4/1 mixture (*v/v*) of benzene and diethyl-amine.

Cyclohexane is another solvent which has received wide attention in Soxhlet extraction of PNA's. An attractive property of cyclohexane in comparison to benzene is its lower capacity for dissolving polar compounds which may cause analytical problems with PAH's (Jackson *et al.*, 1974; Dong *et al.*, 1976). Cyclohexane was virtually just as efficient as benzene in extracting benzo[a]pyrene from particulate matter collected from air samples (Stanley *et al.*, 1967). Yet, in the same report, cyclohexane and benzene were 76 and 95% efficient, respectively, in recovering benzo[a]pyrene from enriched particulate matter. Of interest is a study involving a particulate model in which Cautreels and Van Cauwenberghe (1976) concluded that both benzene and carbon disulfide were superior to cyclohexane in Soxhlet extraction of PAH's. This particulate model consisted of 95% cellulose and 5% carbon black and was enriched with various PAH's. Soxhlet extraction with benzene and carbon disulfide independently for four hr afforded recoveries of 96–106% for phenanthrene, fluoranthene, pyrene, benzo[b]fluorene, and benz[a]anthracene. The use of cyclohexane resulted in recoveries of 31–68% for these compounds.

b. Ultrasonic Extraction   Ultrasonic extraction of PNA's from particulate matter is rapid and may be performed at room temperature. Elevated temperatures which may decompose some PNA's can be avoided in this technique. Ultrasonic extraction of benzo[a]pyrene from atmospheric dust was accomplished satisfactorily with benzene, chloroform, methylene dichloride, and diethyl ether in 30 min (Chatot *et al.*, 1971). Recoveries of less than 1 μg quantities of anthracene, phenanthrene, and benzo[a]pyrene from enriched glass fiber filters were 95–98% after 12 min of ultrasonic agitation with cyclohexane and silica powder (Sawicki *et al.*, 1975).

c. Sublimation   Selected PAH's may be extracted from glass fiber filters by sublimation at an elevated temperature. Benzo[a]pyrene may be extracted from glass fiber filters at pressures of 0.03–0.05 torr and a temperature of 300°C (Schultz *et al.*, 1973). Recoveries of benzo[a]pyrene from enriched glass fiber filters treated at 300°C for one hour in a vacuum sublimation apparatus were 91–103% at levels of 0.0002–20.2 μg (Sollenberg, 1976). In a different study, pyrene, chrysene, benzo[a]pyrene, and benzo[k]fluoranthene were thermally desorbed from enriched glass fibers in a tube maintained at 300°C overnight. During this time nitrogen passed through the tube at 50 mliter/min. The vapors were collected on a precolumn which contained the same stationary phase used in subsequent gas chromatography. Although the mean recoveries from the glass fibers were quantitative, the mean recoveries from the precolumn by thermal desorption ranged from 83 to 93% (Burchfield *et al.*, 1974).

2. *Silver Membrane Filters*

PNA's have been extracted from silver membrane filters by Soxhlet extraction with benzene and also with cyclohexane (Schulte *et al.*, 1975; Searl *et al.*, 1970; Lao *et al.*, 1975). Relevant information on this topic may be found in the discussion on Soxhlet extraction involving glass fiber filters. The technique of ultrasonic agitation in the presence of benzene can be applied to the extraction of PNA's from silver membrane filters (Taylor, 1977).

3. *Filter Papers*

Two methods for extracting PNA's from filter papers have been (1) Soxhlet extraction with acetone (Cooper, 1954) and (2) treatment with cold benzene in a beaker six times (Malý, 1971; Hluchán *et al.*, 1974).

4. *Solid Sorbents*

PNA's may be extracted from Tenax GC by Soxhlet extraction with *n*-pentane for 24 hours. Recoveries were generally over 90% for 10 μg quantities of chrysene, pyrene, perylene, benzo[ghi]perylene, and coronene (Jones *et al.*, 1976).

Benzene-soluble material has been extracted from Chromosorb 102 after air sampling (Jackson and Cupps, 1978). The value of this porous polymer as a sorbent in a sampling and analytical method for various PNA's depends upon extraction efficiencies and other factors.

B.   Low Pressure Liquid Chromatography

Liquid chromatographic techniques at low pressure include thin layer, paper, and column chromatography. Attempts at identification and quantitation may follow all three techniques of separation. However, since extracts from air samples generally are very complex mixtures, separations of the compounds of interest may be incomplete by these techniques alone. Incompletely separated PNA's can lead to errors in analysis. Yet, results of spectroscopic analyses of fractions isolated by these techniques may serve as indicators of certain PNA's or classes of PNA's present. Column chromatography, in particular, may serve as a cleanup step prior to a technique with greater resolution such as gas chromatography or high pressure liquid chromatography.

The extracts of air samples, as from Soxhlet extraction, should be in concentrated form prior to chromatography. Such extracts may be concentrated by carefully evaporating the solvent below 40°C under reduced pressure until the volume is 2–5 mliter. Then, additional evaporation may be accomplished under a stream of dry nitrogen at room temperature or below. The residue and also condensates from vacuum sublimation may be dissolved in a small quantity of solvent such as benzene, cyclohexane, or dichloromethane to facilitate quantitative transfer of the PNA's to the chromatographic medium.

1.   Thin Layer Chromatography

Thin layer chromatography of PNA's may be performed on 20 × 20 cm plates coated with aluminum oxide or acetylated cellulose in layers 0.1–1 mm in thickness. Cellulose with degrees of acetylation ranging from 20 to 40% has been used. Pentane/diethyl ether or hexane/diethyl ether (19/1, v/v) may be used as the mobile phase on the aluminum oxide plates (Pierce and Katz, 1975a,b; Stanley *et al.*, 1967). More polar solvent systems for acetylated cellulose include ethanol/dichloromethane/water (20/10/1, $v/v/v$), *n*-propanol/acetone/water (2/1/1, $v/v/v$), and methanol/diethyl ether/water (4/4/1, $v/v/v$) (Sollenberg, 1976; Pierce and Katz, 1975b; Schultz *et al.*, 1973). Pierce and Katz (1975b) accomplished the TLC separation of 12 PAH's in a two-step process. Thin layer chromatography on aluminum oxide separated the PAH's into three groups: $C_{20}H_{12}$ pentacyclic arenes, $C_{22}H_{12}$ hexacyclic arenes, and $C_{24}H_{14}$ hexacyclic arenes. Thin layer chromatography of the separated groups on acetylated cellulose resulted in the separation of the individual PAH's.

Concentrated extracts from the air samples may be spotted onto the

TLC plates in quantities of 1–2 μliter with the aid of a capillary pipet. Reference standards in quantities of ~2 μg may be spotted on the same plate.

After development of the plates, PNA's may be identified by comparison of R_F values and spectroscopic techniques. The spots may be located by examination of the plates under long wavelength ultraviolet light. Benzo[a]pyrene in quantities of ~1–250 ng may be determined directly on the TLC plate by means of a scanning spectrofluorometer (Sollenberg, 1976; Schultz *et al.*, 1973). Otherwise, the flourescent areas on the plate may be outlined with a stylus, removed from the plate, and treated with diethyl ether or perhaps dichloromethane. [Cellulose, however, should not be treated with dichloromethane because it dissolves in this solvent (Pierce and Katz, 1975b).] After evaporation of the solvent, the residue may be dissolved in a deoxygenated hydrocarbon solvent or in concentrated sulfuric acid for spectrofluorimetric analysis.

Another technique of locating and identifying PNA's on the TLC plate involves spraying of the plate with 1–2% solutions of certain electron-acceptor compounds in acetone or dichloromethane. Many PNA's in quantities of ~10 μg form colored charge-transfer complexes that are visible. The particular colors, together with the R_F values, are useful in identification. Benzo[a]pyrene, for example, forms a gray complex with 2,3-dichloro-5,6-dicyano-1,4-benzoquinone (DDQ), a gray–brown complex with 2,4,5,7-tetranitro-9-fluorenone (TetNF) and gray–green complexes with 2,4,7-trinitro-9-fluorenone (TNF) and 9-dicyanomethylene-2,4,7-trinitrofluorene (CNTNF). Benzo[ghi]perylene forms a green complex with DDQ, a brown complex with TNF, and bluish purple complexes with TetNF and CNTNF (Heacock and Hutzinger, 1975).

2. *Paper Chromatography*

Paper chromatography of airborne PNA's was performed by Malý (1971) and Hlucháň *et al.* (1974) with paper pretreated with a 10% solution of liquid paraffin in petroleum ether. The mobile phase consisted of methanol saturated with liquid paraffin. The spots were located under ultraviolet light and outlined with a pencil. The various colors of the fluorescing PNA's aided in the identifications. The technique of quantitation involved calibration curves of horizontal diameters versus quantities of compounds. The detection limits of benzo[a]-pyrene, 3,4-benzofluoranthene, and 1,2:3,4-dibenzopyrene were ~2–3 ng.

3. Column Chromatography

Low pressure column chromatography of PNA's commonly is performed on alumina in glass tubes with diameters of ~0.5–1 cm and lengths of 25–40 cm or greater. The alumina may be deactivated with 1–3% water. The PNA concentrate is applied to the top of the column, and elution is begun with a nonpolar solvent such as pentane or cyclohexane. As the chromatography proceeds, the polarity of the solvent may be increased with small percentages of diethyl ether or dichloromethane. Generally, the lower molecular weight PNA's such as pyrene and fluoranthene are eluted first (Moore *et al.*, 1967; Schulte *et al.*, 1974). Elution volumes tend to increase with molecular weight and polarity of the PNA's.

Chromatography on silica gel columns may be employed for the purpose of separating aliphatic compounds from the PAH's in the crude extract (Moore *et al.*, 1967; Lao *et al.*, 1973).† This may be performed by initial elution with isooctane. Benzene elutes the PAH's. The benzene fraction may be concentrated and subjected to column chromatography over alumina or to gas chromatography.

Sephadex LH-20, a polar material used in gel permeation chromatography, has been applied to the group isolation of PNA's as a purification step prior to analysis by high pressure liquid chromatography (Novotny *et al.*, 1974). By elution with isopropanol, straight-chain compounds are among the first compounds to be removed from the column. Contrary to a general principle of gel permeation chromatography, PAH's are eluted from Sephadex LH-20 in order of increasing molecular size (Giger and Blumer, 1974).

An interesting purification step of PAH's employed by Giger and Blumer (1974) involved the complexation with 2,4,7-trinitro-9-fluorenone (TNF). A dichloromethane solution of the PAH's and TNF was taken to dryness, and uncomplexed compounds were removed with pentane. Chromatography of a dichloromethane solution of the residue on a column of activated silica gel gave uncomplexed PAH's.

Fractions collected after column chromatography may be analyzed by ultraviolet absorption spectrophotometry or by fluorescence spectrometry. The ultraviolet absorption technique involves scanning the ultraviolet spectra and comparing the sample spectra with spectra of reference compounds. Matching spectra help to identify the PNA's of interest. Peak heights may be determined by a base-line technique (Commins, 1958; Cooper, 1954). Various PAH's in solution in a sample

† An example of the separation on silica gel of saturates from polynuclear aromatic hydrocarbons (PAH's) is given in Volume I, Chapter 16, Fig. 3.

cell with a path length of 1 cm may be quantitated in concentrations of a few micrograms per milliliter (Schulte *et al.*, 1974). However, quantitation of PNA's by UV absorption spectrophotometry is subject to error when several unresolved components in the same solution contribute to the absorbance readings (Giger and Blumer, 1974). Fluorescence spectrometry is a more sensitive analytical technique. Benzo[a]pyrene, for example, may be detected at a concentration of 2 ng/mliter. The relationship between the intensity of fluorescence emission and concentration of various PAH's is linear at sufficiently low concentrations when the sample solution is free of interferences (Pierce and Katz, 1975a,b). The fluorescence emission of benzo[a]pyrene is considerably more specific when the solvent is concentrated sulfuric acid than when the solvent is an organic one. However, the quenching effect by other PNA's in sulfuric acid solution is possible (Jäger, 1977).

C. High Pressure Liquid Chromatography

Polynuclear aromatic compounds may be separated by high pressure liquid chromatography (HPLC) in both the normal-phase and reverse-phase modes. Normal-phase HPLC is performed on a polar stationary phase with a nonpolar mobile phase such as pentane or hexane, while reverse-phase HPLC is performed on a nonpolar stationary phase with a polar solvent system. The polar stationary phases which have been used for PNA analysis include alumina, silica, and porous particles with chemically bonded phases such as an aminosilicone, an aminopropylsilane, or oxypropionitrile (Golden and Sawicki, 1976; Novotny *et al.*, 1974; Wise *et al.*, 1977). The nonpolar stationary phases include octadecylsilane and octadecyltrimethoxysilane chemically bonded to porous particles (Golden and Sawicki, 1976; Dong *et al.*, 1976; Fox and Staley, 1976; Wise *et al.*, 1977; Soedigdo *et al.*, 1975). The polar solvent systems may be methanol/water or acetonitrile/water mixtures.

Principles of separation by normal-phase and reverse-phase modes differ, and each mode is useful in PNA analysis (Sleight, 1973; Wise *et al.*, 1977). The retention of PNA's in a normal-phase system depends substantially on interactions between polar functional groups of the stationary phase and aromatic π electrons of the PNA's. Thus, normal-phase techniques result in class separations of PNA's according to the numbers of aromatic rings. The degree of alkyl substitution may have relatively little effect on the retention of the PNA's during normal-phase liquid chromatography. According to Wise *et al.* (1977), the degree of alkyl substitution has less effect on the retention of PNA's on μBondapak NH$_2$ (a chemically bonded aminopropylsilane) than on silica or alumina during normal-phase liquid chromatography. Separations by reverse-phase techniques depend on differences in solubilities of the PNA's in the mobile phase. An increase in alkyl substitution decreases the solubility of a PNA in polar solvents and increases the retention in the nonpolar, stationary phase. Normal-phase HPLC may serve as a purification or fractionation step prior to analysis by reverse-phase HPLC or by gas chromatography. A favorable characteristic of normal-phase HPLC is that the nonpolar mobile phase may be volatile enough for rapid concentration of the collected fractions.

Polynuclear aromatic compounds emerging from the HPLC column may be detected in a flow-through sample cell with an ultraviolet detector set at a particular wavelength. A variable wavelength detector may permit a wavelength setting for an optimum combination of selectivity and sensitivity. A wavelength may be found for the determination of a particular PNA at which interference by another compound of similar retention time is reduced. Generally, nanogram quantities of completely resolved PNA's can be measured with an ultraviolet detector and a flow-through cell. At a wavelength of 254 nm and a sensitivity setting of 0.16 absorbance unit for full-scale deflection, 95 ng of benzo[a]pyrene and 40 ng of coronene in a 5 μliter sample can be detected following HPLC (Dong *et al.*, 1976). At 250 nm and a sensitivity setting of 0.0625 absorbance unit for full-scale deflection, 19 ng quantities of anthracene, fluoranthene, and benz[a]anthracene in a 190 μliter sample can be detected following HPLC (Strubert, 1973).

A more sensitive and selective detector of PNA's in a flow-through cell is a fluorescence detector. By means of a fluorescence detector equipped with broadband filters, 90 pg of benzo[a]pyrene may be detected with a signal-to-noise ratio of 2:1 when the mobile phase, methanol/water (7/3, *v/v*), is saturated with air. This detection limit may be reduced to 25 pg when the mobile phase is free of dissolved oxygen. With this detector system, the fluorescence response is linear in the 1–100 ng range for a number of PAH's including benz[a]anthracene and benzo[a]pyrene (Fox and Staley, 1976). Das and Thomas (1979) have reported on the use of a quite sensitive fluorescence detector which is capable of detecting subpicogram quantities of fluoranthene, benz[a]anthracene, benzo[k]fluoranthene, and perylene. By optimizing the wavelengths of excitation and emission, a higher degree of selectivity may be achieved. For example, when benzo[e]pyrene and perylene are not resolved on an HPLC column, a fluorescence detector may be used for quantitation of each of the PAH's (Das and Thomas, 1979).

In addition, the detection of a PNA at fixed wavelengths coupled with retention time reduces the ambiguity in identification (Wheals *et al.*, 1975).

The sensitivities and selectivities of the ultraviolet and fluorescence detectors for HPLC may depend upon the complexity of the sample. This concept is applicable to other types of detectors and analyses, also.

D. Gas Chromatography

Gas chromatography is quite useful in the analysis of volatile and thermally stable PNA's because of its power of high resolution and its applicability to small samples. The PNA's analyzed by this technique have ranged up to approximately seven rings in size. Gas chromatographic analysis may follow sample cleanup or fractionation procedures such as solvent partition, column chromatography, thin layer chromatography, or high pressure liquid chromatography (Novotny *et al.*, 1974; Lee *et al.*, 1976a; Brocco *et al.*, 1970). In addition, gas chromatography may be performed directly on the concentrated extract of the air sample prior to mass spectrometric analysis or to UV spectral analysis of collected fractions (Lao *et al.*, 1975; Broddin *et al.*, 1977; Searl *et al.*, 1970).

Both packed columns and capillary columns have been employed in the gas chromatography of PNA's. Although stainless steel columns have been used in PAH analysis, glass columns may give more reproducible results; active surfaces strongly adsorb PAH's and metal surfaces may provide sites for adsorption (Doran and McTaggart, 1974). The packed columns may be 1.8–4.5 m in length with a diameter of 0.32 or 0.64 cm and should be packed with a stationary phase of high thermal stability such as 3–6% Dexsil 300 (a polycarboranesiloxane) or 2% SE-30 (a methylsilicone rubber gum) (Lee and Hites, 1976; Lao *et al.*, 1973; Burchfield *et al.*, 1974; Severson *et al.*, 1976; Searl *et al.*, 1970). Glass capillary columns which are 11–20 m in length with an inner diameter of 0.3 mm have been coated with SE-52 (a methylphenylsilicone) (Lee *et al.*, 1976a, 1977; Giger and Schaffner, 1978). Doran and McTaggart (1974) have reported on the use of a 50 m × 0.25 mm glass capillary column coated with OV-101 (a silicone gum). For the analysis of a complex mixture of PNA's having a wide range of molecular weights, a temperature program in the approximate range of 70°–250°C with a heating rate of 1°–4°C/min may be employed.

Generally, packed columns containing conventional stationary phases such as Dexsil 300 or OV-7 (a methylphenylsilicone) are insufficient for achieving base line separations of certain groups of PAH's of the same molecular weight. These groups include anthracene and phenanthrene; benz[a]anthracene, chrysene, and triphenylene; and benzo[a]pyrene and benzo[e]pyrene (Lane *et al.,* 1973; Janini *et al.,* 1975). However, a 1.8-m × 3.2-mm-i.d. glass column packed with 2.5% N,N'-*bis*[p-butoxybenzylidene]-α,α'-*bi-p*-toluidine (BBBT) can resolve these PAH's with base-line separation. N,N'-*bis*[p-butoxybenzyli-dene]-α,α'-*bi-p*-toluidine is a nematic liquid crystal in the temperature range 188°–303°C (Janini *et al.,* 1976). With the possible exception of triphenylene, glass capillary columns can achieve base-line or near-base-line separation of the above PAH's (Lee *et al.,* 1977; Giger and Schaffner, 1978). [Triphenylene is another PAH which may be found in coke oven emissions (Lao *et al.,* 1975).]

Detectors for gas chromatography which are sensitive to PNA's include the flame ionization detector, the electron-capture detector, and the spectrophotofluorometric detector.

The flame ionization detector (FID) can be used to measure PNA's in the nanogram range. For example, a 1 μliter sample containing pyrene, benz[a]anthracene, and chrysene in approximate quantities of 100 ng each can give rise to sharp peaks following emergence from an appropriate chromatographic column (Janini *et al.,* 1975). The relative responses of the FID to twelve selected PAH's ranging in size from three to five rings are all within the same order of magnitude (Cantuti *et al.,* 1965). Since the FID responds well to many carbon-rich compounds, it is not specific for PNA's.

The response of the electron-capture detector (ECD) is more selective and depends on the electron affinities of the compounds. The relative responses of the ECD to twelve selected PAH's from three to five rings in size span more than two orders of magnitude. Within this group of PAH's, the ECD is most sensitive to benzo[a]pyrene, which can be measured in 1 ng quantities (Cantuti *et al.,* 1965; Davis, 1968). The response of the ECD to many PAH's is temperature dependent, at least when the ECD is operated in the pulsed mode and the carrier gas is 10% methane in argon. For example, a tritium foil ECD operated under the above conditions is more sensitive to pyrene, benzanthracene, and chrysene at a detector temperature of 170°C than at a detector temperature of 240°C (Wentworth and Chen, 1967). Incidentally, an ECD equipped with a nickel-63 source is more desirable than one with a tritium source because of the greater thermal stability (Fenimore *et al.,* 1971; Shoemake *et al.,* 1963). Other factors which affect the sensitivity of the ECD are the magnitude of the applied potential and the cleanliness and geometrical characteristics of the detector.

The spectrophotofluorometric detector (SPFD) is both a sensitive and a selective detector for PNA's. Burchfield *et al.* (1974) have reported detection limits of 3–100 ng for selected PAH's at appropriate wavelength settings. Benzo[a]pyrene in quantities of ~50 ng can be detected at excitation and emission wavelengths of 360 and 404 nm, respectively. At these wavelength settings, 1 μg quantities of benzo[e]pyrene would not interfere (Mulik *et al.*, 1975).

The mass spectrometer is a sensitive detector for the gas chromatography of PNA's and is quite powerful in the identification of these compounds by the detection of characteristic ion fragments (Lao *et al.*, 1973, 1975; Lee *et al.*, 1976b).† The lower quantitation limit for fluoranthene, pyrene, and chrysene may be ~30 pg when the mass spectrometer is operated in the multiple ion detection mode with an electron impact ion source (Van Vaeck and Van Cauwenberghe, 1977). There has been difficulty in identifying particular isomers of PNA's by electron impact mass spectrometry because such compounds can give rise to almost identical mass spectra. If reference standards are available and if the isomers can be resolved on the gas chromatographic column, the retention times can facilitate the identification (Lee *et al.*, 1976a). By means of the technique of mixed charge exchange–chemical ionization mass spectrometry, differences in mass spectra can be detected for PAH's of the same molecular weight. In the cases examined, the ratio of the abundance of the protonated molecular ion, $(M+1)^+$, to that of the molecular ion, $(M)^+$, differs from isomer to isomer. In addition, this ratio tends to increase with an increase in the first ionization potential. A methane–argon mixture may be used as the reagent gas. Benz[a]anthracene, chrysene, and triphenylene, for example, can be distinguished easily by this technique (Lee and Hites, 1977).

E. Miscellaneous Analytical Techniques

Miscellaneous techniques which may be applicable to the analysis of PNA's in fractions collected after an appropriate form of chromatography include nuclear magnetic resonance spectrometry, infrared spectrophotometry, and luminescence spectrometry involving the Shpol'skii effect. A few details of these techniques are presented below. Other analytical techniques for PNA's include spectrophosphorimetry and anodic differential pulse voltammetry (Sawicki and Pfaff, 1965; Coetzee *et al.*, 1976).

---

† Combined gas chromatography/mass spectrometry is discussed in Volume I, Chapter 17, Section II,C, and Volume II, Chapter 21, Section II,B.

1. Nuclear Magnetic Resonance Spectrometry

Fourier-transform proton NMR spectrometry may be used to help identify three- and four-ring methyl PNA's in quantities below ~100 μg. When proton NMR data of reference compounds are available, the chemical shifts of the methyl protons may aid in determining the positions of the methyl groups in the aromatic molecules (Novotny *et al.*, 1974; Lee *et al.*, 1976a,b).

2. Infrared Spectrophotometry

Fourier-transform IR spectrophotometry may be applied to qualitative and quantitative analyses of PNA's. The PNA's which have been sublimed *in vacuo* from a Knudsen cell may be deposited onto a window of cesium iodide in a nitrogen matrix at low temperature. Many of the strong IR bands occur in the 700–900 cm^{-1} region. Various PNA's in quantities of a few micrograms may be identified by this technique (Mamantov *et al.*, 1977).

3. Luminescence Spectrometry Involving the Shpol'skii Effect

The Shpol'skii effect refers to sharp line emission spectra obtained in many cases upon excitation of PAH's embedded in selected *n*-alkane matrices at 77 K. Examples of the *n*-alkane solvents are hexane, heptane, and octane. Emission spectra bearing peaks with half-widths of less than 0.6 nm have been recorded for chrysene, pyrene, benzo[a]pyrene, and benz[a]anthracene, for example, in a cyclohexane-*n*-octane solvent mixture. Detection limits of selected PAH's in appropriate solvents have ranged from 5 to 100 ng/mliter (Kirkbright and de Lima, 1974). In addition, narrow-band excitation spectra having peaks with half-widths of less than 0.8 nm may be recorded for selected PAH's (Farooq and Kirkbright, 1976).

IV. MEASURED LEVELS OF POLYNUCLEAR AROMATIC COMPOUNDS

More than 45 PNA's in coke oven emissions have been detected (Lao *et al.*, 1975; Schulte *et al.*, 1974; Searl *et al.*, 1970; Broddin *et al.*, 1977). Among these compounds are more than 15 methyl PAH's and more than 24 PAH's without alkyl or other substituents. Polynuclear aromatic hydrocarbons detected other than those listed in Table III include fluorene, anthracene, phenanthrene, benzo[a, b, and c]fluorenes, benzo[ghi]fluoranthene, and coronene. Three heterocyclic species detected are acridine, benz[c]acridine, and a benzoquinoline.

Table III presents air concentrations of various PNA's collected on glass fiber filters near coke ovens. All of the compounds in the table are hydrocarbons except benz[c]acridine. Generally, the benzene soluble material extracted from the particulate matter constituted 20–40% of the entire sample by weight (Schulte *et al.*, 1974). The first 13 PNA's in the table accounted for approximately 5% of the benzene soluble fraction, and benzo[a]pyrene accounted for 1% of the extracted material. Results from a coke oven study involving a multistage Andersen cascade impactor indicated that more than 90% of the PAH's analyzed were associated with particles having aerodynamic diameters of less than 3 μm (Broddin *et al.*, 1977).

A factor that should be considered in the air concentrations of PNA's at coke oven plants is the contribution of PNA's from other emission sources. Numerous PNA's have been identified in air samples taken

TABLE III *Measured Levels of PNA's from Coke Ovens*

Compound	Concentrations (μg/m^3) (Schulte *et al.*, 1974)[a]	(Broddin *et al.*, 1977)[b]
Fluoranthene	<1.32–1800	0.33
Pyrene	<0.44–879	0.18
Benz[a]anthracene	0.88–529	
Chrysene	0.90–177	
Benz[c]acridine	<0.22–34	
Benzo[b]fluoranthene	<0.28–250	
Benzo[j]fluoranthene	<0.26–72	
Benzo[k]fluoranthene	<0.61–98	0.50
Benzo[a]pyrene	<1.32–688	
Benzo[e]pyrene	<0.54–95	
Dibenz[a,h]anthracene	<0.50–29	
Benzo[ghi]perylene	<0.98–62	0.14
Anthanthrene	<0.12–28	
Perylene	<0.38–67	0.04
Indeno[1,2,3-cd]pyrene		0.06

[a] Samples were taken with 120 high volume glass fiber filters. Analytical methods consisted of (1) TLC and fluorimetric determination of benzo[a]pyrene, (2) column chromatography and UV determination of 8 PNA's and fluorimetric determination of benzo[a]pyrene, and (3) gas chromatography and UV determination of 13 PNA's in collected fractions.

[b] A sample was collected with an eight-stage Andersen cascade impactor equipped with glass fiber filters. Analyses were performed by gas chromatography–mass spectrometry and results are expressed in terms of sums of all stages.

at locations distant from coke ovens. These PNA's may originate from various other sources of carbonaceous material such as automobiles, coal-fired power plants, domestic chimneys, and forest fires. The concentrations of many of the PAH's listed in Table III which have been collected with particulate matter in selected atmospheric samples generally have been in the approximate range of 0.00001–0.04 $\mu g/m^3$ (Gordon, 1976; Pierce and Katz, 1975b; Fox and Staley, 1976; Sawicki *et al.*, 1962; Dong *et al.*, 1976). This concentration range is based on samples taken at sites located in suburban-rural, suburban, and urban areas of North America. Generally, atmospheric PAH concentrations are lower during the summer season than during the winter season.

REFERENCES

Antonello, C., and Carlassare, F. (1964). *Atti Ist. Veneto Sci., Lett. Arti, Cl. Sci. Mat. Nat.* **122**, 9–19. [*Chem. Abstr.* **64**, 9129e (1966).]

Badger, G. M., and Novotny, J. (1961). *J. Chem. Soc.* pp. 3400–3402.

Badger, G. M., Jolad, S. D., and Spotswood, T. M. (1964). *Aust. J. Chem.* **17**, 771–777.

Bradley, R. S., and Cleasby, T. G. (1953). *J. Chem. Soc.* pp. 1690–1692.

Brocco, D., Cantuti, V., and Cartoni, G. P. (1970). *J. Chromatogr.* **49**, 66–69.

Broddin, G., Van Vaeck, L., and Van Cauwenberghe, K. (1977). *Atmos. Environ.* **11**, 1061–1064.

Burchfield, H. P., Green, E. E., Wheeler, R. J., and Billedeau, S. M. (1974). *J. Chromatogr.* **99**, 697–708.

Burton, R. M., Howard, J. N., Penley, R. L., Ramsay, P. A., and Clark, T. A. (1973). *J. Air Pollut. Control Assoc.* **23**, 277–281.

Campbell, N., and Reid, D. H. (1952). *J. Chem. Soc.* pp. 3281–3284.

Cantuti, V., Cartoni, G. P., Liberti, A., and Torri, A. G. (1965). *J. Chromatogr.* **17**, 60–65.

Capell, L. T., and Walker, D. F., Jr. (1963). "The Ring Index," 2nd Ed., Suppl. I. Chem. Abstr. Serv., Am. Chem. Soc., Washington, D.C.

Cautreels, W., and Van Cauwenberghe, K. (1976). *Water, Air, Soil Pollut.* **6**, 103–110.

Chatot, G., Castegnaro, M., Roche, J. L., Fontanges, R., and Obaton, P. (1971). *Anal. Chim. Acta* **53**, 259–265.

Coetzee, J. F., Kazi, G. H., and Spurgeon, J. C. (1976). *Anal. Chem.* **48**, 2170–2174.

Commins, B. T. (1958). *Analyst (London)* **83**, 386–389.

Cook, J. W., and Hewett, C. L. (1933). *J. Chem. Soc.* pp. 398–405.

Cooper, R. L. (1954). *Analyst (London)* **79**, 573–579.

Das, B. S., and Thomas, G. H. (1979). *In* "Trace Organic Analysis: A New Frontier in Analytical Chemistry" (H. S. Hertz and S. N. Chesler, eds.), Nat. Bur. Stand. (U. S.), Spec. Publ. 519, pp. 41–56. U.S. Government Printing Office, Washington, D.C.

Davis, H. J. (1968). *Anal. Chem.* **40**, 1583–1585.

Dewar, M. J. S. (1952). *J. Am. Chem. Soc.* **74**, 3357–3363.

Dong, M., Locke, D. C., and Ferrand, E. (1976). *Anal. Chem.* **48**, 368–372.

Doran, T., and McTaggart, N. G. (1974). *J. Chromatogr. Sci.* **12**, 715–721.

Epstein, S. S. (1965). *Arch. Environ. Health* **10**, 233–239.

Epstein, S. S., Bulon, I., Koplan, J., Small, M., and Mantel, N. (1964). *Nature (London)* **204**, 750–754.

Farooq, R., and Kirkbright, G. F. (1976). *Analyst (London)* **101,** 566-573.
Fenimore, D. C., Loy, P. R., and Zlatkis, A. (1971). *Anal. Chem.* **43,** 1972-1975.
Fieser, L. F., and Hershberg, E. B. (1937). *J. Am. Chem. Soc.* **59,** 2502-2509.
Fox, M. A., and Staley, S. W. (1976). *Anal. Chem.* **48,** 992-998.
Giger, W., and Blumer, M. (1974). *Anal. Chem.* **46,** 1663-1671.
Giger, W., and Schaffner, C. (1978). *Anal. Chem.* **50,** 243-249.
Golden, C., and Sawicki, E. (1976). *Anal. Lett.* **9,** 957-973.
Gordon, R. J. (1976). *Environ. Sci. Technol.* **10,** 370-373.
Grimmer, G., and Hildebrandt, A. (1965). *J. Chromatogr.* **20,** 89-99.
Heacock, R. A., and Hutzinger, O. (1975). *Mikrochim. Acta* **2,** 101-108.
Hlucháň, E., Jeník, M., and Malý, E. (1974). *J. Chromatogr.* **91,** 531-538.
Huberman, E., and Sachs, L. (1976). *Proc. Natl. Acad. Sci. U.S.A.* **73,** 188-192.
Jackson, J. O., and Cupps, J. A. (1978). *In* "Carcinogenesis" (P. W. Jones and R. I. Freudenthal, eds.), Vol. 3, pp. 183-191. Raven, New York.
Jackson, J. O., Warner, P. O., and Mooney, T. F., Jr. (1974). *Am. Ind. Hyg. Assoc. J.* **35,** 276-281.
Jäger, J. (1977). *Fresenius' Z. Anal. Chem.* **284,** 283-285.
Janini, G. M., Johnston, K., and Zielinski, W. L., Jr. (1975). *Anal. Chem.* **47,** 670-674.
Janini, G. M., Muschik, G. M., and Zielinski, W. L., Jr. (1976). *Anal. Chem.* **48,** 809-813.
Jones, P. W., Giammar, R. D., Strup, P. E., and Stanford, T. B. (1976). *Environ. Sci. Technol.* **10,** 806-810.
Jones, R. C., and Neuworth, M. B. (1944). *J. Am. Chem. Soc.* **66,** 1497-1499.
Keeling, W. O., and Jung, F. W. (1949). *In* "Encyclopedia of Chemical Technology (R. E. Kirk and D. F. Othmer, eds.), Vol. 3, pp. 156-178. Intersci. Encycl., New York.
Kirkbright, G. F., and de Lima, C. G. (1974). *Analyst (London)* **99,** 338-354.
Krafft, F., and Weilandt, H. (1896). *Chem. Ber.* **29,** 2240-2245.
Krstulovic, A. M., Rosie, D. M., and Brown, P. R. (1976). *Anal. Chem.* **48,** 1383-1386.
Kruber, O. (1941). *Chem. Ber.* **74,** 1688-1692.
Lane, D. A., Moe, H. K., and Katz, M. (1973). *Anal. Chem.* **45,** 1776-1778.
Lao, R. C., Thomas, R. S., Oja, H., and Dubois, L. (1973). *Anal. Chem.* **45,** 908-915.
Lao, R. C., Thomas, R. S., and Monkman, J. L. (1975). *J. Chromatogr.* **112,** 681-700.
Lee, M. L., and Hites, R. A. (1976). *Anal. Chem.* **48,** 1890-1893.
Lee, M. L., and Hites, R. A. (1977). *J. Am. Chem. Soc.* **99,** 2008-2009.
Lee, M. L., Novotny, M., and Bartle, K. D. (1976a). *Anal. Chem.* **48,** 1566-1572.
Lee, M. L., Novotny, M., and Bartle, K. D. (1976b). *Anal. Chem.* **48,** 405-416.
Lee, M. L., Prado, G. P., Howard, J. B., and Hites, R. A. (1977). *Biomed. Mass Spectrom.* **4,** 182-186.
Levy, M., and Szwarc, M. (1955). *J. Am. Chem. Soc.* **77,** 1949-1955.
Lewis, I. C., and Edstrom, T. (1963). *J. Org. Chem.* **28,** 2050-2057.
Mailath, F. P., Medve, F., and Morik, J. (1974). *Egeszsegtudomany* **18,** 333-341. [*Chem. Abstr.* **82,** 148426g (1975).]
Malý, E. (1971). *Mikrochim. Acta* pp. 800-806.
Mamantov, G., Wehry, E. L., Kemmerer, R. R., and Hinton, E. R. (1977). *Anal. Chem.* **49,** 86-90.
Mašek, V. (1971). *J. Occup. Med.* **13,** 193-198.
Meyer, H., and Hofmann, A. (1916). *Monatsh. Chem.* **37,** 681-722.
Moore, G. E., Thomas, R. S., and Monkman, J. L. (1967). *J. Chromatogr.* **26,** 456-464.
Morgan, G. T., and Mitchell, J. G. (1934). *J. Chem. Soc.* p. 536.
Moriconi, E. J., Rakoczy, B., and O'Connor, W. F. (1961). *J. Am. Chem. Soc.* **83,** 4618-4623.
Moureu, H., Chovin, P., and Rivoal, G. (1946). *C. R. Acad. Sci.* **223,** 951-952.

Mulik, Cooke, M., Guyer, M. F., Semeniuk, G. M., and Sawicki, E. (1975). *Anal. Lett.* **8,** 511–524.

Murray, J. J., Pottie, R. F., and Pupp, C. (1974). *Can. J. Chem.* **52,** 557–563.

Newman, M. S. (1940). *J. Am. Chem. Soc.* **62,** 1683–1687.

Novotny, M., Lee, M. L., and Bartle, K. D. (1974). *J. Chromatogr. Sci.* **12,** 606–612.

Orchin, M., and Reggel, L. (1947). *J. Am. Chem. Soc.* **69,** 505–509.

Patterson, A. M., and Capell, L. T. (1960). "The Ring Index" (D. F. Walker, asst. ed.), 2nd Ed., Chem. Abstr. Serv., Am. Chem. Soc., Washington, D.C.

Pierce, R. C., and Katz, M. (1975a). *Environ. Sci. Technol.* **9,** 347–353.

Pierce, R. C., and Katz, M. (1975b). *Anal. Chem.* **47,** 1743–1748.

Pupp, C., Lao, R. C., Murray, J. J., and Pottie, R. F. (1974). *Atmos. Environ.* **8,** 915–925.

Richards, R. T., Donovan, D. T., and Hall, J. R. (1967). *Am. Ind. Hyg. Assoc. J.* **28,** 590–594.

Sawicki, E., and Pfaff, J. D. (1965). *Anal. Chim. Acta* **32,** 521–534.

Sawicki, E., Elbert, W. C., Hauser, T. R., Fox, F. T., and Stanley, T. W. (1960). *Am. Ind. Hyg. Assoc. J.* **21,** 443–451.

Sawicki, E., Hauser, T. R., Elbert, W. C., Fox, F. T., and Meeker, J. E. (1962). *Am. Ind. Hyg. Assoc. J.* **23,** 137–144.

Sawicki, E., Belsky, T., Friedel, R. A., Hyde, D. L., Monkman, J. L., Rasmussen, R. A., Ripperton, L. A., and White, L. D. (1975). *Health Lab. Sci.* **12,** 407–414.

Schiller, J. E. (1977). *Anal. Chem.* **49,** 1260–1262.

Schulte, K. A., Larsen, D. J., Hornung, R. W., and Crable, J. V. (1974). "Report on Analytical Methods Used in a Coke Oven Effluent Study," HEW Publ. No. (NIOSH) 74–105. Dep. Health, Educ. Welfare, Nat. Inst. Occup. Saf. Health, Cincinnati, Ohio.

Schulte, K. A., Larsen, D. J., Hornung, R. W., and Crable, J. V. (1975). *Am. Ind. Hyg. Assoc. J.* **36,** 131–139.

Schultz, M. J., Orheim, R. M., and Bovee, H. H. (1973). *Am. Ind. Hyg. Assoc. J.* **34,** 404–408.

Searl, T. D., Cassidy, F. J., King, W. H., and Brown, R. A. (1970). *Anal. Chem.* **42,** 954–958.

Severson, R. F., Snook, M. E., Arrendale, R. F., and Chortyk, O. T. (1976). *Anal. Chem.* **48,** 1866–1872.

Shoemake, G. R., Lovelock, J. E., and Zlatkis, A. (1963). *J. Chromatogr.* **12,** 314–320.

Sleight, R. B. (1973). *J. Chromatogr.* **83,** 31–38.

Soedigdo, S., Angus, W. W., and Flesher, J. W. (1975). *Anal. Biochem.* **67,** 664–668.

Sollenberg, J. (1976). *Scand. J. Work, Environ. Health* **3,** 185–189.

Stanley, T. W., Meeker, J. E., and Morgan, M. J. (1967). *Environ. Sci. Technol.* **1,** 927–931.

Strubert, W. (1973). *Chromatographia* **6,** 205–206.

Taylor, D. G. (1977). "NIOSH Manual of Analytical Methods," 2nd Ed., Vol. 1, Methods P & CAM 183, 184, and 186, DHEW (NIOSH) Publ. No. 77-157-A. Dep. Health, Educ. Welfare, Nat. Inst. Occup. Saf. Health, Cincinnati, Ohio.

Thomas, J. F., Mukai, M., and Tebbens, B. D. (1968). *Environ. Sci. Technol.* **2,** 33–39.

Timmermans, J., and Burriel, F. (1931). *Chim. Ind. (Paris), Spec. No.* pp. 196–197. [*Chem. Abstr.* **25,** 3324 (1931).]

Van Vaeck, L., and Van Cauwenberghe, K. (1977). *Anal. Lett.* **10,** 467–482.

Wentworth, W. E., and Chen, E. (1967). *J. Gas Chromatogr.* **5,** 170–179.

Wheals, B. B., Vaughan, C. G., and Whitehouse, M. J. (1975). *J. Chromatogr.* **106,** 109–118.

Wise, S. A., Chesler, S. N., Hertz, H. S., Hilpert, L. R., and May, W. E. (1977). *Anal. Chem.* **49,** 2306–2310.

Chapter 44

The Production, Management, and Chemistry of Coal Gasification Wastewaters

Richard G. Luthy   *Richard W. Walters*
DEPARTMENT OF CIVIL ENGINEERING
CENTER FOR ENERGY AND ENVIRONMENTAL STUDIES
CARNEGIE-MELLON UNIVERSITY
PITTSBURGH, PENNSYLVANIA

I. INTRODUCTION

The objective of this chapter is to discuss the production, management, and chemistry of coal conversion process wastewater. The importance of water chemistry is emphasized in relation to wastewater characterization and wastewater reuse in coal gasification processes. The chapter begins with a description of coal gasification process water streams and then discusses means by which coal gasification plant

189

water balances are achieved. Coal gasification wastewater sampling and analytical strategies are reviewed, and representative characterization data are discussed in order to illustrate the effect of water chemistry on effluent composition and handling.

Water management and treatment play critical roles in the design of coal gasification facilities. Much of our nation's easily exploitable coal reserves lie in semiarid regions where municipal, agricultural, and industrial water demands compete for a limited water supply. Even in the Appalachian region, local water shortage may prohibit installation of a large high-BTU gasification facility. For these reasons it is necessary that gasification plants be designed to minimize water consumption. A second consideration is that these plants must be environmentally acceptable. This means that maximum attention must be given to treatment of wastewater for reuse in order to eliminate wastewater discharges effectively. Naturally, this goal serves to help minimize water consumption.

The quantity of water consumed in a given gasification plant depends mainly on product quality and on the fraction of unrecovered heat disposed of by wet cooling. Water leaves the plant primarily as vapor from cooling towers, as hydrogen in the methane product, and as occluded water in solid residues. Foul process waters which are treated for reuse, rather than being discharged, do not contribute to this consumption. The extent to which these foul process waters must be treated is dependent on the quality of the effluent and the quality constraints governing reuse options. Consequently, the study of this process water quality is fundamentally related to the evaluation of process water consumption.

Figure 1 is a simplified illustration of the quantity–quality relationships of coal gasification plant influent and effluent streams. In this

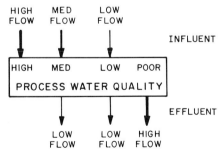

Fig. 1 Representation of influent and effluent water quality and water quantity relationships in high-BTU coal gasification processes.

case influent streams include a high flow of high quality water, e.g., boiler feed makeup; a medium flow of medium quality water, e.g., cooling tower makeup; and a low flow of low quality water, e.g., process water. Effluents include a high volume of poor quality water and two low volume streams of medium and low quality water. Several alternatives for achieving water reuse may exist, depending on the exact nature of the water quality of each stream involved. Reuse can be achieved by upgrading the poor quality effluent to low and medium quality, and the medium and low quality effluents to high quality. Alternatively, it may also be possible to use the low and medium quality effluents directly as low and medium quality feed, respectively, and upgrade the poor quality effluent to meet the high quality need.

In order to provide a framework for discussion of coal gasification water quantity–quality relationships and water chemistry considerations, five coal gasification processes will be considered: (1) Hygas, (2) CO_2-Acceptor, (3) Bi-Gas, (4) Lurgi, and (5) Synthane.

These processes have been chosen on the basis of their involvement in various environmental assessment programs. Each process has completed or is currently involved in pilot scale evaluations which have resulted in the availability of differing amounts of environmental data. Other gasification systems involved in pilot scale evaluations, e.g., the Texaco gasifier, have at best only limited environmental and process data available in the literature on which an evaluation can be based. Although only several gasification processes are being considered, the evaluations, methodologies, and comments made here should be broadly applicable to other coal gasification systems as well as other coal conversion systems such as solvent refined coal and coal liquefaction.

Extensive process descriptions for each of the coal gasification processes listed above can be found elsewhere (Dravo, 1976; Probstein and Gold, 1978). Briefly, each can be characterized by certain distinguishing gasification operating parameters, as listed in Table I. Specific comments regarding configurations in each system include:

(1) Hygas process: The Hygas gasifier operates at high pressure and low temperature (1170 psig, 1850°F) as a fluidized bed in four distinct stages, wherein coal and/or char falls countercurrent to upward gas flows. In the uppermost section (stage 1), hot product gas contacts the slurry coal feed and thereby vaporizes the oil or water used for slurry production. Dry coal falls to stage 2, where initial hydrogasification results from cocurrent contact with hot gases (1800°F) rising from stage 3. Conversion of carbon to methane in this stage is about 20%. Char from this stage falls to stage 3, where it contacts hydrogen rising from

TABLE I   *Operating Parameters for Five Coal Gasification Systems*

		Gasifier			
	Hygas	CO$_2$-Acceptor	Bi-Gas	Lurgi	Synthane
Gasifier type	Fluid bed	Fluid bed	Entrained	Fixed	Fluid bed
Pressure (psig)	1170	150	1000	400	1000
Initial coal devolatilization temperature (°F)	800–1200	1500	1700–2200	400	400–700
Temperature of gasification zone (°F)	1850	1500	~2800	~2500	1800
Coal feed system	Slurry	Lock hopper	Injection with steam	Lock hopper	Lock hopper
Coal size (in.)	0.006–0.079	0.125–0.25	0.0029	0.19–1.5	0.033
Coal residence time	—	—	Zone II, 8–10 s Zone I, 1 s	60 min	—

stage 4 to produce methane and steam to produce carbon monoxide and hydrogen. Carbon conversion in this zone is about 25%. Finally, stage 4 involves steam–oxygen gasification, producing a hydrogen rich gas which rises into stage 3.

(2) CO$_2$-Acceptor Process: In the CO$_2$-Acceptor process, steam and coal react in a low pressure and low temperature (150 psig, 1500°F) fluidized bed. The unique aspect of the system is the use of calcined dolomite (CaO·MgO) in the gasifier which consumes CO$_2$ produced during gasification to form MgO·CaCO$_3$. This consumption of CO$_2$ favors the formation of a hydrogen rich gas, which provides the appropriate hydrogen to carbon monoxide ratio for methanation and thereby eliminates the need for CO shifting.

(3) Bi-Gas Process: The Bi-Gas gasifier operates at high pressure and temperature (1000 psig, 2800°F) in an entrained flow, two-stage gasifier. Steam and coal are injected into the upper stage and are contacted with hot synthesis gas rising from below. This contact results in a rapid heating up of the coal, producing methane at 50% conversion, and char. This entrained char is separated from product gas in a cyclone and recycled to the lower stage of the gasifier. Here, char reacts with steam and oxygen to produce a synthesis gas which rises to the upper stage. Molten slag/ash produced in the lower stage falls to a quench zone and is removed as a slurry.

(4)   Lurgi Process: The Lurgi gasifier operates at low pressure and high temperature (400 psig, 2500°F) in a fixed bed mode. Steam and oxygen are injected into the bottom zone of the fixed bed where combustion occurs. Heat from combustion drives the gasification reaction, which occurs in a zone just above that of combustion. In the uppermost zone, volatilization reactions occur. Ash is collected in a lockhopper located directly below the combustion zone.

(5)   Synthane Process: The Synthane gasifier is a fluidized bed reactor which operates at high pressure and low temperature (1000 psig, 1800°F). Coal is fed to the top of the gasifier, and as it falls to the lower section it contacts steam and oxygen and gasification occurs. The rising product gas contacts the falling feed coal and causes some devolatization to occur. Unreacted char is removed from the bottom of the gasifier.

Gasifier operating parameters have a pronounced effect on the organic content of foul process condensate formed when raw product gas is cooled. Gasifiers having highest devolatilization temperatures produce a cleaner product gas which in turn produces condensates free of tars and oils and other organic contamination. However, other process variables, such as physical gasifier configuration, gas residence time, coal heat up rate, and the extent of gas–solids mixing (Nakles et al., 1975), also influence effluent quality. As discussed in Section IV, effluent quality for a particular gasification system necessarily reflects the net effect of these factors.

II. PROCESS WATER BALANCES

This section describes process water balances and important water chemistry aspects which influence water reuse strategies. Figure 2 shows a general schematic representation of those water streams of importance in the coal gasification process water balances considered. Streams can be classified as either plant influents or effluents. The cooling tower system has a circulating stream as well as three effluent streams (evaporation, drift, and blowdown) and an influent (makeup) stream. The boiler system has a circulating steam/condensate stream. However, since live steam is often used directly in the process, the flow rate of returned steam condensate is usually less than that of steam to the process. Boiler system effluent streams include pressure leak losses and blowdown, with makeup water being supplied as an influent. Additional streams include raw water, process cooling water, process steam, process condensates, treatment blowdowns, and slurry/

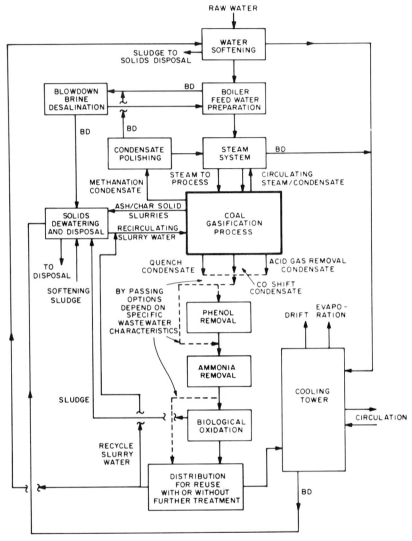

Fig. 2 Schematic representation of major water streams in a coal gasification process water balance.

sludge water. It should be noted that specific process condensate treatment and distribution configurations are strongly dependent on the particular process being considered. Table II shows the quantity of these various water streams for proposed commercial designs of specific coal gasification processes. A discussion of the source and quality of these streams follows.

A. Cooling Water System

1. Quantity Considerations

As shown in Table II, the circulating water in the cooling tower system is typically the single largest stream in the water balance of these coal gasification systems. A large amount of cooling water is sent throughout the process and is returned to the cooling tower at elevated temperature.

A schematic representation of cooling tower water streams is shown in Fig. 3. Heated cooling water from the process enters the top of the cooling tower and is distributed throughout the cross section of the tower as it falls through the packing. The water falls countercurrent to an upward air flow and is cooled both by convective losses to the air and by evaporation.

Assuming a 30°F temperature drop in circulating water and a net evaporation rate of 1400 BTU/lb evaporated (Probstein and Gold, 1978) evaporative loss is 2.1% of circulating water. Besides the loss of water because of evaporation, some water is lost as a result of entrainment

TABLE II *Flow Rates of Major Water Streams for the CO_2-Acceptor and Hygas Processes (Units are water in 1000 lb/hr based on 250 MSCFD SCFD pipeline gas production)*

	CO_2-Acceptor western coal[a]	Hygas western coal[b]	Hygas eastern coal[c]
Cooling water system			
Circulation	8800	10248	142,150
Evaporation	352	410	3243
Drift			283
Blowdown	53	62	368
Makeup	405	472	3894
Boiler feed water system			
Steam consumed	1346	1201	2140
Circulating steam	1890	1896	—
Blowdown	131	169	151
Losses	2	—	111
Process condensates			
Raw gas quench	316	644	755
CO shift		777	
Methanation	313	190	118
Water in disposed solids	39	64	230
Raw makeup water	1161	947	5229

[a] From Braun (1976a).
[b] From Braun (1976b).
[c] From Procon (1978).

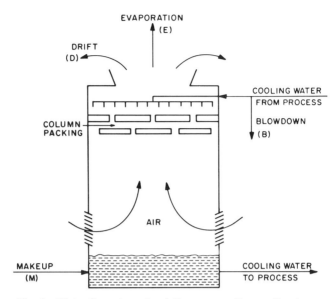

Fig. 3   Water flows in a circulating evaporative cooling tower.

in the air flow as drift loss. This represents a small loss of water relative to the evaporation loss; typical values for drift loss are 0.1–0.2% of the recirculation rate (McCoy, 1974). Since drift is a carryover stream which ultimately is released to the atmosphere and since it contains solids which are released to the atmosphere, drift is both unrecoverable and an emission from the process.

Water which is evaporated from the tower represents a significant water loss from the system. Since this water is essentially free of dissolved solids, evaporation results in the concentrating of solids in the circulating water. A purge, or blowdown, stream is required in order to maintain dissolved solids at a steady concentration. This blowdown also accounts for a net loss of water from the system. To balance losses by evaporation, drift, and blowdown, a makeup stream is necessary.

Mathematically, the water streams are related by

$$M = E + B + D, \tag{1}$$

where, in units of mass of water per time, M is makeup, E is evaporation, B is blowdown, and D is drift.

By including constraints on the solid concentration(s) in the makeup water (Cm) and in the circulating water (Cc), the solids balance becomes

$$M (Cm) = (D + B) Cc. \tag{2}$$

The number of times the makeup water solids have been concentrated to yield the blowdown concentration is

$$X = Cc/Cm = M/(B + D),\qquad(3)$$

where X is the concentration factor.

Combining Eqs. (1) and (3) and rearranging yields:

$$B = [E/(X-1)] - D.\qquad(4)$$

Equation (4) relates the necessary blowdown rate to evaporation, drift, and the desired operating cycles of concentration, X. In practice, the highest value of X for which safe operation can be maintained is sought. From Eq. (3), this value is limited by that component for which the ratio of the circulating concentration constraint to makeup concentration is the smallest.

2. Quality Considerations

Cooling water circulating concentration constraints are related to cooling water chemistry, and reflect limitations imposed by

(1) scaling and fouling because of the buildup of solids,
(2) microbial growth, and
(3) corrosion.

These constraints have led to the operational guidelines presented in Table III. These guidelines reflect the aim to control operational problems by minimizing the concentrations of species causing these problems and by using additives which aid in the prevention of these problems.

As evaporation occurs and solid concentrations build up, there is an increased tendency for insoluble species, e.g., $CaCO_3$ and $CaSO_4$, and to a lesser extent for calcium phosphate, silica, and magnesium silicates, to precipitate and thus form a scale on pipe and exchanger walls of the cooling system. Scale formations ultimately reduce the heat transfer ability in exchangers and therefore should be avoided.

The continuous contact of circulating water with air in the column keeps water well oxygenated, which in turn yields the water supportive to microbial growth. The presence of oxygen and microbial activity may lead to corrosion. Other factors contributing to corrosion include high dissolved solid concentrations owing to the influence on conductivity, and dissolved sour gases, i.e., H_2S, CO_2, NH_3, and HCN. In addition to dissolving oxygen, water also scrubs gases and dust out of the air. Aside from chemical contamination, this can cause plugging and channeling which significantly reduces the efficiency of the tower.

TABLE III   *Guidelines for Circulating Cooling Water Quality*

	Krisher (1978)	Crits and Glover (1975)	Goldstein and Yung (1977)
Langelier saturation index[a]	0.5–1.5		
Ryznar stability index[b]	6.5–7.5		
pH	6.0–8.5	6.5–7.5	
Ca, mg/liter as $CaCO_3$	(20–50)–(300–400)	1200–6000	
Fe, total mg/liter	0.5		
Mn, mg/liter	0.5		
Cu, mg/liter	0.08		
Al, mg/liter	1.0		
S^{2-}, mg/liter	5		
SiO_2, mg/liter	150 (pH < 7.5)	150	
	100 (pH > 7.5)		
SS, mg/liter	100–150	200–400	
TDS, mg/liter	2500		
Conductivity, μmho/cm	4000		
Cl^-, mg/liter			3000
Mg × SiO_2, (mg/liter)^2			3500–60,000
Ca × SO_4, as $(CaCO_3)^2$	500,000		(1.5–2.5) × 10^6
CO_3^{2-}, mg/liter as $CaCO_3$			5
HCO_3^-, mg/liter as $CaCO_3$			300
NH_3			20
PO_4^{3-}			20

[a] $I_{SAT} = pH - pH_S$, where $pH_S = (pK_2 - pK_{SP}) + pCa + pAlk$

[b] $I_{STAB} = 2pH_S - pH$, where $I_{STAB} < 6.0$ scale forming tendency; $I_{STAB} = 7.0$ balanced water; I_{STAB} {7.5–8.5} corrosive.

The treatment of makeup water must be aimed at preventing the problems of fouling, scaling, microbial growth, and corrosion, and should meet the guidelines shown in Table III. At present, there are no well developed guidelines with respect to organic contamination [e.g., chemical oxygen demand (COD) and biological oxygen demand (BOD)] in cooling tower operation. This has resulted in controversy concerning the acceptability of treated wastewater for reuse in cooling towers.

Chemicals can be added to cooling water to prevent the problems described above. From this viewpoint, scaling and fouling is typically controlled through the use of dispersants, antiprecipitants, defloccu-lants, and peptizing agents. Some common additives include acrylam-ide polymers and partially hydrolized polyacrylamide polymers. Halogens, especially chlorine, have been used to control microbial growth. Chromium and phosphate are popular corrosion inhibitors. It should be noted that the use of chlorine and chromium may cause atmospheric emission problems in cooling tower drift and/or evaporation.

B. Boiler Feed Water System

Boiler feed water is the highest quality water required for coal conversion processes. Figure 4 gives a schematic representation of the water streams involved with the boiler feed system. The quantity and quality aspects of this system are discussed below.

1. Quantity Considerations

Water is continuously converted to steam at elevated pressure in either a utility boiler or in waste heat recovery steam generators throughout the process. A portion of the resulting steam goes to the process for heating and for turbine driving and is generally recovered as condensate which is returned to the boiler feed water system. The remaining steam is sent directly into process units (e.g., steam is

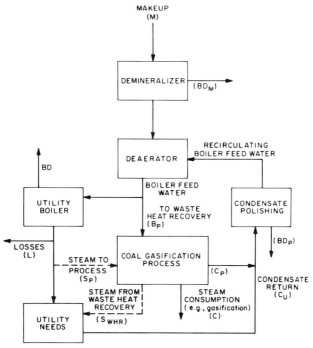

Fig. 4 Major water streams in boiler feed water preparation and distribution systems.

injected directly into the gasifier along with coal and oxygen feeds) and is not recovered as clean condensate. Since steam is generated at elevated pressures, steam losses also occur through pressure leaks.

As water is evaporated in the boilers to produce steam, any dissolved solids are concentrated up in the boiler water inventory. To maintain an acceptable level of solids in the boiler water, a blowdown stream is required. The returned condensate may become contaminated owing to system corrosion and/or leaks of air, gases, or cooling water, and this contamination must also be removed. This condensate is "polished," resulting in a blowdown stream from that treatment system. The amount of steam consumed in the process is

$$C = B_p + S_p - S_{WHR} - C_p, \qquad (5)$$

where, in units of mass of water per time, C is consumption, B_p is boiler feed water to process, S_p is steam to process, S_{WHR} is steam from process waste heat recovery, and C_p is process steam condensate returned to polishing.

As in the case of the cooling tower system, the amount of makeup water required in the boiler feed water system is related to system flow rates by

$$M = C + L + BD_p + BD_m + BD, \qquad (6)$$

where, in units of mass of water per time, M is makeup, C is process consumption, L is pressure losses, BD is boiler blowdown, BD_p is polishing blowdown, and BD_m is demineralizer blowdown.

2. Quality Considerations

Boiler feed water quality is limited owing to problems of (1) scaling, (2) corrosion, and (3) foaming. In order to avoid these problems, the quality guidelines shown in Table IV have been adopted. Data in Table IV show a major difference between cooling water and boiler feed water quality constraints, in that boiler feed water has a strong relationship between water quality and the pressure of steam being generated. The causes of potential scaling and corrosion problems in boilers can be attributed to similar groups of species as in the case of cooling tower water systems. However, in the case of boilers these problems become more severe with higher operating pressures.

Scaling is primarily caused by hardness and suspended metal oxides, especially from iron and copper. Iron and copper may be present as a result of system corrosion. Principal scaling species include SiO_2, Ca^{2+}, Mg^{2+}, HCO_3^-, and SO_4^{2-}, which form $CaSO_4$, $CaCO_3$, $Mg(OH)_2$, Ca

TABLE IV *Guidelines for Water Quality in Modern Industrial Water Tube Boilers*[a]

Operating pressure (psig)	Feedwater				Boiler water	
	Iron (ppm)	Copper (ppm)	Total hardness (ppm CaCO₃)	SiO₂ (ppm)	Total alkalinity (ppm CaCO₃)	Specific conductivity (μmho/cm)
0–300	0.100	0.050	0.300	150	700	7,000
301–450	0.050	0.025	0.300	90	600	6,000
451–600	0.030	0.020	0.200	40	500	5,000
601–750	0.025	0.020	0.200	30	400	4,000
751–900	0.020	0.015	0.100	20	300	3,000
901–1000	0.020	0.015	0.050	8	200	2,000
1001–1500	0.010	0.010	0.000	2	0	150
1501–2000	0.010	0.010	0.000	1	0	100

[a] From Simon (1975).

$(OH)_2$, $Ca(SiO_4)_2$, and magnesium silicates. Corrosion is caused by the presence of dissolved oxygen, acids (e.g., H_2SO_4 and HCl), dissolved sour gases (e.g., CO_2, H_2S, NH_3, and HCN), and chloride. Foaming results from the presence of organic matter, oil, suspended solids, and highly alkaline conditions.

C. Process Water and Liquid Effluent Streams

Unlike the boiler feed water and cooling water systems discussed in the previous two sections, process water needs and effluent streams are highly dependent on the exact nature of the gasification process configuration. Major process specific water streams in the coal gasification process are shown in Fig. 5, which represents a composite of those streams common to one or all of the gasification systems being considered. Process influent water streams generally include

(1)   water for coal slurry feed,
(2)   water for direct contact gas cooling (quenching), and
(3)   water for char solids slurry removal and/or water for ash or slag quench and removal.

Process steam requirements generally include

(1)   steam to gasifier and
(2)   makeup steam to CO shift reactor.

Process effluents can be categorized as follows:

(1)   slag or ash quench water (slurry),
(2)   raw product gas quench condensate,
(3)   CO shift condensate,
(4)   acid gas removal condensate, and
(5)   methanation condensate.

Quantities for these process water and effluent streams are provided in Table II. Since the nature of these streams are highly process dependent, their disposition and source is discussed below for each particular gasification process being considered.

1. Disposition of Process Water Streams

a. CO_2-Acceptor Process   Figure 6 shows a process flow diagram of those process units of concern in the CO_2-Acceptor process.† Low

---

† Organic compounds and inorganic elements in the CO_2-Acceptor process water streams are given in Figs. 10–12 and Table VIII of Chapter 45.

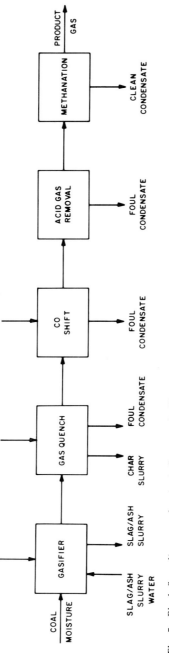

Fig. 5 Block flow diagram of major influent and effluent water streams interacting directly with the high-BTU coal gasification process.

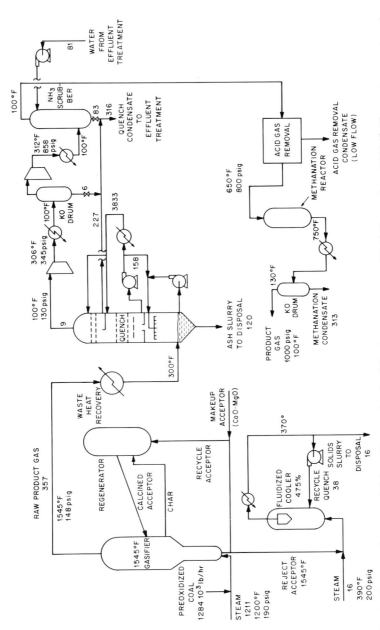

Fig. 6 Conceptual commercial flow diagram for the CO₂-Acceptor process [C. F. Braun, Inc. (1976a)]. Flow rates are given for water in units of 1000 lb/hr, based on 250 MSCFD pipeline gas produced.

pressure steam is consumed in the gasification reaction, and raw product gas and remaining water vapor are sent to the quench tower. The lower section of the quench tower acts effectively as a dry quench, with 38×10^3 lb/hr water being vaporized through contact with the incoming gas. Contact in this section removes mainly gross solids which are comprised of unreacted char and ash.

The raw gas and vapor move into the upper section of the tower. Water circulating in the upper section serves to act as a complete quench where 98% of raw product gas vapor is condensed out of the gas. A large volume of water is recirculated in this upper quench loop, and condensate is removed from the tower to complete a balance. Overhead gas is compressed in two stages to obtain sufficient pressure for pipeline gas distribution and interstage cooling condensate is combined with the high pressure gas scrubbing water. Scrubbed gas proceeds in turn to acid gas removal and to methanation. Water produced as a result of chemical reaction in the methanation process is also recovered.

Water influent streams include low pressure steam (700 psig) to the gasifier and the acceptor cyclone cooler and process water for ammonia scrubbing. Effluents include acceptor slurry, ash quench slurry, quench and acid gas condensates, and methanation condensate. The CO_2-Acceptor process does not require shifting because the correct $CO:H_2$ ratio is obtained in the gasifier. Hence, no shift condensate is produced.

b. Bi-Gas Process   Figure 7 shows a flow diagram of the Bi-Gas process. In contrast to the CO_2-Acceptor system, the Bi-Gas gasifier operates at high pressure (1000 psig). In this scheme, raw gasifier product gas is dry quenched in char cyclones and slightly cooled to 1485°F. The pressurized coal/water slurry feed stream is contacted with this hot dry quenched product gas, which results in cooling of the product gas to 560°F while simultaneously flashing water from the coal slurry feed. Flashed vapor and product gas are separated from coal in the cyclone and proceed to the CO shift reactor. Condensates result from cooling and knocking out of shifted gas effluent, acid gas removal, and methanated effluent streams. High pressure steam is required for the gasification reaction. Circulating slag and char cyclone quench water complete process water needs.

The use of raw product gas to dry the slurried coal feed serves the double duty of cooling the product gas and flashing slurry water. The overall gas cooling process acts as a dry quench—lowering the temperature of the gas while not condensing water vapor. For this reason

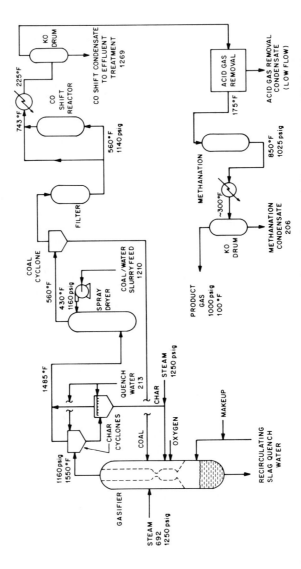

Fig. 7 Conceptual commercial flow diagram for the Bi-Gas process [Air Products and Chemicals, Inc. (1971)]. Flow rates are given for water in units of 1000 lb/hr, based on 250 MSCFD pipeline gas produced.

the process does not produce a quench water effluent stream. The vapor content of the product gas is thus appropriate for the water gas shift reaction without further adjustment.

   c. Hygas Process   Figure 8 shows a flow diagram for a commercial scale Hygas process using western coal. The raw gasifier product gas and solids are separated in a manner similar to that used in the CO_2-Acceptor design. In this case, however, the quench tower works predominately as a dry quench, with less than 2% of the raw product gas vapor content being removed as condensate in the ash/slurry stream. The effluent gas from the quench tower is cooled further, and 52% of the vapor is removed as condensate which is either recycled to the gas quench or sent to water treatment. A portion of the remaining vapor in the product gas is consumed in the CO shift, and the remainder is condensed out in the shift knockout drum. Additional condensates result from downstream acid gas removal and methanation processing.

   Process water requirements in this scheme include circulating water streams for coal slurry feed and ash quench. Quench water needs are conveniently eliminated as a result of the quench tower acting predominately as a dry quench. Water needed for the dry quench is provided by the recycle of condensate from the cooling of the quench overhead gas. As in the case of the Bi-Gas process, high pressure steam is required for the gasification reaction.

   Figure 9 shows variations in the Hygas raw product gas processing which result from utilizing eastern coal. In this design the change in configuration is based on the use of recycle oil rather than water for slurry coal feeding. Gasifier raw product gas and slurry oil vapor enters the quench, which operates in the same manner as for the western coal design. Quenched gas, however, must be cooled to a much lower temperature (100°F rather than 460°F) to achieve complete recovery of the slurry oil. Coincident with this extent of cooling is the near complete removal of water vapor from the product gas stream. This requires a steam input prior to CO shifting to achieve the appropriate $CO:H_2O$ ratio. Although not illustrated in Fig. 9, eastern coal must be pretreated by partial oxidative devolatilization to prevent char agglomeration in the gasifier. Pretreater condensate is comprised mainly of organic compounds and is flared. Other than these considerations, the remaining production process configurations remain essentially unchanged.

   d. Synthane Process   The configuration of the Synthane process is shown in Fig. 10. Here, gasifier raw product gas is quenched in a wet quench where 90% of raw gas vapor content is recovered as condensate. Oils produced in gasification are decanted, and the separated water

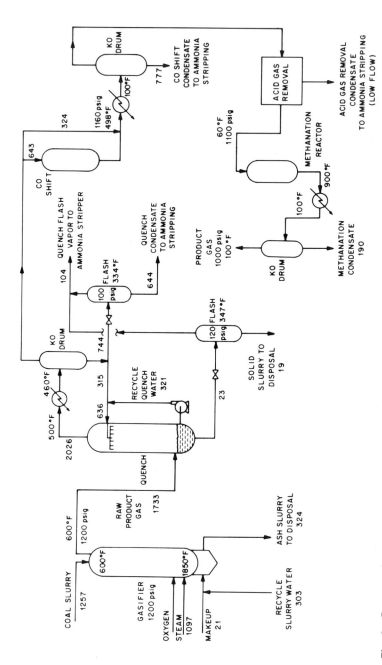

Fig. 8 Conceptual commercial flow diagram for the Hygas process utilizing western coal [C. F. Braun, Inc. (1976b)]. Flow rates are given for water in units of 1000 lb/hr, based on 250 MSCFD pipeline gas produced.

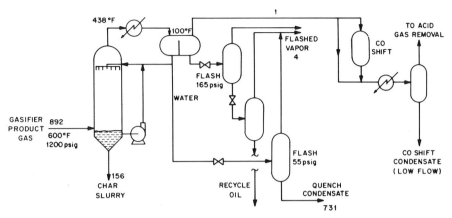

Fig. 9 Conceptual commercial flow diagram for the Hygas process utilizing eastern coal [Procon Inc. (1978)]. Flow rates are given for water in units of 1000 lb/hr, based on 250 MSCFD pipeline gas produced.

stream is split into three streams and distributed as either (1) recycle quench scrubbing water, (2) water to char quench, or (3) quench condensate to water treatment. The near complete quench of vapor from the product gas requires the addition of water to adjust the $CO:H_2O$ ratio prior to CO shifting. Of the 266×10^3 lb/hr of high pressure steam added to the product gas, 29% is provided by quench condensate which is converted to steam using the extreme heat in the char quench. Further downstream processing is typical of those systems already discussed and includes effluent condensates from the CO shift, acid gas removal, and methanation systems.

e. Lurgi Process   Figure 11 shows a flow diagram for the Lurgi gasification process. Processing configurations here are seen to be quite similar to that of the Synthane process in that raw gasifier product gas is quenched with a recirculating water stream, and overhead gas proceeds to downstream processing consisting of CO shift, acid gas removal, and methanation systems. Condensates result from quenching, shift cooling, acid gas removal, and methanation systems. Water requirements include gasifier steam and demineralized water to the acid gas removal system.

2. Quality Considerations

a. Organic and Inorganic Specie Distribution   The exact nature of commercial scale gasification liquid effluent streams from the processes

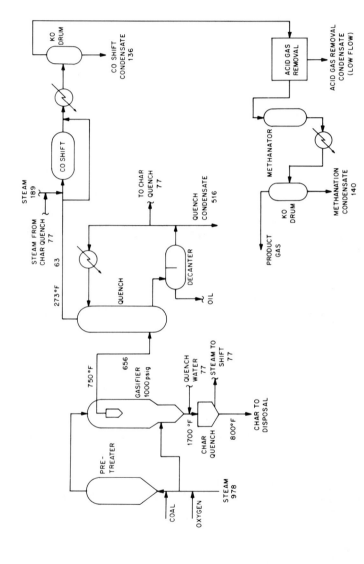

Fig. 10 Conceptual commercial flow diagram for the Synthane process [Probstein and Gold (1978)]. Flow rates are given for water in units of 1000 lb/hr, based on 250 MSCFD pipeline gas produced.

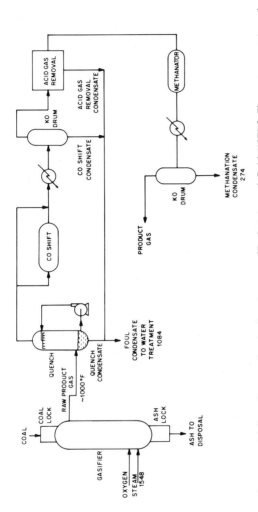

Fig. 11 Conceptual commercial flow diagram for the Lurgi process [Probstein and Gold (1978)]. Flow rates are given for water in units of 1000 lb/hr, based on 250 MSCFD pipeline gas produced.

described above is not known at this time. Many of the coal conversion systems being considered here have at best only been demonstrated at the pilot scale, and various amounts of preliminary pilot plant environmental data are available for each process. Although the Lurgi slagging process is operating at commercial scale in Scotland, the slagging fixed bed process has not been developed beyond the pilot scale. Since pilot plant configurations seldom reflect completely either the structural or operational practices of the commercial version of a process, it is difficult to translate effluent data from a pilot plant to a commercial scale facility. Further, predicting commercial scale effluents based on pilot plant data can be performed accurately only after careful comparison of pilot plant and commercial scale flow configurations and process operating conditions.

In light of available pilot plant data, some general characteristics and comparisons can be made for each of the coal processes being considered. Comparisons on plant effluent quality for organic (i.e., phenol, tar, and oil) and inorganic (i.e., NH_3, HCN, H_2S) species will be made in Sections IV and V. It is constructive to consider here the distributional trends of effluents throughout process water streams. Two designs for the Hygas process will be used to facilitate the evaluation of these trends: the Braun (1976b) design utilizing western coal and the Procon (1978) design utilizing eastern coal. It is important to stress that specific effluent production rates are used only to depict effluent distribution. These effluent production rates are based on a data base which includes various degrees of uncertainty. For this reason these values may not necessarily describe accurate effluent production rates expected in an actual commercial plant.

Figures 12 and 13 show anticipated effluent distribution trends for the western coal and eastern coal versions of the Hygas process, respectively. One basic difference between these designs results from the use of oil for slurrying coal in the eastern design, rather than water which was used for the western coal case. In order to recover this oil, the raw product gas and oil overhead from the quench must be cooled from 440° to 100°F. The contrasting system requires cooling only from 500° to 460°F. As a result, effluent distributions vary tremendously.

The distribution of organics in these designs parallels the temperature relationship governing oil recovery. The oil/water separator in the eastern design operates at 100°F and serves to remove essentially all the phenol content of the raw product gas. Phenol is thus collected in the quench condensate, and little or none is expected to be found in the CO shift condensate. In contrast, the quench effluent knockout drum in the western coal design operates at 460°F, and few phenolics are

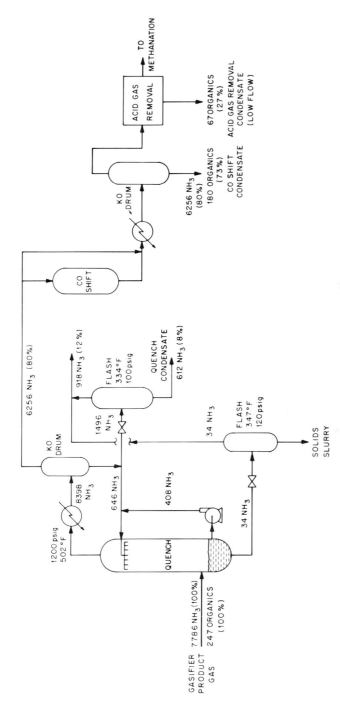

Fig. 12 Organic and inorganic effluent distributional trends for the Hygas process utilizing western coal [C. F. Braun, Inc. (1976b)]. Constituent flow rates are given in units of pounds per hour, based on 250 MSCFD pipeline gas produced.

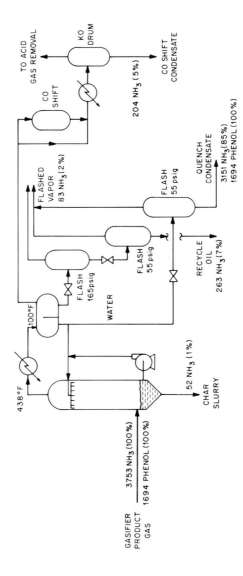

Fig. 13 Organic and inorganic effluent distributional trends for the Hygas process utilizing eastern coal [Procon, Inc. (1978)]. Constituent flow rates are given in units of pounds per hour, based on 250 MSCFD pipeline gas produced.

removed from the gas. It is not until the CO shift knockout drum that organics are separated from the product gas. In this design 73% of the product gas organic content is removed in the CO shift condensate, with the other 27% being removed during acid gas removal.

It is emphasized that oils, tar, phenols, and other organic contamination will be recovered either in one or more of the liquid effluent streams including: product gas quench, CO shift condensate, and acid gas removal condensate. Essentially all organics are removed prior to methanation, hence the methanation condensate is virtually free of organic contamination. The distribution of organics between product gas quench, CO shift, and acid gas removal condensates is fundamentally related to process temperature and configuration.

Most of the inorganic species being considered here (i.e., HCN, NH_3, H_2S, CO_2) are water soluble gases. Their distribution is affected by pressure, temperature, and water phase chemistry. Some of these factors will be generalized by considering the fate of NH_3. The temperature effects which govern the distribution of organic constituents have a similar effect on inorganic species. As shown in Fig. 13 for the Hygas eastern coal design, only 5% of raw product gas ammonia content is not recovered prior to the CO shift reaction. Of the 94% of NH_3 recovered in the oil/water separator, 3% is recycled with the oil, 85% is recovered in the quench condensate, and 2% is flashed and sent to effluent gas treatment. Solids slurry water accounts for the remaining 1%.

In the western coal design, 80% of the raw product gas ammonia content is not removed prior to CO shifting. Of the 20% which is removed, 12% is recovered as flash vapor and is sent to the ammonia stripper and 8% is recovered in the condensate and is also sent to the ammonia stripper. Inorganic species such as NH_3 are removed completely from the product gas in the CO shift condensate, while some organic compounds are carried over and are removed in acid gas condensate.

b. Quality Constraints on Process Water Streams   Since process water usually requires only low-to-medium quality water, most available raw makeup waters can be used without the need for further treatment. For this reason, there are no guidelines reported in the literature for process water quality. However, the important issue here is that closure of plant water balances which emphasize wastewater reuse depend strongly on how dirty water can be, rather than how clean the water can be made. Hence, a strong incentive exists to delineate minimal water quality guidelines for process waters. This will help to achieve complete reuse of plant wastewater effluents while minimizing the treatment necessary to upgrade these wastewaters for reuse. For these reasons, minimum quality criteria required for gasification processes should receive attention from researchers in the future. The configurations of those designs discussed above indicate some potential alternatives for study of water reuse quality.

A common feature of these gasification systems is the use of recirculating quench water for either dry or wet quenching. In the case of dry quenching, water is continuously evaporated and must be supplied as either a recycle or a makeup stream. For the Hygas processes shown in Figs. 8 and 9, quench overhead is cooled, and resulting condensate is recycled to the quench to provide for this need. However, in the case of the Bi-Gas process, no recycle water is available and a makeup stream becomes necessary. The Synthane process (Fig. 10) shows an example of a wet quench in which a recirculating condensate stream accomplishes the quenching. Since the net result of this system is formation of a condensate, water is removed to maintain steady state.

Water quality may be an important concern as it relates to makeup water in the dry quench and to the recirculating water in the wet quench. In the dry quench process, excessive contamination in the makeup water adds to gaseous effluent levels and may burden the removal of these contaminants downstream. This same burden can result if wet quench recirculating water contaminants are allowed to build up to a level which impairs the ability of the quench to act as a scrubber of effluent constituents. Consequently, this may require that a portion of the recirculating stream be treated prior to recycling in order to maintain the scrubbing effect.

In the Bi-Gas and Hygas processes (Figs. 7 and 8, respectively), a coal slurry is used for feeding the coal to the gasifier. In the Bi-Gas design, this water is flashed in the spray dryer and provides the water necessary in the product gas for the CO shift reaction. In the Hygas process, the slurry water is flashed in the uppermost drying zone of the gasifier. In both cases, it is not clear that a traditional clean process makeup water is necessarily the most appropriate makeup water to be used. It may be acceptable to use a partially treated process waste stream or an untreated stream to provide for some or all of this makeup need.

The use of foul raw product gas quench water to produce steam for the CO shift steam requirement in the Synthane process (Fig. 10) suggests another reuse possibility. In this case, high pressure steam ($\sim$1000 psig) is generated by using this foul condensate to quench gasifier char. It is reasonable to suspect that this steam could also be used to help satisfy steam requirements for the gasification reaction. This concept illustrates a means by which a foul condensate can be converted to process steam without having to satisfy boiler feed water quality criteria.

D. Waste Treatment Brines and Solid Slurries

Water balances are made complete by the inclusion of various blow-downs, or brine streams, and solid sludges and slurries. The flows result from water and wastewater processing of the following streams: raw makeup water, boiler feed makeup water, wastewater treatment blowdown, combined blowdown treatment brine, and water from de-watering of ash, char and/or sludge slurries.

As shown earlier in Fig. 2, water treatment units process water and produce an effluent product and a brine. The brine will vary in com-position depending on the input water quality, but is typically sev-eral times more concentrated than that of influent water. In some cases the brine from a very high quality water treatment process may be of better quality than that of a low quality influent stream. For example, brine from blowdown in the boiler feed water system is often suitable for cooling tower water. As a result, brine streams can be manipulated in several ways to complete the water balance. The hierarchy governing the use of brines is directly related to decreasing water quality needs, i.e., the higher quality brine can be used in a step-down process where lower quality water is needed. This strategy results in a few streams ultimately containing all the waste brine solids for disposal.

These most concentrated brine streams may be conveniently dis-posed with dewatered plant solids. Brines may be evaporated to reduce the final water volume destined to the solids disposal unit. Sludges resulting from flocculation, clarification, and biological oxidation of process waters, can be dewatered with slurries from transport of char, ash, slag, and other solids, and combined with these brines. Brines used in this fashion contribute to the water in which solids can be disposed. The sludge dewatering product can be used elsewhere in the process for such purposes as recirculating solid slurry water.

III. SATISFYING PROCESS WATER BALANCES

In order to minimize strategically both environmentally significant effluent discharges and net plant water consumption, all alternatives for satisfying process water input needs by use of process effluents must be considered. These evaluations are made by assessing what each effluent stream is most closely suited for in terms of reuse and what prohibitive problems or expenses are involved in treating these effluents. Current commercial scale designs include detailed process water balances which should reflect these considerations. Four such process water balances will be discussed below. Details concerning their formation provide a means to understand the design intent with regard to water management.

Figure 14 shows a schematic of the CO_2-Acceptor process water balance. Process streams include: (1) raw gas quench and acid gas removal condensates, (2) ammonia stripped recycle quench and ammonia scrubber water, (3) methanation condensate, and (4) quench bottom char/ash slurry. Steam to process consists predominantly (90%) of the gasifier steam requirement.

In this design, ammonia stripped wastewater is divided into three streams: recycle quench water (23%), cooling tower makeup (36%), and solids handling (41%). The solids handling feed water is ultimately recovered as part of the dewatering product, which is then sent to the cooling tower for makeup. The rest of the cooling tower makeup is supplied by boiler blowdown and by raw treated makeup water. The blowdowns from water treatment, viz., lime softening, demineralization, and polishing, are combined with cooling tower blowdown and sent to the evaporator. In this respect, the evaporator represents the ultimate repository for all dissolved solid brines in the process. Here solids are concentrated by a factor of 26, so effectively only 4% of the water in this combined brine is not recoverable. The concentrated brine is removed with dewatered solids.

Water balances for the Hygas process are much more complicated than that for the CO_2-Acceptor process because of the higher system operating pressure and the need for more extensive treatment of the waste streams. Figure 15 shows the water balance for the Hygas process utilizing western coal. Process streams include: (1) raw gas quench condensate, (2) shifted gas condensate, (3) methanation condensate, (4) recirculating ash quench water, and (5) gas quench slurry water. The source of these streams was detailed in Fig. 12. Because of a large variation in the distribution of contaminants in the process effluent water streams, only the shift and acid gas condensates are combined prior to treatment. The quench water stream is ammonia stripped and used as either cooling tower makeup (43%) or as makeup to low-pressure boiler feed water preparation. Shift condensate is ammonia stripped, with the stripped water going to biological oxidation and the overhead gas to ammonia recovery. Biologically treated wastewater is

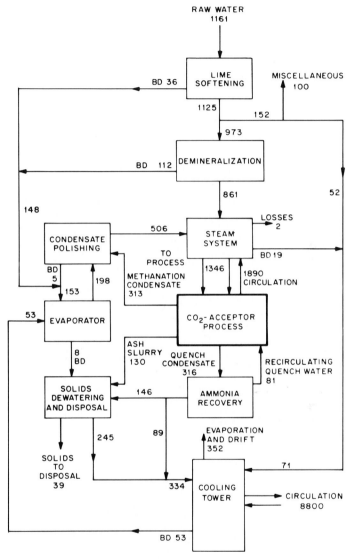

Fig. 14   Conceptual commercial plant water balance for the CO_2-Acceptor process [C. F. Braun (1976a)]. Flow rates are given for water in units of 1000 lb/hr, based on 250 MSCFD pipeline gas produced.

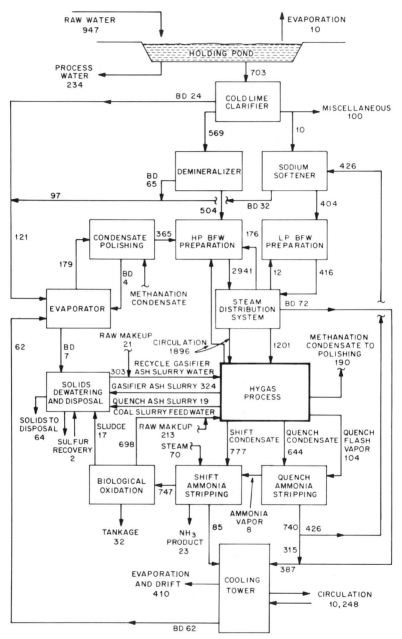

Fig. 15 Conceptual commercial plant water balance for the Hygas process utilizing western coal [C. F. Braun, Inc. (1976b)]. Flow rates are given for water in units of 1000 lb/hr, based on 250 MSCFD pipeline gas produced.

used for coal slurry feed process water makeup. Water separated in ammonia recovery is used for cooling tower makeup. Ash quench water is dewatered, combined with the gas quench slurry, and returned to the process as quench water. Methanation condensate is polished and used as feed to high pressure boiler feed water preparation. As was the case for the CO_2-Acceptor water balance, water treatment blowdowns and cooling tower blowdown are collectively sent to an evaporator, the product of which is polished and used for high pressure boiler feed makeup. Evaporator brine is sent to solids disposal.

Figure 16 shows the process water balance for the Hygas process utilizing eastern coal. Process streams include: (1) quench, CO shift, and acid gas removal condensates, (2) methanation condensate, (3) char slurry from raw gas quench, and (4) ash slurry from gasification.

In this scheme, foul wastewater is treated for phenol and NH_3 removal and recovery, and biological oxidation. Biological process effluent wastewater is combined with raw makeup water and further treated by flocculation and clarification. This water is used for most of the process water makeup needs and is used without further treatment for cooling tower makeup. Methanation condensate is also used directly as cooling tower makeup water. Solid slurries are dewatered with product water going to phenol recovery. Water treatment blowdowns (e.g., from boiler feed water preparation), raw water reverse osmosis blowdown, and the cooling tower blowdown are processed through a reverse osmosis unit, which reduces the brine volume to 10%. The brine volume is reduced further by evaporation, with the evaporator product used for boiler feed makeup. The overall effect of this two-stage brine-handling process is to reduce brine volume from 670×10^3 to 3×10^3 lb/hr, a reduction of more than 99.5% of brine water volume, or a concentration factor of 223.

Cooling tower makeup water consists predominantly of raw makeup water but also includes direct use of methanation condensate. Ninety-seven percent of the raw makeup water comes from flocculation and clarification of fire pond water. The remaining fraction of cooling tower makeup is supplied by reverse osmosis permeate. Boiler feed water comes predominantly from treated fire pond water, though there is a small contribution from the evaporator. It should be pointed out that treated wastewater is not used directly for either cooling tower or boiler feed makeup. Rather, it is combined with raw makeup water and treated as such to satisfy process makeup requirements.

Several modifications of this water balance have been proposed, and most of these modifications have been or will be incorporated in subsequent revisions of the Procon design. High quality methanation

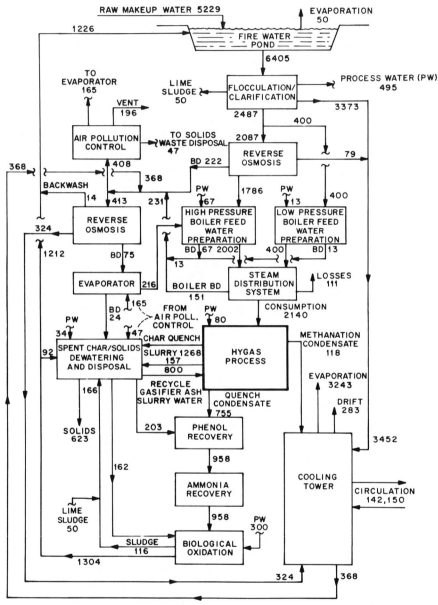

Fig. 16 Conceptual commercial plant water balance for the Hygas process utilizing eastern coal [Procon., Inc., (1978)]. Flow rates are given for water in units of 1000 lb/hr, based on 250 MSCFD pipeline gas produced.

condensate will be polished and used for boiler feed makeup water. Raw clarified water will not be completely treated with reverse osmosis prior to high-pressure boiler feed preparation. This is because product water from reverse osmosis should be of fairly high quality . Since this product is treated further in a mixed bed demineralizer, it is reasonable to anticipate that some portion of the clarified water can bypass reverse osmosis treatment. Treated wastewater from the biological oxidation unit will not be mixed with raw makeup water in the holding pond. The treated process wastewater will in many respects be of much lower quality than that of the raw makeup water. It is not prudent to mix these waters because that would result in a larger volume of lower quality water. Treated wastewater will be processed directly for reuse or will be stored separately in another holding pond.

One final point of interest can be made concerning the water management scheme for the Lurgi process located at Powder River, Wyoming, provided by Panhandle Eastern Pipeline Company (Moon, 1978). As shown in Fig. 17, the system uses several multieffect evaporative distillation units in place of reverse osmosis or other desalination systems. The steam necessary to run such units normally prohibits their use. However, sufficient low pressure steam is made available by recovering heat which normally would have been dissipated by air cooling.

IV. STRATEGIES FOR CHARACTERIZATION OF COAL GASIFICATION PILOT PLANT WASTEWATER EFFLUENTS

The prediction of effluent emissions from coal gasification processes is complicated by the fact that pilot plant discharges are not similar to that which would be discharged from a commercial facility. The central issue is one of data scaling, i.e., how does one take raw pilot plant data and predict commercial scale effluents. Important problems within this issue are generally related to effluent characterization, process characterization, and effluent management. The following discussion is concerned primarily with problems related to effluent characterization. For a more thorough discussion of this topic and issues related to process characterization and effluent management, the interested reader is referred to Luthy *et al.* (1977b), Massey *et al.* (1978), and Proceedings of the Workshop on Health and Environmental Effects of Coal Gasification and Liquefaction Technologies (MITRE Corp., 1979).

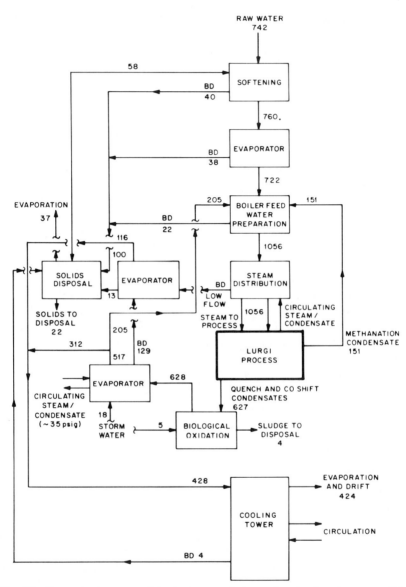

Fig. 17 Conceptual commercial plant water balance for the Lurgi process [Moon (1978)]. Flow rates are given for water in units of 1000 lb/hr, based on 250 MSCFD pipeline gas produced.

A.   Stream Sampling and Measurement Priorities

Because of monetary and temporal constraints, priorities must be decided with regard to the proper emphasis for aqueous, gaseous, and solids sampling during a pilot plant test program. Decisions must also be made with respect to the selection of specific effluent streams for sampling and with respect to ranking by importance various effluent parameters to be analyzed.

Sampling efforts should focus on those streams which: (1) provide baseline effluent production data scalable to larger plant sizes and (2) provide material balances for specific effluent constituents. Priority should be given to those streams which contribute toward the development of a scalable effluent data base. Sampling for material balances serves as a check on the accuracy of the data base. Sampling to determine pilot plant environmental impacts should receive a smaller fraction of the total sampling and characterization effort.

As an example of application of these priorities, consider aqueous effluents from the CO_2-Acceptor and Hygas pilot plants. Figure 18 shows effluent flow patterns in the CO_2-Acceptor pilot plant. As presently configured there are effectively five pilot plant wastewater streams in the CO_2-Acceptor pilot plant. Of these streams, the coal drier–preheater venturi scrubber water and product gas quench condensate provide the best information on wastewater effluent production for a commercial facility. The regenerator off-gas condensate shown in Fig. 18 is an artifact of the pilot plant because commercial designs call for dry heat recovery from this stream.

Figure 19 delineates effluent flow patterns for the Hygas pilot plant. Comparison of Fig. 19 with Figs. 8 and 9 show that, of the nine pilot plant wastewater streams, five are suitable for scalable plant characterization: agglomerating coal pretreater quench water, gasifier ash slurry water, product gas cyclone slurry water, product gas quench condensate, and oil/water/solids interface blowdown. Commercial designs do not call for the production of pretreater quench water. However, coal pretreatment may significantly reduce downstream organic and inorganic loadings.

Selection of parameters for effluent characterization should reflect appropriate concern for monitoring:

(1)   regulated parameters in related industries present at significant levels and

(2)   parameters present at significant levels and needed for design of wastewater treatment facilities.

Richard G. Luthy and Richard W. Walters

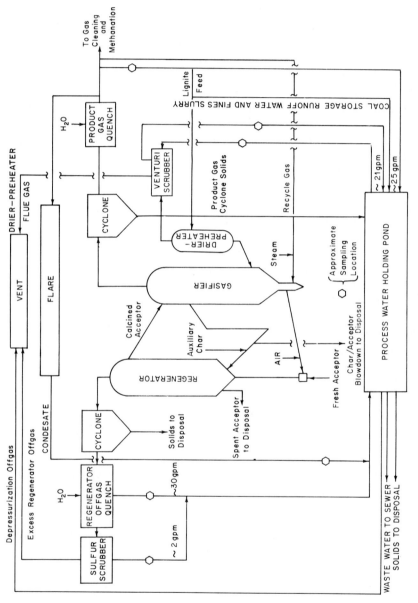

Fig. 18   Effluent flow patterns in the CO₂-Acceptor pilot plant [Luthy, *et al.* (1977b)].

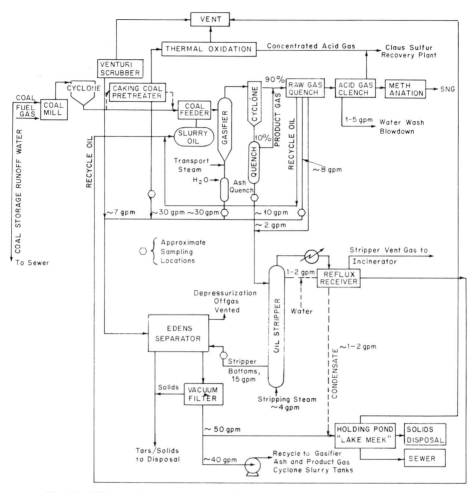

Fig. 19 Effluent flow patterns in the Hygas pilot plant [Luthy *et al.* (1977b)].

A listing of first priority effluent parameters for coal gasification waste-waters may be obtained by examining federally regulated wastewater discharge parameters in related industries, e.g., petroleum refining, iron and steel making, steam electric power, electropolating, ore mining, and coal mining. Table V shows a listing of United States EPA National Effluent Limitations, Guidelines, and Performance Standards in these industries. This list suggests thirteen parameters for characterization of coal gasification wastewaters. Additional parameters required for facility design would include an expanded list of compounds

TABLE V *United States EPA National Effluent Limitations, Guidelines, and Standards in Industries Analogous to Coal Gasification Processing*[a]

Parameters	Petroleum May 9, 1974	Iron and steelmaking June 28, 1974 March 19, 1976	Steam electric power[b] Oct. 8, 1974	Electro-plating April 24, 1975	Ore Mining Nov. 6, 1975	Coal mining May 13, 1976
Physical Properties						
TSS	●	●	●	●	●	●
Inorganic nonmetallics						
pH	●	●	●	●	●	●
F⁻		●		●		
CN⁻		●		●	●	
Oxygen/oxygen demands						
BOD$_5$	●					
COD	●			●		
Organic constituents						
Phenols	●	●	●			
TOC	●	●				
Grease and oil	●					
Nitrogen						
NH$_3$-N	●	●		●		
NO$_3^-$-N		●				
Phosphorus						
PO$_4^{3-}$		●	●			
Sulfur						
S^{2-}	●	●				

[a] Effluent Guidelines (1974a,b,c, 1975a,b, 1976a,b).
[b] Also includes limitations for PCB's, chlorine, corrosion inhibitors, heat.

as appropriate in the categories shown in Table V. Examples of these constituents would be total dissolved solids (TDS), settleable solids, alkalinity, thiocyanate, and biological nutrients .

Trace contaminants in wastewaters may be characterized according to specific constituents, such as heavy metals and priority organic pollutants, or according to specific measures of biological toxicity, such as TLM-50 Ames testing for carcinogenicity/mutagenicity. Federally regulated trace metal wastewater discharge parameters for industries analogous to coal gasification are shown in Table VI. These contaminants should serve as candidates for screening characterization, but survey analyses should be made for other important trace elements which may be more prevalent in gasification wastewaters. For example, boron levels are not included in Table V, but boron has been found at environmentally significant concentrations (≈ 100 mg/liter) in Hygas quench waters.

Lists of priority trace organic pollutants have been promulgated by the EPA. Screening gas chromatographic/mass spectrometric analysis of coal gasification wastewaters is necessary in order to determine wastewater loading of these priority pollutants and removal during treatment. Such an investigation is presently being conducted jointly by Carnegie–Mellon University, Argonne National Laboratory, and the

TABLE VI *United States EPA Regulated Metals For Six Industries in National Effluent Limitations, Guidelines, and Standards*[a]

	Petroleum	Iron and steel-making	Steam electric power	Electro-plating[b]	Ore mining[c]	Coal mining
Al					●	
As					●	
Cd				●	●	
Cr	●	●	●	●		
Cu		●	●	●	●	
Fe		●	●	●	●	●
Hg					●	
Mn		●				●
Ni		●		●	●	
Pb		●		●	●	
Sn		●		●		
Zn		●	●	●	●	

[a] Effluent Guidelines (1974a,b,c, 1975a,b, 1976a,b).

[b] Precious Metals subcategory includes the precious metals Ag, Au, Ir, Os, Pd, Pt, Rh, Ru.

[c] Uranium, radium, and vanadium ores subcategory includes Ra_{225}, U.

Institute of Gas Technology in order to evaluate trace organic concentrations in raw Hygas wastewater and in raw wastewater treated by bench-scale processing.

B. Wastewater Sampling Preservation, and Analytical Methods†

Major factors affecting strategy for effectively characterizing the composition of coal gasification wastewaters are (Luthy *et al.*, 1977b)

(1) the presence of temporal variation in stream compositions owing to either random or systematic influences (e.g., fluctuations in process operating conditions),
(2) the influence of such factors as sample degradation prior to analysis, and
(3) the choice of analytical methods for wastewater analysis.

Tests to evaluate variations in goal gasification stream composition have been reported by Koblin and Massey (1977). In one test, effluent streams from the Hygas process were sampled at 25 min intervals over an 8 hr sampling period. In another test, streams were sampled at 1 hr intervals for 35 to 50 consecutive hours. Depending on location and parameter, the coefficient of variation for ammonia and total organic carbon (TOC) ranged from about 5 to 21%. If the data included non-steady state fluctuations, the coefficient of variation ranged from about 9 to 61%.

For the case of effluent parameters resulting from a homogeneous population having a normal distribution, the number of grab samples required to achieve any desired level of accuracy is

$$n = t^2 s^2 / l^2 \equiv t^2 (cv)^2 / p^2, \tag{7}$$

where n is number of grab samples, t is statistic, s is estimated parameter standard deviation, cv is coefficient of variation, l is desired accuracy, and p is desired accuracy as a fraction of the mean. Figure 20 (Koblin and Massey, 1977) shows graphically that at the 95% confidence level, accuracy is governed by the coefficient of variation of a stream and the number of samples collected.

It was found in the case of the Hygas data that the subset of data reflecting steady state operation approximated a normal distribution as shown in Fig. 21. In this case the use of Eq. (7) is appropriate for

† The sampling and analytical methods for aqueous streams from fluidized-bed combustion processes are described in Volume II, Chapter 36.

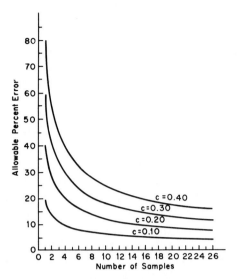

Fig. 20 Relationship between stream coefficient of variation, number of samplings, and the accuracy of a parameter estimate at the 95% confidence level given a normal distribution. C = coefficient of variation = standard deviation/mean. [Koblin and Massey (1977)].

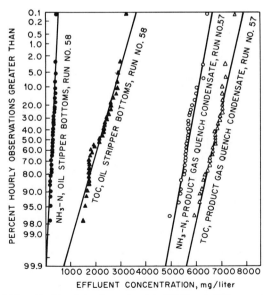

Fig. 21 Probability distribution of steady state TOC and NH_3 concentrations in Hygas pilot plant wastewater streams.

assessing sampling frequency. Oftentimes wastewater effluent data exhibit log–normal distributions rather than normal distributions. In this case Sherwani and Moreau (1975) have shown that an analogous equation suitable for log–normal frequency distributions is

$$n = t^2 \, (\text{antilog } S_g - 1)/p^2, \tag{8}$$

where S_g is the geometric standard deviation.

The degree of sampling accuracy needed to characterize an effluent stream depends on the use of the data. For base line characterization, 25% accuracy may be sufficient. Much greater accuracy is required for evaluation of the effect of minor process variations on effluent production. If the stream has an inherently high coefficient of variation, it may be futile to attempt to reduce accuracy to low values. In this case consideration should be given to evaluation of other strategies, such as moving the sampling location.

Field and laboratory spike and recovery studies have been performed in order to evaluate procedures for wastewater sample preservation and analysis. Screening studies at the CO_2-Acceptor and Hygas pilot plants (Massey *et al.*, 1978) have shown that techniques for sample preservation and analysis required development. Results of identification of suitable sample preservation and analytical techniques are presented in Luthy (1977) and in Luthy *et al.* (1977b). Many of the identifiable problems can be categorized as follows:

(1)   representativeness of wastewater solids sampling,
(2)   influence of ammonia and carbon dioxide on dissolved solids measurements,
(3)   high buffering capacity,
(4)   high concentration of effluent constitutents, and
(5)   complex interactions between sulfide, cyanide, and thiocyanate.

Each of these problems is discussed briefly below.

Wastewater solids sampling is recognized as an important part of an environmental sampling effort. However, many solid containing streams in coal gasification pilot plants exhibit erratic pulse discharge of solids. The standard deviation of total suspended solid levels in grab/composite samples may be as great or greater than mean values. This makes it difficult to assess true values of waste solids loadings in these streams. Another problem with solids analysis is that settleable and suspended solids, tars, and oils may be partitioned nonuniformly during sample collection, preparation, and analysis. Phenol and organic carbon analysis are susceptible to these types of problems. Until this problem can be overcome it is recommended that analysis for constituents which may be affected by the inclusion of solids and suspended material be preceded by membrane filtration. This serves to decouple solid suspended phase constituents from aqueous phase constituents and helps to generate a data base which discerns the nature of various constituents.

It is also preferable to analyze separately for soluble organic carbon and suspended solids, rather than expend effort on measuring total organic carbon because the latter measurement may vary greatly depending on fluctuations in suspended solids loading. In any event the TOC analysis is not particularly useful for design because it does not reflect how much organic carbon may be removed by gravimetric techniques as opposed to how much is dissolved and requires other forms of treatment for removal.

Raw coal gasification quench water contains high levels of dissolved inorganic substances. This results from the relatively high partial pressures of carbon dioxide and ammonia under which these waters are formed. As will be explained later, the most prevalent cation in raw quench waters is the ammonium ion, while the most prevalent anion is bicarbonate. As a result, many quench waters may be modeled as having up to 0.25 to 0.5M/liter NH_4HCO_3.

Because of the high concentration of dissolved gases in raw quench waters, specific conductance values fall in the range of 10,000 to 50,000 μmho/cm. These values do not correlate with TDS measurements because drying for TDS measurements at 180°C to remove occluded and bound water results in thermal decomposition of some salts, especially ammonium bicarbonate. This is different from many natural waters where it has been found that a reasonable estimate of TDS is given by multiplying conductivity by a factor of 0.5–0.7 (A.P.H.A., 1975). Thus, it is recommended that correlations between conductivity and TDS be based on calculated values of total dissolved solids using elemental analysis. Specific conductance could then be used routinely to give an estimate of TDS.

High buffering capacities are prevalent in coal gasification quench effluents because of the high concentration of dissolved ammonia and carbon dioxide. Consequently, the use of standard additions of acids and bases as sample preservatives or reagents during analysis is likely to be insufficient for these effluents. Personnel inexperienced with handling these wastewaters require special instruction in order to appreciate fully the effect of extraordinarily large buffer capacity on the lack of response to standard reagent additions. As will be explained later, the base neutralizing capacity of these wastewaters is greater than the acid neutralizing capacity. Hence preservation procedures which require raising of sample pH often necessitate addition of unusually high quantities of solid sodium hydroxide or lime.

Another effect of high buffering capacity is the liberation from solution of carbon dioxide upon acidification which may cause the sample to effervesce uncontrollably. In the case of certain analyses, e.g., cyanide, this may severely interfere with the analytical procedure (Luthy *et al.*, 1977a).

Doses of preservatives or reagents which react quantitatively with a compound of interest must be checked to ensure that sufficient quantities have been added to react with high concentrations of effluent constituents encountered in coal gasification wastewaters. For example, the recommended preservative dose for sulfide is four drops of 2N zinc acetate per 100 mliter of sample (A.P.H.A., 1975; EPA, 1974). This dose may be insufficient to precipitate all the sulfide present in coal gasification wastewater samples.

The chemistry of interaction between sulfide, cyanide, and thiocyanate in coal conversion wastewaters will be described in the following section. The significance of this interaction is that cyanide in the presence of sulfide and an oxidizing agent (oxygen) will react to form thiocyanate. Hence, it is important to remove sulfide by precipitation at the time of sample preservation in order to prevent degradation of cyanide. Work at Carnegie-Mellon University has shown that cyanide samples which have been preserved by raising pH without prior removal of sulfide will show a decrease in cyanide concentration and an increase in thiocyanate. Low values of cyanide reported for some pilot plant quench waters may reflect improper preservation technique rather than true cyanide production levels.

V.   QUENCH WATER COMPOSITION AND CHEMISTRY

This section describes general characteristics of coal gasification quench waters. Reasons for significant differences in organic contaminant concentration in these effluents are discussed, and general characteristics of coal gasification wastewaters are compared with coke plant weak ammonia liquor. The acid–base chemistry of quench waters will be discussed with regard to foreseeable effects on ammonia recovery systems. Potential chemical interactions among cyanide, sulfur, and thiocyanate are described.

A.   General Characteristics of Coal Gasification Quench Waters

Gasification processes may be divided into two categories with respect to levels of organic contaminants in process condensates: those which produce little or no phenols, oils, and tars, and those which produce substantial quantities of these materials.

Examples of processes showing little or no organic contamination in the product gas are: Bi-Gas, Koppers–Totzek, Combustion Engineering, CO_2-Acceptor, Westinghouse, and U-Gas. Generally, these processes have either entrained flow or fluidized bed gasifiers that operate at temperatures above 1900°F and produce ash as a slag or agglomerate (Bostwick, 1977), but there are exceptions to this generalization. For example, the CO_2-Acceptor process operates at less than 1900°F, and on this basis would be expected to have higher organic production. However, in this case coal is injected to the bottom of the gasifier to yield a higher temperature and residence time for destruction of coal devolatilization products.

Examples of gasification processes which produce effluents with organic contamination are Synthane, Hygas, Lurgi, COGAS, and Wellman–Galusha. Observations regarding hydrocarbon formation in the Synthane process development unit have shown qualitative relationships between processing conditions and the production of phenol, oil, and tars (Nakles *et al.*, 1975; Fillo, 1979). The reduction of heavy tar is promoted by enhanced gas–solids mixing and higher temperatures. Oil production and composition are affected by coal particle size; production increases and composition becomes lighter as particle size becomes smaller. Preliminary studies on phenol decomposition at atmospheric pressure show substantial decomposition above 900°C in less than 4s residence time independent of hydrogen partial pressure. It has been observed that o-cresol decomposes faster than phenol, and that decomposition of phenol is enhanced by the presence of char. Studies of this type indicate that strong, potentially exploitable relationships exist between changes in process variables and reduction of phenols, oil, and tar production.

B. Comparison of High Organic Strength Quench Waters with Coke Plant Ammonia Liquor

Table VII (Luthy and Jones, 1978) compares the chemical composition of coke plant ammonia still effluent with coke plant ammonia liquor (Chapter 38, Fig. 1) and with available data on Synthane process by-product water, Hygas quench water, and H-Coal liquefaction foul water. The data shown in this table are for comparison only, and are not intended to imply that all coal gasification wastewaters are represented by those of either the Synthane or Hygas processes. For example, it is known that effluents from the CO_2-Acceptor (Luthy *et al.*,

TABLE VII   *Comparison of Coal Coking and Coal Conversion Effluents*

Parameter (mg/liter)	Coke plant ammonia still effluent	Coke plant waste ammonia liquor[a]	Synthane process by-product water[b]	Hygas process wastewater[c]	H-coal liquefaction foul water[d]
COD	3400–5700	2500–10,000	15,000–43,000	3000–5100	88,000 (26,500)
Phenol	620–1150	400–3000	1700–6600	560–900	-6800
NH_3-N	22–100	1800–6500	7200–11,000	2600–4600	17,000
NO_3^--N	<0.2	—	—	1–5	<1
Kjeldhal-N	21–27	—	—	4–10	50
P	0.9	<1	—	0.5–1.8	—
CN^-	1.6–6	10–100	0.1–0.6	0.1–0.7	—
SCN^-	230–590	100–1500	22–200	17–45	—
S^{2-}	8	200–500	—	60–220	29,000
SO_4^{2-}	325–350	—	—	60–180	—
Alkalinity (as $CaCO_3$)	525–920	2800–4300[e]	10,000–20,000	9800–15,000	—
Conductivity (μmho/cm)	3500–6000	—	—	30,000	—
pH (units)	9.3–9.8	7.5–9.1	8.5–9.3	7.8–8.0	9.5

[a] Data sources: Rubin and McMichael (1975); Wong-Chong *et al.* (1978); Kostenbader and Flecksteiner (1969); Effluent Guidelines (1974a).

[b] Johnson *et al.* (1977).

[c] Luthy and Tallon (1978). Hygas wastewater was comprised of equal volumes of cyclone and quench effluents.

[d] Reap *et al.* (1977). Stripped foul water with sulfide removed had an average COD of 26,500 mg/liter.

[e] Calculated from data of Jablin and Chanko (1972).

1977b), Bi-Gas, and Koppers–Totzek (Magee *et al.*, 1974) processes contain little or negligible amounts of organic contaminants.

Data in Table VII show that Hygas process wastewater comprised of cyclone and quench effluent has COD and phenol concentrations similar to ammonia still effluent. However, if Hygas cyclone dilution water were discounted, the COD of Hygas wastewater would increase by a factor of approximately two. If the Hygas pilot plant were to operate at a steam-to-coal ratio approximately equal to one, as in the proposed commercial design, rather than at a ratio of two, then quench wastewater COD may increase further. In these respects, the soluble organic contamination of Hygas wastewater in a commercial facility may be significantly higher than that shown in Table VII. However, these effects may be offset by enhanced phenol decomposition in the commercial facility owing to higher temperatures in the gasifier riser. It is interesting to note that the Hygas process does not produce significant quantities of tars and heavy oils as do other processes shown in Table VIII.

The gasification wastewater qualities presented in Table VII illustrate that these wastewaters contain many of the same compounds as coal coking wastewaters. For this reason, the body of literature on processing coal coking wastewaters is useful for gaining insights with regard to processing of coal gasification effluents. However, it must be recognized that some important coal conversion wastewater constituents are present at quite different concentrations than observed in coke plant effluents. For example, based on the limited data available, it appears that cyanide and thiocyanate are at higher concentrations in coke plant ammonia liquors, while sulfide and alkalinity are at higher concentrations in the coal conversion effluents. The differences in relative concentration of cyanide, thiocyanate, and sulfide in coke plant liquor may be explained to a certain extent by the practice of introducing small quantities of oxygen to the carbonization gases during charging operations. This can promote oxidation of sulfide to polysulfide, which in turn can react with cyanide to produce thiocyanate.

C. Acid–Base Chemistry of Coal Gasification Wastewaters

Table VII shows that alkalinity values in Hygas and Synthane wastewaters are noticeably higher than in coke plant ammonia liquor. This results from the presence of relatively high partial pressures of ammonia and carbon dioxide in raw coal gasification product gas. When this gas is cooled (quenched), ammonia and carbon dioxide have a tendency to be dissolved into solution simultaneously because ammonia acts as a proton acceptor for carbon dioxide. This results in a large fraction of ammonia in these wastewaters existing as so-called free ammonia. That is, most if not all of the ammonia in Hygas wastewater can be envisioned as being compensated by a stoichiometric equivalent of dissolved acid gases, which are largely comprised of carbon dioxide. A different situation exists for coke plant ammonia liquor. For example, with approximately 400 mg/liter ammonia–nitrogen and 400 mg/liter alkalinity, only 50 to 60% of the ammonia in the waste liquor is free ammonia. The remaining fraction has been termed fixed ammonia in the industry; it is more difficult to liberate from solution in stripping operations because lime or caustic must be added to the liquor to raise pH and to compensate for acid released as ammonia is liberated.

In theory, most of the ammonia present in Hygas and Synthane wastewater is present as free ammonia. Thus, if the wastewater were stripped with steam or air at elevated temperature, most of the ammonia could be liberated without the addition of lime or caustic for pH adjustment. This has been demonstrated in water reuse investigations with Hygas wastewater, wherein ammonia is liberated by high temperature gas stripping.

The fundamental difference in ammonia stripping characteristics between Synthane and Hygas coal gasification wastewaters, and coke plant ammonia liquor can be explained according to the following definitions (Luthy and Tallon, 1978):

$$\text{free ammonia} \leq HCO_3^- + 2CO_3^{2-} + HS^- + 2S^{2-} + CN^-$$
$$+ \text{ other deprotonated dissolved acid gases,} \qquad (9)$$

$$\text{fixed ammonia} = \text{total ammonia} - \text{free ammonia} \qquad (10)$$

These represent correct chemical definitions for free and fixed ammonia. Free ammonia is that fraction of the total ammonia which is compensated by chemical equivalents of deprotonated dissolved acid gases. This definition removes ambiguities associated with attempts to differentiate between free and fixed ammonia on the bases of contributions of various ammonium salts (Wilson and Wells, 1945). This concept for estimation of fixed ammonia concentration is somewhat analogous to the determination of noncarbonate hardness. This is because free ammonia and noncarbonate hardness are parameters defined on the basis of aqueous phase chemical composition and are related only indirectly to parent species which formed the solution.

D.   Acid–Base Neutralizing Capacity

Acidimetric and alkalimetric titration characteristics aid understanding of acid–base chemistry of gasification quench waters. Figure 22 shows a typical acidimetric titration curve for Hygas wastewater comprised of equal volumes of cyclone and quench wastewaters. The endpoint of the titration is given by the inflection point at a pH of 3.7, which corresponds to a measured alkalinity of 13,000 mg/liter as $CaCO_3$. Figure 23 (Luthy and Tallon, 1978) shows alkalimetric titration characteristics for the same raw wastewater. In this case approximately 0.43 equivalents/liter of lime are required to raise wastewater pH to 10.

These two figures illustrate the significant effect of dissolved carbon dioxide and ammonia on quench water buffering characteristics. At a raw quench water pH of 7.9, most of the alkalinity is present at

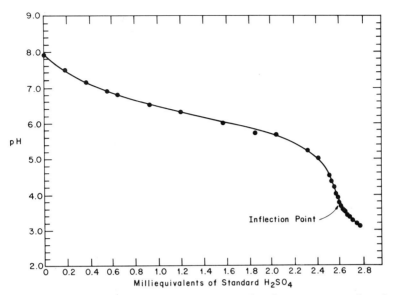

Fig. 22 Acidimetric titration curve for raw Hygas pilot plant wastewater. Sample size = 10 mliter, alkalinity = 13,000 mg/liter as $CaCO_3$.

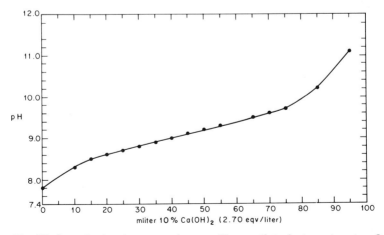

Fig. 23 Alkalimetric titration curve for raw Hygas pilot plant wastewater. Sample Size = 500 mliter. [Luthy and Tallon (1978).]

HCO_3^- and most of the ammonia is present as NH_4^+. Acidimetric titration largely entails the conversion of HCO_3^- to H_2CO_3; alkalimetric titration largely entails the conversion of HCO_3^- to CO_3^{2-} and NH_4^+ to NH_3. Thus, larger equivalent chemical doses are required to raise the pH to a total acidity endpoint than to lower pH to a total alkalinity endpoint.

E.   Interactions among Cyanide, Sulfur, and Thiocyanate

Mechanisms to explain the formation of ammonia and cyanide in product gas during coal coking and coal gasification are understood in a qualitative fashion. Ammonia is believed to be formed as a result of the release of amino-type side chains, e.g., amino or substituted amino groups, from the coal structure. Nitrogen which remains in the coal structure is believed to be contained within multiring compounds and to be much more refractory than aminonitrogen. Under hydrogenation conditions during gasification much more ammonia is formed than under normal coal carbonization conditions, probably as a result of hydrogen attack of the ring bound nitrogen. Hence, the yield of ammonia is greater from coal gasification than from coking (Hill, 1945).

The formation of cyanide during coking is believed to be the result of secondary reactions. Possible pathways for formation of hydrogen cyanide entail reaction of ammonia with C, CH_4, C_2H_4, C_2H_2, and/or CO (Nakles, 1977). Much less is known about the extent of gas phase formation of thiocyanate during gasification.

Regardless of whether or not thiocyanate may be present in gasifier product gas, there is no question that thiocyanate appears in gasification quench waters, coke plant waste ammonia liquors, and coke gas distribution lines, condensates, and deposits (Luthy *et al.*, 1977b; Lowry, 1945). A summary and discussion of possible reaction pathways for aqueous phase production of thiocyanate during coal gasification is presented in Luthy *et al.* (1977a). That paper reviewed possible reaction pathways of cyanide with sulfur species and kinetic studies on the oxygenation of sulfide in aqueous solution. Figure 24 (Luthy *et al.*, 1977a) summarizes a model based upon those reviews and experimental investigations to explain the oxygenation of sulfide in coal gasification wastewaters to yield thiocyanate. The model incorporates features of the sulfide reaction scheme following polysulfide formation as presented by Chen and Morris (1972). Possible reaction pathways for formation of thiocyanate entail reaction of cyanide with polysulfide and thiosulfate. Kinetic investigations of the reaction between cyanide

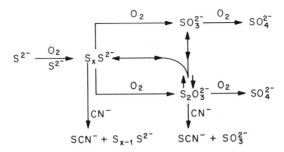

Fig. 24   Reaction pathways of oxygenation of sulfide to yield thiocyanate. [Luthy *et al.* (1977a).]

and polysulfide showed this to be a relatively fast reaction with overall reaction order of 1.54 ± 0.25. The rate constant was found to be 0.24 $(M/liter)^{-0.54}/min$ (Luthy *et al.*, 1977a).

Bartlett and Davis (1958) reported a second-order rate constant of 2.3×10^{-3} liter/M/min for reaction of cyanide with thiosulfate to produce thiocyanate. Thus, the reaction rate of cyanide with polysulfide is several orders of magnitude higher than the reaction rate of cyanide with thiosulfate.

These cyanide–sulfur–thiocyanate interactions have important implications for changes in wastewater quality which may occur during storage or processing. Wastewater oxidation could produce polysulfide from available sulfide and, to the extent that cyanide is present, thiocyanate should form. The fate of cyanide in wastewater processing may effect choice of treatment strategy and reuse options for treated waters. Though thiocyanate is relatively nontoxic as compared to cyanide, thiocyanate can cause severe corrosion and deposition problems in steel process units and lines and may, therefore, play an important role in management of wastewaters for reuse.

VI. SUMMARY AND CONCLUSIONS

Both limited water supply and environmental considerations ultimately lead to motivation for the reuse of gasification process wastewater. Various water management schemes may exist for a given gasification process, all of which depend on the exact nature of the particular waters and on the quality constraints for which the waters will be used. Thus, water quality serves to link water management directly to the study of water chemistry.

From a water management viewpoint, there are many possible configurations of a process water balance. Some aspects of these configurations have become generally accepted as necessary in achieving a process water balance. For example, raw makeup water is inevitably softened and serves as process water, cooling tower makeup water, and feed to the boiler feed water treatment system. Years of experience have lead to the development of several rule of thumb treatment needs for use of raw water in these applications. Although an exact design must be based on specific quantity/quality relationships for water streams in a given plant or process, such a design may be made easily by applying a wealth of treatment data to equations governing the system. In lime softening of raw makeup water for example, lime doses are calculated directly from wastewater levels of hardness and alkalinity.

In contrast, some process water balance configurations unique to the coal gasification process are not well developed, because little or no experience has been obtained in working with these particular waters. As a result, the thrust of the current interest in minimizing water consumption in coal gasification processes is focused on evaluation of process wastewater treatability and reuse.

Effluent water streams from the coal gasification process which are of interest include the raw gas quench, CO shift, and acid gas removal condensates which are poor quality effluents. The methanation condensate is of good quality and is anticipated to be of sufficient quality to be polished and used as feed to the boiler feed water system. Poor quality condensates will generally require some extent of phenol removal, ammonia removal, and/or biological oxidation prior to reuse in the process. However, little is known about (1) the extent to which each of these latter treatments are necessary, (2) the need for additional treatment processes, and (3) uses for which these poor quality waters are most suited.

Although specific processes may differ in overall water management configurations, it appears likely that all high-BTU gasification systems will use some portion of treated wastewater as part of the makeup water requirement for cooling water. The amount of cooling tower makeup water which is needed relative to the volume of poor quality process water depends on the extent to which waste heat is removed by either air or evaporative cooling. Other process specific candidates for the reuse of treated wastewater include water for slurry coal feeding (i.e., Bi-Gas), water for makeup to slag/ash slurry removal, and makeup quench water. The particular distribution of wastewater for treatment and/or reuse depends on specific plant process characteristics and on water conservation requirements.

Assuming that the bulk of process wastewater will be used as cooling water makeup, two issues must be considered. First, quality constraints on circulating cooling water need to be better understood in order to relate to specific unique characteristics of coal gasification wastewaters. Tolerable levels of BOD, phenol, cyanide, thiocyanate, and ammonia in makeup water must be determined. Some experience in the petroleum industry has been obtained concerning organics in cooling water makeup (Mohler and Clere, 1973), which suggests that the cooling tower may be used as an actual treatment system to remove these organics biologically. If such a scheme appears feasible, the chemistry and/or biology of this treatment must receive careful analysis; the extent of removal of various species including potentially toxic substances must be evaluated. Second, in view of such cooling water constraints, the treatability of process wastewaters to meet these constraints must be proven. Several schemes may be envisioned for removal of total dissolved solids (TDS). Reverse osmosis is generally more economical than ion exchange for TDS greater than 2000 mg/liter (Weber, 1974), but both may be limited by residual organics from the biological oxidation treatment. Carbon adsorption may completely remove these organics, but if these organics can be decomposed in the cooling tower, their removal by carbon adsorption is not necessary. Sand filtration has been shown to remove BOD, TOC, and suspended solids effectively (Mohler and Clere, 1977) and may yield an effluent amenable to either reverse osmosis or ion exchange. If TDS levels do not exceed constraints on the cooling tower, the primary issue becomes one of organic contamination. If organic loadings are too high, some extent of sand filtration and/or carbon adsorption may be necessary.

The ability to characterize coal gasification effluents accurately and to predict scalability relationships between the pilot and commercial plants precludes the accurate interpretation of treatability studies on pilot plant wastewaters. Effluent characterization must be geared toward accurately defining effluent production rates and, therefore, requires the use of proven analytical procedures and a carefully evaluated sampling strategy. Relating these effluent production rates to process gasification parameters serves to assist the task of predicting commercial scale effluent production from pilot plant data. An engineering evaluation of pilot and commercial scale process configurations is necessary to predict effluent distribution and loading accurately. These scalability considerations must be combined with results from treatability studies in order to evaluate wastewater reuse options.

Analysis of these problems for reuse of coal gasification wastewater from the Hygas pilot plant are currently under study at Carnegie-

Mellon University. Preliminary screening studies have been completed, and reports summarizing the results of this and future work should be available in 1979.

ACKNOWLEDGMENTS

The information presented in this paper represents various facets of continuing research performed at Carnegie–Mellon University's Center for Energy and Environmental Studies. The authors' research activities have focused largely on characterization and treatment of wastewaters resulting from coal gasification and coal coking. Support for these activities has been provided at various times by the National Science Foundation, the Department of Energy, and the Department of the Interior—Office of Water Research and Technology. Support for the preparation of this manuscript was provided by the Department of the Interior for R. G. Luthy and by the Department of Energy for R. W. Walters.

REFERENCES

Air Products and Chemicals, Inc. (1971). "Engineering Study and Technical Evaluation of the Bituminous Coal Research, Inc. Two-Stage Super Pressure Gasification Process." Rep. No. 60. Allentown, Pennsylvania.

A.P.H.A. (1975). "Standard Methods for Analysis of Water and Wastewater," 14th Ed. Washington, D.C.

Bartlett, P. D., and Davis, R. E. (1958). *J. Am. Chem. Soc.* **80,** Part II, 2513.

Bostwick, L. E. (1977). "Control Technology Development for Fuel Conversion Systems Wastes," Annu. Rep., Contract 68-02-2198. Environ. Prot. Agency, Washington, D.C.

Braun, C. F., Inc. (1976a). "Commercial Concept Design–Western Coal Conoco CO_2-Acceptor Process." Alhambra, California.

Braun, C. F., Inc. (1976b). "Commercial Concept Design—Western Coal IGT Steam-Oxygen Hygas Process." Alhambra, California.

Chen, K. Y., and Morris, J. C. (972). *Environ. Sci. Technol.* **6,** 529–537.

Crits, G. J., and Glover, G. (1975). *Water Wastes Eng.* **12(4),** 45–52.

Dravo Corporation (1976). "Handbook of Gasifiers and Gas Treatment Systems." Rep. No. FE-1772-11. Dep. Energy, Washington, D.C.

Effluent Guidelines (1974a). "Iron and Steelmaking," 39 FR 24113. EPA, Washington, D.C.

Effluent Guidelines (1974b). "Petroleum," 39 FR 16559. EPA, Washington, D.C.

Effluent Guidelines (1974c). "Steam Electric Power," 39 FR 36185. EPA, Washington, D.C.

Effluent Guidelines (1975a). "Electroplating," 40 FR 18129. EPA, Washington, D.C.

Effluent Guidelines (1975b). "Ore Mining and Dressing," 40 FR 51722. EPA, Washington, D.C.

Effluent Guidelines (1976a). "Coal Mining," 41 FR 19831. EPA, Washington, D.C.

Effluent Guidelines (1976b). "Iron and Steel Making," 41 FR 12989. EPA, Washington, D.C.

EPA (1974). "Methods for Chemical Analysis of Waters and Wastes," EPA-625/6-74-003a. Environ. Prot. Agency, Cincinnati, Ohio.

Goldstein, D. J., and Yung, D. (1977). "Water Conservation and Pollution Control in Coal Conversion Processes," EPA Rep. No. 600/7-77-065. Environ. Prot. Agency, Washington, D.C.

Hill, W. H. (1945). *In* "Chemistry of Coal Utilization" (H. H. Lowry, ed.), pp. 1008–1135. Wiley, New York.

Jablin, R., and Chanko, G. P. (1972). *Meet. AICHE, 65th, New York.*

Johnson, G. E., Neufeld, R. D., Drummond, C. J., Strakey, J. P., Haynes, W. P., Mack, J. D., and Valiknac, T. J. (1977). "Treatability Studies of Condensate Water from Synthane Coal Gasification," Rep. No. PERC/RI-77/13. Pittsburgh Energy Res. Cent., Pittsburgh, Pennsylvania.

Koblin, A. H., and Massey, M. J. (1977). *Proc. Pac. Area Chem. Eng. Congr., 2nd, Denver, Colo.*

Kostenbader, P. E., and Flecksteiner, J. W. (1969). *J. Water Pollut. Control* **41,** 199–207.

Krisher, A. S. (1978). *Chem. Eng.* **19,** 79–98.

Lowry, H. H., ed. (1945). "Chemistry of Coal Utilization," Vols. I and II. Wiley, New York.

Luthy, R. G. (1977). "Manual of Methods: Preservation and Analysis of Coal Gasification Wastewaters," FE-2496-2. Dep. Energy, Washington, D.C.

Luthy, R. G., and Jones, L. D. (1978). "Biological Treatment of Coke Plant Wastewater," Cent. Energy Environ. Stud., Carnegie-Mellon Univ., Pittsburgh, Pennsylvania.

Luthy, R. G., and Tallon, J. T. (1978). "Biological Treatment of Hygas Coal Gasification Wastewater," FE-2496-27. Dep. Energy, Washington, D.C.

Luthy, R. G., Bruce, S. G., Jr., Walters, R. W., and Naklles, D. V. (1977a). to appear in *J. Water Pollut. Control,* Sept. 1979.

Luthy, R. G., Massey, M. J., and Dunlap, R. W. (1977b). *Proc. Purdue Ind. Waste Conf., 32nd, Lafayette, Indiana* pp. 518–536.

McCoy, J. W. (1974). "The Chemical Treatment of Cooling Water." Chem. Publ. Co., New York.

Magee, E. M., Jahnig, C. E., and Shaw, H. (1974). "Evaluation of Pollution Control in Fossil Fuel Conversion Processes; Koppers–Totzek Process," EPA-650/2-74-009a. Off. Res. Dev., Environ. Prot. Agency, Washington, D.C.

Massey, M. J., Luthy, R. G., and Pochan, M. J. (1978). "S Synopsis of C-MU Technical Contributions to the DOE Coal Gasification Environmental Assessment Program," FE-2496-39. Dep. Energy, Washington, D.C.

MITRE Corp. (1979). *Proc. Workshop Health Environ. Eff. Coal Gasif. Liquefaction Technol.,* Report No. DOE/HEW/EPA-03, McLean, Va.

Mohler, E. F., Jr., and Clere, L. T. (1973). *Proc. Nat. Conf. Complete Water Reuse, AIChE, New York* pp. 425–447.

Mohler, E. F., Jr., and Clere, L. T. (1977). *Chem. Eng. Prog.* **73**(4), 74–82.

Moon, R. A., Jr. (1978). Personal communication.

Nakles, D. V. (1977). Ph.D. Thesis, Dep. Chem. Eng., Carnegie-Mellon Univ., Pittsburgh, Pennsylvania.

Nakles, D. V., Massey, M. J., Forney, A. J., and Haynes, W. P. (1975). "Influence of Synthane Gasifier Conditions on Effluent and Product Gas Production," PERC/RI-75/6. Pittsburgh Energy Res. Cent., Pittsburgh, Pennsylvania.

Probstein, R. F., and Gold, H. (1978). "Water in Synthetic Fuel Production." MIT Press, Cambridge, Massachusetts.

Procon, Inc. (1978). "Hygas Demonstration Plant Project: Conceptual Commercial Plant Design Process Flow Diagram," DOE Contract No. EF-77-C-01-2618. Des Plaines, Illinois.

Reap, E. J., Davis, G. M., Duffy, M. J., and Koon, J. H. (1977). *Proc. Purdue Ind. Waste Conf., 32nd, Lafayette, Indiana* pp. 929–943.

Rubin, E. S., and McMichael, F. C. (1975). *Environ. Sci. Technol.* **9,** 112.

Sherwani, J. K., and Moreau, D. V. (1975). "Strategies for Water Quality Monitoring," Project No. A-065-NC. Dep. Environ. Sci. Eng. Univ. of North Carolina, Chapel Hill.

Simon, D. E., II (1975). *Proc. Annu. Water Conf., 36th, Eng. Soc. West. Pa., Pittsburgh, Pa.* pp. 65–68.

Weber, W. J., Jr. (1974). "Physiochemical Processes for Water Quality Control." Wiley (Interscience), New York.

Wilson, P. J., Jr., and Wells, J. H. (1945). *In* "Chemistry of Coal Utilization" (H. H. Lowry, ed.), pp. 1371–1481. Wiley, New York.

Wong-Chong, G. M., Caruso, S. C., and Patarlis, T. G. (1978). *AIChE Nat. Meet., 84th, Atlanta, Ga.*

Chapter 45

Environmental Characterization of Products and Effluents from Coal Conversion Processes

Jonathan S. Fruchter   *Michael R. Petersen*
PHYSICAL SCIENCES DEPARTMENT
BATTELLE
PACIFIC NORTHWEST LABORATORY
RICHLAND, WASHINGTON

I. INTRODUCTION

Utilization of coal for power production is currently undergoing a rapid expansion. In addition, the potential for coal to be converted into "synthetic" liquid and gaseous fuels creates another usage for coal in the future. Both conventional coal technology and the newer coal-conversion processes have a considerable potential for environmental and health-related impacts. In both cases, the physical and chemical analysis of materials involved in the processes is an important part of evaluating the potential for these environmental and health effects. Both the total elemental concentrations and the physical and chemical forms are of importance since these forms determine the ultimate fate and effect of elements and species. In this chapter a number of precise

247

and sensitive techniques for both sampling and analysis of coal-conversion plant feedstocks, effluents, products, and by-products for these elements, heavy metal species, gases, and organic constituents are discussed. A multitechnique approach is used. The methods were chosen both for their sensitivity and, where possible, relative freedom from matrix effects in these complex samples.

II. SAMPLING METHODS

The first step in making a meaningful analysis of a plant effluent is to obtain a representative sample. In many cases, this first step is also the most difficult. Experience has shown that before a successful field trip can be undertaken, a certain basic knowledge of the facility is necessary. The basic structure of the process should be studied, and then the particular aspects of the plant design can be examined. Copies of the process flow diagrams are the best source for such data. When one is familiar with the process, the actual sampling points can be more intelligently determined. After further consultation with the plant personnel, the sampling probes, fittings, etc., can be designed.

On arrival at the plant site, much time can be saved by going over the proposed sampling scheme with the plant engineer using the process flow diagrams. Usually, additional sampling points become apparent, and revision of the sampling scheme has to be made on site in order to get more meaningful samples. The key in any sampling scheme is to obtain the best samples possible. This goal involves thoughtful planning, a good working relationship with the plant operators, and close time coordination during the actual sampling so representative samples can be obtained. These thoughts may be intuitively obvious, but judging from previous experience, misleading samples are often collected from such pilot facilities and the timing and coordination can be difficult to execute in practice.†

In the following discussion, the sampling method will be divided into those appropriate for organic and for inorganic analyses. The methods will be further subdivided according to the type of sample.

A. Gases and Particulates—Inorganic

In many ways, the most difficult samples to obtain are those of the gaseous and particulate effluents. This difficulty is in large part because

† Examples of process flow diagrams and sampling points are given for the Synthoil coal hydroliquefaction process in Volume I, Chapter 18, Fig. 1, and for atmospheric fluidized-bed combustion of coal in Volume II, Chapter 35, Fig. 1, and Table I.

of the wide variety of temperatures, flow rates, pressures, chemical composition, and particulate loadings encountered in various plant streams. The difficulty is further compounded by the chemical and physical instability of many of the constituents in these streams. Depending on the nature of the stream and the constituents to be analyzed, several sampling methods may be employed. Depending on the information needed, various traps and filters can be used separately or can be tied together in a sampling train, such as the SASS train described in Volume II, Chapter 36, Fig. 1.

If particulates such as fly ash or coal particles are to be collected in the stream, a filter must be employed. This arrangement can be either a single filter with a pore size appropriate to the smallest particles of interest or a cascade impactor if particle size distribution is of interest. In either case, it is usually necessary to calibrate the particle size cutoffs for the various filters. If representative particulate samples are required, the flow rate through the filter must be isokinetic with the flow in the process streams. It is also helpful to keep the filter isothermal with the gas stream so that volatile constituents including water vapor do not condense. Thus, it is necessary to incorporate the necessary flow and temperature measuring devices and heaters or coolers into the sampling apparatus. The choice of filter material depends on the particulate loading temperature and type of analysis. Membrane-type filters (Nuclepore, Millipore, Fluoropore, etc.) usually have the lowest trace element blanks but cannot stand high temperatures (>140°C, 75°C, 130°C, respectively) and may clog relatively quickly at high particulate loadings. Glass-fiber filters (IPC, etc.) and metal membrane filters† have much greater high temperature tolerance (some up to 500°C) and clog less easily but tend to have high blanks for a number of trace elements. In conducting environmental analyses around a coal-burning or coal-conversion plant, particulate samples of ambient air are needed as well. Although many of the same considerations apply, the generally milder sampling conditions allow the use of additional filter materials, including paper. The literature on particulate sampling is vast, so that the references given are to only a few of the many articles which have been published on the subject (Anderson, 1966; Flesch *et al.*, 1967; Lundgren, 1970; Natusch and Wallace, 1976).

The nature of samplers which follow the filter depend on the constituents to be determined. For many of the regulated priority pollutants (SO_x, NO_x, CO, etc.), individual gas analyzers are available commercially and will not be further discussed. Many of the more

---

† Such as silver membrane filters, Chapter 43, Section II,B.

stable constituents, such as CO_2, N_2, H_2O, can be grab sampled and returned to the laboratory for analysis. Less stable constituents, however, such as H_2S, mercury species, and carbonyl sulfide should be determined on site before substantial alteration of the sample occurs.

For the collection of trace metal vapors in the gas stream, either impingers (gas bubblers) or specialized traps can be used.† Experience with this type of sampling is, as yet, not nearly so complete as with the collection of particulates and aerosols. No general method applicable to all possible species and conditions has yet been developed and validated. When a broad spectrum first look for trace element vapors is desired, a combination of acid, base, and water impingers is generally employed. These general impingers will usually trap some of the constituents in the gas stream, but they are frequently not quantitative and this collection method does not preserve the more unstable species. The lack of quantitative trapping may be because of the high flow rates frequently necessitated by isokinetic particle sampling requirements. Sometimes it is desirable to sample for particles and vapors separately to alleviate this problem. In this case, a filter should be placed in front of the vapor trap to avoid contamination by particles.

Although considerable work remains to be done on general and specific trace element and speciation traps, in some of the more important cases considerable progress has been made. Thus, for example, specific traps for mercury, arsenic, fluoride, ammonia, SO_x, NO_x, etc., have been developed. Figure 1 shows an example of a sampling train for arsenic species used in the authors' studies on coal-fired and coal-conversion plants. The As traps are adapted from those developed by

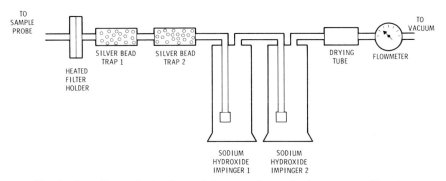

Fig. 1 Sampling train for determining arsenic species in gaseous effluents.

† An example of impinger collection of vapor state metals is given in Chapter 41, Section V,A,3a.

Johnson and Braman (1975). In this sampling train the oxidized forms of arsenic vapor are trapped in the sodium hydroxide impinger after passing through the filter. Arsine and methyl arsines pass through the impingers and are trapped on the silver-coated quartz or pyrex glass beads. Soft glass is avoided in both the beads and the collection apparatus because of the high arsenic blank. A mercury speciation train which works in a similar fashion can also be employed. Each trap successively captures a specific chemical form, and the other forms pass through to subsequent traps (Braman and Johnson, 1974). All forms of mercury studied to date amalgamate at temperatures less than 150°C with the gold beads in the last trap. Hence this trap can also be used by itself to determine total mercury. After sampling, the mercury is removed from each trap by heating at about 400°C and determined by flameless atomic absorption. If large amounts of organics are present in the gas, they may interfere with the mercury determination. In such cases a simple heat transfer of the absorbed mercury to a second gold tube may eliminate the interference. In more difficult cases the vapors must be passed through an alumina trap which removes the organics but passes the mercury.

B. Gases and Particulates—Organic

In gas streams, where the components of interest exceed a concentration of 1 ppm and are reasonably stable, suitable samples are collected using evacuated perfluoroethylene (PFE) coated stainless steel bottles. Such samples can be taken back to the laboratory for gas chromatographic analyses. This collection technique is useful for hydrocarbon analysis but is questionable if the gases are sulfides, nitriles, or other similar type compounds. In such cases, on-line real-time analysis may be the best procedure. We have tried other collection techniques such as cold trapping or adsorption trapping. Open pore polymeric resins are reportedly also suitable for subparts per million concentrations of volatile organic compounds (Pellizzari, 1977). Our experience with these techniques has met with marginal success, and each technique has certain problems. The cold traps require a supply of coolant which, in the field, is difficult to obtain and handle. On thawing, the resulting samples contain considerable water which poses problems for trace determination of volatile compounds. The polymeric adsorbent traps may require long sampling times. Contamination in the polymers themselves limit their usefulness. In addition desorption of the adsorbed sample requires a great deal of finesse and skill for reproducible results. Hence grab sampling has been our mainstay collection technique. In order to increase sensitivity of grab sampling we use up to 2 milliliter of sample for the gas chromatography analysis.

Particulate samples were collected using glass-fiber filters and high volume air samplers. These samplers were placed at various sites around the plants to collect particulate matter for analysis of organic content. The glass-fiber filters were first cleaned by heating at 400°C for several hours and then wrapped and stored in aluminum foil. After collection of the sample, the filter was placed inside the foil again. Some stack type sampling has been done using in-line cyclone separators for the larger size particles. These samples are readily stored under contamination-free conditions in clean press-top cans. Solids were collected as grab samples and stored in the same type cans. Composite samples for time periods up to 24 hr were occasionally collected. All samples, both solid and particulate, were stored at 4°C until the analysis could be performed.

C. Liquid and Solid Samples

The sampling of liquids and solids is considerably easier than gases, although not without some pitfalls.† The major problems in the case of liquids and solids are (1) to get a representative sample from a fairly large amount of material and (2) to avoid contamination and degradation. Many plants of both production and pilot variety have provision for obtaining such samples regularly, as they are frequently needed for process control. These plant samples are often suitable for analysis; otherwise obtaining representative samples may be very time consuming.

For purposes of avoiding contamination in inorganic samples, either PFE or pyrex glass containers are generally best. For the purpose of preventing degradation, liquid samples must usually be split, because the same method does not usually work for all cases. Samples for metals which easily form hydroxides should be acidified; those which form acids should be kept at high pH. Samples should be kept in the dark as much as possible. In the case of solid samples, the final remaining problem is to avoid contamination during grinding, splitting, or other preparation. For organic samples it is best to avoid use of polyethylene or polypropylene since plasticizers continue to leach out of these materials even after cleaning. For certain solid or gas samples, metal is a suitable container, especially if coated with perfluoroethylene polymers. Clean PFE or glass containers are best for collection of trace

† An example of sampling of liquids and solids from fluidized-bed combustion processes is given in Volume II, Chapter 36.

organic samples. However, one must remember that adsorption phenomena may lower the observable organic concentrations in the bulk sample so a thorough solvent rinsing of the container after sample removal is wise. Adsorption of gas components onto container walls should always be checked in any gas analyses. All solvents used in the workshop procedure should be checked for purity and redistilled if necessary.

To minimize degradation we have found that liquid samples are much more stable if kept cold (or frozen) under an inert atmosphere and in the dark. For example, certain light distillates from a coal liquefaction process darken quickly in air at room temperature but appear stable for months when stored as described above. Many organic compounds in water samples are unstable and can be stabilized several ways depending on the compounds present. Dilute solutions of phenols can be biologically degraded, and these samples are preserved by adding a few milliliters of dichloromethane per liter of sample. Other water samples can best be stabilized for about one–two weeks by storing them near 0°C until analysis. Organic liquids were collected in metal (PFE lined) or glass containers filled to the top to exclude air and sealed with PFE-lined caps.

III. ANALYTICAL METHODS

A. Inorganic

Many of the samples obtained from coal-conversion plants are chemically complex, creating the potential for matrix effects in many of the commonly used methods for chemical analysis. Therefore, the techniques employed for inorganic analysis of these materials were chosen when possible for their relative freedom from matrix effects as well as their sensitivity and precision. Because no one method can at this time meet all of these requirements, a multitechnique approach was used as described below.

1. Neutron Activation

In complex substances such as are found in coal wastes and coal-conversion products, neutron activation analysis (NAA) is the method of choice for many trace element analyses because of its relative freedom from matrix effects.

The NAA method can be divided into Instrumental NAA (INAA), and Radiochemical NAA (RNAA). In RNAA, the various neutron-induced products are chemically separated to minimize interferences, whereas in INAA the sample is merely irradiated and counted. Only the instrumental method will be discussed here. There are several comprehensive review papers on INAA published in the literature (Gordon *et al.*, 1968; Soete *et al.*, 1972; Laul and Schmitt, 1973; Abel and Rancitelli, 1975). (Basic INAA procedures are given in Volume I, Chapter 12). The basic parameters controlling sensitivity for a multiele-ment determination are neutron flux, irradiation time, delay interval prior to counting, half-life and γ-ray energy of the induced activity, and efficiency and resolution of the counter. In our scheme, two se-quential irradiations are employed along with counts at various inter-vals to determine radionuclides with various half-lives corresponding to the elements of interest. The final count occurring 40–50 day after the second irradiation is performed on an anticoincidence-shielded Ge(Li) system recently developed in our laboratory.

The anticoincidence counter works in the following way. When an isotope emits two or more γ rays in cascade, if they are simultaneously detected by the NaI(Tl) anticoincidence shield and the Ge(Li) diode, the coincidence events are stored in the first half of a 4096-channel memory while the single γ ray or events in noncoincidence are stored in the second half of the memory. Figure 2 displays coincidence and noncoincidence spectra of the geological standard rock BCR-1. The great advantage in noncoincidence lies in the fact that the Compton continuum (a background interference) is reduced by an order of mag-nitude for low and medium γ-ray energies. Thus, peak/Compton ratios and sensitivities for many elements such as Cr, Sr, Ba, Nd, Rb, Zn, Se, and Sb in noncoincidence spectra, usually low in normal Ge(Li) count-ing, are greatly increased. The 1116 keV peak of Zn was easily measured by the reduced Compton edge of the 1121 keV peak of ^{46}Sc in noncoin-cidence. The 146 keV peak of ^{141}Ce, unlike in normal Ge(Li) counting, contained relatively little contribution from the 143 keV peak of ^{59}Fe. The elements Ni via ^{58}Co and Zr previously not observed in our normal Ge(Li) counting were easily measurable in noncoincidence counting. The 757 keV peak of Zr and the 811 keV peak of Ni have interferences from ^{152}Eu. The contribution amounted to 0.71% for the Zr peak and 1.0% for the Ni peak relative to the 122 keV peak of ^{152}Eu. Overall, the accuracy and relative precision for a number of elements was improved by factors of from three to ten by noncoincidence counting relative to the normal Ge(Li) counting.

2.   X-Ray Fluorescence

Energy dispersive x-ray fluorescence (XRF) spectroscopy with zircon-ium and silver secondary source targets was used to measure certain

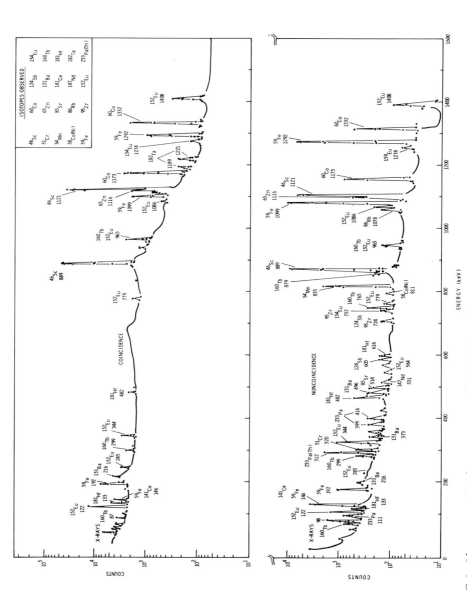

Fig. 2   Coincidence-noncoincidence spectra of U.S. Geological Survey Standard Basalt BCR-1. USGS BCR standard: 0.50g, neutron fluence: 1×10^{17}, decay interval: 50 days, count time: 30 min, Ge(Li) diode 26% coaxial, shielding-Pb plus 30 in. diameter plastic anticoincidence shields.

trace elements where greater sensitivity or accuracy could be obtained
by this method. An example of the difference in spectra obtained by
the two secondary sources is shown for a spent oil shale sample in
Figs. 3 and 4. Some trace elements were measured by both NAA and
XRF to provide a check on the consistency of the two analytical methods. The XRF determination of low atomic number elements is quite
sensitive to the sample matrix since this affects the degree of photon
attenuation. However, the matrix effect diminishes for higher atomic
weight elements. Sample preparation technique for XRF has been described previously by Giauque *et al.* (1973), although some improve-
ments have been made since that publication. For solid samples, the
powdered sample is mixed with cellulose powder and pressed into
uniform pellets. Aqueous samples can be run by freeze or air drying
them with cellulose and pressing them into a pellet or simply evaporating the sample onto a filter paper. Organic liquids can be prepared
by evaporating them on a thin sheet of $\frac{1}{8}$ mil polypropylene. In many
cases, filter samples can be run directly. Otherwise, the filter may be
dissolved or ashed in order to concentrate the particles. The XRF data
are processed using a computer code described by Nielson (1977). This
code uses the coherent and incoherent x-ray scattering peaks to independently calibrate the mass and mass absorption of the sample. (Energy dispersive x-ray fluorescence is discussed further in Chapter 49.)

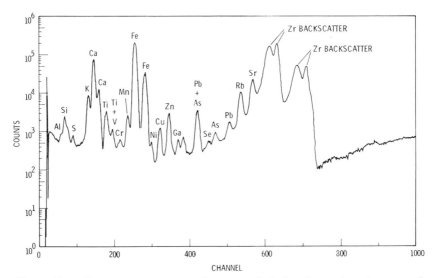

Fig. 3 X-ray fluorescence spectrum of a spent oil shale using a zirconium secondary
source.

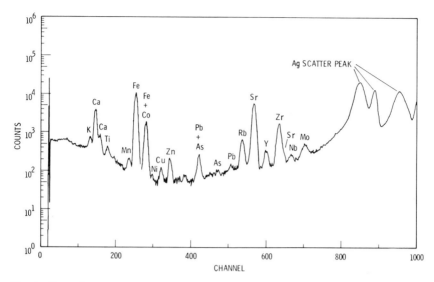

Fig. 4   X-ray fluorescence spectrum of a spent oil shale using a silver secondary source.

3.   *Chemical Speciation Methods*

Our analysis of the trace elements in gaseous effluents is accomplished by collecting the various species on selected sorption beds within a gas-sampling train as discussed in the section on sampling. After collection, the sorption traps containing Hg are analyzed by flameless atomic absorption spectroscopy (see Volume I, Chapter 14, Section II,B,2). Similarly, arsenic as arsine, the oxide and methylated species, can be absorbed on a series of traps. The arsenic compounds can then be removed, reduced, and analyzed by dc plasma emission spectroscopy.

B.   Organic

In our laboratory we have tried various isolation procedures but have settled generally on liquid–liquid extractions for water samples and Soxhlet extractions for solids and particulates. (For some organic liquids and water samples we analyze the sample without any separation or pretreatment.) The extracts can then be further partitioned into acidic, basic, polynuclear aromatic and neutral fractions for individual class analyses. The partitioning scheme uses 1N NaOH, 1N HCl, and di-methylsulfoxide to separate the above compound classes.

1. Gas Chromatography/Mass Spectrometry†

The analytical instrumentation most useful for organic analytical work is gas chromatography/mass spectrometry (GC/MS) and high performance liquid chromatography (HPLC). Both techniques are versatile and allow the analyst to look at volatile and nonvolatile organic compounds. Gas chromatography is used as the initial screening procedure to determine the separation properties of the mixture. Using CG/MS we then identify the compounds of interest and, if the chromatographic separation is good, quantitation is performed on the simple GC without using GC/MS time. The mixtures of organic compounds are often very complex, and the data reduction techniques of the GC/MS data system are called on to identify and quantify the components. Selected ion monitoring, reconstructed mass chromatograms, and other instrumental techniques such as chemical ionization and capillary column chromatography are used to obtain more information about the composition of the mixtures. The GC/MS data system is the most useful current instrumentation in our work with coal conversion processes.

2. High Performance Liquid Chromatography

High performance liquid chromatography (HPLC)‡ is extremely useful for analyses of the more polar compounds such as phenols or acids in samples collected at coal conversion facilities. We have developed a direct analytical technique for phenolic compounds (down to about 10 ppb) in water for use without any sample treatment (Fruchter *et al.*, 1977). In many of the complex samples HPLC is unable to resolve individual compounds but is used to separate fractions which can be further investigated by other techniques such as GC/MS. We feel that suitable HPLC fractionation schemes will eventually replace the bench partitioning schemes now used for polar compound separation. The usefulness of HPLC in identification will expand as new selective detectors are developed and efficient mass spectrometer interfaces evolve.

IV. APPLICATIONS

A. Coal Liquefaction—Refined Solids Process

For the purposes of application we will consider a coal liquefaction process and a coal gasification process. The liquefaction process ac-

† For details on GC/MS see Volume I, Chapter 17, Section II,C, and Volume II, Chapter 21, Section II,B.

‡ Also see Chapter 43, Section III,C.

tually converts high-sulfur coal into a low-sulfur, low-ash fuel, (solvent refined coal—SRC). Depending on the process conditions, the fuel produced may be either a solid (SRC-I) or a liquid (SRC-II). The plant has been successfully operated in both modes for several years producing either a black asphaltic type solid or heavy distillates equivalent to about No. 5 fuel oil. The products have met both the sulfur and ash specifications of the original plant design.

In the process, pulverized bituminous coal is mixed with a process solvent and heated under about 1500 psi hydrogen. Most of the coal dissolves leaving the inorganic material in suspension. The slurry is then filtered to remove the undissolved mineral matter (mineral residue), and the filtrate is flash distilled to remove the solvent leaving a black liquid which on cooling solidifies. In the fuel oil mode some of the slurry is flash distilled leaving the mineral matter in the vacuum bottoms. In both modes the distillate is further fractionated into light, medium, and heavy cuts. The light and medium cuts are treated byproducts. The heavy cuts are either recycled as process solvent (solid product mode) or used as the fuel oil-type product (liquid mode). We have sampled the process many times in both modes of operations and collected various samples of feed coal, the by-product liquids, the mineral residue, solid product, liquid products, airborne particulates, process off gases, and process water and treated effluent water.

Some data which we have collected are shown in Tables I–V. These tables show major and trace element data for samples of feed coal, mineral residue, product solids and liquids, process liquids, particulates, and effluent gases taken from the coal liquefaction plant. In Fig. 5 we show mass balances for six elements in the solid product mode. The balances are only approximate since operating parameters are varied from run to run. Except for mercury, titanium, and bromine, most elements appear to remain with the mineral residue. For the case of bromine, approximately 84% remains with the product, whereas for titanium, approximately 56% remains with the product. In the case of mercury, 89% is unaccounted for in the solid and liquid products and is presumably emitted in the process offgas.

Mercury measurements were made in the offgas stream from the process (liquid mode) and are shown in Table IV. Because of uncertainties in the volume of gas in this process, we are unable at this time to determine whether or not all of the mercury can be accounted for in this stream. The mercury was apparently removed by the sulfur recovery unit at the plant, as no mercury could be detected downstream from this unit or in the sulfur. The scrubber solution was found to contain considerable concentrations of mercury. Several arsenic species

TABLE I Inorganic Element Content of Solid Samples from Coal Liquefaction—Refined Solids Process[a,b]

	Feed coal (solids process)		Mineral residue (solids process)		Solid product		NBS coal std.	
	INAA	XRF	INAA	XRF	INAA	XRF	INAA	XRF
Na	180 ± 4		3150 ± 60		8.8 ± 1.5		420 ± 20	
Cl		276 ± 30		1020 ± 50			850 ± 150	
K	1200 ± 0.02	1000 ± 100		1.16% ± 0.05	115 ± 20		0.27% ± 0.02	
Ca	—	0.069% ± 0.010		1.15% ± 0.05				0.42% ± 0.05
Ti		620 ± 20		2300 ± 200		520 ± 30	0.11 ± 0.02	0.13 ± 0.02
V	29 ± 2	25 ± 3		180 ± 20		10 ± 1	35 ± 4	38 ± 4
Cr	18 ± 2	17 ± 2	150 ± 3	161 ± 10	7.5 ± 2.1	6.0 ± 2.2	19 ± 2	18 ± 2
Mn	35 ± 2						42 ± 6	
Fe	2.29% ± 0.12	2.10% ± 0.15	13.3% ± 0.5	14.5% ± 0.3	270 ± 30	300 ± 40	0.81% ± 0.05	0.78% ± 0.035
CO	5.2 ± 0.2		30 ± 1		0.26 ± 0.04		5.2 ± 0.1	
Ni	21 ± 2	23 ± 1	120 ± 5	115 ± 5	<6	2.1 ± 0.2	16 ± 2	14 ± 2
Cu		10 ± 1		126 ± 10		0.8 ± 0.1		15 ± 2
Zn	26 ± 2	17 ± 2	120 ± 15	138 ± 20	8.1 ± 0.5	7.2 ± 0.8	37 ± 3	32 ± 3
As	19 ± 1		77 ± 5	75 ± 5	3.11 ± 0.15	1.8 ± 0.2	5.7 ± 0.5	6.1 ± 0.5
Se	3.6 ± 0.2		20 ± 2	15 ± 3	0.17 ± 0.03	—	3.3 ± 0.3	3.1 ± 0.4
Br	3.6 ± 0.3	3.9 ± 0.3	4.9 ± 0.4	5.2 ± 0.4	4.7 ± 0.4	4.8 ± 0.4	17 ± 2	17 ± 2
Rb	13 ± 1	15 ± 1	82 ± 10	96 ± 5		<0.4	19 ± 2	18 ± 1
Sr	59 ± 15	140 ± 10	310 ± 30	325 ± 15		<2	170 ± 20	150 ± 20
Sb	1.4 ± 0.1		7.4 ± 0.3		0.96 ± 0.16		3.7 ± 0.3	
Ba	46 ± 5		240 ± 25		0.066 ± 0.009		390 ± 40	
La	8.9 ± 0.5		46.8 ± 1.2		0.14 ± 0.03		10.5 ± 0.2	
Hf	0.44 ± 0.02		2.3 ± 0.2		0.10 ± 0.02		0.97 ± 0.09	
Hg	0.16 ± 0.01		<0.005		0.05 ± 0.005		0.12 ± 0.02	
Pb		8 ± 1		42 ± 3	0.020 ± 0.005	<1		<11
Th	1.9 ± 1		9 ± 1		0.19 ± 0.03		3.4 ± 0.3	
U	1.1 ± 0.1		7.3 ± 0.3		0.54 ± 0.05		1.6 ± 0.2	

[a] SRC-I.
[b] Measurements in parts per million except as noted.

TABLE II *Inorganic Element Concentrations in Solid Samples from Coal Liquefaction (Liquid Product) Process*[a,b]

	Feed coal XRF	Slag XRF
Si	3.1% ± 0.4	5.4% ± 0.5
P	<0.18%	<0.18
S	3.8% ± 0.3	3.2% ± 0.3
Cl	<0.037	0.16% ± 0.2
K	2000 ± 100	3600 ± 300
Ca	0.67% ± .05	1.14% ± 0.08
Ti	800 ± 100	1400 ± 100
Cr	18 ± 3	38 ± 5
Mn	19 ± 3	52 ± 5
Fe	1.3% ± 0.1	3.3% ± 0.2
Ni	10 ± 1	22 ± 2
Cu	6 ± 1	15 ± 1
Zn	17 ± 1	25 ± 2
As	6.9 ± 0.6	15 ± 1
Se	1.3 ± 0.2	2.4 ± 0.3
Br	15 ± 1	25 ± 2
Rb	12 ± 1	24 ± 2
Sr	142 ± 8	268 ± 16
Hg	0.16 ± 0.01	<0.01
Pb	2.5 ± 0.1	3.7 ± 0.7

[a] SRC-II.
[b] Measurements in ppm except as noted.

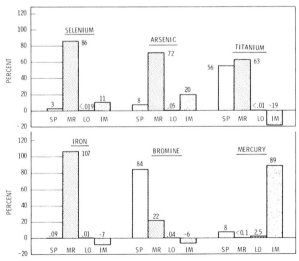

Fig. 5 Mass balances for six elements in a coal liquefaction–refined solid process (SRC-I). SP, solid product; MR, mineral residue; LO, light oil; IM, imbalance.

TABLE III Inorganic Element Content of Liquid Samples from Coal Liquefaction—Refined Solid Process[a]

	Light oil (naphtha) (solid product process)[b]		Process solvent (solid product process)[b]		Light oil (naphtha) (liquid product process)[c]	Heavy oil (liquid product)[c]
	INAA	XRF	INAA	XRF	XRF	XRF
Na	1.7 ± 0.03		0.51 ± 0.01		1600 ± 500	3650 ± 700
S					<55	<55
Cl		<36			<34	102 ± 17
K		<3				36 ± 5
Ca		<3		<67		12 ± 3
Ti	1.8 ± 0.7	2.2 ± 0.15		9.3 ± 3.1	<0.5	6.3 ± 1.8
V		<0.6		<3.2	<6	<1.1
Cr	0.044 ± 0.009	<0.2	0.54 ± 0.02	10.5 ± 1.5	<3	
Mn		0.21 ± 0.06		<1.11	<1.8	660 ± 45
Fe	1.1 ± 0.3	<2	58 ± 2	102 ± 71	<1.6	
CO	0.0026 ± 0.0005		0.0044 ± 0.001			5.0 ± 0.5

Ni	<0.10	<0.05		1.35 ± 0.2	<0.63	0.80 ± 0.30
Cu		<0.07		<0.4	<0.71	1.9 ± 0.2
Zn	0.58 ± 0.04	0.46 ± 0.07	0.86 ± 0.04	0.54 ± 0.14	<1.68 ± 0.35	0.34 ± 0.13
As	0.013 ± 0.001	0.01 ± 0.02	0.105 ± 0.002	<0.23	<0.38	0.31 ± 3.10
Se	0.065 ± 0.020	0.086 ± 0.030	<0.009	<0.15	0.60 ± 0.20	<0.26
Br	0.018 ± 0.002	<0.07	0.25 ± 0.01	0.29 ± 0.05	<0.61	<0.26
Rb	<0.05		<0.06	<0.26	<0.61	
Sr		<1.1		<2.4	<5	<2.4
Sb	<0.002		0.006 ± 0.09			
Ba	<1		<2			
La	<0.001		0.0125 ± 0.008			
Hf	<0.003		<0.003			
Hg	0.05 ± 0.01		<0.01			
Pb		<1		<1.7	<1.7	<76
Th	<0.0009		<0.001			

[a] Measurement in ppm except as noted.
[b] SRC-I.
[c] SRC-II.

TABLE IV *Mercury in Gas Samples from Coal Conversion Processes*

Gasification process		Liquefaction—refined	
	Total Hg		Total Hg
Offgas from coal preheater	10.3 ng/liter	Untreated process gas	20 ng/liter
Untreated product gas	7.1 ng/liter	Scrubbed process gas	<0.4 ng/liter
Scrubbed product gas	<0.2 ng/liter	Scrubber solution	2600 ng/liter

were also measured in the offgas stream. Other data on trace element behavior in this process may be found in Filby *et al.* (1978).

The data presented in Table VI are for the two major fractions, the polynuclear aromatic hydrocarbons (PAH) and the neutral (*n*-alkane) fractions, which are obtained during our partitioning scheme. The concentration of low molecular weight members of both fractions are misleading because their loss in the scheme has not been corrected in this table. As explanation for the table, the light oil, wash solvent, and process solvent are three distillate cuts; the raw process water, mineral residue, solvent refined coal, and particulate filter were extracted before analysis. The particulate filter sample was collected directly over the cooling product before any in-line devices for aerosol control.

Some examples of gas chromatograms of these types of mixtures and their fractions are shown in the following set of gas chromatograms. In Fig. 6 is a chromatogram of a wash solvent sample (380°–480°F). Aromatic compounds are the major components in this distillate cut. Note also that phenolic and aliphatic compounds are also present in the mixture. Diphenyl-ether and some biphenyl are probably contaminants from the process heat exchange liquid.

In Fig. 7 the fraction of the process solvent containing the nitrogen bases is depicted. Pyridines are missing in the fraction because the

TABLE V *Arsenic Speciation Measurements in Gas Samples from a Coal Liquefaction—Refined Solids Process*

	Total arsines	Total vapor As_2O_3	Total vapor methylated arsenic compounds	Total particulate arsenic
Untreated process gas	84 ng/liter	30 ng/liter	<2 ng/liter	<0.2 ng/liter
Scrubbed process gas	0.4 ng/liter	16 ng/liter	<2 ng/liter	<0.2 ng/liter

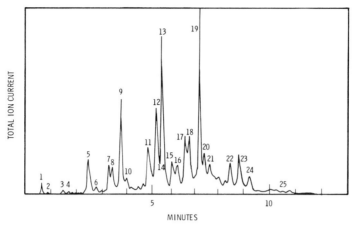

Fig. 6 Composition of wash solvent (bp 380–480°F). (1) Benzene, (2) toluene, (3) xylene, (4) xylene, (5) phenol, (6) C_3-benzene, (7) indan, (8) cresol, (9) cresol, (10) methylindan, (11) C_2-phenol, (12) C_2-phenol, (13) naphthalene, (14) tetralin, (15) C_3-phenol, (16) C_3-phenol, (17) C_3-phenol, (18) dimethylindan, (19) 2-methylnaphthalene, (20) 1-methylnaphthalene, (21) tridecane, (22) biphenyl, (23) diphenyl ether, (24) tetradecane, (25) dibenzofuran.

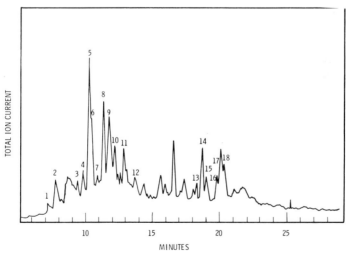

Fig. 7 Composition of the basic fraction and process solvent. (1) C_4-indolene, (2) methyltetrahydroquinoline, (3) C_2-quinoline, (4) methylquinoline, (5) 2-methylquinoline, (6) quinoline, (7) methylquinoline, (8) C_2-quinoline, (9) methylquinoline + ?, (10) C_3-quinoline, (11) 2-ethylquinoline, (12) C_4-indole, (13)–(15), benzoquinolines, (16)–(18), methylbenzoquinolines.

TABLE VI   *Analyses of Samples Collected at Solvent-Refined Coal Plant[a]*

PAH Fraction	Light oil	Wash solvent	Process solvent	Raw process water	Mineral residue	Solvent-refined coal	Particulate filter (concentration $\mu g/m^3$)
xylene	9800	1300					
o-ethylbenzene		1700					
m/p-ethylbenzene		700					
C_3-benzene	3900	1500					
C_4-benzene	4300	500					
indane	510	13000		15	85		
methylindane	180	2500			25		
methylindane	240	1400			55		
methylindane	<5	2300			25		
dimethylindane	330	40			110		
tetralin	<5	4100		<0.1	35		
dimethyltetralin	110	1500		0.5	50		
6-methyltetralin		3200					
naphthalene	1630	32000	100	5	1500	1	3
2-methylnaphthalene	690	32000	3800	2	740	8	16
1-methylnaphthalene	110	12000	930		180	5	4
dimethylnaphthalene	80	13000	11200	0.3	260	6	170
dimethylnaphthalene	70	700	1700		60	3	20
dimethylnaphthalene		4000	4200		150		
dimethylnaphthalene	10	160	650		<2		10
2-isopropylnaphthalene		40	50	2	<1		<0.5
1-isopropylnaphthalene		210	1400	0.7	15		20
C_4-naphthalene		5	50	2	<1		4
cyclohexylbenzene		410			5		
biphenyl	80	10000	5900	0.2	270	2	75
acenaphthylene	2	500	3400	<0.1	45	8	60
dimethylbiphenyl	15	35	2100	0.5	30	9	130
dimethylbiphenyl	21	30	560	0.2	20	7	40
dibenzofuran	8	400	5800	0.6	60	9	160
xanthene	10	30	840	0.1	20	5	40
dibenzothiophene	3	50	4200	1.5	70	30	180
methyldibenzothiophene		15	320	<0.1	8	4	60
dimethyldibenzothiophene	5	15	1200	<0.05	20	13	130
thioxanthene		2	3300	0.1	5	3	120
fluorene	15	250	6600	0.3	80	27	200
9-methylfluorene	15	110	3100	0.3	40	11	150
1-methylfluorene	10	10	3000	0.2	50	18	100
anthracene/phenanthrene	25	130	23000	1.1	500	300	1500
methylphenanthrene	6	15	6200	0.3	100	50	400
1-methylphenanthrene	6		3900	0.2	50	30	300

C₂-anthracene	30	1	10	< 0.05	500	25	6
fluoranthene	700	180	200	0.4	10500	35	15
dihydropyrene	30	1	10	< 0.05	1200	25	6
pyrene	900	280	200	0.6	11200	40	20
Neutral fraction (n-alkanes)				2.3			
n-octane						900	16000
n-nonane						2700	8700
n-decane						5000	9800
n-undecane					50	8300	3900
n-dodecane					80	21000	1400
n-tridecane	**4**	4	90	0.3	30	14000	470
n-tetradecane	12	10	550	0.3	340	11000	170
n-pentadecane	18	8	9100	0.4	1000	4000	60
n-hexadecane	50	7	210	0.3	2000	400	10
n-heptadecane	35	12	80	0.2	3100	120	10
n-octadecane	18	8	50	0.02	920	40	
n-nonadecane	30	3	20		800	500	
n-eicosane	20	3	10		930		
n-heneicosane	35	22	16		600		
n-docosane	55		14		670		
n-tricosane	35		14		980		
n-tetracosane	45		16		900		
n-pentacosane	43		14		740		
n-hexacosane	40		10		450		
n-heptacosane	25	5	8		300		
n-octacosane	28	2	6		150		
n-nonacosane	18	2	5		90		
n-triacontane	22	1	4		60		
n-hentriacontane	15	1	2		40		
n-dotriacontane	11	1	1		10		
n-tritriacontane	7		< 1		5		

[a] Concentration in parts per million.

boiling range for the process solvent is 480°–850°F, but indolenes, quinolines, acridines, indoles, and benzoquinolines are present. Some PAH compounds have been inadvertently carried over in this fraction.

In Fig. 8 a chromatogram of a methylene chloride extract of the raw process water is shown. The major extractable components are phenol, cresols, and xylenols. However, many hydrocarbons, such as the aromatic compounds found in the process liquids, are also present.

In Fig. 9 a chromatogram of an extract of a filter particulate collected in the product cooling building is depicted. The major components are volatile aromatic compounds which evolve from the hot product. Note that phenanthrene is a major constituent in these aerosols.

The gases evolved from the process are separated from the slurry as the slurry is depressurized. In a commercial plant these gases, after cleanup, would supply process heat, but in the pilot operation after removal of the acid gases, the mixture is flared. The major organic components of these gases consist of aliphatic compounds, both straight and branched members, methylated cyclohexanes, benzene, toluene, xylenes, C_3-benzenes. We have also tentatively identified carbon disulfide and pyrrolidine in the gases.

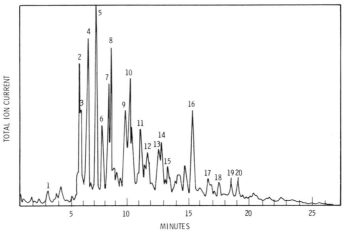

Fig. 8 Extractable organic compounds found in process water before any treatment. (1) indan, (2) phenol, (3) tetralin, (4) cresol + naphthalene, (5) cresol, (6) xylenol, (7) methylnaphthalene, (8) xylenol, (9) biphenyl + C_2 naphthalene, (10) phenyl ether, (11) acenaphthene, (12) dibenzofuran, (13) fluorene + C_4 naphthalene, (14) methylfluorene, (15) xanthene, (16) phenanthrene, (17) methylphenanthrenes, (18) carbazole, (19) fluoranthene, (20) pyrene.

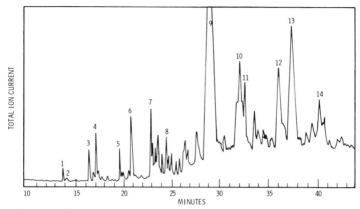

Fig. 9 Composition of aerosols from cooling refined solid product. (1) 2-Methyl-naphthalene, (2) 1-methylnaphthalene, (3) biphenyl, (4) phenyl-ether, (5) acenaphthalene, (6) dibenzofluran, (7) fluorene, (8) dimethylbiphenyl, (9) phenanthrene, (10) methyl-phenanthrene, (11) methylphenanthrene, (12) fluoranthene, (13) pyrene, (14) methylpyrene.

B.   Applications—Coal Gasification Process

The gasifier process studied is a second generation gasification process (carbon dioxide acceptor process) designed to produce synthetic natural gas from lignite rank coals. Pulverized coal is fed into a fluidized bed gasifier containing calcined dolomite (or limestone). During gasification, the dolomite accepts the carbon dioxide and some of the sulfur species, and the ash fuses onto acceptor particulates. The dolomite bed is constantly recirculating to a regenerator where the acceptor is recalcined for recirculation back into the gasifier.†

We participated in several sampling trips to this plant. Various samples were obtained during two runs. These samples included (1) quench waters from the regenerator and gasifier quench systems, preheater venturi scrub water, and waste pond water, (2) gas samples from before and after the gasifier quench, from after the regenerator quench, and from preheater vent gases, and (3) solid samples of gasifier overhead fines, airborne particulates, and spent acceptor.

Trace and major element data for samples of the feed coal, gasifier fines, regenerator fines, spent acceptor, and three scrubber waters are shown in Tables VII and VIII. Most of the mineral matter in the coal ends up in the spent acceptor. However, some of the more volatile

† Flow diagrams for the CO_2-Acceptor process are given in Chapter 44, Figs. 6 and 18, the latter showing sampling locations.

TABLE VII *Inorganic Element Concentrations in Solid Samples from CO_2-Acceptor Coal Gasification Process[a]*

	Feed coal		Gasifier fines		Lime (Spent acceptor)		Regenerator Fines	
	INAA	XRF	INAA	XRF	INAA	XRF	INAA	XRF
Na	0.40% ±0.01		0.24% ±0.01		640 ±30		0.41% ±0.01	
Si		5800 ± 1650		—		<1.4%		<2.7%
P		<1300		1.05% ± 0.19		3.4%		2.8% ± 0.4
S		0.60% ± 0.05		1.65% ± 0.14		0.91% ± 0.10		3.7% ± 0.3
Cl		<160		<380		<500		1000 ± 400
K		55 ± 15		470 ± 60		<300		3500 ± 300
Ca		1.61% ± 0.10		12.7% ± 0.9		47%		35.9% ± 24
Ti		550 ± 40		2590 ± 180		1200 ± 90		4300 ± 280
V				100 ± 30				114 ± 25
Cr	9.5 ± 0.2		32 ± 1	28 ± 2	28 ± 7		82 ± 5	85 ± 10
Mn		31 ± 1		170 ± 13		202 ± 15		180 ± 17
Fe	0.27% ± 0.01	0.290% ± 0.020	2.07% ± 0.05	2.28% ± 0.14	1.15% ± 0.01	1.34% ± 0.09	1.7% ± 0.2	1.8% ± 0.1
CO	1.4 ± 0.1		9.6 ± 0.2		4.2 ± 0.1		24.8 ± 0.4	
Ni	4.9 ± 0.6	5.7 ± 0.5	42 ± 5	44 ± 3	53 ± 8	60 ± 5	68 ± 8	67 ± 6
Cu		7.8 ± 0.3		67 ± 5		50 ± 4		69 ± 7
Zn		5.7 ± 0.3		38 ± 3		59 ± 4		58 ± 4
As	1.2 ± 0.1	1.3 ± 0.1	22.5 ± 0.5	8.6 ± 1.4	4.6 ± 0.1	5.2 ± 0.8	8.3 ± 0.3	7.6 ± 0.7
Se		1.91 ± 0.06		3.0 ± 0.4		2.8 ± 0.6		13.5 ± 1.1
Br	1.09 ± 0.08	1.0 ± 0.1	5.7 ± 0.3	4.8 ± 0.5	3.1 ± 0.1	2.4 ± 0.5	27.1 ± 0.4	28 ± 2
Rb	<2	1.3 ± 2	5.1 ± 1.0	5.7 ± 0.6		3.4 ± 0.7		17 ± 1
Sr		296 ± 5		940 ± 50		840 ± 40		1540 ± 70
Sb	0.26 ± 0.04		1.8 ± 0.3		0.32 ± 0.06		0.88 ± 0.15	
Ba	320 ± 400		1080 ± 120		300 ± 40		1350 ± 110	
La	24.8 ± 0.2		18.8 ± 0.2		9.2 ± 0.2	24.8 ± 0.3		
Hf	5.2 ± 0.3		2.2 ± 0.1		0.68 ± 0.04			
Pb		3.1 ± 0.2		8.6 ± 1.4		<3.8		2.6 ± 0.1
Th	1.5 ± 0.1		6.9 ± 0.1		3.2 ± 0.1		9.8 ± 0.2	27 ± 2
U	4.4 ± 0.4		2.6 ± 0.3		2.8 ± 0.2		4.4 ± 0.4	

[a] Measurement in parts per million by weight except as noted.

TABLE VIII   *Inorganic Element Concentrations in Water Samples from CO_2-Acceptor Coal Gasification Process*

	Gasifier quench water XRF	Regenerator scrubber water XRF
Si	<100	<100
P	<28	<22
S	<26	210 ± 20
Cl	3900 ± 150	1040 ± 100
K	8 ± 1	3 ± 1
Ca	320 ± 30	440 ± 30
Ti	1.7 ± 0.2	3.5 ± 0.3
V	0.16 ± 0.02	<0.27
Cr	<0.1	<0.1
Mn	0.5 ± 0.2	0.7 ± 0.2
Fe	18 ± 1	22 ± 1
Ni	0.05 ± 0.01	0.09 ± 0.01
Cu	0.10 ± 0.02	0.16 ± 0.02
Zn	0.15 ± 0.02	0.20 ± 0.02
As	0.020 ± 0.005	<0.01
Se	<0,01	<0.01
Br	<0.044 ± 0.005	0.14 ± 0.01
Rb	<0.02	<0.02
Sr	3.1 ± 0.1	4.0 ± 0.2
Pb	0.04 ± 0.02	0.16 ± 0.02

elements such as arsenic and bromine tend to partition onto the gasifier fines. In addition, the gasifier fines are somewhat higher in most of the elements measured, because it is not diluted by the acceptor which tends to be low in these elements. The feed coal measured had already been through a preheater where water and other volatiles including some mercury were driven off. As shown in Table IV, measurable quantities of mercury were detected in the preheater offgases.

Chromatograms of extracts of the above waters are given in Figs. 10–12. The scrub water from the preheater venturi contained an interesting class of rosin acid compounds which probably were occluded in the raw coal and escaped on heating. Abietane, abietatriene, ferruginol, retene, and an unknown related compound are major components of the extractable fraction of organics from this scrub water. The homologous n-alkane series is present as well as some phenolic compounds. The gasifier quench water contains naphthalene as the major organic compound at about 10 ppm. A variety of polynuclear aromatic hydro-

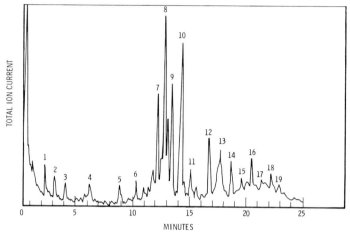

Fig. 10   Extractable organic compounds in preheater scrub water at CO_2-Acceptor coal gasification plant. (1) Phenol, (2) cresol, (3) xylenol, (4) C_6-phenol, (5) cadalene, (6) C_3-cyclohexane, (7) C_9-phenol, (8) abietane, (9) abietatriene, (10) unknown, (11) retene, (12) ferruginol, (13)-(19) *n*-alkanes.

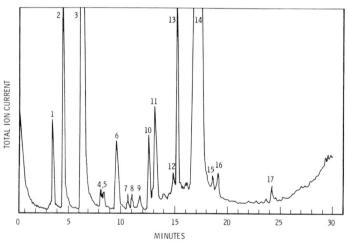

Fig. 11   Extractable organic compounds in gasifier quench water at CO_2 Acceptor coal gasification plant. (1) Indene, (2) unknown (perhaps tritholane), (3) naphthalene, (4) (5) methylnaphthalenes, (6) C_2 naphthalene, (7) biphenylene, (8) biphenyl, (9) C_3-naphthalene, (10) methylacenaphthylene, (11) phthalate contaminant, (12) dibenzothiophene, (13) phenanthrene, (14) sulfur (S_8), (15) fluoranthene, (16) pyrene, (17) bisphenol.

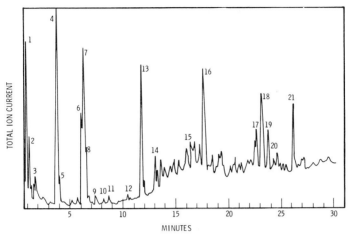

Fig. 12 Extractable organic compounds in regenerator quench water at CO_2-Acceptor coal gasification plant. (1) xylene, (2) cyclohexanone, (3) cyclohexenone, (4) chlorocyclohexanol, (5) salicylaldehyde, (6) phenol, (7) naphthalene, (8) benzothiophene, (9) cresol, (10) methylnaphthalene, (11) ethylphenol, (12) acenaphythylene, (13) dodecanol, (14)-(21) plasticizer contaminants, (20) bisphenol.

carbons (PAH) were identified as well as bisphenol. Sulfur, present in colloidal form, was also extracted from the water and can be seen in the chromatogram. Grab samples of the raw gas directly from the gasifier contained CO, CO_2, and CH_4 as the major components. Grab sampling did not allow any minor gaseous species to be identified. The regenerator quench water contained a variety of oxygenated compounds not present in other parts of the process. Ketones, aldehydes, phenols, and alcohols as well as some PAH compounds have been identified. Plasticizer contamination was present only in the regenerator water sample, but whether these contaminants are sampling related or process related we cannot say with certainty.

Because of the complexity of effluents from coal conversion facilities and the variety of compounds which may belong exclusively to these process wastes, much development in organic analysis is needed for future work. In our work the emphasis has been essentially on surveying what compound bases are present and some estimates of the concentrations of major constituents. As demonstration units are built, a more thorough study of composition will be needed. The ultimate question of what the biological impact will be is a much broader area of investigation and the analytical chemist must be ready to assist in answering that question by being able to find which compounds, or at least classes, are present in effluents.

V. CONCLUSIONS

The growing concern for environmental considerations in this country has greatly increased the need for chemical analyses of all types of plant effluents, products, and by-products. Because of the physical and chemical complexity of many of these samples, making these analyses with sufficient sensitivity and accuracy is not a trivial problem. In fact, it is one of the greatest challenges facing present-day analytical chemists. There is a need to consider many factors that are not stressed in many books and courses because of the chemical complexity of these samples. Many of the early studies of chemical and power plant effluents are in fact neither accurate nor representative. In this chapter we have tried to point out some of these considerations which include taking representative samples, interfacing with the plant and its personnel, avoiding sample contamination and degradation, and choosing analytical techniques which minimize matrix effects. Because this field is a new and rapidly growing one, we cannot give any final solutions. We can only encourage continuing research in this general case. We also encourage each chemist faced with sampling and analyzing plant effluents to consider carefully the individual characteristics and requirements of the situation before proceeding.

REFERENCES

Abel, K. H., and Rancitelli, L. A. (1975). *Adv. Chem. Ser.* **141,** 118.
Anderson, A. A. (1966). *Am. Ind. Hyg. Assoc. J.* **27,** 160–165.
Braman, R. S., and Johnson, D. L. (1974). *Environ. Sci. Technol.* **12,** 996.
Filby, R. H., Shah, K. R., Hunt, M. L., Khalil, S. R., and Sautter, C. A. (1978). "Fossil Energy," R & D Rep. No. 53, Interim Rep. No. 26, Vol. III, Part 6: NTIS FE/496-717. Dep. Energy, Washington, D.C.
Flesch, J. P., Norris, C. H., and Nugent, A. E., Jr. (1967). *Am. Ind. Hyg. Assoc. J.* **28,** 507–516.
Fruchter, J. S., Petersen, M. R., and Ryan, P. W. (1977). "Characterization of Substances in Products, Effluents and Wastes from Synthetic Fuel Development Process," Rep. BNWL-2224. Battelle, Pacific Northwest Lab., Richland, Washington.
Giauque, R. D., Goulding, F. S., Jaklevic, J. M., and Pohl, R. H. (1973). *Anal. Chem.* **45,** 641.
Gordon, G. E., Randle, K., Goles, G. G., Corliss, J. B., Beeson, M. H., and Oxley, S. S. (1968). *Geochem. Cosmochim. Acta* **32,** 369–396.
Johnson, D. L., and Braman, R. S. (1975). Chemosphere **4,** 333.
Laul, J. C., and Schmitt, R. A. (1973). *Proc. Lunar Sci. Conf., 4th,*   pp. 1349–1367.
Lundgren, D. A. (1970). *Air Pollut. Control Assoc.* **20,** 603–608.
Natusch, D. F. S., and Wallace, J. R. (1976). *Atmos. Environ.* **10,** 315–324.

Nielson, K. K. (1977). *Anal. Chem.* **49,** 641.

Pellizzari, E. A. (1977). "Analysis of Organic Pollutants by Gas Chromatograph and Mass Spectroscopy," EPA-600/2-77-100. Environ. Prot. Agency, Washington, D.C.

Soete, D. O., Gijbel, R., and Hoste, J. (1972). "Neutron Activation Analysis." Wiley, New York.

Part XI

SPECIAL INSTRUMENTAL TECHNIQUES FOR ANALYSES OF COAL AND ITS PRODUCTS

Chapter 46

Instrumental Activation Analysis of Coal and Coal Ash With Thermal and Epithermal Neutrons

E. Steinnes

INSTITUTT FOR ATOMENERGI
ISOTOPE LABORATORIES
KJELLER, NORWAY

I. INTRODUCTION

The burning of coal is recognized as one of the major sources of atmospheric pollution. Concern about toxic elements in the environment has therefore led to increased attention to the elemental composition of coal and has emphasized the need for reliable and sensitive methods for the determination of trace elements in samples of coal and coal ash. One of the analytical techniques showing a great potential in this respect is neutron activation analysis, which has played a significant role in the earth sciences during the last 20 years. The general aspects of neutron activation analysis have been discussed in several

279

textbooks (e.g., Bowen and Gibbons, 1963; Lyon, 1964; Kruger, 1971; De Soete *et al.*, 1972), and its application to fossil fuels is the subject of a previous chapter (Volume 1, Chapter 12) and Chapter 47.

Although other neutron sources are used in some cases, the nuclear reactor is by far the dominant irradiation source for neutron activation analysis. The most remarkable feature of reactor activation analysis is probably its high sensitivity, which depends on the nuclear properties of the elements concerned and the neutron flux. At a flux of 2×10^{13} n cm^{-2} s^{-1}, which is available in a great number of research reactors, more than 50 elements can be determined in amounts of 1 ng or less. In most cases it is necessary to employ radiochemical separations in order to reach the ultimate sensitivity, which usually involves a significant work load. Very sensitive determinations may, however, be carried out even by measurements directly on the irradiated samples. This purely instrumental neutron activation analysis (INAA) is quite simple and rapid and often allows the simultaneous determination of a considerable number of elements at trace concentrations.

In INAA, because of contribution from interfering radionuclides, the sensitivity for each element depends on the concentrations of other elements in the sample. The low neutron activation cross sections of the matrix elements carbon, oxygen, and hydrogen make coal a favorable case for this technique, and several workers have demonstrated the feasibility of INAA for the multielement analysis of coal and coal ash (Block and Dams, 1973; Abel and Rancitelli, 1975; Sheibley, 1975). Selective irradiation with epithermal neutrons offers advantages for a number of elements in INAA (Steinnes and Rowe, 1976; Rowe and Steinnes, 1977). In this chapter the possibilities offered by INAA in the elemental analysis of coal and coal ash at the present state of the art are reviewed. The application of epithermal neutron activation as compared to the conventional irradiation with the whole energy spectrum of reactor neutrons is discussed on a theoretical and experimental basis.

II.   INSTRUMENTAL NEUTRON ACTIVATION ANALYSIS

Activation analysis can be described as consisting of two separate steps:

(1)   production of a radioactive nuclide from the element in question by some kind of nuclear reaction;

(2)   measurement of the induced radioactivity.

In the case of reactor neutron activation, the nuclear reaction concerned is usually of the (n,γ) type, which means that the radionuclide produced is an isotope of the target element. The activity D of the induced radionuclide at the end of the irradiation can be expressed by

$$D = N\sigma\phi a[1 - \exp(-0.693t/t_{1/2})], \tag{1}$$

where N is the number of atoms of the element concerned present in the sample; σ is the cross section of the reaction producing the nuclide; ϕ is the neutron flux in the irradiation position; a is the relative abundance of the target isotope; t is the duration of the irradiation; $t_{1/2}$ is the half-life of the induced radionuclide.

Thus, the sensitivity for the determination of an element at a given neutron flux level is proportional to the product of the isotopic abundance and activation cross section of the target isotope, and is also dependent on the length of the irradiation in relation to the half-life of the induced activity. When γ-spectrometry is used for the activity measurements, the sensitivity is furthermore proportional to the fraction of the radioactive atoms that decay with emission of the particular γ-ray measured.

Although the amount of the desired element can be calculated from Eq. (1), provided that the detection efficiency of the actual radionuclide with the equipment used is known, the most precise and accurate results are obtained by irradiating a standard containing a known amount of the element along with the sample and calculating the result from the ratio of measured activity in sample and standard.

A quite extensive treatment of practical aspects of activation analysis, such as the preirradiation and postirradiation treatment of samples, equipment, and procedures for activity measurements and processing of counting data in the analysis of fossil fuels is given in Volume I, Chapter 12 and need not be repeated in detail here. The choice of irradiation time depends on the half-life of the induced radionuclide, and the procedure to be used is dependent on the irradiation time as well as conditions in the irradiation position such as neutron flux, temperature, and energy distribution of neutrons. After the irradiation, it is often convenient to let the samples decay for some time before measurements in order to reduce interfering radioactivity. In most INAA multielement schemes for coal, one short (few minutes or less) and one long irradiation (several hours or more) are employed. Examples on procedures for multielement analysis by INAA using thermal or epithermal neutrons are given in the experimental section.

γ-spectrometric measurements are carried out by means of Ge(Li) solid-state detectors coupled to multichannel analyzers. Peak areas corresponding to characteristic γ-energies in the spectra are calculated according to methods with a variable degree of sophistication, usually with the aid of a computer. Additional information may be obtained if measurements with a conventional Ge(Li) detector are supplemented by γ-spectrometry using a germanium low-energy photon detector (LEPD). The LEPD is superior to the ordinary Ge(Li) detectors at energies below about 150 keV. These detectors are shown in Fig. 1, Chapter 12, Volume I.

Relevant nuclear data for the determination of 66 elements in coal and coal ash by INAA are listed in Table I. In many cases two or more isotopes of an element may be used for the determination. Only those γ-emitting radionuclides, formed by the (n,γ) reaction, which are of interest for the analysis of coal and coal ash are listed. Activities with half-lives of 10 s or less are not included since their use in INAA requires special facilities. Some additional elements, such as P and Tl, possess adequate sensitivity for determination at very low levels by neutron activation, but the induced radioisotopes only emit β^--radiation that cannot be measured without performing a radiochemical separation. In some cases γ-rays other than those given in Table I may be more advantageous, because of spectral interferences.

The limit of detection by neutron activation can vary by several orders of magnitude from element to element. While elements such as Sc, Au, Ir, and several rare earths can be determined by INAA at the sub-ppm level even in very complex matrices, the sensitivity for other elements such as Mg, K, Ca, and Fe is rather poor. It is possible, however, to obtain values for those elements along with other data when they are present as major components, such as in coal ash.

III. ACTIVATION WITH EPITHERMAL NEUTRONS

As indicated before, the sensitivity for the determination of a particular element by INAA often depends on the concentration levels of other elements in the sample giving rise to interfering activities in the γ-spectrometry measurements. The effects of such interfering radiation can in some cases be minimized by using an anticoincidence shielded Ge(Li) spectrometer (Abel and Rancitelli, 1975) or a low-energy photon detector, instead of the ordinary coaxial-type Ge(Li) detector. A different approach is to enhance the formation of a radionuclide relative to interfering ones by an energy-selective neutron activation. This can be achieved by activation with the epithermal component of the reactor neutron spectrum.

TABLE I *Relevant Nuclear Data for Elements Suitable for Determination by Instrumental Activation Analysis Using Thermal or Epithermal Neutrons from a Nuclear Reactor*

Element	Stable isotope	Abundance (%)	Neutron activation cross section (barn) Thermal $(\sigma)^a$	Epithermal $(I)^b$	$\frac{I}{\sigma}$	Radionuclide formed	Half-life	Most useful γ-ray Energy (keV)	Abundance (%)
F	^{19}F	100	0.0098	0.04	4.1	^{20}F	11.4 s	1634	100
Na	^{23}Na	100	0.528	0.31	0.59	^{24}Na	15.0 hr	1368	100
Mg	^{26}Mg	11.2	0.0038	0.030	7.9	^{27}Mg	9.5 min	844	70
Al	^{27}Al	100	0.232	0.17	0.73	^{28}Al	2.3 min	1779	100
S	^{36}S	0.015	0.15	—	—	^{37}S	5.1 min	3105	90
Cl	^{37}Cl	24.5	0.43	0.17	0.41	^{38}Cl	37.3 min	2168	100
K	^{41}K	6.8	1.48	1.28	0.86	^{42}K	12.4 hr	1525	18
Ca	^{46}Ca	0.0033	0.7	0.32	0.46	^{47}Ca	4.6 d	1297	74
—	—	—	—	—	—	→ ^{47}Sc	3.4 d	160	73
—	^{48}Ca	0.185	1.1	0.90	0.82	^{49}Ca	8.8 min	3084	89
Sc	^{45}Sc	100	10	—	—	^{46m}Sc	20 s	143	46
—	—	—	25	11	0.44	^{46}Sc	84 d	889	100
Ti	^{50}Ti	5.3	0.179	~5	~30	^{51}Ti	5.8 min	320	95
V	^{51}V	100	4.90	2.7	0.55	^{52}V	3.75 min	1434	100

TABLE I (Continued)

| Element | Stable isotope | Abundance (%) | Neutron activation cross section (barn) | | | Radionuclide formed | Half-life | Most useful γ-ray | |
			Thermal $(\sigma)^a$	Epithermal $(I)^b$	$\frac{I}{\sigma}$			Energy (keV)	Abundance (%)
Cr	^{50}Cr	4.3	16.0	8.5	0.53	^{51}Cr	27.8 d	320	9
Mn	^{55}Mn	100	13.3	14	1.1	^{56}Mn	2.58 hr	847	99
Fe	^{58}Fe	0.31	1.14	1.2	1.1	^{59}Fe	45 d	1099	56
Co	^{59}Co	100	19.9	—	—	^{60m}Co	10.4 min	59	2.1
—	—	—	37.5	75.0	2.0	^{60}Co	5.2 yr	1333	100
Ni	^{58}Ni	67.8	—	—	—	^{58}Co^c	71 d	811	99
—	^{64}Ni	1.16	1.50	0.9	0.6	^{65}Ni	2.55 hr	1482	25
Cu	^{63}Cu	69.1	4.4	5.0	1.1	^{64}Cu	12.8 hr	511	38
—	^{65}Cu	30.9	2.20	2.5	1.1	^{66}Cu	5.2 min	1039	9
Zn	^{64}Zn	48.9	0.82	1.6	2.0	^{65}Zn	247 d	1116	49
—	^{68}Zn	18.6	0.07	0.24	3.4	^{69m}Zn	13.9 hr	439	95
Ga	^{71}Ga	39.5	4.7	25	0.53	^{72}Ga	14.1 hr	834	96
Ge	^{70}Ge	20.6	3.5	—	—	^{71}Ge	11.4 d	Ga x rays	—
—	^{74}Ge	36.7	0.52	0.9	1.7	^{75}Ge	83 min	265	11

As	^{75}As	100	4.4	63	14	^{76}As	26.4 hr	559	43
Se	^{74}Se	0.90	55	500	9.1	^{75}Se	120 d	265	60
—	^{76}Se	9.0	21	42	2.0	^{77m}Se	17.4 s	162	50
Br	^{79}Br	50.5	2.6	—	—	^{80m}Br	4.4 hr	37	36
—	—	—	11.0	125	11	^{80}Br	17 min	616	7
—	^{81}Br	49.5	3.0	50	17	^{82}Br	35.3 hr	777	83
Rb	^{85}Rb	72.1	0.45	7	16	^{86}Rb	18.7 d	1079	8.8
—	^{87}Rb	27.9	0.12	2.3	19	^{88}Rb	17.7 min	1836	21
Sr	^{84}Sr	0.55	0.7	7.5	11	^{85}Sr	65 d	514	100
—	^{86}Sr	9.9	0.8	4	5	^{87m}Sr	2.8 hr	389	80
Y	^{89}Y	100	0.0010	0.89	890	^{90m}Y	3.1 hr	202	97
Zr	^{94}Zr	17.4	0.0075	0.38	51	^{95}Zr	65 d	757	49
—	—	—	—	—	—	→ ^{95}Nb	35 d	766	100
—	^{96}Zr	2.8	0.05	5.0	100	^{97}Zr	17.0 hr	743	99
—	—	—	—	—	—	→ ^{97}Nb	72 min	658	98
Nb	^{93}Nb	100	0.15	8.0	53	^{94m}Nb	6.3 min	Nb x rays	—
Mo	^{98}Mo	23.8	0.14	7.5	54	^{99}Mo	67 hr	740	12
—	—	—	—	—	—	→ ^{99m}Tc	6.0 hr	140	90

TABLE I (*Continued*)

Element	Stable isotope	Abundance (%)	Neutron activation cross section (barn)			Radionuclide formed	Half-life	Most useful γ-ray	
			Thermal $(\sigma)^a$	Epithermal $(I)^b$	$\frac{I}{\sigma}$			Energy (keV)	Abundance (%)
—	^{100}Mo	9.6	0.20	3.9	20	^{101}Mo	14.6 min	2089	16
—	—	—	—	—	—	$\rightarrow$ ^{101}Tc	14.0 min	307	91
Ru	^{96}Ru	5.5	0.21	5.5	26	^{97}Ru	2.8 d	215	91
—	^{102}Ru	31.6	1.3	4.2	3.2	^{103}Ru	39.5 d	497	88
—	^{104}Ru	18.8	0.47	5.2	11	^{105}Ru	4.4 hr	724	48
—	—	—	—	—	—	$\rightarrow$ ^{105}Rh	35.9 hr	319	19
Rh	^{103}Rh	100	139	1100	7.9	^{104m}Rh	4.4 min	51	47
Pd	^{108}Pd	26.8	12	186	16	^{109}Pd	13.5 hr	88	5
Ag	^{109}Ag	48.6	4.7	55	12	^{110m}Ag	255 d	658	96
Cd	^{114}Cd	28.9	0.30	20	67	^{115}Cd	53.5 hr	528	26
—	—	—	—	—	—	$\rightarrow$ ^{115m}In	4.5 hr	336	50
In	^{113}In	4.3	7.8	~180	~20	^{114m}In	50 d	190	17
—	^{115}In	95.7	161	2600	16	^{116m}In	54 min	1293	80
Sn	^{112}Sn	0.95	0.71	27	38	^{113}Sn	117 d	In x rays	—

—	—	—	—	—	—	→¹¹³ᵐIn	100 min	391	64
—	¹¹⁶Sn	14.2	0.006	0.49	82	¹¹⁷Sn	14.0 d	158	87
—	¹²²Sn	4.7	0.15	0.83	5.5	¹²³Sn	40 min	160	84
—	¹²⁴Sn	6.0	0.14	9	64	¹²⁵Sn	9.6 min	332	97
Sb	¹²¹Sb	57.3	6.2	180	29	¹²²Sb	2.8 d	564	66
—	¹²³Sb	42.7	4.0	120	30	¹²⁴Sb	60 d	1691	50
Te	¹³⁰Te	34.5	0.2	0.4	2	¹³¹Te	25 min	150	68
—	—	—	—	—	—	→¹³¹I	8.0 d	365	82
I	¹²⁷I	100	6.2	150	24	¹²⁸I	25.0 min	443	14
Cs	¹³³Cs	100	2.6	30	12	¹³⁴ᵐCs	2.90 hr	127	14
—	—	—	30.0	465	16	¹³⁴Cs	2.05 yr	797	99
Ba	¹³⁰Ba	0.12	11	270	25	¹³¹Ba	12.0 d	496	48
—	¹³⁴Ba	2.6	0.16	24	150	¹³⁵ᵐBa	28.7 hr	268	16
—	¹³⁸Ba	70.4	0.35	0.3	0.9	¹³⁹Ba	83 min	166	23
La	¹³⁹La	100	9.0	11	1.2	¹⁴⁰La	40.2 hr	1597	96
Ce	¹⁴⁰Ce	88.5	0.58	0.49	0.84	¹⁴¹Ce	33 d	146	48
—	¹⁴²Ce	11.1	1.1	1.40	1.3	¹⁴³Ce	33 hr	293	46
Nd	¹⁴⁶Nd	17.2	1.4	2.8	2.0	¹⁴⁷Nd	11.0 d	91	28

TABLE I (*Continued*)

| Element | Stable isotope | Abundance (%) | Neutron activation cross section (barn) | | | Radionuclide formed | Half-life | Most useful γ-ray | |
			Thermal $(\sigma)^a$	Epithermal $(I)^b$	$\frac{I}{\sigma}$			Energy (keV)	Abundance (%)
—	148Nd	5.7	2.5	17	6.8	149Nd	1.9 hr	211	27
—	150Nd	5.6	1.3	14	11	151Nd	12 min	116	40
—	—	—	—	—	—	→ 151Pm	28 hr	340	21
Sm	152Sm	26.7	210	2900	14	153Sm	47 hr	103	28
—	154Sm	22.6	5.5	27	4.9	155Sm	22 min	104	73
Eu	151Eu	47.8	2800	11400	4.1	152mEu	9.3 hr	963	12
—	—	—	5300	3800	0.72	152Eu	12.5 yr	1408	22
—	153Eu	52.2	400	1500	3.8	154Eu	16 yr	1274	37
Gd	152Gd	0.21	1100	3000	2.7	153Gd	242 d	98	55
—	158Gd	24.8	3.5	80	23	159Gd	18.0 hr	363	9
Tb	159Tb	100	25	400	16	160Tb	72 d	299	30
Dy	164Dy	28.1	2100	—	—	165mDy	1.26 min	Dy x rays	—
—	—	—	2600	800	0.31	165Dy	2.32 hr	95	4
Ho	165Ho	100	62	660	11	166Ho	27.0 hr	81	5.4

Er	170Er	14.9	6	35	6	171Er	7.5 hr	308	63
Tm	169Tm	100	106	1700	16	170Tm	130 d	84	3.3
Yb	168Yb	0.14	3200	25000	7.8	169Yb	31 d	198	35
—	174Yb	31.7	65	31	0.48	175Yb	4.2 d	396	6.0
—	176Yb	12.6	5.5	7	1.3	177Yb	1.9 hr	150	16
Lu	175Lu	97.4	18	600	33	176mLu	3.7 hr	88	10
—	176Lu	2.6	2050	~1100	~0.5	177Lu	6.7 d	208	6.1
Hf	178Hf	27.2	52	—	—	179mHf	18.6 s	214	94
—	180Hf	35.2	12.6	28	2.2	181Hf	43 d	482	81
Ta	181Ta	100	22	700	32	182Ta	115 d	1222	34
W	186W	28.4	37	420	11	187W	24.0 hr	686	27
Re	185Re	37.1	110	1700	15	186Re	3.7 d	137	9
—	187Re	62.9	75	310	4.1	188Re	16.8 hr	155	10
Os	190Os	26.4	16	39	2.4	191Os	15 d	129	25
Ir	191Ir	38.5	925	4000	4.3	192Ir	74 d	468	49
—	193Ir	61.5	110	1390	13	194Ir	17.4 hr	328	10
Pt	196Pt	25.2	0.8	~6	~8	197Pt	19 hr	77	20
—	198Pt	7.2	3.7	53	14	199Pt	30 min	543	24

TABLE I (*Continued*)

Element	Stable isotope	Abundance (%)	Neutron activation cross section (barn) Thermal $(\sigma)^a$	Epithermal $(I)^b$	$\dfrac{I}{\sigma}$	Radionuclide formed	Half-life	Most useful γ-ray Energy (keV)	Abundance (%)
—	—	—	—	—	—	→ ^{199}Au	3.15 d	158	37
Au	^{197}Au	100	98.8	1550	15.7	^{198}Au	2.69 d	412	95
Hg	^{196}Hg	0.15	3000	~1200	~0.4	^{197}Hg	2.70 d	78	18
—	^{202}Hg	29.8	4.9	4.3	0.88	^{203}Hg	47 d	279	77
Th	^{232}Th	100	7.4	82	11	^{233}Th	22.3 min	Pa x rays	—
—	—	—	—	—	—	→ ^{233}Pa	27 d	312	38
U	^{235}U	0.72	580^d	275^d	0.47	Fission products	Various	Various	—
—	^{238}U	99.3	2.72	280	103	^{239}U	23.5 min	75	51
—	—	—	—	—	—	→ ^{239}Np	2.35 d	278	14

[a] Values from Sher (1974).
[b] Values from Albinsson (1974).
[c] Produced by reactor fast neutrons.
[d] Fission cross section.

A. General Background

The neutron energy spectrum in a nuclear reactor can be divided into three different components:

(a)   Unmoderated fission neutrons, usually denoted reactor fast neutrons, exhibiting energies mainly in the range 0.1–10 MeV, with the most probable energy at about 1 MeV. The most probable nuclear activation reactions are of the (n,p), (n,α), and (n,2n) types.

(b)   Neutrons of intermediate energy, which are in the process of slowing down in the moderator. This component is called epithermal, or resonance, neutrons and covers an energy range from below 1 eV up to about 1 MeV. In this region the (n,γ) and (n,n') reactions are the predominant types of interest.

(c)   Neutrons in thermal equilibrium with the moderator atoms, showing an energy distribution following the Maxwell distribution law. With a few exceptions the thermal neutrons produce radionuclides only by (n,γ) reaction. In a well-moderated reactor most of the neutrons have energies below 1 eV.

As a general rule, the cross section of an (n,γ) reaction is inversely proportional to the neutron velocity (v). Most nuclides follow this "1/v law" in the thermal neutron region, and some follow the law quite closely in the epithermal region as well. Many nuclides, however, show large resonances for neutron absorption in the epithermal region. If the thermal neutron component is excluded by a suitable filter such as 0.7 mm cadmium foil, which gives an effective "cutoff" at about 0.4 eV, a selective activation of nuclides having large neutron resonances is obtained relative to those following the 1/v law.

The epithermal activation properties of a nuclide can be conveniently expressed by means of the *cadmium ratio*:

$$R_{Cd} = (\phi_{th}\sigma + \phi_e I)/\phi_e I, \qquad (2)$$

where ϕ_{th} is the thermal and ϕ_e the epithermal component of the neutron flux in the irradiation position concerned; σ is the thermal neutron activation cross section; and I the corresponding activation cross section for epithermal neutrons (resonance activation integral, including the "1/v tail"). If the cadmium ratios of a nuclide of interest (D) and a nuclide interfering with the measurement (d) are known in a particular irradiation position, the benefit to be obtained by introducing activation under a cadmium cover can be quantitatively expressed by means of an "advantage factor" (Brune and Jirlow, 1964):

$$Fa = (R_{Cd})_d/(R_{Cd})_D. \qquad (3)$$

If the cadmium ratios cannot easily be obtained, the ratio I/σ may also be a good indicator of the feasibility of epithermal activation. For a nuclide following the $1/v$ law this ratio is about 0.4, and the activity produced by (n,γ) reaction in a well thermalized flux comes almost entirely from thermal neutrons. Many nuclides, however, show I/σ ratios of 10 or more, and in this case a very significant part of the induced activity may be because of the epithermal neutrons. Values for the thermal neutron cross section σ, resonance activation integral I, and the I/σ ratio are listed in Table I for the nuclides concerned. If the I/σ ratio associated with the formation of the nuclide of interest is significantly higher than that of the nuclides giving rise to major interfering activities, the use of epithermal activation with a cadmium cover would appear to be advantageous.

A more extensive survey of epithermal activation analysis discussing the advantages and limitations of the technique is given elsewhere (Steinnes, 1971).

B.   Application to Coal and Coal Ash

The major element composition of coal ash is somewhat similar to that of silicate rocks. It has been pointed out earlier (Brunfelt and Steinnes, 1969; Steinnes, 1971) that most of the nuclides giving rise to major activities in common silicate rocks on neutron activation have low I/σ ratios. This applies for instance to ^{23}Na, ^{27}Al, ^{45}Sc, ^{50}Cr, ^{151}Eu, ^{164}Dy, and also to some extent to ^{41}K, ^{46}Ca, ^{55}Mn, ^{58}Fe, and ^{139}La. On the other hand, a large number of elements normally present in trace concentrations in silicate materials show quite high I/σ ratios. In the determination of these elements in rocks by INAA, the use of epithermal activation offers a clear advantage, and the same applies to samples of coal ash (Steinnes and Rowe, 1976; Rowe and Steinnes, 1977).

In the case of coals, the major elements in the organic fraction are not appreciably activated either by thermal or epithermal neutrons, and the advantage of epithermal activation is therefore similar to the case of ash.

Some practical considerations regarding epithermal irradiations under a cadmium cover should be mentioned at this point. Cadmium is a very strong neutron absorber, therefore a certain limitation may exist regarding the maximum amount of the element that can be irradiated at all times. This may restrict one to a small sample size if many samples are to be irradiated simultaneously. For the same reason, the safety instructions in force may put limitations on the use of cadmium in connection with pneumatic tube systems for short irradiations. During the irradiation a considerable buildup of radioactivity in the cadmium takes place (mostly ^{115}Cd-^{115m}In), and a hot-cell facility should preferably be available for remote unpacking of irradiated batches.

IV. EXPERIMENTAL PROCEDURES

This section gives a brief description of the INAA procedures used in a feasibility study involving a comparison of thermal and epithermal irradiation in the analysis of coal and coal ash. The procedures were also applied to a study of mass balances of 38 elements at three coal-fired power plants (Zubovic *et al.*, 1978). The thermal neutron activation procedures are quite similar to those used by other workers, while those used with epithermal activation were designed especially for this project. A more detailed description is given elsewhere (Steinnes and Rowe, 1976; Rowe and Steinnes, 1977).

A summary of the irradiation and counting procedures used are given in Table II. Irradiations were carried out in the JEEP-II reactor (Kjeller, Norway) at a thermal neutron flux of about 1.5×10^{13} n cm^{-2} s^{-1} and a cadmium ratio of 3.0 (^{197}Au) or in the RT-3 tube of the NBS-reactor (National Bureau of Standards, Gaithersburg, Maryland) at a thermal neutron flux of about 5×10^{13} n cm^{-2} s^{-1}. After each irradiation the samples and standards were subjected to γ-spectrometry with coaxial Ge(Li) detectors at two different decay intervals. Spectra for the short irradiations were processed on a NORD-I minicomputer. Spectra for the long irradiations were collected on magnetic tape and processed on an IBM 370 computer using the program "SPECTRA" (Baedecker, 1976).

Some details on irradiation and sample preparation are given below.

A. Short Irradiation

Samples of about 100 mg were weighed and sealed in small envelopes of polyethylene foil for thermal neutron irradiation in the JEEP-II reactor. Standards were made by absorbing 50 μliter of an appropriate solution on to filter paper and sealing in the same sort of polyethylene envelopes. Two samples or standards were irradiated in each "rabbit." After 20 min delay, measured with a stopwatch, the samples were counted simultaneously with two Ge(Li) detector systems interfaced to a NORD-I computer. The counting time was 10 min, and the interval between subsequent irradiations was 15 min to allow sufficient time

TABLE II *Summary of Irradiation and Counting Procedures Used in the Comparison of Thermal and Epithermal Activation of INAA of Coal and Coal Ash*

Irradiation time	Cd cover	Reactor	Number of samples per irradiation	Decay interval 1	Decay interval 2	Counting time	Elements determined
30 min	Yes	JEEP-II	6	30–60 min	1 d	10 min	Ga, As, Br, Sr, In, Cs, Ba, La, Sm, Ho, W, U
1–2 d	Yes	JEEP-II	15–20	6–7 d	20–25 d	1 hr	Sc, Cr, Fe, Co, Ni, Zn, As, Se, Br, Rb, Sr, Zr, Mo, Sb, Cs, Ba, La, Ce, Sm, Tb, Yb, Hf, Ta, W, Th, U
1 min	No	JEEP-II	1	20 min	1 d	10 min	Na, Mg, Al, Cl, K, Ca, Ti, V, Mn, Cu, Br, Sr, I, Ba, La, Eu, Dy, U
4 hr	No	NBS	15–20	6–7 d	20–25 d	1 hr	Sc, Cr, Fe, Co, Zn, As, Se, Br, Rb, Sr, Zr, Sb, Cs, Ba, La, Ce, Nd, Sm, Eu, Tb, Yb, Lu, Hf, Ta, Th

for concentration printouts and changing of samples. After 24 hr, the samples were recounted, in the same sequence, to avoid time-consuming calculation of decay corrections.

For the epithermal activation, samples of about 100 mg were wrapped in 30 × 30 mm sheets of aluminum foil for irradiation. Standards were prepared by evaporating aliquots of appropriate solutions on to the aluminum foil, drying at room temperature, and folding. Six samples and one set of standards were wrapped together in another aluminum foil and placed inside a 1-mm thick cadmium box (14-mm internal diameter, 10-mm internal height). The cadmium container was wrapped in aluminum foil, placed in a polyethylene "rabbit," and irradiated in the pneumatic tube system of the JEEP-II reactor. Insertion took place 30 min before reactor shutdown and the "rabbit" was retained in the reactor for another 20 min before return to the laboratory. After unpacking, the samples were weighed and wrapped in inactive aluminum foil for counting.

B. Long Irradiation

Samples of about 100 mg were used. A multielement standard, made by grinding an obsidian sample from Horse Mountain, Oregon, to <200 mesh and spiking with appropriate concentrations of Ba, Co, Cr, Cs, Rb, Sb, Ta, Zr, Sc, La, Eu, and Tb, was used as a monitor for most of the elements studied. The multielement standard was calibrated against pure solutions and U.S. Geological Survey standard rocks. For Ni, As, Se, Br, Sr, Mo, W, and U, standards were made by evaporating aliquots of dilute solutions on to 30 × 30 mm sheets of aluminum foil, drying at room temperature, and folding.

For the epithermal neutron activation in the JEEP-II reactor about 15 samples and a set of standards were packed in the same sort of cadmium box as used in the short irradiation. Before counting, the samples were transferred to inactive aluminum foils and weighed. Samples and standards for thermal neutron activation were weighed into polyethylene minivials (5 mm-internal diameter, 10-mm depth), which were heat-sealed and irradiated in the NBS reactor. For γ-spectrometry measurements the irradiated minivials were placed inside larger polyethylene vials.

V. COMPARISON OF RESULTS OBTAINED BY THERMAL AND EPITHERMAL ACTIVATION

The National Bureau of Standards reference materials of coal (SRM 1632) and fly ash (SRM 1633) were selected as test material for this study. Results are given in Table III, along with certificate values and literature data, where available. These values may be compared with those in Table VI, Chapter 12, Volume I. Each value is the mean of five separate aliquots, and the error limits are based on the standard deviation of a single value. For each element the irradiation mode giving the best result and the corresponding radionuclide are indicated. In cases where more than one technique was found to give an acceptable result, the preferred mode is underlined.

In the following, the possibilities offered by instrumental activation analysis using thermal or epithermal neutrons are discussed on the basis of the results obtained from the present study of standard ma-

TABLE III Elemental Concentrations in NBS Coal (SRM 1632) and NBS Fly Ash (SRM 1633) Determined by Instrumental Activation Analysis Using Thermal or Epithermal Neutrons[a]

Element	Radio-nuclide used	Short thermal	Short epi-thermal	Long thermal	Long epi-thermal	SRM-1632 Present value[b]	SRM-1632 NBS certified value	SRM-1632 Literature values[c] range	SRM-1633 Present value[b]	SRM-1633 NBS certified value	SRM-1633 Literature values[c] range
Na	24Na	X				380 ± 25		347–414	2830 ± 140	3070	3200–3400
Mg%	27Mg	X				0.17 ± 0.03		0.15–0.25	1.78 ± 0.20	1.98	1.52–1.8
Al%	28Al	X				1.74 ± 0.04		1.76–1.90	12.4 ± 0.3		12.5–14.3
Cl	38Cl	X				844 ± 37		846–1066	—		
K%	42K	X				0.30 ± 0.02		0.28–0.29	1.80 ± 0.13	1.72	1.61–1.8
Ca%	49Ca	X				0.35 ± 0.03		0.43–0.44	4.69 ± 0.14		3.92–4.7
Sc	46Sc			X̲		3.80 ± 0.05		3.5–4.5	26.7 ± 0.7		21–32
Ti	51Ti	X				0.089 ± 0.005	0.080[e]	0.084–0.104	0.70 ± 0.03		0.61–0.74
V	52V	X				35.0 ± 2.9	35 ± 3	33–40	237 ± 20	214 ± 8	208–240
Cr	51Cr			X		20.8 ± 0.8	20.2 ± 0.5	18.9–21.6	129 ± 3	131 ± 2	113–180
Mn	56Mn	X				41 ± 4	40 ± 3	40–47	488 ± 14	493 ± 7	460–540
Fe%	59Fe			X̲	X̲	0.90 ± 0.02	0.87 ± 0.03	0.84–0.93	6.20 ± 0.05		6.08–6.69
Co	60Co			X̲	X	5.70 ± 0.12	6[e]	5.1–6.2	40.3 ± 0.4	38	35–46
Ni	58Co				X	16	15 ± 1	12.1–18.4	99 ± 9	98 ± 7	97–109
Cu	64Cu	X				15.0 ± 1.2	18 ± 2	18	115 ± 8		133
Zn	65Zn			X̲	X	39 ± 6	37 ± 4	30–38	201 ± 6	210 ± 20	208–216
Ga	72Ga		X			5.8 ± 0.4		5.2–8.5	40.7 ± 1.2		49
As	76As		X	X		6.3 ± 0.2	5.9 ± 0.6	4.6–8.9	61.5 ± 2.4	61 ± 6	54–68
Se	75Se			X		3.0 ± 0.3	2.9 ± 0.3	2.4–3.4	10.8 ± 0.8	9.4 ± 0.5	9.1–10.2
Br	82Br		X	X	X̲	19.6 ± 0.4		14.2–19.3	9.2 ± 0.6		6–12
Rb	86Rb				X	18.3 ± 1.6		16.3–24	108 ± 4	112	111–125
Sr	85Sr				X						
	87mSr		X̲			161 ± 9		1.02–161	1430 ± 60	1380	1301–1700
Zr	95Zr				X	—			310 ± 70		301–500

Element	Isotope									
Mo	99mTc	X			3.1 ± 0.1		0.20-3.4	25.3 ± 1.6		1.52
In	116mIn	X			0.017 ± 0.001		0.07-0.23	0.128 ± 0.008		0.28-0.32
Sb	122Sb	X	X		3.0[d]		3.1-4.5	6.0 ± 0.2		6.9-7.8
	124Sb	X	X							
I	128I	X			2.9 ± 0.3		2.8-6.6	—		
Cs	134mCs	X								
	134Cs	X	X		1.52 ± 0.11		0.35-2.6	8.4 ± 0.2		0.63-9.4
Ba	131Ba	X	X		306 ± 20		280-405	2550 ± 30		2490-2780
	139Ba	X	X		338 ± 14			2540 ± 50		
La	140La	X	X	X	10.3 ± 0.5		7.9-11.3	81 ± 3		64-82
Ce	141Ce	X	X		19.5 ± 0.7		18.5-23.3	150 ± 2		146-169
Nd	147Nd	X			8.7 ± 1.0		10.7	58 ± 2		61-81
Sm	153Sm	X	X		1.41 ± 0.06		1.6-1.8	12.1 ± 1.4		12.4-15
Eu	152mEu	X								
	152Eu	X			0.38 ± 0.04		0.21-0.41	2.69 ± 0.09		2.5-3.1
Tb	160Tb	X	X		0.27 ± 0.01		0.23-0.5	2.01 ± 0.06		1.8-3.1
Dy	165Dy	X	X		1.12 ± 0.06		1.3-1.4	9.4 ± 0.5		10.2-10.9
Ho	166Ho	X			0.24 ± 0.03			1.94 ± 0.13		
Yb	169Yb	X								
	175Yb	X	X	X	0.84 ± 0.07		0.69-0.8	6.6 ± 0.4		5.5-7
Lu	177Lu	X	X		0.109 ± 0.011		0.1-0.15	0.94 ± 0.09		1.0-1.2
Hf	181Hf	X	X		0.83 ± 0.06		0.89-1.10	6.7 ± 0.3		7.5-10.8
Ta	182Ta	X			0.273 ± 0.006		0.17-0.3	2.00 ± 0.06		1.6-2.2
W	187W	X	X		0.71 ± 0.07		0.74-0.75	3.9 ± 0.4		4.6
Th	233Pa	X	X	X	3.12 ± 0.10	3.0[e]	1.28-3.65	24.0 ± 0.5	24	24.8-32.2
U	239U	X	X		1.46 ± 0.02	1.4 ± 0.1	1.2-1.46	12.7 ± 0.5	11.6 ± 0.2	11.3-12.0
	239Np	X			1.45 ± 0.04			12.2 ± 0.5		

[a] Values are given in ppm unless otherwise indicated.

[b] The uncertainty listed is the standard deviation of a single value.

[c] Ondov et al. (1975), Nadkarni (1975), Klein et al. (1975), Ruch et al. (1975), Chattopadhyay and Jervis (1974), and Millard and Swanson (1975).

[d] Inhomogeneous distribution of Sb in SRM-1632.

[e] Information value only.

terials as well as from experience gained from the investigation of samples of coal, fly ash, and bottom ash from various coal-fired power plants (Zubovic *et al.*, 1979). It may be added that some improvement might be expected for elements such as some of the rare earths by the use of a low-energy photon detector. Another element that may be determined by that means is mercury, by thermal neutron activation and measurement of the 77 keV γ-ray of ^{197}Hg after about 7 d. That, however, requires irradiation in sealed quartz ampoules in order to avoid loss of Hg by volatilization.

A. Precision and Accuracy

Using the combination of thermal and epithermal activation, data for the content of 41 elements were recorded both in SRM-1632 and SRM-1633. In addition, data for Cl and I were obtained for the coal, while results for Zr could be obtained for the fly ash. The precision is in general slightly better in the case of the fly ash and is of the following order of magnitude for the elements concerned:

a. Coal

±5% or better:		Al, Sc, Cr, Mn, Fe, Co, As, Br, Sm, Tb, Ta, Th, U
±5–10%	:	Na, Cl, K, Ca, Ti, V, Cu, Ga, Se, Rb, Sr, Mo, In, Sb, Cs, Ba, La, Ce, Eu, Dy, Yb, Lu, Hf, W
±10–20%	:	Mg, Ni, Zn, I, Nd, Ho

b. Fly Ash

±5% or better:		Na, Al, Ca, Sc, Cr, Mn, Fe, Co, Ga, As, Sr, Mo, Sb, Cs, Ba, La, Ce, Nd, Eu, Tb, Hf, Ta, Th, U
±5–10%	:	K, Ti, V, Cu, Zn, Se, Br, Rb, In, Sm, Dy, Ho, Yb, Lu, W
±10–20%	:	Mg, Ni, Zr

It appears from Table III that epithermal activation offers more precise results for 19 elements, while the conventional thermal neutron irradiation is to be preferred for the other 25 elements concerned. As expected, the advantage of epithermal activation is most apparent in cases with relatively high I/σ ratios. An exception is iodine, in which case the determination by epithermal activation was made impossible by volatilization loss from the standard during irradiation.

By using the optimal combination of the two irradiation techniques it appears that almost 40 elements may be determined in coal and coal ash with a precision of ±10% or better, provided that the sample is homogeneous at the 100 mg level, which may not always be the case for coal samples.

Evaluating the accuracy of the results shown in Table III is not so easy because certified values are available only for about 15 elements in SRM-1632 and SRM-1633. The agreement is good in cases where certified values exist. For the other elements the present values in most cases fall within the range of data reported in the literature, which is, however, quite wide in many cases. For elements such as Ga, Mo, In, and Ta it appears that epithermal activation analysis is superior to the techniques used to produce the literature data.

B. Sensitivity

As stated before, the sensitivity for the determination of an element by instrumental activation analysis depends to a great extent on spectral interferences that reflect the matrix composition. An exact figure for the detection limit therefore cannot be given for a group of materials such as coal where the composition may vary considerably. Using the combination of thermal and epithermal activation, the following may be generally assumed as detection limits in coal:

<0.01 ppm	:	In, Ta
0.01–0.1 ppm	:	Sc, Co, As, Br, Sb, Cs, Eu, Tb, Yb, Lu, Hf, Th, U
0.1–1 ppm	:	Cr, Ga, Se, Mo, I, La, Ce, Sm, Ho, W
1–10 ppm	:	Na, V, Mn, Ni, Cu, Zn, Rb, Ba, Nd
10–100 ppm	:	Cl, Sr, Zr
>100 ppm	:	Mg, K, Ca, Ti, Fe

Some rare elements not included in this study, such as Au and Ir, would also show detection limits of 0.1 ppm or less in coal by instrumental methods.

The detection limits in an ash sample are poorer than in the coal because of the higher concentration of most elements contributing significantly to the matrix activity. Since the concentrations of most trace elements are correspondingly higher, however, the conditions for determination become slightly better, as indicated by the improvement in precision observed in SRM-1633 compared with SRM-1632 for many elements. This is in part because volatile elements, such as bromine, contributing significantly to the matrix activity in coal, are to a great extent released on burning the coal.

C. Sources of Systematic Error

Sources that could lead to significant systematic errors in instrumental activation analysis are of three types:

(1) Spectral interferences. The γ-rays that could interfere with the measurement of a particular γ-ray in instrumental activation analysis are in most cases well known, and do not represent a great problem with the present state of the art in detector technology and data reduction facilities. Such interferences should, however, always be considered in order to introduce corrections whenever necessary.

(2) Errors associated with different neutron exposure to sample and standard. This may occur if the sample contains strong neutron-absorbing elements as major constituents. This is not the case for coal and coal ash neither for thermal nor epithermal neutrons. Since carbon is a good neutron moderator, one might expect some moderation of epithermal neutrons in a coal sample, which might introduce possible systematic errors in epithermal activation analysis. Such effects, however, are shown to be insignificant with the present sample size of 100 mg (Steinnes and Rowe, 1976).

(3) Interfering nuclear reactions, the most important of which are the thermal neutron fission of uranium and fast-neutron induced reactions of the (n,p) and (n,α) type. Fission products of uranium may interfere seriously with the determination of elements such as Zr, Mo, Ba, and Ce in samples with a high uranium content. Appropriate corrections can be introduced by means of a simultaneously irradiated uranium standard. In epithermal activation the fission-product interference become less serious in many cases because of the low I/σ ratio of the fissioning isotope ^{235}U. Concerning fast-neutron induced reactions, these are in most cases insignificant in thermal neutron activation. They become far more important in the case of epithermal activation because their reaction rate is not reduced by the presence of cadmium. Nevertheless, they do not constitute a large source of error in most cases. In the present work, minor corrections had to be introduced to the observed results from epithermal activation in the case of Sc and Cr. If Na, Mg, Al, or K were to be determined after epithermal activation, this interference would be more serious. These elements are, however, determined more conveniently after thermal neutron activation, where the interference is of no great concern for the samples discussed here.

In general, the systematic errors associated with instrumental neutron activation analysis of coal and coal ash are known and can be adequately accounted for. The accuracy of the analysis may therefore be nearly the same as the precision.

VI. CONCLUSION

Instrumental neutron activation analysis with a nuclear reactor is a convenient and sensitive technique for the simultaneous determination of a number of elements in coal and coal ash. Nearly 40 elements may be detected by thermal neutron activation at the concentrations in which they are present in coal, and of these about 30 elements may be determined quantitatively in most samples of coal and coal ash with a satisfactory result. The introduction of epithermal activation using a cadmium cover increases the number of elements that can be satisfactorily determined to 40 or more. According to the experience presented here, the following 21 elements are more favorably determined by the epithermal irradiation technique:

Ni, Ga, As, Se, Br, Rb, Sr, Zr, Mo, In, Sb, Cs, Ba, Sm, Tb, Ho, Hf, Ta, W, Th, U.

Elements determined with a better precision and sensitivity using the conventional thermal neutron activation are:

Na, Mg, Al, Cl, K, Ca, Sc, Ti, Vi, Cr, Mn, Fe, Co, Cu, Zn, La, Ce, Nd, Eu, Dy, Yb, Lu.

If the samples are homogeneous at the 100 mg level, most of these elements may be determined with a precision of ±10% or better in coal and coal ash. Since many of the elements concerned are trace elements that are difficult to determine by other analytical techniques, the instrumental activation analysis is likely to play a significant role in future studies of trace elements in coal and their behavior in the burning of coal.

REFERENCES

Abel, K. H., and Rancitelli, L. A. (1975). *In* "Trace Elements in Fuel" (S. H. Babu, ed.), pp. 118–138. Am. Chem. Soc., Washington, D.C.

Albinsson, H. (1974). *In* "Handbook on Nuclear Activation Cross-Sections," Tech. Rep. Ser. No. 156, pp. 15–86. IAEA, Vienna.

Baedecker, P. A. (1976). *In* "Advances in Obsidian Glass Studies" (R. E. Taylor, ed.), pp. 00–00. Noyes Press, Park Ridge, New Jersey.

Block, C., and Dams, R. (1973). *Anal. Chim. Acta* **68,** 11–24.

Bowen, H. J. M., and Gibbons, D. (1963). "Radioactivation Analysis." Oxford Univ. Press (Clarendon), London and New York.

Brune, D., and Jirlow, K. (1964). *Nukleonik* **6,** 242-244.

Brunfelt, A. O., and Steinnes, E. (1969). *Anal. Chim. Acta* **48,** 13-24.

Chattopadhyay, A., and Jervis, R. E. (1974). *Anal. Chem.* **46,** 1630-1639.

De Soete, D., Gijbels, R., and Hoste, J. (1972). "Neutron Activation Analysis." Wiley (Interscience), New York.

Klein, D. H., Andren, A. W., Carter, J. A., Emery, J. F., Feldman, C., Fulkerson, W., Lyon, W. S., Ogle, J. C., Talmi, Y., and Bolton, N. (1975). *Environ. Sci. Technol.* **9,** 973-979.

Kruger, P. (1971). "Principles of Activation Analysis." Wiley (Interscience), New York.

Lyon, W. S. (1964). "Guide to Activation Analysis." Publ. under the auspices of Div. Tech. Inf., USAEC, Princeton, New Jersey.

Millard, H. T., and Swanson, V. E. (1975). *Trans. Am. Nucl. Soc.* **21,** 108-109.

Nadkarni, R. A. (1975). *Radiochem. Radioanal. Lett.* **21,** 161-176.

Ondov, J. M., Zoller, W. H., Omez, I., Aras, N. K., Gordon, G. E., Rancitelli, L. A., Abel, K. H., Filby, R. H., Shah, K. R., and Ragaini, R. C. (1975). *Anal. Chem.* **47,** 1102-1109.

Rowe, J. J., and Steinnes, E. (1977). *Talanta* **24,** 433-439.

Ruch, R. R., Cahill, R. A., Frost, J. K., Camp, L. R., and Gluskoter, H. J. (1975). *Trans. Am. Nucl. Soc.* **21,** 107-108.

Sheibley, D. W. (1975). *In* "Trace Elements in Fuel" (S. P. Babu, ed.), pp. 98-117. Am. Chem. Soc., Washington, D.C.

Sher, R. (1974). *In* "Handbook on Nuclear Activation Cross-Sections" Tech. Rep. Ser. No. 156, pp. 1-13. IAEA, Vienna.

Steinnes, E. (1971). *In* "Activation Analysis in Geochemistry and Cosmochemistry" (A. O. Brunfelt and E. Steinnes, eds.), pp. 113-128. Universitetsforlaget, Oslo.

Steinnes, E., and Rowe, J. J. (1976). *Anal. Chim. Acta* **87,** 451-462.

Zubovic, P., Steinnes, E., and Rowe, J. J. (1979). *U.S. Geol. Surv. Bull.* (in press).

Chapter 47

Fast-Neutron Activation Analysis for Oxygen, Nitrogen, and Silicon in Coal, Coal Ash, and Related Products

Alexis Volborth†
NUCLEAR RADIATION CENTER AND
DEPARTMENT OF GEOLOGY
WASHINGTON STATE UNIVERSITY
PULLMAN, WASHINGTON

† Former address: Chemistry Department, University of California, Irvine; and Chemistry and Geology Departments, North Dakota State University, Fargo, North Dakota.

303

I. INTRODUCTION

Nuclear activation analysis is a well established technique suitable for qualitative and quantitative chemical analysis of nearly all elements. As early as 1936, von Hevesy and Levi were able to determine one part of dysprosium in a thousand parts of yttrium by irradiating the sample with low-energy neutrons produced in a radium–beryllium source (von Hevesy and Levi, 1936). The use of neutron activation analysis (NAA) did not become widespread until the end of the 1950s because of the scarcity of the isotopic neutron sources and the relatively low fluxes generated by such devices. With the development of nuclear reactors and accelerator-type neutron generators, the NAA has become one of the most widely useful analytical techniques.

The perfection of scintillation detectors and the multichannel analyzers (MCA) in the 1960s and especially of the lithium drifted germanium detectors Ge (Li) during the last decade have transformed NAA into one of the most sensitive, precise, accurate, selective, rapid, and reliable methods of analytical chemistry. It is now entirely on a par with x-ray fluorescence (XRF) and atomic absorption (AA), and perhaps it is even more useful in terms of the total number of elements that can be determined satisfactorily and the relative simplicity of application.

While nuclear activation in the broad sense encompasses all activation or extitation of the nucleus of the atom that can be effected by such distinct nuclear particles as neutrons, protons, deuterons, tritons, alpha particles, and, also, in some light elements, by hard gamma rays, the most commonly used exitation is that with neutrons of varying energies.

A. Fast-Neutron Activation versus Thermal or Reactor Neutron Activation

The interaction between neutrons and atomic nuclei of the elements is called neutron activation (NA). Depending on the energy of the neutron and the nature of the target nucleus, a number of nuclear transmutations can occur. The most commonly used types of neutron interaction are the so-called n gamma (n,γ) and (n,p) reactions. The first is typical of neutrons of low energy or so-called "slow neutrons" or "thermal neutrons" generated primarily in a nuclear fission reactor and possessing a broad spectrum of neutron energies. The second reaction is typical of high energy or so-called "fast-neutrons" that can also form a part of the neutron energy spectrum of a reactor but can be specifically produced in Cockroft–Walton-type accelerators. The nuclear chemist distinguishes accordingly between thermal neutron activation (THNA) and fast-neutron activation (FNA).

Two typical nuclear reactions are presented below:

$$^{51}_{23}V + ^{1}_{0}n = ^{52}_{23}V + \gamma, \qquad ^{27}_{13}Al + ^{1}_{0}n = ^{27}_{12}Mg + ^{1}_{1}H.$$

The first reaction can be written in an abbreviated form:

$$^{51}V(n, \gamma)^{52}V.$$

This is a typical (n,γ) reaction where a thermal neutron interacts with vanadium isotope 51 producing vanadium isotope 52 and gamma radiation characteristic of the product. Thus (n,γ) reactions produce heavier isotopes of the same element being irradiated. Since this isotope is radioactive, it can be identified by its characteristic gamma rays and the half-life of its decay.

The second reaction can be abbreviated:

$$^{27}Al(n, p)^{27}Mg.$$

It is an (n, p) reaction occurring when a fast neutron (n) knocks off a proton (p) from aluminum 27, forming magnesium-27 and a proton particle.

In general, nuclear reactions can be written:

$$X + a = Y + b$$

or in abbreviated form:

$$X(a,b)Y,$$

where X is the target nucleus, Y is the product nucleus, a is the projectile, and b is the emitted light particle, or photon, in (n,γ) reactions.

In the above cases, a neutron capture occurs and, depending on the energy of the projectiles the product particle, is either an isotope of the target element or another element one position lower on the periodic chart of the elements. A neutron capture in general results in a compound nucleus, usually excited, because an addition of a nucleon (n or p) increases the energy of the product by 8 million eV (1 eV is the energy acquired by a charged particle falling through a potential of 1 V). Other types of nuclear transmutation reactions are known, for example $(n,2n)$ type reactions where the capture of a fast neutron causes the emission of two neutrons making a chain reaction possible. (n,α) reactions are also known. Using alpha particles, protons, and deuterons, such reactions as (α,γ), (d,n), (d,p), (d,γ), and (p,d) are possible. Combined, all these and other types of different nuclear reactions, when used for determinative purposes, constitute nuclear activation analysis.

The main difference between fast-neutron activation and thermal neutron activation is in the energy of neutrons used and, therefore, in the sources of the neutrons. The fast neutrons are produced in Cockroft–Walton accelerators that use the so-called "(d–t)" reaction. This can be written as:

$$\ _1^3\text{H} + \ _1^2\text{H} = \ _2^5\text{He}^* = \ _2^4\text{He} + \ _0^1\text{n} + 17.6 \text{ MeV}, \quad \text{or abbreviated} \quad \text{T(d,n)}^4\text{He},$$

signifying that when the tritium target is bombarded by deuterium (accelerated to above the 110 kV energy threshold) a very fast-decaying compound nucleus He-5 is formed, which decomposes immediately into a helium-4 particle and a fast neutron (of approximately 14.7 MeV energy), causing considerable surplus thermal energy to be emitted.

This reaction produces monoenergetic neutrons (actually of a very narrow energy spectrum depending on acceleration voltage and the difference between the energies of the forward and the backward scattered neutrons), whereas the neutrons produced in a reactor have a very broad spectrum reaching out into the "fast" region.

B. Advantages of Fast-Neutron Activation

Because the probability of neutron and target interaction depends on the energy of the neutron projectile and the type of the target nucleus (cross section), having a source of monoenergetic neutrons is of great advantage to the nuclear chemist. Whereas the gamma-ray spectrum of a sample irradiated in a reactor is highly complex, the "fast-neutron" irradiated sample spectrum is much simpler. In addition, some light elements, for example oxygen, nitrogen, and silicon, produce strong, distinctive, and simple gamma-radiation spectra when irradiated by fast neutrons, whereas when they are irradiated in a reactor with other elements hardly any gamma activity is apparent. Such spectra are, in addition, highly complicated by the overlapping spectra of the heavier elements that generally produce little or no gammas when irradiated by fast neutrons.

Further advantages of FNA are the relatively short half-lives of product nuclei that result in fast decay of induced radioactivity of the sample irradiated. This also means more rapid analysis and thus better precision because repeated determinations are easy to perform and samples are no longer "hot" only minutes after irradiation, simplifying storage.

One main advantage of using fast neutrons is the availability of relatively inexpensive generators that today cost about $40,000 versus a necessary investment of at least $500,000 and up to $1,000,000 for a modern pulsed reactor such as the Triga Mark I and subsequent models built by the Gulf-General Atomic Corporation. Also, the relative ease of operation of an accelerator that can be switched off and on when needed, and that contains only tritium as the radioactive target material, which is only a weak β-emitter and thus easy to shield, must be taken into account when compared to the necessity of permanently employing licensed reactor operators and securing licenses, which are relatively difficult to obtain, if one were to consider the use of a nuclear reactor. Thus, while nuclear reactors are best used by consortia or pooled groups centrally financed, the small accelerator is easily acquired and put into operation with a minimum of expense and time lost. It in fact costs less than a modern XRF machine and about as much as a sophisticated AA apparatus. If intended for training students in neutron activation analysis, the fast-neutron generator output neutrons can be thermalized with paraffin or water so that most of the thermal neutron activation reactions can be reproduced in class with the only difference that the flux will be about three orders of magnitude less than in a reactor. Thus, except for reactor operator training, the accelerator can be used to simulate most reactions produced in a reactor and thus be a useful tool for training in neutron activation analysis. Modern Cockroft–Walton accelerators can produce fast-neutron fluxes of about 10^{10} n/cm²/sec with typical neutron yields of $>10^{11}$ n/sec. This is sufficient for the detection of numerous elements at minor concentrations, and is especially suitable for the determination of several elements at major concentrations. In fact, FNA as an analytical method applied to major element analysis in terrestrial rocks, meteorites, lunar rocks, and coal (O, Si, Al, N) has proved to have significant advantages when compared with other methods (Volborth and Banta, 1963; Volborth, 1966; Volborth and Vincent, 1967; Volborth *et al.*, 1975, 1977b,c,d; Vincent and Volborth, 1967; Vogt and Ehmann, 1965; Ehmann and Morgan, 1970; Ehmann *et al.*, 1974; Hamrin *et al.*, 1975; Morgan and Ehmann, 1970; James *et al.*, 1975). Bibby and Sellschop (1974) did an interference and accuracy study in geological materials. A review of methods used in FNAA is given by Nargolwalla and Przybylowicz (1973).

Accelerator activation in general compliments reactor neutron activation. Its application is even broader than indicated above because particles other than deuterium and targets other than tritium can be selected (protons, for example, can be used to excite x rays, PIXE), and the acceleration voltage can be further increased to cause a multitude of nuclear reactions. Here, however, we are in the realm of nuclear physics, and the practical applications to analytical problems become more difficult and expensive. In this chapter we confine our description to FNAA methods specifically developed for coal and coal related industrial products.

C. Advantages of Nondestructive Analysis of Coal

Coal, lignite, bituminous coal, anthracite, and industrial organic materials produced by the upgrading processes of these solid fuels are usually of a highly complex chemical nature. In terms of elemental composition, these substances are composed mainly of carbon, oxygen, hydrogen, nitrogen, sulfur, silicon, iron, and aluminum in major concentrations and contain a multitude of trace elements, some of which are main sources in environmental pollution. The organic "coal molecule" is highly complex and unstable when heated or reacted with industrial solvents. Highly volatile gases such as methane, carbon dioxide, carbon monoxide, nitrogen, hydrogen, and especially water are present in different forms and escape at different temperatures. Grinding and storage of coal, drying processes, and most sample preparation and diminution cause major changes in the constitution of these products. The interpretation of all these processes requires ideally the knowledge of the total composition during each significant step of handling or industrial upgrading. The classical coal analysis methods consist of time consuming procedures and have as their main deficiency the inability to determine oxygen other than by "difference." The so-called "moisture" determination of the proximate coal analysis is also a bulk determination of the "weight loss" when drying, which includes to a significant extent the gases listed above (Volborth, 1976a,b,c).

The advantage of truly nondestructive techniques lies in the fact that they permit the determination of elemental concentrations without first destroying or altering the substance in any way, thus avoiding possible loss, dilution, and contamination by foreign substances. An analytical system that avoids essentially all changes in composition during preparation for analysis, if well calibrated, should then give us the best possible balances or summations. Methods must be, therefore, as much as possible interference free, rapid, and inexpensive. In addition, an industrial laboratory always needs the most straightforward methods that can be perfected and used by technician level personnel after a reasonable period of training. The introduction of FNAA seems to provide such methods for oxygen, silicon, and nitrogen. The main advantage is in the ability to determine oxygen accurately within about 1% standard deviation. Applied to all coals and coal derived products, this ability permits independent checks of moisture related oxygen

(Volborth, 1976c), of oxygen in coal ash (Volborth, 1976d; Volborth *et al.*, 1977e), and it also enables one to produce more meaningful summations in material balances than was possible when using "oxygen by difference" data.

In addition to nondestructive FNAA determination of oxygen, silicon, and nitrogen (Volborth and Miller, 1977; Chyi *et al.*, 1976), the combination of these methods with thermal neutron activation for trace elements (Ruch *et al.*, 1977) could, when fully developed, result in a better and faster overall method for coal analysis that will give more information than classical coal analysis.

II. INSTRUMENTATION

A. Generators of 14 MeV Neutrons

Several types of fast-neutron generators exist (Volborth, 1972). One distinguishes two different designs. The older Cockroft–Walton-type generators were of the "drift-tube" type (Fig. 1). These consist of an ion source with a deuterium gas supply and a series of accelerating electrodes in a so-called "lens assembly," all placed into a metallic dome called the "accelerator head." In this liquid cooled dome deuterium ions are produced by radio-frequency exitation or other methods, leaked into the lens assembly under high vacuum driven with an ion

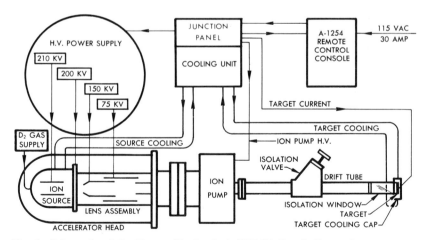

Fig. 1 Schematic of the Kaman Nuclear's A-1254 "drift-tube" type fast-neutron generator. Output 2.5×10^{11} 14.3 MeV Neutrons per second. (Courtesy of Kaman Nuclear, 1700 Garden of the Gods Road, Colorado Springs, Colorado 80907.)

pump and accelerated toward the negative end by stepwise decreasing voltages. The accelerated deuterons are focused to enter a metallic tube of varying length called the "drift tube," hence, the term drift-tube accelerator. At the end of this tube a metallic target usually of titanium foil, saturated with tritium and cooled by liquid freon and water, is placed. The fast neutrons are produced at the target whence they are scattered in all directions, depending on the collision angle of the deuterium projectile with the tritium atom, and thus acquiring some- what different energies. The nuclear reaction that causes the fast neu- trons to form is the so-called deuterium–tritium reaction, "(d–t)" re- action, as given previously.

The recently developed "sealed fast neutron sources or tubes" (Fig. 2) work on the same principle as the drift-tube accelerators with the difference being that the housing is permanently sealed and the parts are miniaturized. In addition, these tubes contain the necessary amount of tritium and deuterium, which are accelerated together so that the target while producing neutrons is reimpregnated by tritium and thus is not depleted as rapidly as in a drift-tube generator.

For the analyst, the sealed type neutron generator is more convenient to operate, lasts longer, and produces a steady neutron flux plateau that does not fall off as it does in the drift-tube-type accelerator. An additional convenience is that in the sealed tube, tritium is perma- nently confined to the sealed chamber, therefore making the cumber- some tritium smear and urine tests required when working with the drift tubes superfluous. Because of its design, a sealed tube can produce neutrons for 100 to 400 hr depending on the nature of use, whereas, a drift tube device needs a tritium target change every 2–4 hr.

B. Signal Processing Systems as Applied to Short- Lived Radioactivity and Fast–Neutron Activation

Nuclear detectors and multichannel analysis with data acquisition systems in NAA are covered in Volume I, Chapter 12. Here only the dual system of single channel analysis and reasons for such an elaborate system in FNAA will be discussed.

Short-lived radioactivity typically utilized in FNAA causes several related problems in NAA and in counting that are difficult to correct. At high counting rates, the built-in capability of modern multichannel analyzers to count for the real time period by recording the lost time or "dead time," which is because of pulse overlap, and automatically counting longer, is of little use. This is because the pulse rate decreases

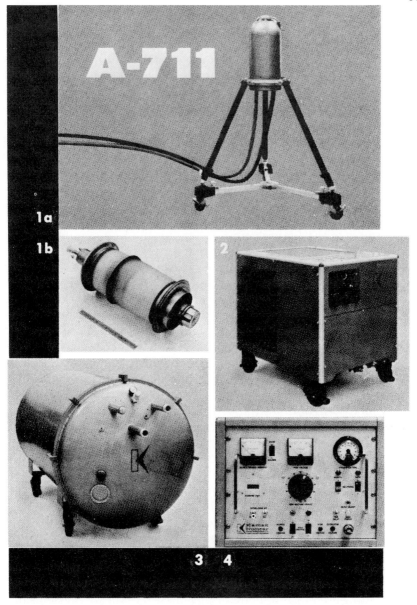

Fig. 2 Components of the Kaman Nuclear's "sealed tube" fast-neutron generator A-711. Output $>10^{11}$ 14.3 MeV neutrons per sec. (1) The accelerator head with the sealed neutron generating tube enclosed in a pressurized sulfur hexafluoride SF_6 atmosphere (a) including the Penning ion source (b) utilizes mixed-beam of tritium and deuterium. (2) The cooling system consisting of three subsystems: the closed loop target cooling tap water system, the Freon 113 cooling system for the ion source, a heavy duty commercial-type refrigeration unit to cool the target and the source cooling systems. (3) High voltage power supply pressurized with SF_6. (4) The remote control unit console including a 50 ft cable. (Courtesy of Kaman Nuclear.)

so rapidly that the counts accumulated at the end of the counting interval, which last typically four half-lives, would constitute 93.8% of all counts obtainable, thus leaving only a few additional counts to be collected if proceeding to count for the lost period of time. The result is that one merely increases the "background" counts relative to counts resulting from the nuclide of interest. This means that for accurate work some other means of achieving true ratios between the standard and the unknown sample must be found. This especially applies to analytical systems based on sequential comparison of simultaneously irradiated samples, in which the sample or the standard are counted first and compared to counts collected for the other sample after one-half of the lifetime is counted. In such systems decay corrections must be performed mathematically, and if there is considerable dead time effect because of the higher counting rate for the first-counted sample, these decay corrections are insufficient because of the different dead time corrections required. Simultaneous irradiation and simultaneous counting solve this problem well, especially if both the standard and the unknown sample are selected so as to contain a similar quantity of the element analyzed for. Such a system has been suggested by Volborth and Banta (1963), built with National Science Foundation support, and discussed by Volborth (1966). It was further perfected in 1972 (Volborth *et al.*, 1972), and in 1975 at the University of California, Irvine (Volborth *et al.*, 1975), it was especially adapted to the analysis of oxygen, nitrogen, and silicon in coal in 1977 (Volborth and Miller, 1977). A block diagram of this system is given in Fig. 3.

A dual irradiation system, in addition to making the cumbersome decay corrections as applied by Vogt and Ehmann (1965) superfluous, minimizes the effect of dead time when count ratios of two samples with similar counting rates are taken, and can also be further adapted for "crisscross" irradiation and counting of samples as shown in Fig. 3. If samples are switched so that they are irradiated and counted in opposite positions and rotated in addition along their axes, two sequential counts, if added, will represent data corrected for neutron flux variation, spottiness of the tritium target if samples are not rotated, and the possible drifts of the counting system electronics (Volborth, 1969).

The superiority and simplicity of such dual "crisscross" systems have been demonstrated in several papers listed previously. Advantages in industrial-type applications, in coal analysis especially, were demonstrated by Volborth and Miller and co-workers (Volborth *et al.*, 1977e, 1978; Miller and Volborth, 1976).

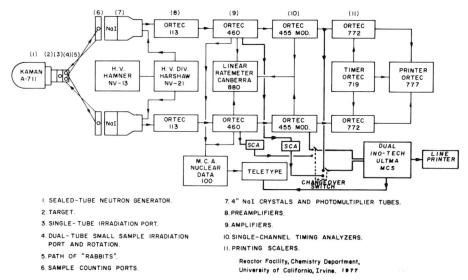

I. SEALED-TUBE NEUTRON GENERATOR.

2. TARGET.

3. SINGLE-TUBE IRRADIATION PORT.

4. DUAL-TUBE SMALL SAMPLE IRRADIATION
   PORT AND ROTATION.

5. PATH OF "RABBITS".

6. SAMPLE COUNTING PORTS.

7. 4" NaI CRYSTALS AND PHOTOMULTIPLIER TUBES.

8. PREAMPLIFIERS.

9. AMPLIFIERS.

10. SINGLE-CHANNEL TIMING ANALYZERS.

11. PRINTING SCALERS.

Reactor Facility, Chemistry Department,
University of California, Irvine. 1977

Fig. 3 A block diagram of the dual fast-neutron activation, transfer and data handling system for the sequential determination of oxygen, silicon, and nitrogen, etc., at the University of California, Irvine. In addition to the Kaman sealed tube generator on the left with two permanent transfer systems and the Ortec Electronics (Ortec, Inc., 100 Midland Road, Oak Ridge, Tenn. 37830) indicated on the drawing, this system also includes the INOTEC ULTIMA II analyzer, built by INO-TECH, INC., Madison, Wisconsin 53719. This latter expanded system is equipped with two analog to digital converters of 8192 resolution permitting simultaneous collection of data from two detectors utilizing the "dual" self-correcting "criss-cross" principle used in the earlier constructed system with Ortec electronics described above.

C. Gamma-Ray Spectrum Produced after Fast-Neutron Irradiation

The gamma-ray spectra of samples irradiated by fast neutrons are generally much simpler than spectra of the same samples irradiated in reactors. This is because of the fact that fewer elements are excited by FNA. This is the result of a virtually monoenergetic spectrum of FNA sources as compared to a very broad neutron energy spectrum in reactors. This greatly helps in the interpretation of results and the counting and reduces the background significantly. A typical FNA spectrum of a rock (granite) is given in Fig. 4. The only strong peaks seen here are, from left to right, the 0.511 MeV because of positron annihilation gammas, the peaks from iron, aluminum, magnesium, and silicon (most prominent), as well as the 6.1 and 7.1 MeV-wide peaks with

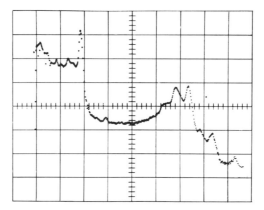

Fig. 4 Gamma ray spectrum of a granitic rock activated for 20 sec and counted for 20 sec; on logarithimic scale. The first wide peak on left is the 0.511 annihilation gamma peak, the sharp tallest peak is the ^{28}Al = 1.78 MeV, originating from ^{28}Si. The four wide peaks on the right are all due to oxygen-16 (^{16}N = 6.1, 7.1 MeV) including escape peaks. The total spectral width in this figure is about 8 MeV.

escape peaks on the right caused by oxygen-16 (nitrogen-16). This spectrum was taken about 4 sec after irradiation counting for 20 sec.

Corresponding Ge(Li) spectra are shown in Volume I, Chapter 12, Figs. 2, 7, and 8, and results are discussed in Chapter 46. It should be emphasized that while our Fig. 4 encompasses the whole gamma-ray spectrum from 0 to about 10 MeV energy, the previously cited spectra are only short segments of Ge(Li) spectra. The use of thallium-activated sodium iodide scintillation crystals, NaI(Tl), for the detection of the short-lived activities of the FNA product nuclides is of an advantage because of an order of magnitude higher efficiency as compared with Ge(Li) semiconductor detectors. Large NaI(Tl) crystals 4, 5, or 6 in. in diameter are preferred, especially when oxygen is determined, because such crystals result in smaller escape peaks. To avoid counting rates over 10,000 counts/sec and dead time effects depending on the concentration of the element analyzed, the neutron output can be adjusted down by lowering the milliamperage of the generator. We have found that adjusting the generator output so that each count produces about 50,000 counts for oxygen in the sample after a 20 sec accumulation period results in no detectable dead time loss with a 7.3 sec half-life and a modern double-delay-line amplifier and single-channel analyzer combination.

D. Shielding and Safety Requirements of a Fast-Neutron Activation Laboratory

It was mentioned previously that the generation of fast neutrons is relatively inexpensive when compared to reactor neutron activation. Nevertheless, added costs result from some critical shielding problems. These determine the location of an accelerator that has to be chosen in a way to prevent most of the 14 MeV neutrons from reaching any working area around and above the neutron source. The best location is, therefore, the basement of a building that is in a corner where two walls are facing clay or other permanent fill and that has a thick concrete wall and ceiling. Preferably, the room should be reserved for the counting and remote control equipment in such a way that the operator and personnel are placed as far from the source target as possible. From solid concrete bricks one builds thick walls with a narrow "labyrinth" walk so that no straight path for neutrons exists.

Another frequently used approach is to build a silo with 35 in. diam and 110-in. deep as shown in Fig. 5. From the standpoint of shielding efficiency and space requirement, which in this case is relatively small, this arrangement maybe is best. However, we have found that in laboratories with industrial demands handling hundreds of samples per day, the disadvantage of such an installation is a poor, and time consuming access to the accelerator head in case repairs become necessary, especially since some sample containers ("rabbits") tend to become stuck and break up at the irradiation port, necessitating cleaning. This difficulty is partially because of a relatively sharp bend of the sample transfer system plastic tubing and because of the fact that some samples, if improperly sealed or too wide, can be caught in the rotating mechanism and be broken up occasionally. Lifting the whole accelerator dome with all the cooling and electric cables by a winch is a rather time-consuming operation. Therefore, we prefer a "walk-in" arrangement.

An important consideration is the thickness of concrete or concrete block wall used for shielding. Here the rule of thumb is that at least 5.5 ft of solid concrete is required to bring the fast-neutron level to the required minimum safety standard level of 10 neutrons per cm^2 per second (10 $n/cm^2/sec$), which is the maximum permissible dose for a 40 hr week working period.

Figure 6 gives the fluxes of fast neutrons versus thickness of concrete. In the three installations performed by this author, 6 to 7 ft of concrete were used in order to bring the fast-neutron level below 5 to 0.5 $n/cm^2/$ sec.

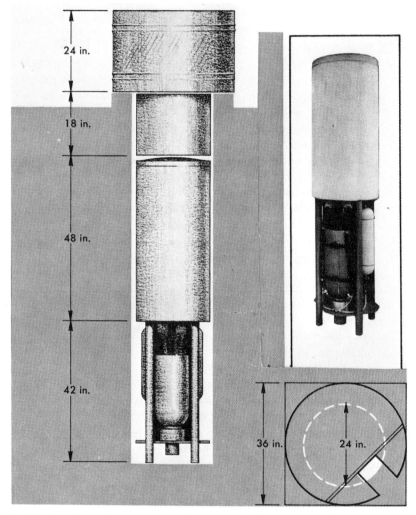

Fig. 5 A typical well-mounting facility similar to the UCI facility for the sealed tube A-711 Kaman nuclear neutron generator. An additional 2-ft-high tank filled with polyethylene beads is placed above the well for shielding. The whole assembly can be lowered into a well by a manual or electric hoist. (Courtesy of Kaman Nuclear.)

In addition to proper shielding, which can also be done with tanks or containers filled with water or blocks of paraffin, safety requires built-in interlock systems that prevent entry. External warning devices, such as red lights that indicate when the accelerator is in operation and prevent accidental entry, are mandatory. Emergency KILL switches

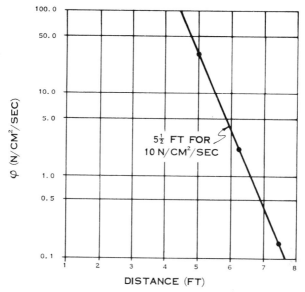

Fig. 6   Fast-neutron flux versus distance for concrete. (Courtesy of Kaman Nuclear.)

must be located in positions accessible when working with the machine. Kaman Nuclear Corporation (1966) Technical Bulletin No. 104 is a good guide to shield considerations. Nevertheless, it is imperative to measure the neutron fluxes with calibrated monitors during and after the installation of a neutron generator. It is important to remember that thermalized neutrons are also present, that the machine also produces hard x rays and α-particles, and that the cooling water oxygen is also activated making the pump housing also radioactive during operation. Distance from the source is an important consideration, too. After use, the metallic parts near the source are strongly radioactive because of the aluminum and iron in the steel parts, and after prolonged operation, one should preferably wait for 24 hr before working for extended periods at the target end of the machine. At least 2 hr waiting before entry is recommended. One always has to wear safety monitors and badges when working with these accelerators. In general, shielding and operations must follow the Atomic Energy Commission Regulation 10 CFR 20. A useful review on high-intensity neutron sources has been published by the U.K. Atomic Energy Authority [see Olive *et al.* (1962)].

III. METHODS

A. Sample Preparation Techniques for Fast-Neutron Activation of Coal, Coke, Coal Ash, Oils, Humic Substances, and Asphaltenes

One of the major considerations in preparation of coal and related substances for analyzing is their instability and partial volatility of some light hydrocarbons, methane, etc., and other gases such as carbon monoxide, carbon dioxide, and adsorbed nitrogen, as well as the many forms of water distinguished by the coal chemist depending on the method of determination. The procedures used in coal sampling and subsequent storage in closed containers are well defined in ASTM standards (D2234-72 and D197-30 "reapproved" 1971) (ASTM, 1975). Sieve analysis and preparing coal samples for analysis and grinding are also well covered (ASTM standards D410-38 "reapproved" 1969, and D2013-72) (ASTM, 1975). Nevertheless, recently, questions have been raised as to the adequacy of some of the grinding procedures because of oxidation and loss of moisture in some coals during dimi- nution if exposed to air, and "cryocrushing" in liquid nitrogen was proposed by Solomon and Mains (1977) to avoid this effect when coal has to be crushed for accurate scientific tests.

Assuming that the sampling and storage as well as crushing and sieving, if any, have been performed properly according to the ASTM standards, the analyst still is confronted with a substance that can contain as much as 40% water, will oxidize when dried in air, and could lose considerable moisture when exposed to the dry atmosphere in a laboratory (Volborth, 1976c,d; Volborth *et al.*, 1977d,e, 1978). Sim- ilar considerations apply to coke, ash (high-temperature ash and low- temperature ash) (HTA and LTA), oils and tars, which often are emul- sions with considerable amounts of water, asphaltenes, humic acid concentrates, etc.

In order to permit a well normalized basis for our work with all these substances, we use a unified procedure when preparing our samples, which consists of putting aside the powders or volatile substances for the period of sample packing and weighing and determining the per- cent loss or added weight (if samples have been dried). This loss or gain is then computed assuming it is all water (not true in light oils). This permits a normalization relative to oxygen of the escaped or ad- sorbed water and results in more meaningful data than if the escaped or added moisture were not considered (H_2O = 88.9% O). Concentra- tion and dilution effects for all major constituents are corrected for by this method in a simple computer program. The magnitude of these corrections can be seen from Table I where losses of moisture of several percent are recorded with some coals exposed to laboratory ambient air and moisture for only from 5 to 15 min. Obviously, if these losses remained unrecorded and uncorrected the whole coal analysis could not have been balanced in a stoichiometric sense. This effect is difficult to avoid because weighing and packing of coal into containers or "rabbits" for activation takes varying periods of time and demands hand manipulation, compacting, and sometimes repeated opening of the containers. This method has also permitted us to determine approximately how much moisture loss occurred in the coal during each subsequent weighing. It was shown that in relatively coarse wet coal each opening of the container over periods of years does sometimes cause significant losses that can be monitored only after homogenizing and redetermining the moisture content. On the other hand, very dry,

TABLE I *Losses and Gains of Coal Powder Samples during the Period of Weighing and Packing Approximate Duration of Manipulations 5 to 15 Min*

Sample	Description	Loss of wt.% (as received)	Gain of wt.% (dried at 105°C)	Total moisture % by Brabender method
PSOC 416	Lignite; Titus Co., TX	−6.14	0.33	29.22
PSOC 420	Subbit C; Milan Co., TX	−5.20	0.39	31.90
PSOC 92	Lignite; Richland Co., MT	−3.68; −4.01	1.87	34.00
PSOC 86	Lignite; Mercer Co., ND	−3.48; −2.34	1.33	28.60
PSOC 88	Lignite; Mercer Co., ND	−2.82; −3.98	1.89	31.80
PSOC 140	Lignite; Harrison Co., TX	−1.78	1.20	35.95
PSOC 139	Subbit C; Harrison Co., TX	−1.77	1.04	18.60
PSOC 154	HVC; Kane Co., UT	−0.82	0.63	12.80
PSOC 289	HVB: Grundy Co., IN	−0.71	0.57	17.70
PSOC 280	HVC; Sullivan Co., IN	−0.57	0.46	10.60
PSOC 313	HVB; Carbon Co., UT	−0.55	0.34	5.80
PSOC 305	HVB; Jefferson Co., OH	−0.20	0.18	4.05
PSOC 340	HVA; Jefferson Co., PA	−0.12	0.14	2.50
PSOC 259	Med Vol; Clearfield Co., PA	0.00	0.15	0.70
PSOC 142	Med Vol; LeFlore Co., OK		0.14	1.10
PSOC 134	Med Vol; Jefferson Co., AL		0.07	0.80
PSOC 157	Med Vol; Pitkin Co., CO	0.00	0.00	0.75
PSOC 129	Low Vol; Cambria Co., PA	0.00	0.03	1.00
PSOC 318	Low Vol; Somerset Co., PA		0.08	1.25
PSOC 319	Low Vol; Somerset Co., PA		0.07	1.85
PSOC 255	Med Vol; Clearfield Co., PA (+)	0.03	0.04	0.50

fine coal powder adsorbs considerable water very rapidly upon opening the bottle. Another solution used for some samples was to equilibrate them under ambient conditions. But this process takes time and is best applicable to coals dried in an air drying room (see U.S. Bureau of Mines, 1967, p. 2) and not to coals "as received" nor to coals "oven dried" at 105°C.

In coal ash, irrespective of whether these are HTA or LTA, the hygroscopicity problem is less serious but amounts to 0.1–0.4 wt.% for the former and may be as high as 1–2% with the latter, depending on the type of clay mineral and other minerals ashed by radio-frequency induction heating.

In oils, the percentage of dissolved or suspended ash varies as well as that of moisture. The difficulty is in the selective evaporation of the lighter hydrocarbons, which cannot be controlled. Asphaltenes and humic acid mixtures are very difficult to standardize and pack because of their high viscosity. We have found that heating was necessary in some cases, which may remove, selectively, water and low-boiling point hydrocarbons if present and has frequently caused further solidification of these gluelike substances making packing into the plastic "rabbits" impossible or very difficult. The method, in brief, consists of homogenizing the sample in the receiving container by tumbling for several minutes or mixing with a glass rod and of opening the container and putting about 500 mg of powder or the substance on a flat tared dish by the balance for the period of packing and sealing the "rabbit." Immediately after the "rabbit" is packed and sealed, the dish is weighed again and the weight loss or increase determined and reported in weight percent.

B. Special Data Reduction Processes

Calculation and reporting of classical coal analyses present problems because of the nature of coal. Generally, analysis of coal is performed on "air-dried" samples. At the U.S. Bureau of Mines (1967) Coal Laboratories in Pittsburgh, air drying consists of two consecutive operations at 30°–35°C, lasting 20–24 hr each, and subsequent weighing. If the difference between the weighings does not exceed 3.5%, the sample is considered having achieved equilibrium with the air in the laboratory. From the "air drying loss" the *"as received"* values for moisture, volatile matter, fixed carbon, and ash are calculated. This requires data on proximate analyses. Thus, for example:

$$\text{Moisture as received} = \text{moisture} \times \frac{(100\text{-air drying loss})}{100}$$

$$+ \text{ air drying loss}$$

$$\text{Volatile matter as received} = \text{volatile matter} \times \frac{(100\text{-air drying loss})}{100}$$

$$\text{Fixed carbon as received} = \text{fixed carbon} \times \frac{(100\text{-air drying loss})}{100}$$

$$\text{Ash as received} = \text{ash} \times \frac{(100\text{-air drying loss})}{100}$$

Data on the ultimate coal analysis are also recalculated to the "as received" basis. For example:

$$\text{Hydrogen as received} = H \times \frac{(100\text{-air drying loss})}{100}$$

$$+ 1/9 \text{ air drying loss}$$

$$\text{Carbon as received} = C \times \frac{(100\text{-air drying loss})}{100}$$

$$\text{Nitrogen as received} = N \times \frac{(100\text{-air drying loss})}{100}$$

$$\text{Oxygen as received} = O \times \frac{(100\text{-air drying loss})}{100}$$

$$+ 8/9 \text{ air drying loss}$$

Since total moisture is often performed on "as received" coal at 105°C in specially designed ovens or on "air dry" coal, recalculations to dry coal, on a "dry" or *moisture free* basis of values determined on air dried coal samples become necessary. For example:

$$\text{Volatile matter in dry coal} = \text{volatile matter} \times \frac{100}{100\text{-moisture}}$$

$$\text{Fixed carbon in dry coal} = \text{fixed } C \times \frac{100}{100\text{-moisture}}$$

$$\text{Ash in dry coal} = \text{ash} \times \frac{100}{100\text{-moisture}}$$

$$H \text{ in dry coal} = (H\text{-}1/9 \text{ moisture}) \times \frac{100}{100\text{-moisture}}$$

$$O \text{ in dry coal} = (O\text{-}8/9 \text{ moisture}) \times \frac{100}{100\text{-moisture}}$$

In addition, calculations from "air dry," "as received," or "moisture free" data to "moisture (dry) and ash free" basis (daf) are required for better comparison of coal analyses when the ash resulting from the inorganic matter (minerals in coal) as well as the greatly varying moisture are removed. For example:

$$\text{Volatile matter, daf} = \text{volatile matter} \times \frac{100}{100\text{-(moisture + ash)}}$$

$$\text{Fixed C, daf} = \text{fixed C} \times \frac{100}{100\text{-(moisture + ash)}}$$

$$\text{H, daf} = \text{(H-1/9 moisture)} \times \frac{100}{100\text{-(moisture + ash)}}$$

$$\text{C, daf} = \text{C} \times \frac{100}{100\text{-(moisture + ash)}}$$

$$\text{O, daf} = \text{(0-8/9 moisture)} \times \frac{100}{100\text{-(moisture + ash)}}$$

Similar recalculations are performed by the coal analyst for nitrogen, sulfur, and calorific values depending on how the coal sample has been pretreated.†

When it became obvious that most coal samples received for our tests had had varying histories, some packed "as received" others as "air dried" or as "dry" (at 105°C), and had been exposed to unknown steps during handling, packing, transport, and storage, it was decided to write a computer program that would treat all samples "as received" and, on the basis of our "moisture" determination and the moisture reported by the sample source, recalculate the rest of the source data to the concentrations at the time our sample was removed for oxygen analysis. (Volborth *et al.*, 1976, 1977*a,e*, 1978). This approach has permitted us to normalize all data for comparison and calculate the coal analysis on an actual "as received" basis.

The computer printout, Table II, demonstrates the result of such computation. The first column describes the nature of the analytical procedure and the elements determined, the second gives source data on an "as received" basis, the third gives the recalculated "as received" data describing the computed approximate composition of the sample as it was received in our laboratories, the fourth column gives the "DRY" moisture-free percentages of constituents based on our moisture determination, the "DAF" column gives recalculated data on the

---

† A detailed discussion of procedures for computing ultimate analyses, volatile matter, and calorific value on a dry mineral-matter-free (dmmf) basis is given in Volume II, Chapter 20, Section V.

TABLE II Oxygen Determination by Fast-Neutron Activation Analysis in Coals University of California, Irvine—Data and Recalculations

Proximate Analysis, %	Pennsylvania, Cambria County—low volatile bituminous coal				The Pennsylvania State University PSOC-126 LTA or M.M. Basis		
	Source data as received	As received	Dry	DAF	As received	Dry	DLTAF
Moisture	0.64	0.99 ± 0.1			0.99		
Ash	6.97	6.95	7.01				
LTA (or MM)					9.36	9.45 (LT)	
Volatile	18.50	18.43	18.62	20.02	17.95	18.13	20.02
Fixed Carbon	73.89	73.63	74.37	79.98	71.70	72.42	79.98
Summations	100.00	100.00	100.00	100.00	100.00	100.00	100.00
Ultimate analysis, %							
Hydrogen	4.50[a]	4.53[a]	4.46	4.79	4.53[a]	4.46	4.92
Carbon	82.37	82.08	82.90	89.15	82.08	82.90	91.55
Nitrogen	0.85	0.85	0.86	0.92	0.85	0.86	0.94
Organic sulfur	0.44	0.44	0.44	0.48	0.44	0.44	0.49
Other sulfur	0.65	0.65	0.65	0.65	0.65	0.65	0.00
Chlorine	0.04	0.04	0.04	0.04	0.04	0.04	0.04
Carbon dioxide	0.00	0.00	0.00	0.00	0.00	0.00	0.00
Oxygen (by diff)	(4.18)	(4.48)	(3.63)	(4.61)	(2.07)	(1.20)	(2.05)
Oxygen (by FNAA) as is	7.49	7.77 ± .05	6.96	4.01[b]	7.77	6.96	3.20[b]
Oxygen (by FNAA) dried	7.32[c]	7.61[c]	6.80 ± .08	3.84[b]	7.61	6.80	3.02[b]
Sum. Inc. FNAAO (as is)	96.34	96.35	96.31	100.05	96.35	96.31	101.15
Sum Inc. FNAAO (dried)	96.18	96.19	96.15	99.88	96.19	96.15	100.98
% Cations, etc. as ash	3.76[d]	3.75	3.79		5.33[e]	5.39	
Summations (as is)	100.10	100.10	100.10	100.05	101.68	101.70	101.15
Summations (dried)	99.94	99.94	99.94	99.88	101.52	101.54	100.98

[a] Includes H from moisture.
[b] Excludes estimated oxygen in ash.
[c] Based on moisture content.
[d] Estimated as 54% of ash content.
[e] Estimated 57% of LTA content.

sample dry and ash free. The last three columns signify similar recalculations when instead of HTA, the LTA data, if available, are introduced into the computations in order to calculate better the "organic oxygen" of the coal molecule. In all cases "OXYGEN BY DIFFERENCE" has been retained in order to compare and detect discrepancies between these values and the "OXYGEN DETERMINED" and normalized, depending on the method of reporting of the analysis. Oxygen determined on the wet "as received" sample is also given as well as oxygen determined in the "OVEN DRIED" sample of the same coal.

The space restrictions do not permit us to go into more detail on these programs here, but it may be sufficient to point out that where good balances are achieved and the data on oxygen determined compare well with "oxygen by difference" (after such normalizations), one may accept the whole coal analysis as sufficiently well done. Discrepancies, on the other hand, should indicate specific inadequacies or errors and may point toward proper solutions. The multitude of data that have to be recalculated in this case makes the use of a computer program mandatory.

C. Determination of Oxygen

Coal sample, irrespective of grain size, is packed into an oxygen-free polyethylene "rabbit" of about 2.5 mliter capacity (see Fig. 7), or larger, pressing to compactness if it is a fine powder. The container is closed, weighed, and sealed with a hot rotating platinum rod, and the seam is examined with a binocular microscope for possible holes. The "rabbit" and a "standard rabbit" containing either oxalic acid ($\sim$76% O), benzoic acid ($\sim$26% O), or triphenylmethanol ($\sim$6.3% O) (all cross calibrated and with known moisture content) are transferred by a fast pneumatic system and irradiated for 20 sec at a flux of approximately 10^8 n/cm²/sec, returned in front of two shielded 6-in. NaI(Tl) crystals, counted for 20 sec, returned to start position, switched into opposite transfer tubes automatically, returned after a timed period of about 2 sec (just enough for the switch), and irradiated again. This can be repeated 2, 4, 6, or 8 times depending on the precision desired, each time adding the counts collected from the two opposite ports (see Fig. 3). The selection of the type of standard depends on the quantity of oxygen in the coal sample. The closer the oxygen contents in both samples, the less will be the "dead time" effect, if any. The ratios of total accumulated counts of both samples are taken and multiplied by the known content of oxygen in the standard to obtain the total oxygen in the unknown. Usually the oxygen content in the empty container is

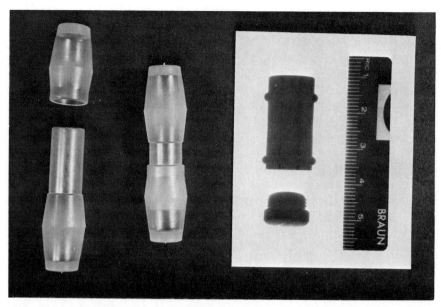

Fig. 7 Polyethylene containers of different capacities and shapes for solid and liquid samples for FNAA.

so small that correction is not necessary, however, when the oxygen content in the substance analyzed becomes very small, it has to be taken into account. This has been solved by a small computer program where the background counts are always fed in and subtracted. A Nova computer built into our new University of California at Irvine system (Fig. 3) is so programmed that background correction, oxygen percentages determined, and counting, as well as actual precision based on the number of individual determinations, are output. Figure 8 shows a similar arrangement installed by the author at Dalhousie University in Halifax, Canada, in 1972 (Volborth *et al.*, 1972; Volborth, 1969).

The nuclear reaction used in this case is

$$^{16}O(n,p)^{16}N.$$

The nitrogen-16 decays emitting 6.1 and 7.1 MeV gammas (see Fig. 4) that are counted using a wide window in a single-channel analyzer encompassing the region between 4.5 and 7.5 MeV, or simply integrating with an open window with a 4.5 MeV threshold. The main interferences arise from fluorine reaction,

$$^{19}F(n,\alpha)^{16}N,$$

Fig. 8 Fast-neutron activation laboratory built by the author at the Dalhousie University, Halifax, Canada, at the Canadian National Research Council Atlantic Regional Laboratory.

which is a primary interfering reaction. Fortunately, in coals and related substances, the fluorine concentration rarely exceeds 0.1%. The correction factor for F determined in our UCI laboratory is approximately $0.6 \times \%F = \%O$. Boron also interferes, but the correction factor there is $0.1 \times \%B = \%O$, which may be neglected unless major boron is present (Volborth and Miller, 1977).

D. Determination of Nitrogen

The same coal sample selected for the determination of oxygen is irradiated for 5 to 10 min, transferred in front of the NaI(Tl) detectors, and, after a 5 to 10 min cooling period to permit the activity caused by oxygen, silicon, and aluminum to decay, it is counted by accumulating, for 5 to 10 min, the gammas at 0.511 MeV caused by positrons emitted in the reaction $^{14}N(n,2n)^{13}N$. The resulting annihilation gammas have a half-life of 10 min. Simultaneous counting of standard and unknown

(Volborth and Miller, 1977) eliminates the decay corrections, necessary in sequential methods, described by others (Hamrin *et al.*, 1975; James *et al.*, 1975, 1976; Block and Dams, 1974; Bibby and Champion, 1974). Because of long decay (2.6 hr) of Mn-56 produced in the $^{56}Fe(n,p)^{56}Mn$ reaction, it was found that the next activation is preferably performed after at least a 10 hr wait, or, in practice, next day. If coincidence counting is available, the elimination of other radiation that does not result from positron emission would further improve the results. Such a technique, however, without the advantages of simultaneous counting, was employed earlier by James *et al.* (1975, 1976).

Several interferences have to be eliminated or their effect reduced to a lowest possible minimum. These are the primary interferences from reactions: $^{12}C(p,\gamma)^{13}N$, $^{13}C(p,n)^{13}N$, and $^{16}O(p,\alpha)^{13}N$. The incident protons are produced by the "knock-on" neutron–proton collisions within the coal sample and the organic container whenever hydrogen is present. Geisler *et al.* (1971) have demonstrated that the resulting quantity of ^{13}N from these reactions is a complex function of the H, C, and O content.

Additional interferences result from other reactions that produce positrons, e.g., $^{39}K(n,2n)^{38}K$ and $^{31}P(n,2n)^{30}P$. The former interference is especially troublesome because the ^{38}K activity has a half-life of 7.7 min. The 2.5 min activity of ^{30}P will sufficiently decay during a 10 min waiting period.

Taking the above discussed difficulties into consideration, a timing sequence shown in Fig. 9 was selected with repeat irradiation performed in opposite positions next day. In order to correct for the background caused by the varying amounts of the inorganic matter reported as ash, a corresponding quantity of ash for each coal can be activated separately, after mixing with graphite, for the same period, and the total background subtracted after identical irradiation procedure and waiting and counting times. Another approach is to activate pure ash from each coal and calculate the background knowing the percentage of this ash in the coal. This also has been done. Results of these studies have only been partially published (Volborth and Miller, 1977). This work was done with Dr. R. R. Ruch of the Illinois State Geological Survey (Ruch *et al.*, 1978). Selected data from this work are shown in Table III in sequence with decreasing ash content.

E. Determination of Silicon

Determination of silicon in complex substances such as meteorites and rocks is shown to be accurate and precise in publications by Wing

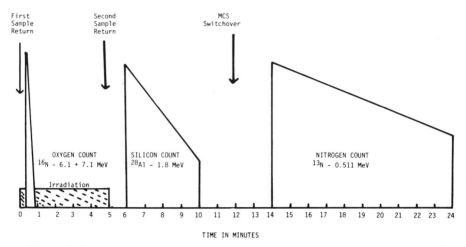

Fig. 9 Schematic timing diagram for sequential dual fast-neutron activation analysis for oxygen, silicon, and nitrogen in coal adapted to the INOTEC ULTIMA II analyzer facility partially built on Department of Energy Contract E(04-3) 34 PA 241, the reactor facility of the University of California in Irvine.

(1964), Vogt and Ehmann (1965), Vincent and Volborth (1967), and in coal and coal ash by Block and Dams (1974).

The reaction

$$^{28}Si(n,p)^{28}Al$$

produces a 2.3-min 1.78-MeV gamma emitting aluminum-28 isotope. This reaction has a large cross section resulting in a most prominent gamma peak as shown in Fig. 4.

The main interfering reaction is the $^{31}P(n,\alpha)^{28}Al$, which is primary, necessitating the knowledge of the concentration of phosphorus in the sample. Since the distribution of phosphorus in coals seems to peak at about 20–30 ppm and seldom exceeds 100 ppm (Gluskoter *et al.*, 1977, p. 65) this interference may be ignored. More serious interference comes from iron that is abundant in coal (peaking between 1 to 3%) (Gluskoter *et al.*, 1977, p. 69). It produces Mn-26 in the reaction $^{56}Fe(n,p)^{56}Mn$. The Mn-56 decays, emitting 1.81 MeV gammas with a 2.58 hr half-life. This means that the background under the 1.78 Al-28 peak will increase with repeated irradiations. For very accurate work, a waiting period longer than 10 hr is advisable. However, because of the short irradiation time used for oxygen, and a good cross section, silicon can still be determined after a 90 sec waiting period and the activation repeated up to 4 times. The ideal timing arrangement is

TABLE III *Nitrogen Determinations Based on Graphite/Ash Mixture Background Correction*

Sample number[a]	Type	Reported[a]			UCI determined[b]		
		HTA, %	N, %	H₂O, %	H₂O, %[c]	N, %(I)	N, %(II)
C-19487	Mineral residue SRC	63.9	0.48	2.3	2.05	0.45, 045	
C-18940	Residue, H coal filter cake	25.2	1.21	5.4	17.6[c]	1.29, 1.24	
C-19196	Still bottoms residue	17.14	1.57	0.1	0.13	1.76, 1.68	
C-18903	Illinois No. 6 100 mesh	10.7	1.17	2.5	2.9	1.19, 1.30	1.29
C-19141	SR coal	9.5	1.87	1.2	2.36	1.47, 1.33	2.17, 2.70
C-19740	Whole coal, Tenn.	8.7	1.64	1.5	1.57	1.42, 1.35	1.51, 1.46
C-19197	H-coal liquid	0.2	1.17	2.1	16.0[c]	1.35, 1.14	1.15, 1.12
C-19143	SR coal	0	2.43	0.1	0.16	2.31, 2.16	2.89
C-19486	SR coal	0.1	2.09	0.5	0.41	1.74, 2.09	2.11, 209
C-18937	Graphite	0	0.58	0.01	3.09	0.21, 0.21	0.37, 0.25
PSOC-106	HVB, Indiana	14.38	0.47			1.79, 2.09	1.82
PSOC-108	HVA, Penn.	9.50	1.11			1.09, 1.38	0.95, 1.30
PSOC-140	Lignite, Texas	9.37	1.14			1.23, 1.14	
PSOC-141	Lignite, Texas	9.02	1.23			1.46, 0.94	
PSOC-143	Low Vol, Okla.	8.82	1.36			1.71, 1.24	
PSOC-137	Med Vol, Ala.	7.05	1.42			1.50, 1.38	
PSOC-126	Low Vol, Penn.	7.01	0.86			1.47, 1.78	1.69, 1.76
PSOC-121	HVA, W.Va.	2.55	1.75			1.73, 1.78	
PSOC-136	Med Vol, Ala.	2.42	1.70			1.57, 1.73	

[a] C-19487, etc., Illinois State Geological Survey, Kjeldahl Nitrogen, on "as received" basis; PSOC-106, etc., Pennsylvania State University Sample Bank Data. Reported on oven dried (105°C).

[b] On samples, packed "as received."

[c] Loss of volatile matter at 105°C included.

shown in Fig. 9, where silicon is determined the next day, with "rabbits" in reversed counting ports. The Al²⁸ activity in silicon-rich samples is so high that fast double-delay-line amplifiers are advisable to minimize dead time.

Some silicon data on different coals received from Dr. A. Davis of The Pennsylvania State University and Dr. R. R. Ruch of the Illinois State Geological Survey are compared in Table IV with chemical data. Considering that these results were obtained in two days in 10 to 20 min counting intervals and that the method can be incorporated into a sequential industrial on-line procedure, including the determination of oxygen, silicon, and nitrogen, the advantages of this approach are clear. A more complete study of this method is in progress.

Accurate data on total silicon in coal and coal ash give important information on the type of mineral matter in coal. Block and Dams (1974) have shown that particulate matter polluting the air around coal burning furnaces consists mainly of silica. Silica thus can be considered the main matrix in analyses of airborne particles in industrial centers burning coal. It may be used as an indicator when the effect of major exhaust gas scrubbing installation has to be monitored.

F.   Interferences in Fast-Neutron Activation Analysis for Oxygen, Nitrogen, and Silicon

Whereas the dual activation and simultaneous counting system corrects automatically for electronic drifts, neutron flux irregularities, short term electric surges, uneven distribution of tritium in the target, and makes the normally required timing records and decay corrections unnecessary, a few background corrections and interferences because of nuclear reactions have to be considered for accurate work.

In Table V, the main interfering nuclear reactions have been tabulated for oxygen, nitrogen, and silicon, and the gamma energies and half-lives given.

Of practical interest to the analyst are the equivalent interferences relative to the percentages of the interfering elements, which are frequently known, expressed preferably in terms of the percentage of the determined element. This permits the correction of the total intensity of the element sought, expressed in percent, by simple subtractions. In Table VI the equivalent interferences determined by us experimentally, as well as interferences reported in the literature, are compiled.

The data on fluorine interference on the oxygen determination vary significantly depending on the system, especially, (a) on the position and distance of the sample in relation to the neutron source or target and the beam axis of the accelerator, (b) the scattering of neutrons from components near the target such as shielding, target cooling fluids, and thermalization effects because of the polyethylene "rabbit" containers, and (c) the accelerating potential of the primary ion beam that may also partially consist of neutrons with kinetic energies of about 2.5 MeV resulting from the $^2H + ^2H = ^3He + {}^1n$ reaction.

The correction factors for fluorine reported in Table VI vary from the factor of 0.415 derived from Cuypers and Cuypers (1968) work and referred to frequently in the literature, our work in Canada (0.514)

TABLE IV *Silicon Determinations, University of California, Irvine*

Sample number[a]	Type	Reported[b,c] HTA, %	Reported[b,c] Si, %	UCI Determined[b,c] Si, %(I)	UCI Determined[b,c] Si, %(II)
C-19142	Residue SRC process	66.96	10.0	—, 11.7	11.3, 11.6
C-19198	Filter-cake H-coal	36.32	7.2	6.88, 7.64	
C-18940	Residue H-coal filter cake	25.17	5.8	5.04, 5.12	
C-19698	Whole coal, Ky.	17.9	3.9	4.33, 4.25	
C-18941	Residue, still btms, H-coal	15.84	3.0	3.58, 3.19	
C-19739	Whole coal, Tenn.	13.5	3.4	3.38, 3.32	3.32, 3.37
C-18903	Illinois No. 6, 100 mesh	10.59	2.7	2.07, 2.13	2.65, 2.59
C-19700	Whole coal, Tenn.	10.0	2.0	2.41, 2.23	2.28, 2.17
C-19740	Whole coal, Tenn.	8.7	2.0	1.77, —	1.71, 1.76
C-19197	H-coal liquid	0.2	0	<0.010	<0.0010
C-19143	SR coal	0.01	0.04	0	
PSOC-139	Lignite, Texas	15.35	4.2	4.01, 3.65	
PSOC-106	HVB, Indiana	14.38	3.9	3.55, 3.76	3.67
PSOC-142	Med vol., Okla.	10.18	1.0	0.72, 0.88	
PSOC-108	HVA, Penn.	9.50	1.5	1.35, 1.37	1.34, 1.40
PSOC-137	Med vol., Ala.	7.05	0.9	0.89, 0.85	
PSOC-126	Low vol., Penn.	7.01	1.7	1.31, 1.35	1.35, 1.41
PSOC-133	Med vol., W.Va.	6.10	1.0	1.16, 1.08	
PSOC-109	HVA, Penn.	5.15	1.2	1.99, 1.78	1.80, 1.91
PSOC-121	HVA, W.Va.	2.55	0.2	0.15, 0.18	
PSOC-136	Med vol., Ala.	2.42	0.5	0.42, 0.48	

[a] C-19142, etc., Illinois State Geological Survey Samples; PSOC-139, etc., Pennsylvania State University Samples.

[b] On Illinois State Geological Survey Samples, reported on "as received" basis.

[c] On Pennsylvania State University Samples, reported on oven dried (105°C) basis and determined on dried samples.

TABLE V *Fast-Neutron Activation Reactions Used in Analysis of Coal and Coal Ash*

Oxygen	Nitrogen	Silicon
$^{16}O(n,p)^{16}N$	$^{14}N(n,2n)^{13}N$	$^{28}Si(n,p)^{28}Al$
$\gamma = 6.1; 7.1$ MeV; 7.2 sec	$\gamma = 0.511$ MeV; 10 min	$\gamma = 1.78$ MeV; 2.24 min
	Main interferences	
$^{19}F(n,\alpha)^{16}N$	$^{13}C(p,n)^{13}N$	$^{31}P(n,\alpha)^{28}Al$
	$^{16}O(p,\alpha)^{13}N$	$^{27}Al(n,\gamma)^{28}Al$
$^{11}B(n,p)^{11}Be$	(Recoil protons)	$^{27}Al(n,\alpha)^{24}Na$
	$^{31}P(n,\alpha)^{28}Al$	1.37, 2.8 MeV; 15 h
$\gamma = 8.0; 6.8; 5.9; 4.7$ MeV;	$^{39}K(n,2n)^{38}K$	$^{24}Mg(n,p)^{24}Na$
13.6 sec		$^{56}Fe(n,p)^{56}Mn.$
		0.8; 1.8; 2.12 MeV; 2.6 h

TABLE VI   *Equivalent Interferences in Fast-Neutron Activation*

Oxygen	Nitrogen[a]	Silicon[b]
100% B = 10% O[c]	100% C = ~0.1% N	100% P = 49.4% Si
100% F = 41.5% O[d]	100% ¹⁶O = ~0.8% N	100% Al = 0.25% Si
100% F = 51.4% O[e]		100% Mg = 0.23% Si
100% F = 60 % O[c]		100% Fe = 0.21% Si

[a] Rison *et al.* (1967).
[b] Bibby (1975).
[c] Determined by Volborth at UCI (1976 unpublished data).
[d] Cuypers and Cuypers (1968).
[e] Volborth *et al.* (1972), determined at Dalhousie University.

(Volborth *et al.*, 1972), and our work at UCI (0.6), unpublished. The large differences clearly indicate the necessity of determination of this correction factor for each installation and system and of checking it everytime after the distance of the sample from the target is changed. Our recent work at UCI (unpublished) confirms the importance of this correction when very accurate oxygen data are required. Bibby and Guinn (1976) have shown that the threshold energy difference in FNA can be used also to estimate the fluorine content in oxygen bearing substances.

G.   Precision and Accuracy of Fast-Neutron Activation Analysis of Major Elements

Neutron activation analysis is presently regarded as an analytical tool for predominantly trace element research at ppm and ppb levels. At these concentrations a relative standard deviation of 10% is perfectly acceptable. It is frequently not realized that this technique is as well suited for the determination of many elements at major concentrations. This is especially so with the FNA type of neutron activation. It is in such cases that the precision required has to be an order of magnitude higher and preferably within 1% or better.

When major concentrations of elements are to be determined, one of the main considerations is also the effect of "dead time", and the "pile up" of pulses, which phenomena cause losses of counts and may broaden and even shift the photopeaks upward as the count rates increase. The fact that in FNA only relatively few peaks are present simplifies these problems, permitting the selection of wider windows in the single channel analyzers or the MCA's. Also, the use of the dual irradiation and simultaneous counting, with similar standards selected as comparison pairs, improves significantly the accuracy because of equivalent losses when based on count ratios.

The high precision and accuracy of our systems have been described and proven in work done at the University of Nevada in Reno, the Dalhouse University in Halifax, Canada, and the University of California at Irvine (Volborth, 1966, 1969, 1972; Volborth and Vincent, 1967; Volborth *et al.*, 1972, 1975, 1977b,d,e; Vincent and Volborth, 1967; Volborth and Miller, 1977).

In these references it is shown that the standard deviation of the mean for oxygen is usually 0.2 to 0.4% at 95% confidence. Because in most cases no interferences are present, the oxygen results are highly accurate, with deviations from the theoretical value of the same order as the standard deviation of the mean. This was shown with such well-standardized, stable, and pure compounds as oxalic acid, benzoic acid, and crystal quartz for oxygen, the latter also for silicon.

For the determination of nitrogen, mixtures of graphite and amygdalin (3.06% N), piperine (4.91% N), 1-nitroso-2-naphthol (8.09% N), adenine (51.82% N), and other stable nitrogen-bearing compounds were prepared. The pure compounds were also analyzed. In all cases, similar concentrations of the element sought were prepared to bracket the concentrations of the unknown in the sample. Because of relatively small concentrations and complex correction procedures, the relative standard deviation in our work on nitrogen varied in the region of 0.2 to 10%. In many cases, the inaccuracy resulted from incomplete knowledge of the concentration of the interfering elements and the composition of the ash used as blank.

The data on silicon show a high precision of 0.1 to 0.4% relative, because of the high number of counts accumulated. However, in the case of significant phosphorus, and because of an inadequate waiting period between irradiations, disproportionate "dead time," and high background, the accuracy of the silicon determination by FNAA is not as good as it is for oxygen.

IV. DISCUSSION AND CONCLUSIONS

Fast-neutron activation analysis for oxygen, nitrogen, and silicon is an accurate, rapid, and inexpensive method that can be used to supplement the classical coal analysis. This method helps to further quantify the old methods and permits a more meaningful, and from the scientific standpoint, more accurate and informative interpretation of results. Fast-neutron activation analysis can detect errors in the classical coal analysis. It can give stoichiometric balances if other data are correct and thus help the chemist to provide the coal upgrading engineer with necessary information rapidly while the tests are being conducted. In the case of oxygen, the question of whether oxidation or reduction of any interim products occurs during a run can be answered in a few minutes after the sample has been submitted for analysis. Calculations of BTU values may be improved when one finds that not all "moisture" is water but also consists of some combustible gases. Processes and rates of changes occurring in coal during storage can be easily studied. Organic oxygen can be better estimated than previously. Calculations and material balances are improved and result in more informative and scientifically more useful data. An approximate double check of total sulfur in sulfur-rich coals can be done. Liquefaction, gasification, solvent refining, in situ burning of coal and similar coal utilization in industrial processes should be easier to follow and to interpret by checks of "true" oxygen versus the quite unreliable and often erroneous "oxygen by difference."

Instead of waiting for time-consuming proximate and ultimate analyses, in many cases a single accurate oxygen determination will tell the coal chemist and engineer more than the whole analysis, once the correlations of true oxygen with the different steps of analysis and industrial processes are available. It is this type of data that is still lacking and is presently being accumulated at our laboratories at Washington State University in Pullman, and the University of California in Irvine.

ACKNOWLEDGMENT

The author wishes to acknowledge the invaluable help received from Dr. G. E. Miller, Reactor Supervisor and Lecturer in Chemistry at the University of California at Irvine during the development of the methods and systems described in this chapter. Coal samples marked PSOC have been kindly provided by Dr. A. Davis from The Pennsylvania State University Coal Data Bank. Samples marked C-XXXXX have been kindly supplied by Dr. R. R. Ruch of the Illinois State Geological Survey.

REFERENCES

ASTM (1975). "1975 Annual Book of ASTM Standards. Part 26: "Gaseous Fuels; Coal and Coke; Atmospheric Analysis." Am. Soc. Test. Mater., Philadelphia, Pennsylvania.
Bibby, D. M. (1975). *Anal. Chim. Acta* **79**, 125–137.
Bibby, D. M., and Champion, H. M. (1974). *Radiochem. Radioanal. Lett.* **18**(4), 177–184.
Bibby, D. M., and Guinn, V. P. (1976). *J. Radioanal. Chim.* **29**, 125–127.

Bibby, D. M., and Sellschop, J. P. F. (1974). *J. Radioanal. Chem.* **20**(2), 677–693.

Block, C., and Dams, R. (1974). *Anal. Chim. Acta* **71**, 53.

Chyi, L. L., James, W. D., Ehmann, W. D., Sun, G. H., and Hamrin, E. E. (1976). *Sci. Ind. Appl. Small Accel., 4th Conf.* pp. 281–287.

Cuypers, M., and Cuypers, J. (1968). *J. Radioanal. Chem.* **1**, 248–264.

Ehmann, W. D., and Morgan, J. W. (1970). *Proc. Apollo 11 Lunar Sci. Conf.* **2**, 1071–1079 (*Geochim. Cosmochim. Acta* Suppl. No. 2).

Ehmann, W. D., Miller, M. D., Maw-Suen Ma, and Pacer, R. A. (1974). *Lunar Sci. V* **1**, 203–205.

Geisler, M., Maul, E., and Panse, H. (1971). *Radiochem. Radioanal. Lett.* **8**(6), 349–355.

Gluskoter, H. J., Ruch, R. R., Miller, W. G., Cahill, R. A., Dreher, G. B., and Kuhn, J. K. (1977). *Ill. State Geol. Surv., Circ. No.* 499, 1–154.

Hamrin, C. E., Maa, P. S., Chyi, L. L., and Ehmann, W. D. (1975). *Fuel* **54**, 70.

James, W. D., Ehmann, W. D., Hamrin, C. E., and Chyi, L. L. (1975). *Trans. Am. Nucl. Soc.* **21**, Suppl. No. 3, 30.

James, W. D., Ehmann, W. D., Hamrin, C. E., and Chyi, L. L. (1976). *J. Radioanal. Chem.* **32**, 195.

Kaman Nuclear Corporation (1966). Tech. Bull. No. 104. Colorado Springs, Colorado.

Miller, G. E., and Volborth, A. (1976). *Sci. Ind. Appl. Small Accel., 4th Conf.* pp. 288–292.

Morgan, J. W., and Ehmann, W. D. (1970). *Anal. Chim. Acta* **49**, 287.

Nargolwalla, S. S., and Przybylowicz, E. P. (1973). "Activation Analysis with Neutron Generators." Wiley, New York.

Olive, G., Cameron, J. F., and Clayton, C. G. (1962). *U.K. At. Energy Auth., Res. Group, Rep.* **AERE-R-3920**.

Rison, M. H., Barber, W. H., and Wilkins, P. E. (1967). *Radiochim. Acta* **7**, 196.

Ruch, R. R., Cahill, R. A., and Frost, J. K. (1977). *Trans. Am. Nucl. Soc.* **27**, 159.

Ruch, R. R., Miller, G. E., and Volborth, A. (1978). unpublished

Solomon, J. A., and Mains, G. J. (1977). *Fuel* **56**, 302–304.

U.S. Bureau of Mines (1967). *Bull.* No. 638, 1–82.

Vincent, H. A., and Volborth, A. (1967). *Nucl. Appl.* **3**, 753–757.

Vogt, J. R., and Ehmann, W. D. (1965). *Int. J. Appl. Radiat. Isot.* **16**, 573–580.

Volborth, A. (1966). *Fortschr. Mineral.* **43**(1), 10–21.

Volborth, A. (1969). "Elemental Analysis in Geochemistry." Elsevier, Amsterdam.

Volborth, A. (1972). *In* "The Encyclopedia of Geochemistry and Environmental Sciences," pp. 779–789. Van Nostrand-Reinhold, New York.

Volborth, A. (1976a). "Fossil Energy," ERDA COO-2898-2. U.S. Gov. Print. Off., Washington, D.C.

Volborth, A. (1976b). "Fossil Energy," ERDA COO-2898-3. U.S. Gov. Print. Off., Washington, D.C.

Volborth, A. (1976c). "Fossil Energy," ERDA COO-2898-4. U.S. Gov. Print. Off., Washington, D.C.

Volborth, A. (1976d). "Fossil Energy," ERDA COO-2898-7. U.S. Gov. Print. Off., Washington, D.C.

Volborth, A., and Banta, H. E. (1963). *Anal. Chem.* **35**, 2203–2205.

Volborth, A., and Miller, G. E. (1977). *Trans. Am. Nucl. Soc.* **27**, 160–161.

Volborth, A., and Vincent, H. A. (1967). *Nucl. Appl.* **3**, 701–707.

Volborth, A., Dayal, R., McGhee, P., and Parikh, S. (1972). *Am. Soc. Test. Mater., Spec. Tech. Publ.* No. 539, 120–127.

Volborth, A., Miller, G. E., and Garner, C. K. (1975). *Am. Lab.* **7**(10), 87–98.

Volborth, A., Miller, G. E., Garner, C. K., and Jerabek, P. A. (1976). "Fossil Energy," ERDA COO-2898-8. U.S. Gov. Print. Off., Washington, D.C.

Volborth, A., Miller, G. E., Garner, C. K., and Jerabek, P. A. (1977a). "Fossil Energy,"
    DOE COO-2898-10. U.S. Gov. Print. Off., Washington, D.C.
Volborth, A., Miller, G. E., and Garner, C. K. (1977b). *Chem. Geol.* **20,** 327-331.
Volborth, A., Miller, G. E., and Garner, C. K. (1977c). *Chem. Geol.* **20,** 85-91.
Volborth, A., Miller, G. E., Garner, C. K., and Jerabek, P. A. (1977d). *Fuel* **56,** 204-208.
Volborth, A., Miller, G. E., Garner, C. K., and Jerabek, P. A. (1977e). *Fuel* **56,** 209-215.
Volborth, A., Miller, G. E., Garner, C. K., and Jerabek, P. A. (1978). *Fuel* **57,** 49-55.
von Hevesy, G., Levi, H. (1936). *Math. Fys. Medd., Inst. Theor. Phys., Univ. Copenhagen*
    14(5), 3.
Wing, J. (1964). *Anal. Chem.* **36,** 559-564.

Chapter 48

Electron Probe Microanalyzer in Coal Research

Robert Raymond, Jr.   *Ron Gooley*
LOS ALAMOS SCIENTIFIC LABORATORY
GEOLOGICAL RESEARCH GROUP
LOS ALAMOS, NEW MEXICO

I. HISTORY

The technique of x-ray spectrochemical analysis has resulted from Moseley's (1913) discovery that the wavelength (or energy) of emitted x radiation is a function of atomic number and is characteristic of the emitting element. Applications of this technique were pursued through the 1920s and 1930s; most notable was the discovery and identification of hafnium from its x-ray wavelengths by Coster and Von Heresy (1923). In the late 1940s, while electron microscopy was developing rapidly, a patent was issued to Hillier (1947) that described the concept of electron probe microanalysis (EPM). However, Hillier did not pursue the concept. At the 1949 Electron Microscope Conference in Delft, Netherlands, Castaing and Guinier (1949) presented their first report describing specimen excitation by an electrostatically focused electron beam and measurement of characteristic emitted x radiation by an x-

337

ray spectrometer. Castaing subsequently improved the instrument by introducing electromagnetic beam focusing lenses in place of the electrostatic lenses. In his brilliant doctoral thesis (Castaing, 1951), he discussed relationships between specimen composition and x-ray intensity, and he formulated the basis for many current data reduction algorithms.

Geoscientists were slow to begin using the instrument; by 1964, only 60 geoscience papers had been published using EPM data (Keil, 1973). Since that time, the instrument has revolutionized petrology and geochemistry. Almost all major geoscience departments in academic, industrial, and governmental laboratories now use the EPM extensively, and the papers dealing with EPM use in the geosciences number well over 1000. At the time of this writing, though, less than a dozen papers had been published that note use of EPM data in coal research.

Many excellent detailed descriptions of EPM hardware and data reduction software have been published. The authors offer a rather brief description of the instrument and its capabilities here before describing its applicability to coal analysis. A bibliography of some of the better literature is given at the end of the chapter if more detailed instrument descriptions are desired.

II.   ELECTRON PROBE MICROANALYSIS

The EPM uses a finely focused electron beam that impinges on a polished sample, generally at 15–20 keV, producing x rays characteristic for elements present in the sample. The electron column and basic instrumentation of the EPM is very similar to that of a scanning electron microscope (SEM). The major difference between the two instruments, and modes of operation, is that in general, the EPM is used to determine chemical composition, while the SEM defines surface morphology of the specimen, i.e., produces a magnified image of the specimen (see Chapter 42, Figs. 1–4).

Both instruments use a finely focused electron beam that impinges on the sample with accelerating voltage generally at 15 or 20 keV. Most modern EPMs can function in a SEM mode, and some SEMs, when equipped with x-ray spectrometers, are virtually the same as EPMs. In fact, modern electron beam instruments are so versatile that the distinction between an EPM and a SEM becomes rather vague. For the purposes of our discussion, we shall define an EPM as an electron beam instrument that uses wavelength dispersive spectrometers (WDS) for x-ray analysis and an optical microscope for specimen viewing during analysis.

Modern SEMs are commonly equipped with energy dispersive spectrometers (EDS) to permit chemical analysis. Until recently, EDS was limited to qualitative analysis (i.e., identifying only what elements are present, but not quantitative composition). In the last two or three years, significant progress has been made, and several manufacturers now offer quantitative data reduction software that produces results approaching those of WDS analysis. Some pros and cons of WDS and EDS are discussed in later sections, and the reader is referred to more complete literature in the bibliography at the end of the chapter.

A. Basic Instrument

1. *Electron Column*

Figure 1 shows a schematic representation of the basic components of an EPM. The electron gun is of triode design, with a tungsten filament, a grid cylinder, and an anode. Thermionic emission from the heated filament is the source of free electrons. The electrons are accelerated down the column through a hole in the anode, which operates

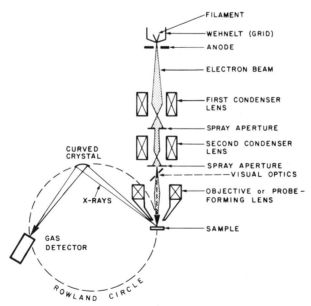

Fig. 1 Schematic representation of the EPM showing the electron column configuration and the placement of the sample, curved crystal, and gas detector on the Rowland circle.

at a high positive potential (generally 15 or 20 keV) with respect to the filament. The grid, or Wehnelt cup, which is operated at a negative potential with respect to the filament, controls emitted current. Commonly, a feedback circuit from a beam limiting aperature lower in the column is used to adjust the grid bias to regulate beam current. The grid also actually forms an electrostatic lens and decreases the beam diameter.

The beam is further shaped and demagnified by apertures and electromagnetic lenses. Older EPMs commonly used one condenser lens and one final probe-forming (objective) lens; SEMs and more modern EPMs generally use two condenser lenses and the final lens. The electron beam is focused to a diameter of about 1 μm for EPM use; beam diameter for SEM use may be as small as 50 Å. The EPM beam also may be defocused or rastered at the point of sample impingement to perhaps as much as 50 μm across, i.e., the spot size or rastered area of analysis is enlarged. This practice is especially common with beam-sensitive samples, or when it is desirable to obtain an "averaged" analysis over a broader area, if the sample is compositionally inhomogeneous on a micrometer-size scale. With WDS work caution must be exercised with "broad beam" analysis to be certain that the spectrometers are in x-ray focus across the entire width of the beam. Energy dispersive detectors "see" a much larger sample area than wavelength dispersive spectrometers, so this potential problem is not as critical.

2. Wavelength Dispersive Spectrometry (WDS)

Excellent wavelength resolution, high peak-to-background intensity ratios, low detection limit (<0.01% for many elements), and the ability to analyze x rays emitted from low Z elements (B, C, N, O) are the prime merits of WDS. These devices employ a curved crystal that is mechanically moved to diffract x rays into a gas proportional detector at positions where Bragg's law is satisfied:

$$N\lambda = 2d \sin \theta, \qquad (1)$$

where N is an integer, λ is the x-ray wavelength, d is the crystal lattice spacing, and θ is the angle of x-ray incidence on the crystal. The wavelength range for which a particular crystal is suitable is limited and is dependent on its d spacing. A variety of crystals are available to cover all elements from $Z = 5$ (B) to $Z = 92$ (U). Perhaps the most common crystals now used in modern EPMs are LiF (lithium fluoride), PET (pentaerythritol), TAP (thallium acid phthalate), and ODPb (lead orthodecate), the latter for B, C, N, and O.

Detector output pulses are shaped, amplified, digitized, and (on automated instruments) stored on the computer for on-line data reduction. Most EPMs are equipped with multiple spectrometers, each with a different crystal. For example, on an instrument with three spectrometers, one might use a TAP crystal on one spectrometer, a PET on the second, and a LiF crystal on the third. Spectrometers are commonly equipped with automatable crystal changers. On automated instruments, each spectrometer is computer driven first to the peak position for each selected element within its wavelength range, then to a preselected position off of each peak for background determinations. Peak and background intensities are determined at each position by a preselected counting time or number of counts. After x-ray intensities are determined on the analytical standards, a typical ten element analysis takes approximately 2.5 min using an efficient automation-data reduction software system.

Figure 1 schematically shows positioning of the sample, crystal, and detector on the Rowland circle. Mechanisms for moving the crystal and detector as the spectrometer is scanned through its wavelength range are fairly complex and must be well engineered. The sample, crystal and detector must always remain on the Rowland circle, and the x-ray takeoff angle must remain constant.

3. Energy Dispersive Spectrometry (EDS)

Energy dispersive x-ray spectrometry makes use of a semiconductor (solid state) x-ray energy detector and a multichannel analyzer (MCA) for energy analysis. X-ray photons impinging on a lithium drifted silicon (SiLi) semiconductor detector create electron-hole pairs. (A "hole" is a lattice site that carries an effective positive charge by the loss of an electron that is promoted to a higher energy level, the conduction band shared by the entire crystal). When a potential is applied across the crystal in its ground state, little or no current flows, but x-ray photon absorption creates an amount of "free" charge that reflects the energy of the incident photon. The applied potential carries the charge (electrons and holes) to a charge-sensitive preamplifier that converts the charge pulse to a voltage pulse. The voltage pulse is amplified, shaped, and passed to a multichannel analyzer. The basis for energy spectroscopy is the relationship between charge and absorbed x-ray photon energy.

The MCA sorts the incoming signal according to variations in pulse amplitude (corresponding to variations in photon energy). It counts and stores in memory the number of pulses that fall into channels (pulse amplitude windows) and displays the resulting energy spectrum on a cathode-ray tube (see Fig. 2). Most modern multichannel analyzer systems provide at least 1024 channels and have a selective operational range (e.g., 0–10, 0–20, 0–40 keV, etc.). For 1024 channels and an operating range of 0–10 keV, each channel covers ~10 eV. Some type of cursor, or channel marker, that can be electronically moved to a selected channel is displayed on the CRT, as is the number of the

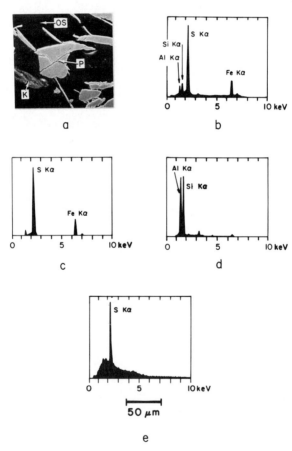

Fig. 2 (a) SEM micrograph of a coal specimen containing pyrite and kaolinite and EDS spectra of that sample. (b) The rastered area spectrum represents elements present in the whole specimen, while the three additional spectra represent small analytical sites as defined in the SEM micrograph: (c) pyrite (P), (d) kaolinite (K), and (e) organic sulfur in the coal (OS).

selected channel. It is convenient with inexpensive MCAs to operate with 1024 channels at 0–10 keV full range, because when the curser is moved to an x-ray peak, the channel number corresponds (with proper placement of a decimal point) to the x-ray line energy. For example, the energy of the Si Kα line is 1.787 keV, and the channel number will read 179. More sophisticated MCAs commonly are programmed to display the element symbol, the x-ray line, and other pertinent information.

4. Light Optics

A light optical microscope allows for specimen viewing at high magnification, so that the electron beam can be accurately positioned on the area of interest. For proper x-ray focusing into the detector, the sample, diffracting crystal, and detector must lie on the perimeter of the Rowland circle at all times during analysis. The focal point of the microscope is coincident with the focal point of the crystal spectrometer so that correct sample positioning for x-ray focus is assured.

B. Analytical Standards

Quantitative analysis by EPM is conducted by comparing the x-ray intensity produced from a sample to that produced on a standard of known composition. Therefore, standard quality is of utmost importance in producing acceptable quantitative data. Acquiring good standards is a particular problem in geosciences because of compositional complexity and inhomogeneity of most minerals. Observed x-ray intensities (those "seen" by the spectrometers) differ from the intensities of x rays actually generated by electron bombardment of the specimen because of x-ray absorption, flourescence, and the atomic number effect (discussed further in Section II,D). These effects depend directly on specimen composition, thus the ideal standard is one whose composition is indentical to the unknown. Complete sets of well-analyzed, homogeneous mineral standards, either natural or synthetic, spanning large compositional ranges, are not available. The analyst, therefore, generally must resort to a standard that one hopes is close in composition to the specimen. The very nature of microanalysis places serious restrictions on standards; they must be homogeneous on a micrometer size scale. Most laboratories accumulate a large selection of standards in time, but seldom is a standard available in sufficient abundance for broad distribution among many laboratories. Very little interlaboratory calibration has therefore occurred.

C. Sample Preparation

Types of samples used for EPM analysis are commonly the same as for optical microscopy. Standard polished sections like those used in metallurgy, or polished petrographic thin sections (with no cover slip) are the most common samples. X-ray production by electron beam excitation occurs within approximately 2 μm of the sample surface. The intensity of x rays emerging from beneath the surface is dependent upon their path length through sample materiaļs. The x-ray path length out of the sample is increased considerably if a 1-μm-diam electron beam is positioned in the bottom of a 2 μm scratch on the sample surface. Therefore, a polished flat sample that is as free of scratches and surface relief as possible should be used.

Most geologic materials are poor electrical conductors; coal is no exception. Therefore, to prevent both charge buildup on the sample surface at the point of electron impact and resulting beam deflection, the sample must be coated by a thin layer of conductive material before analysis. A vacuum evaporator is used to deposit the conductive layer, and for several reasons carbon is generally used instead of materials such as Be, Al, Cu, and Au. In thin film, carbon is essentially optically transparent, facilitating specimen viewing during analysis. Carbon is nonpoisonous (the problem with BeO), and does not appreciably absorb electrons or soft x rays.

A potential problem, especially in coal analysis, is the fact that most geologic materials are poor thermal conductors. This means that heat generation under an intense electron beam may seriously affect the sample and the analysis by vaporizing volatile sample material and by element migration. (An additional problem with vaporizing sample material is contamination of the electron column.) Thermal problems can be reduced by defocusing the electron beam to produce a larger spot size, operating in a beam raster mode, or lowering beam intensity by reducing accelerating voltage and/or beam current.

D. Electron Beam–Sample Interactions

When the electron beam impinges on a sample, several things happen (Fig. 3), a few of which are pertinent to the present discussion. Electrons penetrate below the sample surface (about 1–3 μm at 15 keV, depending on sample composition and structure) and undergo a series of elastic and inelastic interactions with electrons of sample atoms. Some inner shell sample electrons are ejected; this process is followed by outer shell electrons cascading in to fill the vacancies. The energy

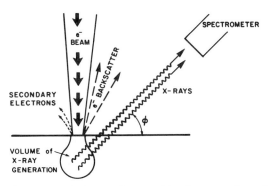

Fig. 3 Electron beam–sample interaction resulting in generation of x rays, backscattered electrons, and secondary electrons. Note that the volume of the sample energized to produce x rays is indeed much larger than suggested by the small area defined by beam impingement on the sample.

difference between the inner shell (vacancy) and the outer shell (original site of the transposed electron) is manifested by emission of a photon. Electron energy levels are discrete and characteristic for each element; thus, emitted x rays are characteristic in energy (or wavelength) for each element in the sample, neglecting the background white spectrum present due to the large number of electron interactions. This is the basis for x-ray analysis.

Some electrons are elastically backscattered and are lost for x-ray production; these must be considered during x-ray data reduction. Other beam electrons undergo inelastic collisions, ejecting sample electrons and losing some energy to them. Some ejected sample electrons (secondary electrons) have low energy (typically ~50 eV) and are easily reabsorbed by the sample. Some secondary electrons produced very near the surface escape; these are utilized for imaging in the SEM.

1. Sample Matrix Effects

The relationship between the abundance ratio of the unknown and standard for element A and the ratio of x-ray intensities emitted from the unknown and standard by element A is given by

$$C_U^A/C_S^A = I_U^A/I_S^A, \quad \text{or} \quad C_U^A = (I_U^A/I_S^A)C_S^A, \tag{2}$$

where C is concentration, I is x-ray intensity, U and S denote unknown and standard, and A implies element A. However, this relationship is valid if and only if the unknown and standard are identical in composition and structure. Observed x-ray intensity is affected by specimen absorption, secondary x-ray fluorescence, electron backscattering and electron stopping power. All of these factors are functions of sample composition and structure and are commonly referred to as matrix or ZAF effects (Z = atomic number, A = absorption, F = fluorescence). Data reduction algorithms calculate and correct for these effects by computer.

a. Atomic Number Effect Sample composition and structure affect electron interactions, therefore x-ray production, in the sample in two ways: electron backscatter and electron stopping power. Both of these factors increase with increasing electron density of the sample or with increasing average atomic number Z. Backscattered electrons result from single and multiple elastic interactions between incident (beam) and sample electrons and are lost to x-ray production. Castaing (1960) obtained an expression to describe electron penetration (therefore stopping power) into a sample:

$$\rho R(x) = 0.033\,(E_o^{1.7} - E_c^{1.7})A/Z, \tag{3}$$

where ρ is sample density in grams per cubic centimeter, $R(x)$ is distance in meters, and E_o and E_c, both in kiloelectron volts, are electron beam energy and the critical excitation potential of the target, below which no x-ray production can occur. Others have revised and tested this equation experimentally; a brief review is given by Goldstein and Yakowitz (1975). Electron penetration for x-ray production into common silicate minerals at 15 keV accelerating potential is less than 2 μm; into coal it would be less than 4 μm.

b. Absorption As depicted in Fig. 3, x-ray production occurs beneath the sample surface. As x rays pass through sample materials, some are absorbed, so that the intensity seen by the detector is reduced with respect to primary x-ray production in the sample. The amount of absorption that occurs is dependent on sample composition and the path length through the sample, which is proportional to the cosecant of the x-ray takeoff angle (ϕ in Fig. 3). It follows then, that absorption is greater with instruments employing a low x-ray takeoff angle.

c. Fluorescence Characteristic x radiation produced by electron excitation of sample atoms can excite other sample atoms, producing a secondary fluorescence effect. This phenomenon has the opposite effect of absorption and may cause observed radiation for some elements to be enhanced. Generally, x-ray absorption affects observed intensity to a greater degree than fluorescence, and heavier elements are absorbers of light element radiation. In compounds or alloys where two elements close in atomic number are present, the opposite may happen. For example, consider an Fe–Ni alloy with equal amounts of each element. Fluorescence of Fe $K\alpha$ radiation by Ni $K\alpha$ radiation dominates over absorption, and Ni $K\alpha$ radiation is strongly absorbed by Fe. As absorption decreases with increasingly higher x-ray takeoff angle, fluorescence becomes increasingly important.

2. Instrumental Effects

Data reduction must also consider (a) background counts, (b) instrumental drift, and (c) detector deadtime. These factors are commonly called instrumental effects. Background results in two areas: (1) stray pulses or signals because of electronic noise and (2) the x-ray continuum because of the almost infinite variety of electron–electron and electron–photon interactions. Small changes may occur in electron beam intensity between standard and sample x-ray intensity measurements due to variations in the instrument's electronics. Therefore beam current is commonly monitored, and x-ray intensities adjusted by a normalization factor whenever this drift is significant. Most wavelength dispersive x-ray detectors employ gas proportional counters. These devices have a finite pulse resolving time, commonly on the order of 2 μs, which produces a nonlinear response at high counting rates. Thus a correction factor for this detector "deadtime" must be applied, especially when there are significant differences in contents of the element of interest between sample and standard.

E. Data Reduction

Reduction of raw x-ray intensity data to meaningful composition data is now almost always done by computer. Modern EPMs generally use a dedicated minicomputer for on-line data reduction and instrument automation. Beaman and Isasi (1970) critically reviewed available data reduction programs in 1970. Basic techniques in data reduction algorithms have changed little since that time, but considerable progress has been made in combined automation–data reduction schemes and also in faster, more efficient software routines. A prospective EPM buyer should investigate the field thoroughly before purchase.

III. APPLICATIONS

Considering the heterogeneous and fine-grained nature of inorganic and organic constituents of coal, EPM capabilities offer a variety of applications valuable to coal research. Therefore, the paucity of current literature concerning use of the EPM in coal studies reveals an important broad research field that as yet has received insufficient attention. In this section, we describe some studies that serve as examples of broader areas of EPM application.

A. Major/Minor Element Analysis

For the purposes of this discussion, we define a major element as one present in abundance greater than 2 wt %, a minor element between 100 ppm and 2 wt %, and a trace element as present at less than 100 ppm. In general, EPM results should be accurate and reproducible to ±2% of the amount present for most major elements, provided the standards are reliable. Relative accuracy decreases with decreasing elemental concentrations. For element concentrations about 1 wt %, relative accuracy should be within ±5%. The following text shows where the EPM has been used for quantitative analysis of organic sulfur in coal (Raymond and Gooley, 1978); this study is an example of minor element analysis.

The American Standards for Testing and Materials (ASTM) procedures for sulfur analysis in coal call for the analytical determination of values for total, pyritic, and sulfatic sulfur. Organic sulfur is calculated by subtracting pyritic and sulfatic sulfur from the total. The procedures are aimed at providing rapid, inexpensive, and reproducible data for coal utilization. The pyritic, sulfatic, and organic sulfur contents reported by the procedures adequately reflect the amount of sulfur that can be removed by sizing, specific gravity separation, and hindered settling techniques. But any error in total, pyritic, or sulfatic sulfur determination shows up as an error in organic sulfur determination. Reasons for error in pyritic sulfur determination have been reported (Greer, 1977; Raymond and Gooley, 1978). See also Volume II, Chapter 20, Section IV, C, 2. For coal conversion processes such as liquefaction, the need for accurate organic sulfur analysis appears to be very critical, approaching an accuracy of 0.2 wt. %—an accuracy that the ASTM methods do not claim to ensure. Results to date indicate such accuracy is obtainable with the EPM.

To prepare for an EPM analysis of organic sulfur, coal samples ground to −20 mesh are potted in epoxy and mounted on glass slides. The samples are polished and mosaics of portions of their surfaces are prepared at approximately 400× magnification. Coal macerals present in the mosaics are identified by oil immersion reflected light microscopy.† Mosaics are necessary since the samples must be carbon coated for conductance during electron bombardment, and once coated, reflectance levels are obscured and maceral identification is extremely difficult. Commonly EPMs have a high magnification optical system (250–400×), and using morphology of the various coal grains in conjunction with the mosaics, the points of EPM analysis can be exactly located.

Standardization datum is determined as an average of background corrected intensities on ten carbon standard beads containing 4.1 wt. % sulfur. After each maceral analysis the computer prints x-ray intensities and wt % sulfur as determined by comparison with the x-ray intensity of the carbon bead standard. Matrix (ZAF) corrections are not made because the standard and unknown are very similar in composition, both are essentially carbon with minor sulfur. Errors that may be introduced without ZAF corrections are less than analytical precision, i.e., less than 5% of the amount present at the 1% sulfur level.

X-ray intensity of iron is continually monitored during analysis, and any intensities in excess of background levels are assumed to be contamination by pyrite and those results are rejected. At the same time, EDS spectra are monitored for any elements that might suggest the presence of other minerals, especially sulfides or sulfates. Therefore, reported sulfur contents are most likely organic. Another possibility is elemental sulfur, but it is unlikely that sulfur would be homogeneously distributed throughout each maceral in elemental form.

The results of a sample analysis include a number of organic sulfur measurements for all maceral types present in the sample. An example of the analysis of a high volatile A bituminous coal is given in Table I. The mean sulfur content for each maceral type was determined and was multipled by the weight percent of that maceral in the dry coal sample. The weight percent of individual macerals in the dry coal were determined by the Pennsylvania State University Coal Section by multiplying maceral density by volume percent in the coal. Volume percent was determined by point counting 1000 macerals. The sulfur contents contributed to the total organic sulfur content of the coal by the individual macerals is also shown in Table I. The summation of the contributions of the individual macerals gives the total organic sulfur concentration as determined by EPM. As can be seen for the example in Table I, this number for total organic sulfur corresponds very closely to that determined by ASTM methods. It should be noted that in the

† For examples, see Volume I, Chapter 1, Figs. 1–3, and Chapter 2, Fig. 1.

TABLE I   *EPM Analysis of a High-Volatile A Bituminous Coal[a]*

Maceral	EPM Analyses (wt % S)					$\bar{X}$ wt % S	Maceral wt %	S contri- bution (wt %)
Vitrinite	0.59	0.58	0.70	0.58	0.60	0.61	52.8	0.322
	0.63	0.55	0.67	0.56	0.64			
Pseudovitrinite	0.49	0.60	0.62	0.57	0.52	0.56	16.4	0.092
Fusinite	0.22	0.36	0.19	0.37	0.19	0.27	6.2	0.016
Semifusinite	0.56	0.33	0.44			0.44	6.2	0.028
Sporinite	0.63	0.60	0.59	0.70	0.66	0.64	6.0	0.038
Micrinite	0.55	0.62				0.59	2.8	0.016
Macrinite	0.38	0.52	0.52	0.60	0.51	0.51	0.7	0.004
EPM Organic S=0.52								
ASTM Organic							TOTAL	0.516
S=0.57								

[a] After Raymond and Gooley (1978).

case of the example shown, the breakdown of maceral constituents follows that scheme used by the Pennsylvania State University Coal Section.

In samples that contain no pyrite, organic sulfur must equal total sulfur less sulfatic sulfur since no error may develop as a result of inaccurate pyrite analysis. In samples that contain very little pyrite, the error that would be transposed to organic sulfur as a result of incorrect pyritic sulfur analyses would be small. The analytical approach discussed above was applied to 18 coal samples that range in rank from subbituminous C through low volatile bituminous. The samples come from 10 states within the contiguous United States, include 18 different coal seams, and range in age from Pennsylvanian (upper Carboniferous) to Paleocene. Results for seven of these samples are shown in Table II. For these samples, sulfatic sulfur as determined by ASTM methods equals zero and pyritic sulfur as observed by optical microscopy is negligible. Since pyritic sulfur content is small, the erorr between the EPM and ASTM methods should also be small. Such is the case, for the EPM analyses are nearly equal to those obtained by ASTM methods. Therefore, it is reasonable to assume that the EPM method for determination of organic sulfur in coal is valid.

In addition to the seven coals just discussed, 11 coals were analyzed that contain an appreciable amount of pyrite as observed by optical microscopy. In all cases the discrepancies between EPM and ASTM analyses were greater than those seen for coals containing nearly no pyritic sulfur. Furthermore, with increasing amounts of fine-grained pyrite, the EPM and ASTM calculations for organic sulfur were increasingly different.

B. Trace Element Detection

For this discussion we defined trace elements as those present in abundances less than 100 ppm. This concentration should be considered as the minimum detection limit for routine analysis. However, much lower detection limits are attainable through special procedures (e.g., Buseck and Goldstein, 1969). Additionally, it should be remembered that the concentration of a trace element for EPM detection is its concentration at the analytical point under the electron beam, which may be quite different than its concentration on a whole sample basis. For example, the bulk sample concentration of Ti may be 10 ppm for some particular coal, but if the Ti is present in the sample as minute grains of rutile, its concentration within those grains is about 60 wt. %.

The work of Wewerka *et al.* (1978) has focused on structural relationships and associations among trace elements and major minerals in coal wastes. In this project, the EPM has been used in a qualitative mode in conjunction with an ion microprobe. Whereas the electron microprobe may measure an area as small as 2 μm across, the ion microprobe used in this coal waste research was limited to measuring elemental concentrations in larger areas, in many cases up to 20 μm across. But, where the EPM has a maximum sensitivity for most elements of around 100 ppm, the ion microprobe is often sensitive to concentration as low as 1 ppm. These two instruments were applied concurrently.

TABLE II *ASTM and EPM Sulfur Determinations (wt %) for Coals Containing Little or No Pyrite Visable under Reflected Light Microscopy*[a]

Coal/rank	ASTM Total S	ASTM Pyritic S	ASTM Organic S	EPM Organic S	Organic S | ASTM– EPM |
(1) Upper Elkhorn No. 3/hvAb	0.65	0.01	0.64	0.78	0.14
(2) Ohio No. 5/subC	1.02	0.01	1.01	0.96	0.05
(3) Lower Elkhorn/hvAb	0.60	0.03	0.57	0.52	0.05
(4) Hazard No. 7/hvAb	0.67	0.04	0.63	0.61	0.02
(5) Upper Sunnyside/hvAb	0.69	0.06	0.63	0.66	0.03
(6) Blind Canyon/hvAb	0.48	0.14	0.35	0.39	0.04
(7) Dietz No. 3/subB	0.29	0.14	0.16	0.22	0.06

[a] After Raymond and Gooley (1978).

One major concern in this project was the trace element compositions of pyrite grains. Using the EPM, Wewerka *et al.* (1978) identified four trace elements (Tl, Mn, As, and Cu) that were occasionally associated with the pyrites of the Illinois Basin refuse samples, but they suspected other chalcophile elements might also be present. By using the ion microprobe with its greater sensitivity, several additional trace elements were observed. Most were associated with ever present clay impurities, but some (Se, Hg, and Mo) were occasionally associated with the pyrite regions. But what were the relationships of these trace elements to the pyrite regions? Were they incorporated within the pyrite mineral structure, did they in fact show up in the ion microprobe analyses only because they coexisted within the larger pyrite grains as micromineral inclusions, or were they part of the clay intimately associated with the pyrite regions?

If the trace elements were included in the pyrite as micromineral inclusions, they should have been present in detectable quantities and the EPM would have identified them. But even though the ion microprobe was able to show that these trace elements were present, the EPM could not explicitly locate them. This allowed a conclusion to be drawn, that since such elements are present, they must be distributed more or less uniformly as contaminants in the pyrite structure or associated clay, rather than as micromineral inclusions. In this case the absence of elements within the detectable limits of the EPM justified drawing a substantial conclusion that could impact greatly on how the various coal waste materials must be treated.

C. Elemental Distribution Mapping

A useful technique commonly employed using the EPM as a qualitative tool is element mapping.† This technique is especially useful when combined with secondary electron imaging. Figure 4 is a secondary electron image and a series of element distribution maps of pyrite crystals in a kaolinite groundmass. Element maps can be made with either WDS or EDS; these were made with WDS in the following manner. The electron beam is focused to a small spot and put into a raster mode with the beam scan on the sample synchronized with the scan on an oscilloscope. A spectrometer is peaked on an x-ray line of the element of interest, and detector output is recorded on the screen of the oscilloscope. Thus, x-ray pulses from the element of interest are

---

† See Volume II, Chapter 27, Figs. 4–6 for elemental maps by electron microprobe for seven elements in lignite, and Chapter 42, Fig. 5 for an elemental map of iron in hematite from ash.

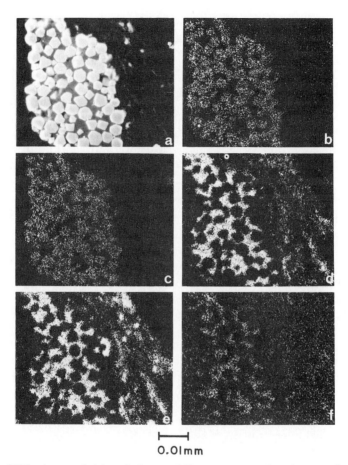

0.01mm

Fig. 4 SEM micrograph (a) and elemental K_α x-ray maps of pyrite crystals within a kaolinite groundmass located in a coal specimen. Note how the sulfur (b) and iron (c) define the pyrite crystals, while the aluminum (d), silicon (e), and oxygen (f) define the kaolinite groundmass supporting the pyrite as well as other kaolinite in the coal matrix.

displayed on the oscilloscope as bright spots. The oscilloscope image is photographed for a permanent record. High peak-to-background ratio and ability to detect elements of low Z are the advantages of WDS over EDS for element mapping. The primary disadvantage of WDS is if areas over about 100 μm across are scanned, crystal spectrometers become defocused at the edges of the scanned area. Since EDS detectors "see" a larger sample area than WDS detectors, this problem is minimized.

Harris and Yust (1978) have combined optical microscopy and the backscattered mode of scanning electron microscopy with x-ray mapping capabilities of the EPM to differentiate and identify the various inorganic constituents within a carbonaceous plug found in a solvent refined coal pilot plant. Harris and Yust realized that because of the complexity of the microstructure of the plug and the difficulty involved in identifying constituents, neither optical micrographs (Fig. 5) nor scanning electron micrographs would alone be of much use. In contrast, a backscattered electron micrograph would clearly distinguish between organic and inorganic constituents within the sample (Fig. 5). The high Z minerals (pyrrhotite, quartz, and kaolinite) would appear as bright, well-defined areas while the lower atomic numbered constituents (the

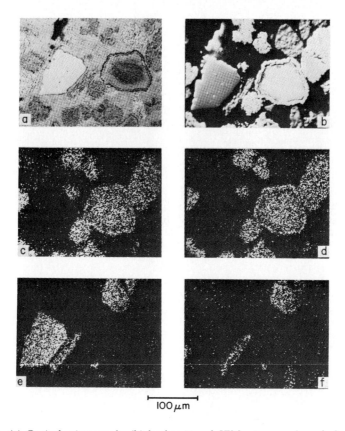

100 μm

Fig. 5 (a) Optical micrograph, (b) backscattered SEM micrograph and elemental K_α x-ray maps of a portion of a SRC Pilot Plant carbonaceous plug. The backscattered SEM micrograph locates the inorganic mineralogy, while the sulfur (c), iron (d), silicon (e), and aluminum (f) define the presence of pyrrohtite ($Fe_{(1-x)}S$), quartz (SiO_2), and kaolinite ($Al_2Si_2O_5(OH)_4$). (After Harris and Yust, 1978.)

hydrocarbons present in a semicoke phase) would appear dark black. Once the specific area of mineralogic interest had been thus defined, the inorganic constituents could be readily located and defined by making Fe, S, Al, and Si $K\alpha$ x-ray maps with the EPM as shown in Fig. 5. Furthermore, the x-ray maps showed relationships between these minerals that in turn provided interesting knowledge concerning the formation of the carbonaceous plug.

IV. CONCLUSION

The capability of nondestructive quantitative analysis on micrometer-size points of a sample is a characteristic unique to electron beam instruments. In most cases, the EPM is superior to the SEM for quantitative analysis. The heterogeneous and fine-grained nature of coal constituents therefore make the EPM uniquely adaptable to many aspects of coal characterization. At the same time, as with other types of instruments, it has limitations that restrict it to specific problems. For example, it is not particularly adaptable for on-line "quality control" applications in large scale production or utilization plants. It is unsuitable for trace element analysis due to minimum detection limits on the order of 100 ppm. Spacial resolution for qualitative analysis is about 1 μm; for good quantitative analysis it is about 5 μm. Despite these limitations, the EPM offers potential solutions to many problems concerning organic and inorganic chemistry of coal, especially if used in conjunction with other techniques such as x-ray diffraction, Mössbauer spectroscopy, scanning and transmission electron microscopy, and optical microscopy. Use of the EPM in coal research is in its infancy and is certain to increase significantly in the near future.

ACKNOWLEDGMENT

This research was supported by the Department of Energy, Division of Basic Energy Sciences under contract W-7405-ENG-36.

REFERENCES

Beaman, D. R., and Isasi, J. A. (1970). *Anal. Chem.* **42,** 1540–1568.
Buseck, P. R., and Goldstein, J. I. (1969) *Geol. Soc. Am. Bull.* **80,** 2141–2158.
Castaing, R. (1951). Ph.D. Thesis, Univ. of Paris.
Castaing, R. (1960). *Adv. Electron. Electron Phys.* **13,** 317–386.
Castaing, R., and Guinier, A. (1949). *Proc. Conf. Electron Microsc. Delft* pp. 60–63.
Coster, D., and Von Heresy, G. (1923). *Naturwissenchaften* **11,** 133.
Goldstein, J. I., and Yakowitz, H. (1975). "Practical Scanning Electron Microscopy." Plenum, New York.

Greer, R. T. (1977). *In* "Scanning Electron Microscopy" (O. Johari, ed.), Vol. 1, pp. 79–93. IIT Res. Inst., Chicago, Illinois.

Harris, L. A., and Yust, C. S. (1978). *In* "Scanning Electron Microscopy" (O. Johari, ed.), Vol. 1, pp. 537–542. Scanning Electron Microscopy, Inc., AMF O'Hare, Chicago, Illinois.

Hillier, J. (1947). U.S. Pat. No. 2, 418,029.

Keil, K. (1973). *In* "Microprobe Analysis" (C. A. Anderson, ed.), pp. 189–239. Wiley (Interscience), New York.

Moseley, H. (1913). *Philos. Mag.* **26,** 1024–1034.

Raymond, R., Jr., and Gooley, R. (1978). *In* "Scanning Electron Microscopy" (O. Johari, ed.), Vol. 1, pp. 93–107. Scanning Electron Microscopy, Inc., AMF O'Hare, Chicago, Illinois.

Wewerka, E. M., Williams, J. M., Vanderborgh, N. E., Harmon, A. W., Wagner, P., Wanek, P. L., and Olsen, J. D. (1978). LA-7360-PR. Los Alamos Sci. Lab., Los Alamos, New Mexico.

SUGGESTED BIBLIOGRAPHY

Anderson, C. A., ed. (1973). "Microprobe Analysis." Wiley (Interscience), New York.
An excellent book for instrumentation, theory, quantitative analysis, applications to geology and other disciplines, and other instruments including laser microprobe and ion microprobe.

Birks, L. S. (1963). "Electron Probe Microanalysis." Wiley (Interscience), New York.
A bit dated as to modern instrumentation and techniques, but good for concepts and theory.

Goldstein, J. I., and Yakowitz, H. (1975). "Practical Scanning Electron Microscopy." Plenum, New York.
Although the title indicates SEM, this book also includes a sizeable amount of information about and pertaining to EPM applications, and a final chapter on the ion microprobe.

Woldseth, R. (1973). "X-Ray Energy Spectrometry." Kevex Corp., Burlingame, California.
Written for EDS analysis, but useful information to aid the reader in x-ray theory, including electron excitation.

Chapter 49

X-Ray Fluorescence Analysis of Trace Elements in Coal and Solvent Refined Coal

John W. Prather

CIBA-GEIGY CORPORATION
MCINTOSH, ALABAMA

James A. Guin and Arthur R. Tarrer

DEPARTMENT OF CHEMICAL ENGINEERING
AUBURN UNIVERSITY
AUBURN, ALABAMA

I. INTRODUCTION

A. Background

In view of the predicted shift to coal and coal derived products as energy sources and concern for the environment, increased attention has been directed to the area of producing a clean burning fuel from coal that will be suitable for use in electric power generating plants. Most of these plants burn massive amounts of coal each minute, and any viable process that will feed these power plants with clean burning coal must have a similar throughput. Many analyses will be necessary to monitor both the feed coal and the refined product.

The research described herein was undertaken to ascertain the ability of energy dispersive x-ray fluorescence (EDXRF) analysis to determine

357

trace element content of solvent refined coal (SRC). Rapid simultaneous quantitative analyses of a large number of elements is possible with EDXRF. In this research some 17–20 elements are quantitatively analyzed in both raw feed coal and product SRC from the Wilsonville SRC Pilot Plant, Wilsonville, Alabama.

B. Analysis of Trace Metals

The content of trace metals in SRC product and in feed coal is important for several reasons: Certain metals, particularly iron, catalyze the hydrogenation and hydrodesulfurization reactions that occur in the SRC process (Guin *et al.*, 1978; Morooka and Hamrin, 1977). If the SRC product is to be used to fire gas turbines, then trace metals such as Ca, Na, and V present potential corrosion or fouling problems. When the feed coal has a high oxygen (~17%) and low sulfur (~0.7%) content as many of the western sub-bituminous coals do, trace metals such as calcium need be monitored since they react with carbon dioxide formed in the SRC dissolver. The carbonates thus formed tend to collect in the dissolver, eventually resulting in plugging problems. Trace metals, such as lead in the SRC product, present potential environmental problems. Also, metals such as iron and aluminum are present in sufficient quantity in the feed coals that SRC mineral residue could represent an attractive source of these metals when current sources have been depleted.

The analysis of trace metals in coals has been the subject of several earlier papers (Berman and Ergun, 1968; Sweatman *et al.*, 1963; Kiss, 1966; Ondov, *et al.*, 1975). Analytical methods, such as atomic absorption spectroscopy (Schultz *et al.*, 1977; Coleman *et al.*, 1977; Chapter 14, Volume I) have been presented for determining the content of trace metals in clean coal products. However, in view of the large number of analyses required to maintain reasonable quality control of the SRC product, alternate methods that are more rapid and less expensive than those currently available are needed. Application of EDXRF to assays of feed coals and SRC products is attractive because it allows simultaneous analysis of trace elements. This analysis can then be used to consolidate three analyses: sulfur content, ash content, and yield of liquefied coal, that at present must be carried out independently of one another. Energy dispersive x-ray fluorescence also lends itself well to automated routine analyses resulting in further time savings over the three independent methods for the same analytical task.

II. ANALYTICAL PROCEDURES

A. Equipment

All EDXRF analyses were performed on a Kevex 0810RW x-ray fluorescence system (Kevex Corporation, Foster City, California 94404) consisting of a 0810A x-ray subsystem, 5100 x-ray spectrometer, Rigaku Gigerflex 3kW x-ray generator (60 kV and 180 mA) Digital Electronics Corporation PDP 11/03 computer with RX01 dual floppy disk bulk storage. The system uses a high power x-ray tube (Ag target) to produce x rays from a series of selectable secondary targets (Ti, Ge, Mo, and Sn). The spectrometer uses a solid state detector of 30 mm^2 active area that has resolution of <165 eV at 5.9 KeV.

Atomic absorption experiments for iron were done on a Perkin-Elmer model 305A atomic absorption spectrophotomer (Perkin-Elmer Corporation, Norwalk, Connecticut 06856) using an acetylene/air flame and a wavelength setting of 249 nm. Sulfur determinations were made using a Leco model 521-500 sulfur analyzer specially fitted for low-level sulfur detection (Laboratory Equipment Corporation, St. Joseph, Michigan 49085).

Carbon, hydrogen, and nitrogen analysis were carried out on a Perkin-Elmer model 240 CHN analyzer (Perkin-Elmer Corporation, Norwalk, Connecticut 06856) that was fully automated and controlled through a Tektronix 31 programmable calculator.

Ash determinations were made by ashing powdered coal in a Lindburg type 123-4 tube furnace at 900°C. A flow of 50 mliter/min oxygen through the tube was utilized to insure complete combustion of the coal samples.

B. Samples and Sample Preparation

1. Coals and SRCs

The coals used in this work were obtained from several sources. Samples of Kentucky No. 9/14, Illinois No. 6, Pittsburgh No. 8, Monterey, Wyodak (Amax), Rosebud, Big Horn, and Emery coals were obtained as lump coal from the SRC pilot plant at Wilsonville, Alabama. These coals were used to produce SRC product in the plant. Four samples were obtained from the Pennsylvania State University, Coal Research Section, University Park, Pennsylvania 16802. These samples were cataloged as PSOC-166 (Kentucky No. 11 Hopkins County) and

PSOC-273 (Kentucky No. 11 Muhlenburg County). One coal sample was obtained from the United States Department of the Interior, Bureau of Mines. The sample coal was from the Lower Kittanning bed with Bureau of Mines Laboratory No. K-64364. One sample was obtained from the National Bureau of Standards (Ondov et al., 1975; von Lehmden et al., 1974) as NBS-1632. This coal is actually a mixture of several coals.

Solvent refined coal samples were obtained from Catalytic, Inc., and Southern Services, Inc., Wilsonville, Alabama. The five samples were produced using Western Kentucky No. 9/14, Illinois No.6, Pittsburgh No. 8, Illinois (Monterey mine) and Wyodak (Amax) coals with laboratory samples numbers 16572–16576, respectively. These SRCs are specification grade product produced between August 1, 1974, and November 22, 1975.

2. Procedure

Approximately 10 g samples of all coals and SRCs were first ground so that the entire sample passed through a 325 mesh screen (<45 μm). The samples were then dried in a vacuum oven at 105°C for at least 3 hr. From these samples 1.25 in. pellets were made using a boric acid backing and were analyzed by EDXRF under a vacuum of 5×10^{-4} torr. Also, fractions of the same samples were used to perform Leco Sulfur, C, H, and N, and atomic absorption analyses.

C. Analysis of Data

The EDXRF data were analyzed using the Kevex matrix correction program, and adaptation of the Shell EXACT (Otvos et al., 1976) program. This program employs a fundamental parametric method that accounts for matrix interactions because of absorption and enhancement for elements in the sample. The EXACT model is basically similar to models presented by Sherman (1955), Shiraiwa and Fujino (1966), and Criss and Birks (1968) with simplifications.

III. RESULTS AND DISCUSSION

To confirm the analytical procedures used in this work, EDXRF analysis was performed on an NBS standard coal sample, No. 1632. The results (Table I) are compared to the literature values that were obtained by extensive analyses in 13 different laboratories. The results for the most part, compare very well with the reported analyses.

TABLE I *EDXRF Analysis of NBS Standard Coal No. 1632*

Element	Weight percent This work		Literature (Ondov et al., 1975; von Lehmden et al., 1974)	
Al	1.98	± 0.10	1.85	± 0.18
Si	3.19	± 0.10	3.2	
P	0.12	± 0.01	—	
S	1.52	± 0.05	—	
Cl	0.0811	± 0.005	0.0890	± 0.0125
K	0.241	± 0.007	0.28	± 0.03
Ca	0.442	± 0.012	0.43	± 0.05
Ti	0.116	± 0.005	0.110	± 0.01
V	0.005	± 0.001	0.0036	± 0.0003
Cr	0.0009	± 0.0002	0.0020	± 0.0009
Mn	0.0037	± 0.004	0.0043	± 0.0004
Fe	0.890	± 0.024	0.84	± 0.04
Ni	0.0083	± 0.0007	0.0018	± 0.0004
Cu	0.0024	± 0.0003	—	
Zn	0.0037	± 0.0003	0.0030	± 0.001
Ga	0.0008	± 0.0001	—	
Ge	0.0005	± 0.0001	—	
Pb	0.0020	± 0.0002	0.0030	
Se	0.0006	± 0.0001	—	
Br	0.0019	± 0.0001	0.00193	± 0.0002
Rb	0.0011	± 0.0001	0.0021	± 0.0002
Sr	0.0128	± 0.0003	0.0161	± 0.0016
Y	0.0007	± 0.0001	—	

Table II gives the EDXRF, C, H, N, and ash analyses for the 14 coals used in this study. Table III gives the same set of analyses for the five SRC products studied in this work. As expected, smaller amounts of the elements analyzed for by EDXRF were detected in the SRC products than in the respective feed coals. However, the relative amounts of the elements in the feed coals is also reflected in the SRC product, i.e., Wyodak and Pittsburgh No. 8 SRC show high-calcium content with respect to the other SRC's just as is found in the Wyodak and Pittsburgh No. 8 feed coals. It would appear that feed coal elemental content is the controlling factor with respect to the elemental content of the SRC product.† This trend is evident in sulfur content of the SRC product as

† For comparison with Synthoil feed coal and product see Table II, Chapter 18, Volume I.

TABLE II EDXRF, C, H, N and Ash Analyses for Coal (Moisture Free Basis)

Element	NBS No. 1632	Emery	Weight percent Wyodak (AMAX)	Bighorn	Rosebud	Lower Kittanning	West Kentucky No. 9/14
Ash	12.17	9.40	7.21	5.70	5.51	13.33	7.05
C	71.7		64.68	69.05	68.42	72.60	70.37
H	4.57	5.40	4.95	5.31	5.48	4.70	5.18
N	1.98	1.50	1.35	1.53	1.58	1.30	1.56
O (by diff)	10.08	19.80	21.05	18.78	19.55	3.0	10.46
Al	1.98	0.842	0.849	0.868	0.716	1.291	0.903
Si	3.19	1.561	1.085	0.917	0.857	1.938	1.469
P	0.12	0.0923	0.095	0.106	0.132	0.279	0.2043
S	1.520	1.205	1.016	0.809	1.526	3.595	3.030
Cl	0.080	—	0.004	0.004	—	0.289	0.0142
K	0.241	0.0256	0.0384	0.0241	0.0326	0.1186	0.1073
Ca	0.442	1.142	1.697	0.885	1.083	0.1617	0.1356
Ti	0.116	0.0797	0.0813	0.0661	0.0465	0.0861	0.0515
V	0.005	0.0120	0.0466	—	0.0156	0.0169	0.0077
Cr	0.0009	0.0065	0.0042	0.005	0.0134	0.0107	0.0019
Mn	0.0037	0.0038	0.0043	0.0033	0.0085	0.0062	0.0031
Fe	0.890	0.5779	0.4239	0.4026	0.7204	3.304	1.203
Ni	0.0083	0.0011	0.0008	0.0016	0.0023	0.0007	0.0003
Cu	0.0024	0.0021	0.0079	0.0077	0.0023	0.0016	0.0014
Zn	0.0037	0.0024	0.0029	0.0018	0.0006	0.002	0.0023
Br	0.0019	0.0002	0.0002	0.0002	0.0011	0.0045	0.0005
Rb	0.0010	—	—	—	—	0.008	—
Sr	0.0128	0.0066	0.987	0.0095	0.0115	0.0054	0.0022
Pb	0.0020	0.0008	0.0011	0.0013	0.0007	0.0034	0.0023

Element	Illinois No. 6	Pittsburgh No. 8	Weight percent Monterrey (Illinois)	PSOC-166 (Kentucky No. 11)	PSOC-216 (Kentucky No. 14)	PSOC-220 (Kentucky No. 11)	PSOC-273 (Kentucky No. 11)
Ash	8.01	8.72	12.00	13.68	10.99	12.37	10.44
C	74.74	73.61	68.35	65.61	72.84	71.98	64.14
H	5.44	5.39	5.03	4.70	5.29	5.32	4.57
N	1.51	1.69	1.54	1.08	1.61	1.33	1.87
O (by diff)	7.81	7.71	12.12	9.22	7.48	5.02	8.41
Al	1.053	0.9025	1.429	2.394	1.1005	1.768	0.8599
Si	1.855	1.563	3.024	3.916	1.455	2.536	1.556
P	0.1949	0.1721	0.4108	—	0.2697	—	0.2869
S	3.514	2.888	5.852	4.001	4.086	5.215	4.652
Cl	0.0231	—	0.0332	—	0.0557	0.0140	0.0056
K	0.1243	0.1062	0.2029	0.2128	0.0779	0.1459	0.1071
Ca	0.2698	1.560	0.3681	0.0782	0.1521	0.3992	0.1941
Ti	0.0679	0.0681	0.0778	0.1512	0.0559	0.0813	0.0597
V	0.0138	0.0107	0.0107	0.0212	0.0062	0.0090	0.0093
Cr	0.0053	0.0033	0.0078	0.0123	0.0045	0.0329	0.0118
Mn	0.0045	0.0054	0.0117	0.0074	0.0108	0.0148	0.0064
Fe	0.8697	0.7496	1.8803	1.5355	3.4433	2.6430	2.9486
Ni	—	—	0.0072	0.0045	0.0087	0.0197	0.0024
Cu	0.0024	0.0049	0.0046	0.0436	0.0095	0.0490	0.0231
Zn	0.0018	0.0036	0.0238	0.0026	0.0047	0.0154	0.0036
Br	0.0011	—	0.0003	—	0.0009	0.0004	—
Rb	—	—	—	0.0020	0.0006	—	—
Sr	0.0018	0.0150	0.0018	0.0023	0.0090	0.0027	0.0013
Pb	0.0011	—	0.0016	0.0024	0.0039	0.0013	0.0032

TABLE III   *EDXRF, C, H, N, and Ash Analysis of SRC (Moisture Free Basis)*

Element	Western Kentucky	Illinois No. 6	Pittsburgh No. 8	AMAX	Monterrey
		Weight percent			
Ash	0.110	0.093	0.135	0.268	0.261
C	74.99	73.24	77.44	75.89	71.89
H	5.96	6.10	5.93	6.12	6.36
N	1.86	1.93	1.95	2.03	1.87
O(by diff)	16.51	18.06	13.69	15.84	18.74
Al	—	—	—	—	—
Si	—	—	—	0.0386	0.0971
P	0.0903	0.0961	0.0760	0.0415	0.1164
S	0.8066	0.8531	1.1237	0.1800	1.3368
K	0.0012	—	—	—	—
Ca	0.0153	0.0040	0.0247	0.0407	0.0038
Ti	0.0108	0.0217	0.0099	0.0429	0.0159
V	0.0041	0.0041	0.0053	0.0118	0.0038
Cr	—	0.0026	0.0028	0.0089	—
Mn	0.0021	0.0031	0.0009	0.0065	—
Fe	0.0582	0.0282	0.0251	0.0317	0.0623
Ni	—	0.0011	0.0009	0.0014	—
Cu	0.0004	0.0017	0.0007	0.0021	0.0012
Zn	0.0006	0.0008	0.0008	0.0006	0.0005
Br	0.0002	0.0004	0.0004	—	0.0004
Sr	—	—	—	0.0001	0.0009
Pb	0.0002	0.0002	0.0005	0.0003	0.0003

well; with all cases showing approximately 75% reduction in sulfur content of the feed coal by the SRC process.

With respect to ash content of the coals and SRC products, an interesting relationship was observed between ash found by ashing the samples in a tube furnace and the sum of the weight percents of metallic elements (Al, Si, K, Ca, Ti, V, Cr, Mn, Fe, Ni, Cu, Zn, Sr, and Pb) determined by EDXRF (Prather *et al.*, 1977). Figures 1 and 2 show that as expected a linear relationship exists between these two sets of data. In the case of the coals analyzed a least square linear regression gives the following equation:

$$\text{Percent ash} = (1.73) \, \Sigma \, \text{wt.\% EDXRF elements.} \qquad (1)$$

The coefficient r was 0.951 for the linear fit.

Analysis of the data on SRC resulted in the following equation:

$$\text{Percent ash} = (1.43) \, \Sigma \, \text{wt.\% EDXRF elements.} \qquad (2)$$

The coefficient of correlation r was 0.975 for the same regression

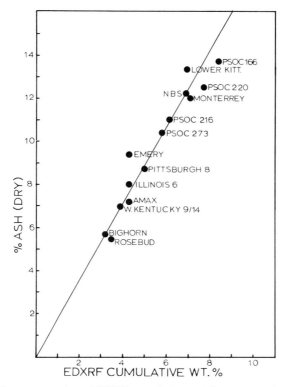

Fig. 1 Weight percent ash vs. EDXRF cumulative weight percents for several coals.

analysis performed on the coal data. In both cases it seems this empirical relationship shows a sufficient correlation among the data to enable one to calculate percent ash in either feed coal or SRC product with the accuracy needed for this type analysis. It is conceivable that nondestructive ash analysis could be carried out on coal and SRC product in approximately 1 and 1.5 hr, respectively, the difference being that longer data acquisition times are necessary for SRC product because of the lesser amounts of the elements under analysis. This contrasts with the 3–24 hr time periods required for ash determination by conventional procedures.

It was thought that if such good relationships existed between ash and EDXRF data, then one should be able to calculate exactly the ash content of coal and SRC product by assuming that the elements found by EDXRF should exist in the ash as their most common oxides and that negligible metallic elemental loss was incurred during high temperature ashing. Figures 3 and 4 show plots of the observed percent

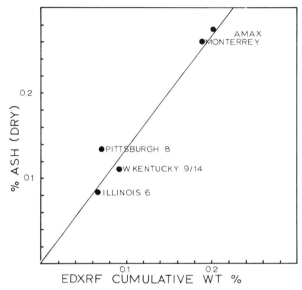

Fig. 2 Weight percent ash (moisture free basis) vs. EDXRF cumulative weight percents for SRC products.

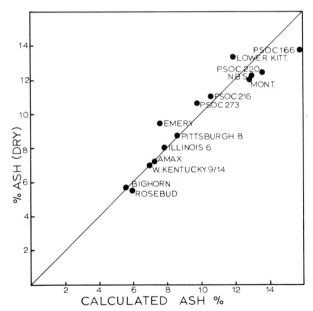

Fig. 3 Experimental weight percent ash (moisture free basis) vs. weight percent ash calculated from EDXRF data for several coals.

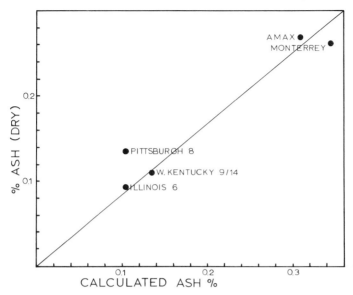

Fig. 4 Experimental weight percent ash (moisture free basis) vs. weight percent ash calculated from EDXRF data on SRC products.

ash versus theoretical ash calculated from EDXRF data. In both cases a satisfactory correlation was found as seen on the figures which show that the measured ash (ordinate) is in essential agreement with the calculated ash (abcissa) for most of the samples.

These findings lend confidence to the ability of EDXRF to predict ash content of coal and SRC and show that high temperature ashing results in negligible metallic elemental loss.

Ash content is important in the solvent refining of coal in aspects other than pollution abatement. Ash content is often used in calculating yield or conversion of coal to liquefied coal (Neavel, 1976). Generally, the percent ash in the feed coal is compared to the ash content of the residue that remains after liquefaction and a conversion is calculated. It was thought that if EDXRF data and ash content showed a high degree of correlation, then it would be possible to use the EDXRF data to calculate conversion based on some elemental balance rather than an ash balance. If this were true, then only the elements present in substantial amounts would need to be monitored by EDXRF and these used in the calculation. Table IV gives the results of two such elemental based conversion calculations compared to corresponding ash based calculations. It can be seen that conversion calculated on an iron basis

TABLE IV *Comparison of Liquefaction Yield DATA Calculated on Ash and EDXRF Basis*

Basis	Yield	Coal
Ash[a]	64.17	Wyodak
Fe (EDXRF)	64.00	
Si (EDXRF)	65.12	
Ti (EDXRF)	64.70	
Ash[a]	86.91	Western Kentucky 9/14
Fe (EDXRF)	84.90	
Si (EDXRF)	82.80	
Ca (EDXRF)	87.70	

[a] Ash yield data taken from Stino (1977).

would result in conversion data of equal quality with results obtained on an ash basis. Conversion calculated from other elemental bases are also given for comparison.

IV. CONCLUSION

It has been shown that EDXRF is capable of analyzing coal and SRCs for elemental content. These analyses provide simultaneous results for several important elements, namely, sulfur, iron, and calcium, as well as percent ash and liquefaction yield. These analyses presently take well over 2 hr per sample to complete, but by using EDXRF this time can be reduced to about 1 hr per sample. Furthermore, EDXRF has an added feature in that it may be automated, leading to even shorter analysis times per sample. Thus, EDXRF can be concluded to be an effective tool that can provide very accurate and rapid analyses of trace elements in coals and SRC product.

ACKNOWLEDGMENTS

The authors are grateful to the University Activities Division of the U.S. Department of Energy for support of this work under Contract No. EX-76-(S-01)-2454. They also wish to thank the personnel of Southern Services, Inc., who kindly supplied coal and SRC samples for this work. The authors also wish to acknowledge the technical assistance of G.A. Thomas and W.R. Glass of the Auburn University Coal Conversion Laboratory.

REFERENCES

Berman, M., and Ergun, S. (1968). *Fuel* **47**, 285.

Coleman, W. M., Szabo, P., Wooten, D. L., Dorn, H. C., and Taylor, L. T. (1977). *Am. Chem. Soc., Div. Fuel Chem., Prepr., 173rd Natl. Meet., New Orleans, La.* **22**, 226.

Criss, J. W., and Birks, L. S. (1968). *Anal. Chem.* **40**, 1080.

Guin, J. A., Tarrer, A. R., Prather, J. W., Johnson, D. R., and Lee, J. M. (1978). *Ind. Eng. Chem., Process Des. Dev.* **17**, 118.

Kiss, L. T. (1966). *Anal. Chem.* **38**, 1731.

Morooka, S., and Hamrin, C. E., Jr. (1977). *Chem. Eng. Sci.* **32**, 125.

Neavel, R. C. (1976). *Fuel* **55**, 237.

Ondov, J. M., Zoller, W. H., Olmez, I., Aras, N. K., Gordan, G. E., Rancitelli, L. A., Abel, K. H., Filby, R. H., Shah, K. R., and Ragaini, R. C. (1975). *Anal. Chem.* **47**, 1102.

Otvos, J. W., Wyld, G. E. A., and Yao, T. C. (1976). *Annu. Denver X-Ray Conf., 25th.*

Prather, J. W., Tarrer, A. R., and Guin, J. A. (1977). *Am. Chem. Soc., Div. Fuel Chem., Prepr., 174th Natl. Meet., Chicago, Ill.* **22**, 72.

Schultz, H., Gibbon, G. A., Hattman, E. A., Booher, H. B., and Adkins, J. W. (1977). PERC/RI-44-2. Pittsburgh, Pa.

Sherman, J. (1955). *Spectrochim. Acta* **7**, 283.

Shiraiwa, J., and Fujino, N. (1966). *Jpn. J. Appl. Phys.* **5**, 886.

Stino, S. (1977). M.S. Thesis, Auburn Univ., Auburn, Alabama.

Sweatman, T. R., Norrish, K., and Durie, R. A. (1963). *CSIRO Misc. Rep.* No. 177.

von Lehmden, D. J., Jungers, R. H., and Lee, R. E., Jr. (1974). *Anal. Chem.* **46**, 239.

Chapter 50

Mössbauer Analysis of Iron-Bearing Phases in Coal, Coke, and Ash

F. E. Huggins   *G. P. Huffman*
U. S. STEEL CORPORATION
RESEARCH LABORATORY
MONROEVILLE, PENNSYLVANIA

371

I. INTRODUCTION

Much interest has been directed toward increased commercial utilization of the extensive coal reserves of the United States. In particular, it is likely that steam coal production will be greatly expanded in the near future. As a result, many new analytical techniques have recently been applied to coal to search for answers to problems posed by increased utilization of coal. Since many of the problems, especially for steam coal utilization, arise from the mineral matter included in coal, it is not surprising that many of these techniques are applied specifically to obtain a better characterization of the inorganic constituents of coal and coal products.

One such technique is Mössbauer spectroscopy, which is ideally suited to the characterization of the iron-bearing phases in coal and coal products. For reasons that will be discussed later, Mössbauer spectroscopy is limited to the investigation of the phases in coal that contain iron; however, this limitation is not too severe because of the detailed information provided by Mössbauer spectroscopy and because iron oxide is usually the third or fourth most abundant oxide in coal ash. Furthermore, the iron-bearing minerals in coal play a significant role in problems of coal utilization. For example, the importance of pyrite in the pollution of the atmosphere by sulfur-bearing gases is well known, and there are several studies currently underway investigating the removal of pyrite and other iron-bearing minerals from coal. Another problem of great economic importance is the behavior of ash and slag in commercial combustion equipment. It is well known that iron oxides derived from the iron-bearing minerals in coal play a pivotal role in determining this behavior. Other areas is which detailed analyses of iron-bearing phases, particularly pyrite, in coal should contribute significantly include environmental studies (such as acid-mine drainage, coal-refuse disposal, coal weathering, and oxidation), paleoenvironment studies, and problems associated with mineral matter in coal-conversion technologies.

In this chapter, a brief description of the theory and practice of Mössbauer spectroscopy will be presented, along with a detailed review and discussion of the characterization by Mössbauer spectroscopy of iron-bearing minerals in coals and their transformation products after carbonization and combustion. Methods of making quantitative determinations of pyritic sulfur in coal, and geological, environmental, and other applications of Mössbauer analysis of coal and coal products will also be discussed.

II. MÖSSBAUER SPECTROSCOPY

A. Physical Description of the Mössbauer Effect

Mössbauer spectroscopy, or nuclear γ-ray spectroscopy as it is often called in Eastern Europe, utilizes the Mössbauer effect to obtain information in such diverse fields of study as physics, metallurgy, biology, and mineralogy. The Mössbauer effect is the name given to the phenomenon of recoilless resonant emission and absorption of γ rays by nuclei that was first demonstrated by Mössbauer in 1957 (Mössbauer, 1958). For his discovery, Mössbauer was awarded the Nobel prize for Physics in 1961.

Not all nuclei exhibit an observable Mössbauer effect, for a nucleus must comply with a number of restrictions that will be discussed later. To date, the Mössbauer effect has been observed in 85 isotopes of 44 elements (Stevens and Stevens, 1976). Of the major elements commonly found in coal ash, only potassium and iron exhibit Mössbauer effects. Since Mössbauer spectroscopy with potassium is experimentally difficult and chemically uninformative, it seems likely that only iron Mössbauer spectroscopy will find widespread application in the area of coal science. Consequently, subsequent discussion in this section will be limited to the Mössbauer effect in iron.

For iron, only the nuclide of mass 57 exhibits a Mössbauer effect. The most convenient method of generating the necessary ^{57}Fe γ rays is to use the radioactive decay of ^{57}Co shown in Fig. 1. ^{57}Co decays by electron capture to an excited nuclear state of ^{57}Fe, which subsequently decays to its ground state by successive emission of a 123 keV γ ray and a 14.4 keV γ ray; the 14.4 keV γ ray is the one used for Mössbauer spectroscopy. When such a γ ray interacts with a second ^{57}Fe nucleus in its ground state, this nucleus can be excited to the 14.4 keV level, provided the energy of the incident γ ray precisely matches the energy difference between the ground and excited states of the absorbing nucleus. This process is the resonant emission and absorption criterion. The lifetime of the excited state determines the precision with which the energies must be matched. For ^{57}Fe, the half-life of the 14.4 keV level is approximately 10^{-7} s, so that according to the Heisenberg

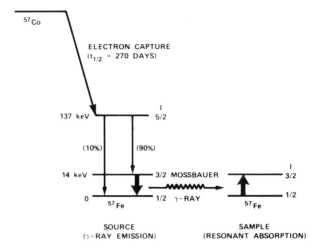

Fig. 1 Nuclear energy levels involved in the generation and resonant absorption of ^{57}Fe Mössbauer γ-rays. The symbol I denotes the nuclear spin quantum number.

uncertainty principle the agreement between the energies has to be better than 1 part in 10^{12}. Prior to Mössbauer's discovery, it was generally assumed that the emitting and absorbing nuclei behaved essentially as billiard balls, recoiling with momenta equal and opposite to that of the outgoing or incident γ rays. It is readily shown that the recoil energy E_R in this situation is given by

$$E_R = E_\gamma^2/2Mc^2, \tag{1}$$

where E_γ is the γ-ray energy, M is the nuclear mass, and c is the speed of light. The fractional energy loss for a 14.4 keV γ ray emitted by a ^{57}Fe nucleus would be about 1 part in 10^7, which violates considerably the criterion that the incident γ-ray energy must agree with the transition energy of the absorbing nucleus to better than 1 part in 10^{12}.

Mössbauer's critical observation was to note that if the emitting and absorbing nuclei are both imbedded in solids, recoil energy can only be transmitted to the lattice by exciting a lattice vibration or phonon. When the recoil energy of Eq. (1) is less than or comparable to typical phonon energies for the solid in question, there is a significant probability that the emission and absorption processes will occur without the excitation of a phonon. Momentum is then conserved by recoil of the entire crystal, and the recoil energy is completely negligible so that resonant, recoil-free emission, and absorption processes can occur.

The fraction of the γ rays emitted or absorbed without recoil by nuclei in a particular lattice may be derived from the theory of lattice vibrations. In the simple Debye approximation, it can be shown (Wertheim, 1964) that at a temperature T, which is small compared to the Debye temperature θ_D, the recoil-free fraction f, is given by

$$f = \exp\left[-\frac{E_R}{k\theta_D}\left(\frac{3}{2} + \frac{\pi^2 T^2}{\theta_D^2}\right) \right]. \tag{2}$$

Although Eq. (2) is strictly valid only for monatomic solids, it clearly demonstrates, together with Eq. (1), what factors affect the recoil-free fraction. It is seen that f will decrease with increasing γ-ray energy and temperature and increase with increasing values of M and θ_D.

For ^{57}Fe, these factors, along with the relatively long half-life of the excited nuclear state and the convenience of ^{57}Co as a source material, are combined in such a way that iron is by far the most popular element for Mössbauer studies. This popularity is especially marked for earth science applications because iron is also an element of major geochemical abundance.

B. Experimental Technique

The most common experimental arrangement used for Mössbauer spectroscopy is shown schematically in Fig. 2a. For ^{57}Fe spectroscopy, the radioactive source normally consists of from 10 to 100 mC of ^{57}Co in a metallic matrix with a relatively high recoil-free fraction in which

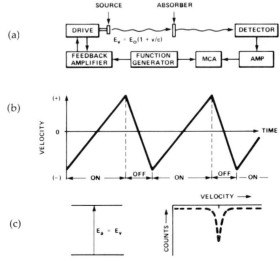

Fig. 2 Schematic representation of a Mössbauer experiment. (a) Block diagram of a Mössbauer spectrometer, (b) source velocity as a function of time, (c) a simple spectrum.

the nuclear levels are not split by hyperfine interactions (see Section II,C); Pd, Rh, Cr, and Cu are common choices for the source matrix. The absorber may be any solid containing iron, and it is usually in the form of a powder sample or thin foil that presents a thin aspect to the γ-ray beam to minimize self-absorption effects. Generally, areal densities of about 5 to 20 mg/cm² of iron are found to be optimum. Although the iron content of coal is relatively low (typically ~1–2 wt %), the organic component of coal is essentially transparent to 14.4 keV γ rays. Consequently, relatively thick powder samples with total areal densities of coal of about 0.5–1.0 g/cm² can be used, yielding areal densities of iron in the optimum range. The total amount of coal sampled by a typical Mössbauer measurement is about 0.5–2.0 g.

As will be discussed further in Section II,C, hyperfine interactions between the nuclei and the electrons in solids cause the nuclear energy levels to shift and split in several ways. In order to sweep the energies of the emitted γ rays through the various transition energies of the ^{57}Fe nuclei in the absorber, the source is vibrated back and forth over a small velocity range (typically of order ±1 cm/s) by an electromechanical drive. This motion causes the γ-ray energy to be varied because of the Doppler effect. Most commonly, the constant-acceleration driving mode is used, so that the source velocity is a linear function of time as shown in Fig. 2b. The Doppler-shifted energy of γ rays emitted when the source is moving at a velocity v is

$$E_v = E_0(1 + v/c), \qquad (3)$$

where c is the speed of light and E_0 is the energy of the γ ray emitted from a stationary source. The γ rays passing through the absorber are detected, amplified, and fed to a multichannel analyzer. A pulse from the function generator starts the analyzer at the beginning of each velocity cycle, and each channel stores the total number of γ rays transmitted by the absorber at a particular source velocity or at a particular Doppler-shifted energy, E_v.

When $E_v = E_a$, the energy difference between the ground and excited states of the nuclei in the absorber, the γ rays are resonantly absorbed. For a cubic, nonmagnetic absorber with a single type of iron site, a plot of the total number of counts accumulated over a period of time versus the source velocity will then exhibit a single Lorentzian peak as illustrated in Fig. 2c.

C. Hyperfine Interactions

The three principal types of interaction between electrons and nuclei in solids are the electrostatic, quadrupole, and magnetic hyperfine interactions. The effects of these three interactions on the nuclear energy levels and the Mössbauer spectra for ^{57}Fe nuclei are shown in Fig. 3. ^{57}Fe nuclei in a cubic, nonmagnetic environment will exhibit a single peak as illustrated in Fig. 3a. If the crystalline environment of the absorbing nuclei is less than cubic, an electric field gradient (EFG) interacts with the quadrupole moment of the ^{57}Fe excited state; for nonmagnetic absorbers, this causes the single peak to split into two peaks (Fig. 3b). In a magnetically ordered absorber, a large internal magnetic field interacts with the nuclear magnetic moments, causing the Mössbauer pattern to exhibit six symmetrical absorption peaks (Fig. 3c). The positions of the absorption peaks on the velocity scale define three Mössbauer parameters that are used to characterize these hyperfine interactions; the isomer shift (δ), the quadrupole splitting (ϵ), and the magnetic hyperfine field (H).

1. Isomer Shift

The electrostatic interaction between the electronic and nuclear charge distributions causes both the ground- and excited-state energy levels to shift; it is characterized by the Mössbauer parameter known

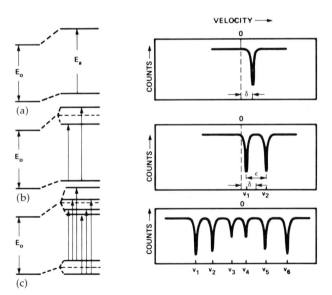

Fig. 3 Schematic illustration of the shifts and splittings in the nuclear levels and resulting Mössbauer spectra caused by interactions between the nucleus and the electrons: (a) isomer shift, (b) quadrupole splitting, (c) magnetic hyperfine splitting. One should note that the magnitudes of the shifts and splittings of the nuclear levels are actually very small ($\sim 10^{-9}$–10^{-7} eV) compared to E_0 (14.4 keV).

as the isomer shift (δ). For a single-peak spectrum (Fig. 3a), the isomer shift is simply the shift of the absorption peak away from zero on the velocity scale. The isomer shift for multipeak spectra is defined as the position of the centroid of the spectrum. For a quadrupole doublet (Fig. 3b),

$$\delta = (v_1 + v_2)/2, \tag{4}$$

whereas for a magnetic hyperfine spectrum (Fig. 3c),

$$\delta = (v_1 + v_2 + v_5 + v_6)/4. \tag{5}$$

where v_i is the velocity of the ith peak, counting from low velocity. The isomer shift of metallic iron is normally assigned the value of zero, and the isomer shifts of all other absorbers are measured relative to that of metallic iron. Variations in isomer shift values arise from differences in the electron density at the iron nucleus caused by the character of the chemical bonding in which the iron atom or ion is involved. Generally, it is found that the isomer shift has characteristic values for the different ionic states of Fe (Walker et al., 1961). Metallically bonded iron atoms have small isomer shifts (~ -0.1–0.1 mm/s), whereas high-spin Fe^{3+} and Fe^{2+} ions normally have medium-sized (~ 0.2 to 0.5 mm/s) and large (~ 0.9–1.3 mm/s) isomer shifts, respectively. Other factors, such as different spin states or unusual coordination numbers, will give rise to values outside these ranges.

2. Quadrupole Splitting

The quadrupole splitting arises from an asymmetric charge distribution about the ^{57}Fe nucleus that produces an electric field gradient (EFG) at the nuclear site. For nonmagnetic compounds, the interaction of the EFG with the quadrupole moment of the excited state of ^{57}Fe partially removes the degeneracy of that level, causing it to split into two levels; the resultant Mössbauer spectrum is known as a quadrupole doublet (Fig. 3b). The quadrupole splitting (ϵ) for a quadrupole doublet is simply the separation of the two doublet peaks in millimeters per second. For magnetically ordered compounds, the effect of a quadrupole interaction is to shift the peaks labeled 2–5 in Fig. 3c relative to those labeled 1 and 6. The quadrupole parameter is defined as

$$\epsilon = [(v_6 - v_5) - (v_2 - v_1)]/4. \tag{6}$$

This expression is equal to the change induced in the excited-state nuclear energy levels by the quadrupole interaction only in the case of axial symmetry (Wertheim, 1964). However, because of its ease of measurement, the parameter defined by Eq. (6) is commonly used to characterize the strength of the quadrupole interaction in magnetically ordered compounds, and this practice will be adopted in this chapter.

In general, there are two contributions to the EFG (Ingalls, 1964): one arises from an asymmetric distribution of the surrounding atoms or ligands (ligand contribution); the other is the contribution because of a nonspherical distribution of valence electrons over the 3d orbitals of the iron atom or ion (valence contribution). The valence contribution, which is larger and temperature dependent, is absent in high-spin ferric compounds and low-spin ferrous compounds because the distribution of valence electrons over the five 3d oribtals is always symmetric in these cases. Hence, high-spin Fe^{3+} and low-spin Fe^{2+} compounds have small, temperature-independent quadrupole splittings in nearly all cases. High-spin Fe^{2+} compounds have much larger, temperature-dependent, quadrupole splittings and can be distinguished easily from other compounds.

3. Magnetic Hyperfine Splitting

In magnetically ordered compounds (ferromagnetic, antiferromagnetic, or ferrimagnetic), the uncompensated 3d electron spins interact with and produce spin polarization of the s electrons. The spin-polarized s electrons produce a magnetic hyperfine field at the nuclear site that, for ^{57}Fe, is normally negative (directed oppositely to the atomic magnetic moments) and quite large ($\sim$1 to 5 $\times$ 10^2 kG). [See, e.g., Watson and Freeman (1961) for discussion of origin, sign, and magnitude of ^{57}Fe hyperfine fields.] The Zeeman interaction of the nuclear magnetic moment with the magnetic hyperfine field removes all degeneracy of both the excited- (spin 3/2) and ground- (spin 1/2) state energy levels. Because of nuclear selection rules, only six transitions are normally allowed between the Zeeman split levels, and a six-peak Mössbauer spectrum results (Fig. 3c). For thin absorbers in which the magnetic moments are randomized (for example, in a pulverized powder sample), the six-line spectra exhibit relative peak intensities of approximately 3:2:1:1:2:3. The magnitude of the magnetic hyperfine field, H, may be simply defined in terms of the separation of the outermost peaks in Fig. 3c;

$$H = 30.98 \, (v_6 - v_1), \tag{7}$$

where H is in kilogauss ($\equiv 10^{-1}$ T) and v_6 and v_1 are in millimeters per second. The magnetic hyperfine field has been a parameter of great interest in fundamental studies of magnetism (Wertheim, 1964). In this chapter, however, it will be used only for identification of the magnetically ordered phases in coal, coke, and ash.

D. Mössbauer Analysis of Multiphase Samples

Every iron-bearing phase exhibits a Mössbauer spectrum with a characteristic set of Mössbauer parameters (δ, ϵ, and H). Mössbauer spectra obtained from multiphase specimens, such as coal and coal derivatives, may frequently consist of a mixture of several quadrupole, magnetic hyperfine, and single-peak components, and there is often considerable overlap of some of the Lorentzian absorption peaks arising from different phases. Nevertheless, the spectral components arising from different phases can normally be resolved by least-squares computer analysis, and the Mössbauer parameters (δ, ϵ, and H) determined from such analyses can be used to identify the iron-bearing phases that are present. Furthermore, the percentage of the total iron contained in a particular phase can be determined from the areas of the peaks contributed by that phase. It can be shown (Hien and Shpinel, 1963; Huffman and Louat, 1967) that the peak area A is related to a parameter known as the effective Mössbauer absorption thickness by the expression

$$A(X_{N,j}) = CX_{N,j} [\exp(-X_{N,j}/2)] [I_0(X_{N,j}) + I_1(X_{N,j})], \tag{8}$$

where C is a constant determined by the radioactive source, I_0 and I_1 are modified Bessel functions, and $X_{N,j}$ is the effective absorption thickness associated with the jth peak contributed by the Nth phase. The total effective absorption thickness for a phase N that contributes M_N peaks to the spectrum is directly proportional to the number of iron atoms n_N contained in that phase per unit volume of sample and is given by

$$X_N = \sum_{j=1}^{M_N} X_{N,j} = n_N f_N \sigma_0 a_{57}, \tag{9}$$

where f_N is the recoil-free fraction of phase N, σ_0 is the maximum cross section for resonant absorption (2.56×10^{-18} cm^2), and a_{57} is the isotopic abundance of ^{57}Fe (0.0219). In the analysis of multiphase samples, it is frequently assumed that the recoil-free fractions f_N do not differ significantly for the various phases present. In this approximation, the percentage of the total sample iron contained in phase N is given by

$$P_N = 100 \, X_N / \sum_K X_K = 100 \, X_N / X_{\text{tot}}, \tag{10}$$

where the sum includes all K iron-bearing phases present and X_{tot} is the total effective absorption thickness for the sample. Although the assumption of near-equality of the recoil-free fractions has not been adequately tested for the many iron-bearing phases that occur in coal and coal products, Eq. (10) probably gives a reasonably accurate approximation to the iron phase percentages in most instances. It should be noted that many investigators assume the percentage of iron in a given phase simply to be proportional to the total area under the peaks contributed by that phase. In the opinion of the authors (Huffman *et al.*, 1974; Huffman and Huggins, 1978), the effective thickness approach is more accurate since, as seen from Eqs. (8) and (9), peak areas do not depend linearly on the number of iron atoms contained in different phases. For most coal samples, the effective thicknesses are usually small, and normally the difference between the iron phase percentages determined by the two approaches will also be small. However, the analysis of Eqs. (8)–(10) is easily performed during the least-squares computer analysis of the Mössbauer data and is preferable for coals of high ash content or for coal ash. Additionally, the effective thickness may be used to make absolute determinations of the weight percentages of iron contained in particular phases in coal, as discussed further in the section of this chapter dealing with pyritic–sulfur determinations.

E. Additional Points of Interest

Before closing this section, several additional aspects of Mössbauer spectroscopy that are of general interest should be briefly mentioned.

1. *Temperature Dependence of Mössbauer Spectra*

More conclusive identification of the phases in complex samples can sometimes be accomplished by obtaining Mössbauer spectra at different temperatures, particularly at cryogenic temperatures. Generally, the quadrupole splitting and magnetic hyperfine splitting are temperature dependent, and the relative change in peak positions with temperature will often lead to better resolution of different components in the spectrum and, hence, more reliable assignments. In addition, transitions from paramagnetic to magnetic states may be induced by lowering the temperature of the sample. The observation of a magnetic state will generally distinguish between an iron-rich and iron-poor phase, for example, between illite and chlorite or between ankerite and siderite. Furthermore, the presence of superparamagnetic phases can usually be conclusively established only by obtaining Mössbauer spectra over a range of temperatures.

2. Reemission or Scattering Mössbauer Spectroscopy

For samples that cannot be conveniently prepared in thin absorber form, an alternative experiment can be utilized that entails accumulation of Mössbauer spectra by detection of radiation reemitted from the absorber. The 14.4 keV level of ^{57}Fe decays 90% of the time by a process known as internal conversion. The systematics of the decay are such that for every 100 14.4 keV γ rays resonantly absorbed, 10 14.4 keV γ rays, 90 7.3 keV internal conversion electrons, 65 5.4 keV Auger electrons, and 35 6.5 keV K_α x rays are emitted. By appropriate choice and positioning of the detection system, these emitted species can be selectively detected to obtain γ-ray, electron, or x-ray reemission Mössbauer spectra. Such spectra appear inverted to normal transmission spectra but otherwise contain identical information. As detailed elsewhere (Spijkerman, 1971; Huffman, 1976), reemission Mössbauer spectroscopy is particularly useful for surface studies and has been used in one coal-related study to follow the weathering reactions of pyrite (Baker, 1972).

3. General Mössbauer References

Much more complete descriptions of the theory, practice, and applications of Mössbauer spectroscopy can be found in the books by Wertheim (1964), Greenwood and Gibb (1971), Bancroft (1973), and others. One other reference work should also be mentioned, namely, the Mössbauer Effect Data Index (Stevens and Stevens, 1976), which is virtually a complete guide to the literature of Mössbauer spectroscopy. This index, originally published as annual volumes, but now as a monthly journal (Stevens *et al.*, 1978), not only attempts to list all literature references utilizing Mössbauer spectroscopy, but also contains many useful review articles on specific aspects of Mössbauer spectroscopy. Many of the data discussed in this chapter were obtained through this index.

III. MÖSSBAUER ANALYSIS OF IRON-BEARING MINERALS IN COAL

More than 50 different minerals have been found in coal; however, most have been found in trace amounts only. For example, Mackowsky in Stach *et al.* (1975) lists 45 different minerals, of which only 13 contributed more than 5% to the mineral matter in some coals. Other authors generally agree with Mackowsky's list concerning the more common minerals (see Volume II, Chapter 26, Table I); however, a number of minor minerals, especially sulfates, can be added (e.g., Gluskoter, 1975; Palache *et al.*, 1951) and micromineralogical techniques using scanning electron microscopy (Finkelman and Stanton, 1978) are likely to add many more rare minerals to the list. Table I lists all common and many rare minerals in coal and classifies them according to their iron content. This table shows that iron-bearing minerals constitute approximately half of the minerals known to occur in coal at the present time. Mössbauer data for these minerals will be discussed according to the groups shown in Table I. Of the common minerals in coal, only quartz and kaolinite are essentially iron free.

It should be noted that the compilation of Mössbauer data in Tables II–VI on different minerals in coals is based on many spectra, and in some cases different spectra of coal from the same seam, but sampled at different locations, illustrate different points. Such spectra are not distinguished in these tables, and it should not be assumed that spectral data in one table can be correlated with those in another. A summary of the iron distribution between different minerals is given in the appendix for samples from many coal seams.

A. Sulfides

Two iron-bearing sulfides are of importance; the first is pyrite because it is of common occurrence and a main contribution to pollution of the atmosphere by sulfurous gases when coal is burned; the second is marcasite because it is believed (e.g., Balme, 1956) that the marcasite content may be a paleoenvironment indicator related to pH. However, the occurrence of marcasite in many coals is negligible so that its use as such an indicator is restricted. The interest in such paleoenvironment indicators is because of the fact that coals rich in pyrite and sulfur are commonly those that have been influenced by marine environments. The actual form of pyrite in the coal, whether as blebs, euhedral crystallites, framboids, or dendritic growths,[†] may also be related to paleoenvironment conditions (Reyes-Navarro and Davis, 1976) and is probably a more general indicator than iron sulfides as marcasite. Pyrite and marcasite are dimorphous and have essentially identical compositions, FeS_2.

† See Chapter 42, Fig. 3, and Chapter 56, Figs. 1 and 2.

TABLE I Partial List of Minerals Found in Coal, Classified According to Their Iron Content

Iron in mineral is:	Essential (>5%)		Minor (1 to 5%)		Absent (<1%)	
Occurrence of the mineral in coal is:	Common[a]	Rare[a]	Common	Rare	Common	Rare
Sulfides	Pyrite	Marcasite Chalcopyrite Arsenopyrite Pyrrhotite Melnikovite		Sphalerite		Galena
Clays/silicates		Chlorite Biotite, etc.	Illite	Montmorillonite Muscovite, etc.	Kaolinite	Feldspars Zircon, etc.
Carbonates	Siderite Ankerite		Calcite	Dolomite		
Oxides/hydroxides		Hematite Magnetite Goethite Lepidocrocite Limonite			Quartz	Rutile Diaspore
Sulfates		Szomolnokite Rozenite Melanterite Roemerite Coquimbite Jarosite etc.				Gypsum Barite, etc.
Others		Humboldtine, etc.				Apatite Halite, etc.

[a] Common—usually >5%; rare—usually <5% of total mineral matter.

1. Pyrite

Pyrite is the most common sulfide mineral in coal and is usually the dominant iron-bearing phase in Mössbauer spectra of coal. Huffman and Huggins (1978) detected pyrite in all 40 coals they investigated and found it to be the major iron-bearing mineral in 70% of the samples. Except for the exploratory study by Lefelhocz *et al.* (1967), in which the iron content was often close to the limit of detectability of their method, all other Mössbauer spectra of coal have shown the presence of pyrite (Montano 1977; Levinson and Jacobs, 1977; Jacobs *et al.*, 1978; Smith *et al.*, 1978). Pyrite was also found to be the dominant iron-bearing mineral in peat (Gamayunov *et al.*, 1975). Typical Mössbauer spectra of pyrite-rich coals are shown in Fig. 4. The Mössbauer parameters obtained from pyrite in various coals and from pure samples are listed in Table II. Studies of the Mössbauer parameters of pyrite in an applied magnetic field at cryogenic temperatures (Montano and Seehra, 1976) have confirmed that iron is in the low-spin, divalent state in pyrite.

2. Marcasite

Marcasite has been recognized in some coals by optical microscopy techniques; however, it is always subordinate to pyrite and, given the similarity of Mössbauer parameters of marcasite and pyrite (Table II), it is not surprising that no Mössbauer study of coal has found unequivocable evidence for marcasite in coal. It is possible that low values

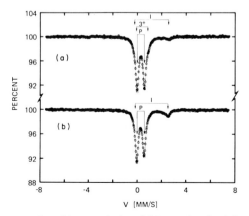

Fig. 4 Mössbauer spectra of two coals in which nearly all of the iron is in the form of pyrite. (a) coal from Herrin No. 6 seam, Ill. (b) coal from Pittsburgh seam, Pa. The arrows labeled P, I, and 3+ indicate Mössbauer peaks due to pyrite, illite, and the ferric sulfate, jarosite, respectively.

TABLE II Mössbauer Parameters[a] for Iron-Bearing Sulfides and Pyrite-Rich Coals

Mineral and formula	% Fe	δ	ϵ	H	Reference
Pyrite, FeS_2	100	0.31	0.62	—	This work
	100	0.25	0.62	—	Montano and Seehra (1976)
	100	0.33	0.61	—	Morice et al. (1969)
	100	0.28	0.61	—	Lefelhocz et al. (1967)
Marcasite, FeS_2	100	0.29	0.50	—	Morice et al. (1969)
	100	0.25	0.50	—	Lefelhocz et al. (1967)
	100	0.26	0.51	—	Temperley and Lefevre (1966)
Pyrrhotite, $Fe_{1-x}S$ (monoclinic, $x = 0.125$)	27	0.68	0.02	304	Huffman and Huggins (1978); Hucl et al. (1975); Vaughan and Ridout (1970)
	14	0.68	0.09	292	
	31	0.66	0.06	251	
	28	0.66	0.07	227	
Pyrrhotite, $x \neq 0.125$	Very complex				Schwartz and Vaughan (1972)
Chalcopyrite, $CuFeS_2$	100	0.20	0.00	350	Vaughan and Burns (1972)
	100	0.26	0.05	315	Raj et al. (1968)
Arsenopyrite, FeAsS	100	0.27	1.16	—	Goncharov et al. (1970)
Sphalerite, (Zn,Fe)S	8	0.61	0.62	—	Vaughan and Burns (1972)
	to	to	to		
	44	0.67	0.67		

Coal seam and state	% Fe_{pyr}[b]	δ_{pyr}	ϵ_{pyr}	Reference
Rosebud, Mont. (Sbb)	99	0.32	0.62	This work
Herrin No. 6, Ill.	99	0.30	0.62	This work
Redstone, W. Va.	97	0.31	0.60	Montano (1977)
Pittsburgh, W. Va.	92	0.31	0.60	Montano (1977)
Pittsburgh, Pa.	87	0.29	0.59	This work
Average of 31 coals	—	0.303	0.614	Huffman and Huggins (1978)

[a] All parameters measured at room temperature; δ—isomer shifts relative to metallic iron; ϵ—quadrupole splittings; H—magnetic hyperfine splittings. In this and subsequent tables, values for δ and ϵ in mm/s, H in kG.

[b] % Fe_{pyr}: percentage of total iron in the coal in the form of pyrite.

for the Mössbauer parameters of pyrite in certain coals may be because of the presence of a significant amount of marcasite (Huffman and Huggins, 1978).

3. Other Sulfides

A number of other iron-bearing sulfides have been reported in coal; these include pyrrhotite ($Fe_{1-x}S$, where $0 < x < 0.125$), arsenopyrite (FeAsS), chalcopyrite ($CuFeS_2$) and iron-bearing sphalerite (ZnS). The Mössbauer parameters for these phases, given in Table II, are based on pure phases, as none of them have been observed in coal by Mössbauer spectroscopy. Pyrrhotite, a ferrimagnetic phase, is of interest because it may be generated at the surfaces of pyrite grains during pulverization of the coal and be responsible for the enhancement of the magnetic properties of pyrite for high-gradient magnetic separation of sulfur from coal (Jacobs *et al.*, 1978). Depending on the number and ordering of vacancies, the structure of pyrrhotite may be of monoclinic, orthorhombic, or hexagonal symmetry, and the complexity of the Mössbauer spectra of pyrrhotites reflects the influence of the nonstoichiometry (Schwartz and Vaughan, 1972; Hucl *et al.*, 1975; Power and Fine, 1976). As discussed in Section VI, iron sulfides formed by the reduction of pyrite are common constituents of coke and of char formed during coal liquefaction.

B. Clays and Other Silicates

In the Mössbauer spectra of many coals (e.g., Fig. 5b), a quadrupole doublet with an isomer shift of 1.12 mm/s and a quadrupole splitting of 2.65 mm/s has been frequently observed (Lefelhocz *et al.*, 1967; Jacobs *et al.*, 1978; Huffman and Huggins, 1978). Originally, this doublet was tentatively attributed to "organic iron" (Lefelhocz *et al.*, 1967). Subsequent work by the same group and by the authors has shown that this assignment is incorrect, primarily because this doublet persists unchanged during low-temperature ashing, which would radically alter any organically bound iron. In addition, on the basis of a number of different tests, Huffman and Huggins (1978) showed that these peaks cannot arise from a ferrous sulfate. The remaining alternative is to attribute this doublet to a ferrous silicate. As will be discussed in later sections, this assignment is consistent with Mössbauer results for coke and coal ash, which show that this ferrous doublet arises from a phase in coal that transforms to either a ferrous (coke) or a ferric (ash) glass.

Many different silicates have been reported to be present in coals (e.g., Mackowsky in Stach *et al.*, 1975), however, only the layer-silicate

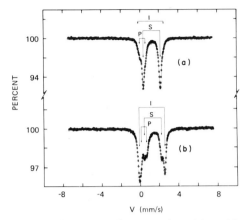

Fig. 5 Mössbauer spectra of an iron-carbonate-rich coal from (a) Gholson   Ala. and of a coal rich in ferrous-bearing clays from (b) Pocahontas No. 3 seam, W. Va. The arrows P, S, and I indicate Mössbauer peaks due to pyrite, siderite, and ferrous-bearing clays (illite and chlorite), respectively.

minerals, clays, chlorites, and possibly micas, have been found consistently in coals. These phases are listed in Table III in approximate decreasing order of occurrence.

Because of the very low amounts of FeO commonly found in kaolinite, it is unlikely that this mineral makes any significant contribution to the ferrous silicate doublet, despite its common occurrence. Montmorillonite and muscovite are too low in FeO, are of less common occurrence, and have Mössbauer parameters that are too different from those of the doublet in question to be responsible.† The same arguments apply to biotite, but with less certainty. It is likely then that illite or chlorite (particularly, the variety known as chamosite) or both are the main candidates for this doublet, which was observed in 90% of the coals studied by Huffman and Huggins (1978), and was the dominant iron-bearing phase in 20%. Because of the similarity of the parameters for these phases, other techniques will have to be used to distinguish the relative contributions from each mineral. For example, seven coals in which this doublet was observed (Huffman and Huggins, 1978) were also investigated by scanning electron microscopy equipped with an energy-dispersive x-ray spectrometer (Lee *et al.*,

† Only one coal of those examined so far has had an absorption consistent with the presence of significant amounts of montmorillonite. About 11% of the iron in coal from the Somerset C seam, Colorado, was in a quadrupole doublet with parameters, $\delta = 1.12$, $\epsilon = 2.79$.

TABLE III *Mössbauer Parameters for Iron-Bearing Silicates and Iron-Silicate Rich Coals*

Mineral	Typical wt % FeO range[a]	Maximum wt % FeO + Fe$_2$O$_3$[a]	δ^b	ϵ^b	Reference
Kaolinite	0.0-0.2	2.0	—	—	Fe^{2+} not reported
Illite	1.0-4.0	20.0	1.08-1.20	1.7-2.9	Coey (1975)
			1.12	2.62	Huffman and Huggins (1978)
Montmorillonite	0.0-1.0	10.0	1.13	2.87	Huffman and Huggins (1978)
Chlorite[c]	1.0-40.0	50.0	1.13	2.52	Huffman and Huggins (1978)
			1.12-1.18	2.60-2.66 ⎱	Coey (1975)
			0.98-1.26	2.43-2.53 ⎰	
Chamosite[d]	20.0-40.0	50.0	1.08-1.20	2.49-2.65	Coey (1975)
Muscovite[c]	0.0-3.0	10.0	1.09-1.17	2.87-3.09 ⎱	Coey (1975)
			1.10-1.16	2.14-2.28 ⎰	
Biotite[c]	1.0-20.0	30.0	1.06-1.16	2.54-2.72 ⎱	Coey (1975)
			1.04-1.14	2.08-2.28 ⎰	
Coal seam, state	% Fe$_{sil}$[e]		δ_{sil}	ϵ_{sil}	Reference
Pocahontas No. 3, W. Va.	48		1.13	2.66	This work
Anthracite, unknown	44		1.12	2.65	This work
Black Creek, Ala.	34		1.12	2.64	This work
Pratt, Ala.	25		1.15	2.65	This work
Sewell, W. Va.	24		1.11	2.65	This work
Average of 29 coals	—		1.12	2.65	Huffman and Huggins (1978)

[a] Based on analyses listed by Grim (1968) and Deer *et al.* (1962).

[b] Parameters for Fe^{2+} components only. Values quoted as ranges derived from many studies as reviewed by Coey (1975).

[c] Two distinct Fe^{2+} doublets are observed in some samples.

[d] Low-temperature, iron-rich form of chlorite.

[e] % Fe$_{sil}$: percentage of total iron in coal in the form of iron silicate.

1978, and unpublished data); illite was found to be a common constituent of all seven coals, whereas Pocahontas No. 3 Seam, West Virginia, and Gholson Seam, Alabama, also showed the presence of significant amounts of a phase rich in Fe, Al, and Si, a composition consistent with chamosite.

C. Carbonates

The third iron-bearing phase commonly observed in coal and Mössbauer spectra of coal is an iron-bearing carbonate. This doublet was observed in about half of the coals investigated by Huffman and Huggins (1978) and typically has an isomer shift around 1.23 mm/s and a quadrupole splitting around 1.81 mm/s at room temperature. The doublet dominated the spectrum in a few cases (e.g., Fig. 5a).

Most carbonate minerals in coal contain some Fe^{2+}, but the two in which iron is a major and essential component are siderite ($FeCO_3$) and ankerite [$Ca(Fe,Mg)(CO_3)_2$]. The Mössbauer parameters for these two phases are given in Table IV, along with data for a number of coals in which iron carbonates were abundant. The parameters obtained for coals are closest to siderite; although ankerite is known to occur in the Gholson Seam, Alabama, (Lee *et al.*, 1978), its contribution must be of less significance as the quadrupole splitting is only slightly reduced from that of pure siderite.

D. Oxides and Oxyhydroxides

Iron-bearing oxides and oxyhydroxides are not common in fresh, unweathered coals, and their presence will usually be indicative of weathering and oxidation processes. The oxides, magnetite and hematite, are easy to recognize in Mössbauer spectra of coals because they give rise to six-peak, magnetically split spectra. The most intense peaks of these spectra are well removed from the peaks for pyrite, siderite, and iron-bearing clays, and identification of very small amounts of these oxides is straightforward. Figure 6 shows the spectrum of coal refuse from the Pittsburgh seam, Pennsylvania. Both hematite and magnetite are present. Table V lists Mössbauer parameters

TABLE IV   *Mössbauer Data for Iron in Carbonates and for Carbonate-Rich Coals*

Mineral, formula		δ	ϵ	Reference
Siderite, $FeCO_3$		1.23	1.81	Walker *et al.* (1971)
		1.21	1.78	Lefelhocz *et al.* (1967)
Ankerite $Ca(Fe,Mg)(CO_3)_2$		1.20	1.50	Lefelhocz *et al.* (1967)
Coal Seam, State	% Fe_{sid}[a]			
Somerset B, Colo.	89	1.24	1.80	This work
Gholson, Ala.	79	1.23	1.77	This work
Hartshorne, Ark.	76	1.23	1.77	This work
Bulli, Australia	67	1.23	1.83	This work
Somerset C, Colo.	58	1.24	1.82	This work
Average of 19 coals	—	1.23	1.79	Huffman and Huggins (1978)

[a] % Fe_{sid}: percentage of total iron in coal in the form of iron carbonates.

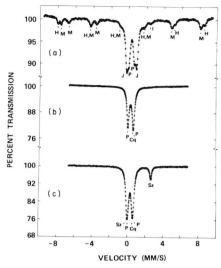

Fig. 6   Mössbauer spectra of coal refuse (a) (Pittsburgh seam, Pa) and of low-temperature ashes [Herrin No. 6 seam (b) and Davis seam, Ill (c)] containing significant amounts of iron sulfates or iron oxides. The arrows labeled H, M, J, P, I, Cq and Sz indicate Mössbauer peaks due to hematite, magnetite, jarosite, pyrite, illite, coquimbite, and szomolnokite, respectively.

TABLE V   *Mössbauer Data for Oxides and Oxyhydroxides and Coals Containing Such Phases*

Mineral, formula	% Fe	δ	ε	H (kG)	Reference
Hematite, $\alpha\text{-Fe}_2\text{O}_3$	100	0.37	−0.10	515	Greenwood and Gibb (1971)
Magnetite, Fe_3O_4	67	0.67	0.0	463	Daniels and Rosencwaig
	33	0.31	0.0	498	(1969)
Goethite, $\alpha\text{-FeOOH}$	100	0.37	0.25	0–390[a]	Govaert *et al.* (1976)
Lepidocrocite, $\gamma\text{-FeOOH}$	100	0.39	0.55	0	Greenwood and Gibb (1971)
Limonite—fine grained intergrowth of hematite, goethite, and ferrous phosphate					Janot *et al.* (1968)

Coal seam, location	% Fe	δ	ε	H	Phase	Reference
Pittsburgh (refuse), Pa.	10	0.39	−0.08	514	$\alpha\text{-Fe}_2\text{O}_3$	
	14	0.66	0.00	461		This work
	7	0.27	0.00	491	Fe_3O_4	
Sewell, W. Va.	19	0.65	0.00	453	Fe_3O_4	
	9	0.30	0.00	491		This work

[a] Goethite at room temperature may be antiferromagnetic ($H = 390$) or superparamagnetic ($0 < H < 390$ kG) depending on grain size (Govaert *et al.*, 1976).

for iron oxides and oxyhydroxides and also for some of these phases in coals.

As the iron oxyhydroxide phases have parameters very similar to pyrite, unambiguous identification of these phases in pyrite-rich coals by Mössbauer spectroscopy at room temperature is not possible, except in the case of magnetically ordered goethite. Hence, low-temperature spectra may be useful since all these phases will become magnetically ordered as the temperature is lowered.

E. Iron Sulfates

A number of iron sulfates have been reported in various coals (Palache et al., 1951; Gluskoter, 1975), particularly in low-rank coals. It is likely that these phases are found in coal as a result of weathering and oxidation of pyrite, although it is conceivable that some of the phases could be primary in certain coals. Mössbauer data for the iron sulfates listed by Gluskoter (1975) and for certain iron-sulfate-bearing coals are shown in Table VI. In Fig. 6 are shown Mössbauer spectra of a jarosite-rich coal (top), a coal containing szomolnokite (bottom), and a coal known from XRD data to contain coquimbite (center). The identification of jarosite and szomolnokite by Mössbauer spectroscopy is fairly straightforward, even though the Mössbauer parameters of szomolnokite are similar to those of the ferrous clay component of coal (compare with clay data in Table III). However, the presence of coquimbite and other ferric sulfates can only be inferred on the basis of the asymmetry of the pyrite absorption peaks (Fig. 6b). It is possible that such asymmetry can arise from other phases, for example iron oxyhydroxides, and usually other evidence will be needed to confirm such assignments.

The Mössbauer spectra of Illinois coals shown by Smith et al. (1978) appear to contain jarosite and other sulfates; however, as their spectra were recorded at 91 K and as they did not assign the spectra completely, direct comparison is not possible.

F. Organoiron and Other Compounds

Many elements are believed to be in coal in varying degrees of inorganic and organic combination (Zubovic, 1966; Gluskoter et al., 1977). Iron has been found to have largely inorganic affinities in most coals (Gluskoter et al., 1977), and Mössbauer studies of iron in coal also argue for a strong inorganic affinity for iron as no convincing evidence for iron in organic combination has been found. As noted earlier, Lefelhocz et al. (1967) originally suggested that the ferrous silicate peaks

TABLE VI *Mössbauer Data for Iron Sulfates and Sulfate-Bearing Coals*

Mineral, formula	% Fe	δ	ε	Reference
Melanterite, $FeSO_4 \cdot 7H_2O$	100	1.31	3.20	Vertes and Zsoldos (1970)
Rozenite, $FeSO_4 \cdot 4H_2O$	100	1.23	3.17	Vertes and Zsoldos (1970)
Szomolnokite, $FeSO_4 \cdot H_2O$	100	1.26	2.71	Huffman and Huggins (1978)
Roemerite, $FeSO_4 \cdot Fe_2(SO_4)_3 \cdot 14H_2O$ (USNM R16094)[a]	26 / 27 / 47	1.27 / 0.53 / 0.38	3.27 / 0.25 / 0.37	This work
Coquimbite, $Fe_2(SO_4)_3 \cdot 9H_2O$ (USNM R16094)[a]	32 / 68	0.51 / 0.43	0.00 / 0.26	This work
Kornelite, $Fe_2(SO_4)_3 \cdot 7H_2O$ (USNM 104277)[a]	100	0.44	0.15	This work
Jarosite, $XFe_3(SO_4)_3(OH)_6$				
$X = K^+$	100	0.40	1.15	Hrynkiewicz et al. (1965)
$X = Na^+$	100	0.40	1.10	
$X = H_3O^+$	100	0.40	1.00	
$X = NH_4^+$	100	0.40	1.15	
$X = 1/2Pb^{2+}$	100	0.40	1.15	

Coal seam, state	% Fe_{coal}	δ	ε	Phase	Reference
Lignite, unknown	71	0.37	1.15	Jarosite	Huffman and Huggins (1978)
Pittsburgh (refuse), Pa.	39	0.36	1.14	Jarosite	Huffman and Huggins (1978)
Herrin No. 6, Ill.	21	0.38	1.10	Jarosite	This work
Pocahontas No. 4, W. Va.	55	1.25	2.71	Szomolnokite	Montano (1977)
Pittsburgh, W. Va.	8	1.27	2.71	Szomolnokite	Montano (1977)
Davis, Ill. (C18105)[b]	14	1.27	2.71	Szomolnokite	This work
	6	0.55	—	Coquimbite	
Herrin No. 6, Ill. (C18903)[b]	>12	—	—	Coquimbite	This work

[a] Sulfate samples supplied by J. S. White, Jr., Smithsonian Institution, Washington, D.C.

[b] Low-temperature ash samples supplied by H. J. Gluskoter, Illinois State Geological Survey, for which XRD data also show presence of sulfates identified.

in Mössbauer spectra of certain coals might be organically bound, but they subsequently ruled out this possibility. Studies of peat indicate that most of the iron is inorganically bound (Gamayunov *et al.*, 1975) even as coalification begins. However, a study of an Australian brown coal does indicate that if any iron were associated with carboxyl groups, it could be altered very rapidly upon exposure to air and so organic iron in this form might not be found in laboratory samples, even if it were present in the coal seam (Schafer, 1977). Except for possible loss of organic iron by such alteration, the evidence appears strongly in favor of most, if not all, of the iron being in inorganic combination.

At least one organoiron mineral has been identified in coal, namely the oxalate, humboldtine, $FeC_2O_4 \cdot 2H_2O$ (Palache et al., 1951). The Mössbauer parameters of the pure phase ($\delta = 1.17$ mm/s and $\epsilon = 1.73$ mm/s; Caric et al., 1975) are close to those of siderite, and absorption because of this phase could be hidden by siderite absorption; however, it should be noted that this phase has not been observed in Mössbauer spectra of siderite-free coals. Iron salts of other organic acids could well occur in coal but have not been identified.

Other iron-bearing minerals that could possibly be found in coal are detrital minerals, resistant to weathering, such as chromite ($FeCr_2O_4$), ilmenite ($FeTiO_3$), and iron-bearing silicates such as almandine garnet and tourmaline. Nondetrital minerals that might be found include iron phosphates such as vivianite [$Fe_3(PO_4)_2 \cdot 8H_2O$], iron-bearing silicates such as greenalite, minnesotaite, and stilpnomelane, and additional iron sulfates. However, none of these phases is likely to be common, as unusual, and restricted conditions are probably necessary for the observation of any of these minerals.

IV.   GEOLOGICAL AND ENVIRONMENTAL INFORMATION FROM MÖSSBAUER DATA

A.   Geological Factors Influencing Iron Mineral Formation

There has been some discussion in the literature concerning the factors that determine iron-mineral formation during coalification. For instance, the question of whether siderite replaces pyrite, or vice versa, as the rank of the coal increases has been discussed at length. As the literature survey by Smyth (1966) indicates, arguments have been presented to support both points of view, based mainly on subjective interpretations of microscopic observations of pyrite–siderite intergrowths in coal.

Mössbauer spectra of coals provide quantitative information on the distribution of iron among sulfides, silicates, and carbonates; these data (see appendix) on the distribution of iron are shown as a function of rank of the coal in Fig. 7. Although the number of samples may be too few to be an adequate test of general applicability, particularly for

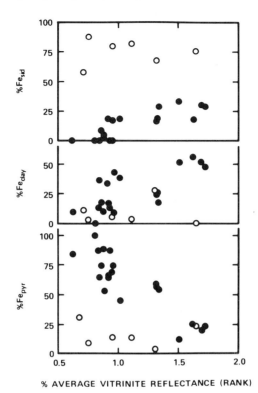

Fig. 7 Distribution of iron between pyrite, siderite, and clay minerals plotted as a function of percent average vitrinite reflectance, a parameter indicative of the rank of the coal. Open circles indicate coals rich in iron carbonates.

nonbituminous coals, a trend of increasing iron in silicates and carbonates at the expense of iron in pyrite with increasing rank can be discerned. Based on the trends shown in Fig. 7, it might be concluded that pyrite is a mineral associated more with low-rank coals, whereas iron-bearing clays and siderite are found more abundantly in high-rank coals. However, it is not clear whether the relationships shown in Fig. 7 are because of increasing coalification or whether they are reflecting differences in depositional environments or position in the stratigraphic column (i.e., geologic age) between the peats that gave rise to the coals of different ranks. Moreover, the possible different modes of formation of minerals in coal (e.g., Mackowsky, 1968) have not been taken into account. A more systematic test of the relationship between iron mineralogy and rank would be to find a single seam that shows a large variation in rank so that variations in depositional conditions, age, and mineral formation would be minimized.

Certainly, paleoenvironmental factors will influence the distribution of iron between different minerals, and such an influence is suggested by the group of coals that are unusually rich in iron carbonates (open symbols in Fig. 7). These coals are restricted in the United States for the most part to southern and western states (see appendix) and such a geographical restriction is most likely because of differences in depositional environments. Freshwater rather than marine environments are believed to be conducive for siderite formation (Kemezys and Taylor, 1964), whereas pyrite formation is favored by brackish-marine conditions (Reyes-Navarro and Davis, 1976). The conditions for mineralization of lignites are discussed in Volume II, Chapter 27.

In theory, the combination of Mössbauer data on the relative abundances of the different iron-bearing phases and Eh–pH diagrams as developed by Garrels and Christ (1965) might be used to deduce redox (Eh) and acidity (pH) conditions of the aqueous medium associated with coal, and also the relative concentrations of dissolved species such as sulfide and carbonate anions and amorphous silica. Iron-bearing mineral assemblages (pyrite–chamosite–siderite–ankerite–carbonaceous matter) similar to those found in coal are also found in sedimentary iron formations (e.g., Klein and Bricker, 1977) and have been extensively interpreted in terms of Eh–pH diagrams.

An example of the mineral equilibria expected at certain concentrations of dissolved CO_3^{2-}, S^{2-} anions and in solutions saturated with respect to SiO_2 is shown in Fig. 8 (Garrels and Christ, 1965, Fig. 7.23). If Fig. 8 were to apply directly to iron minerals in coal, it would indicate that for pyrite, siderite, and an iron-bearing silicate to be in equilibrium, the aqueous fluid associated with coal would have to be reducing (Eh = -0.2---0.3 V) and slightly alkaline (pH $\sim$ 8.0). It should be noted that dissolved CO_3^{2-} is 10^6 greater than S^{2-} for this particular diagram.

Such diagrams also indicate why magnetite and pyrrhotite are rarely observed as primary minerals in coal. For magnetite to be found, the fluid must be undersaturated with respect to silica and very low in dissolved sulfide and carbonate species, whereas for pyrrhotite to be found, extreme reducing conditions and very low dissolved sulfur and carbonate species in the fluid are necessary.

Although general information of this type can be obtained from Eh–pH diagrams, more detailed information will be hard to obtain, given the number of variables and the difficulty of demonstrating that the minerals in coal are actually in equilibrium.

From the brief discussion in this section, it is clear that there are

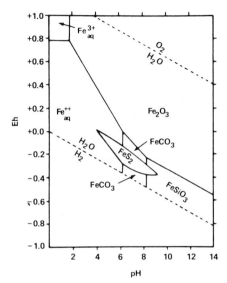

Fig. 8 Example of an Eh–pH diagram (Garrels and Christ, 1965, Fig. 7.23) showing stability relations among iron oxides, carbonates, sulfides, and silicates at 25°C and 1 atm total pressure in the presence of water. Other conditions: total $CO_2 = 10°$; total sulfur $= 10^{-6}$; amorphous silica is present.

many geological factors that might influence the iron distribution among minerals in coal. These factors include the paleoenvironmental conditions of peat formation, the influence of the aqueous medium associated with coal, and possibly external factors such as the degree of metamorphism. Obviously, more work and data are needed to establish whether the iron distribution, as determined by Mössbauer spectroscopy, can be used to obtain specific geological information.

B. Weathering

Coals exposed to the atmosphere often undergo drastic changes in the character of both organic and inorganic components. As the iron-bearing minerals are particularly sensitive to weathering, Mössbauer spectra can be used to diagnose the extent and degree of the oxidation and hydration reactions and to indicate possible reaction sequences. Such reactions are of considerable interest because of the problems posed by acid-mine drainage, coal-refuse disposal, useless coking properties of oxidized coal, etc., as a result of weathering.

Weathering products such as hematite, magnetite, and various iron sulfates can be recognized relatively easily in Mössbauer spectra; an example of the spectrum of weathered coal refuse is shown in Fig. 6. As a brief illustration, Table VII summarizes data on the distribution of iron for the original coal and for two mesh-size fractions of the refuse. It is apparent that jarosite and iron oxides are the stable iron-bearing phases in the refuse and that the finer mesh-size fraction is more completely weathered.

Baker (1972) utilized Mössbauer spectroscopy in the back-scattering mode to investigate oxidation and hydration reactions at the surfaces of pyrite crystals in water seeded with bacteria. His results indicated that both ferric sulfate and ferric oxyhydroxide were formed at the surface of the pyrite crystal after 87 days exposure.

Similar Mössbauer studies would appear to have potential in the areas of acid-mine drainage, coal-refuse disposal, and other environmentally sensitive issues resulting from weathering of coal.

V. QUANTITATIVE DETERMINATION OF PYRITIC SULFUR IN COAL BY MÖSSBAUER SPECTROSCOPY

A very attractive feature of Mössbauer spectroscopy in the area of coal science is its potential for the direct, nondestructive determination of the pyritic-sulfur content of coal. For a number of reasons, the technique is unlikely to supplant the current ASTM standard test, No. D2492, for pyritic sulfur (ASTM, 1976), even though comparison of the two techniques does suggest failings of the ASTM method that ought to be avoided whenever possible. In this section, three methods of making quantitative Mössbauer determinations of pyritic sulfur in coal

TABLE VII *Illustration of Weathering of Refuse of Coal From Pittsburgh Seam as Function of Mesh Size*

Coal seam, state	Mesh fraction	Pyrite	Siderite	Clay	% Fe in Hema-tite	Magne-tite	Jaro-site	% Fe in oxides and Sulfate
Pittsburgh, Pa., fresh	−48 + 200	87	3	6	0	0	4	4
Pittsburgh, Pa., refuse	−48 + 200	41	0	5	23	11	21	55
Pittsburgh, Pa., refuse	−200	29	0	2	9	21	39	69

are discussed and compared. The relative merits of the ASTM and Mössbauer methods are also discussed.

A. Quantification of the Mössbauer Technique

Three different methods have been proposed for quantifying Mössbauer spectroscopy for the determination of pyritic sulfur: the first, suggested by Montano (1977), is by direct calculation from the spectral areas for pyrite; the second, suggested by Levinson and Jacobs (1977), is to calculate pyritic sulfur from Mössbauer data for the coal and the total iron content of the coal ash; and the third, suggested by Huffman and Huggins (1978), is by means of a calibration curve for the Mössbauer data. These methods are discussed individually.

1. *Method of Montano*

In the method outlined by Montano (1977), the area of the pyrite peaks is related to the number of iron atoms contained in pyrite in the coal sample by means of a theoretical expression for the area of Mössbauer peaks derived by Lang (1963). This method requires a value for the recoil-free fraction of the phase in question. In Montano's study, this value was determined by measuring the spectral area contributed by pyrite in a particular coal at two different temperatures, and using these areas to determine a recoil-free fraction from the Debye model (see Section II). As will be discussed further, it appears that Montano's estimate of the recoil-free fraction (0.60 ± 0.15) is too large since the pyritic sulfur contents determined by his method are systematically lower than those of either the ASTM or the other two Mössbauer methods.

2. *Method of Levinson and Jacobs[†]*

By combining Mössbauer data on the percentage of iron contained in pyrite with the measurement of the weight percent of total iron in the coal by chemical means, the weight percent of pyritic iron and hence of pyritic sulfur can be calculated (Levinson and Jacobs, 1977). This method avoids the theoretical assumptions of the Debye model employed by Montano (1977), but requires additional measurements, namely, the weight percent of iron in the coal. By extending this

[†] Levinson and Jacobs (1977) implied that this method should only be used for coals in which most of the iron is in the form of pyrite. They advocate using calibration curve methods for general Mössbauer determinations of pyritic sulfur in coals (Levinson, personal communication).

method to the general case, the following expression can be written:

$$\text{wt\% } S_{pyr} = (64.12/55.85)(\% \text{ Fe}_{pyr} \cdot \% \text{ Fe}_{ash} \cdot \% \text{ ash})/10^4, \qquad (11)$$

where % Fe_{pyr} is the percentage of iron in the form of pyrite determined from the Mössbauer spectra, % Fe_{ash} is the weight percent of iron in the ash, and % ash is the weight percent of ash in the coal.

This method is likely to be accurate for coals in which pyrite is the dominant iron-bearing mineral; however, for coals in which pyrite is not dominant, any significant differences between the recoil-free fractions of the iron-bearing mineral phases will introduce systematic errors.

3. Method of Huffman and Huggins

In the method described by the authors (Huffman and Huggins, 1978), a calibration curve (Fig. 9) for the determination of pyritic sulfur was obtained from Mössbauer measurements on a series of carefully prepared mixtures of pyrite and coconut charcoal of accurately known pyritic-sulfur content. As discussed in Section II,D, the effective Mössbauer absorption thickness for pyrite is determined in the least-squares analysis of the spectra, and division of this parameter by the areal density d of the Mössbauer absorber (total sample weight divided by the cross-sectional area of the sample holder) defines a Mössbauer mass-absorption coefficient X_{pyr}/d which has the form

$$X_{pyr}/d = f_{pyr} \cdot a_{57} \cdot \sigma_0 \cdot n_{pyr}, \qquad (12)$$

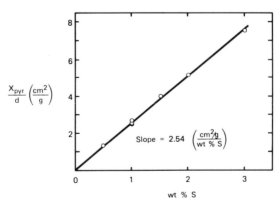

Fig. 9 Mössbauer mass absorption coefficient for pyrite (X_{pyr}/d) vs. wt % sulfur for coconut charcoal–pyrite mixtures. This calibration plot is used to obtain values for wt % pyritic sulfur from Mössbauer data.

where f_{pyr} is the effective recoilless fraction of pyrite, n_{pyr} is the number of iron atoms contained in pyrite per gram of coal, and the remaining terms have been previously defined for Eq. (9). This expression can readily be written in terms of the weight percentage of pyritic sulfur, $\% S_{pyr}$, as follows:

$$X_{pyr}/d = (f_{pyr} N_0 a_{57} \sigma_0/2 \times 32.06 \times 100) \% S_{pyr} = C \cdot \% S_{pyr}, \quad (13)$$

where N_0 is Avogadro's number and 32.06 is the atomic weight of sulfur. As indicated, the term in parentheses is a constant, and is equal to the slope of the linear calibration curve of Fig. 9;

$$C = 2.54 \ (cm^2/g/wt\% \ sulfur), \quad (14)$$

therefore

$$\% S_{pyr} = 0.394 \cdot X_{pyr}/d, \quad (15)$$

where the Mössbauer mass absorption coefficient, X_{pyr}/d, has the units of square centimeter per gram and is readily determined from the least-squares computer analysis of the spectrum and the weight of the sample.

Two minor criticisms can be directed at this method:

(1) Nonresonant absorption of γ rays by mineral matter is not taken into account in establishing the calibration curve. Unless the coal is very low in pyrite and high in other mineral matter, this problem is unlikely to be serious, as pyrite itself is one of the most efficient minerals in coal for nonresonant absorption of the γ radiation, and the calibration method already allows for this absorption by pyrite. The silicate minerals, quartz, kaolinite, illite, etc., are hardly more absorbing than the carbonaceous matter; however, the carbonate minerals, calcite ($CaCO_3$) and siderite ($FeCO_3$), because of the presence of higher atomic-weight elements, may be a more serious problem in this respect and the method may need slight modification for high mineral-matter, carbonate-rich coals.

(2) The method is probably somewhat specific to a given laboratory. Different methods of computer fitting and data reduction, as well as different experimental designs, could cause slightly different calibration curves to be derived in different laboratories.

B. Comparison of Mössbauer Methods and ASTM Standard Method for Pyritic Sulfur

It is well known that the ASTM Standard method, No. D2492 (ASTM, 1976) (see Volume I, Chapter 6, Section VII), for forms of sulfur may often underestimate the amount of pyritic sulfur in a coal, even if all possible precautions and care are taken. The Mössbauer spectra taken by Levinson and Jacobs (1977) at the distinct stages of the ASTM test document this fact very well. These spectra are reproduced in Fig. 10 and show that some of the pyrite is actually dissolved during HCl treatment, which is designed to remove just iron-bearing and other sulfates from coal, and that some of the pyrite remains undissolved after HNO_3 treatment, which is intended to dissolve pyrite from coal. Optical and scanning-electron microscopic examination of HNO_3-treated coal has also shown that not all the pyrite is removed (Edwards *et al.*, 1964; Greer, 1977). A detailed discussion of the incomplete extraction of pyrite with HNO_3 is given in Chapter 56, Section V,B,2,a. Hence, we would expect that the Mössbauer estimate of pyritic sulfur should normally be equal to or greater than the ASTM estimate because the Mössbauer method can be performed directly on the coal without the need for separation of mineral matter and should sample all pyrite in the coal.

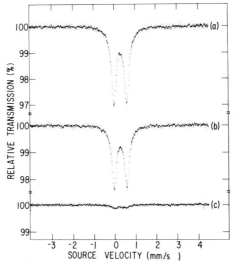

Fig. 10 Mössbauer spectra of a pyrite-rich coal (Pittsburgh seam, W.Va.) at various stages of the ASTM standard test D2492 for forms of sulfur: (a) untreated, (b) after HCl treatment, (c) after HNO_3 treatment. The spectral area is proportional to the wt % pyrite in the coal. Hence, these spectra indicate (1) dissolution of pyrite during HCl treatment [area of (b) less than that of (a)] and (2) minor amounts of pyrite remain in coal after HNO_3 treatment. (From Levinson and Jacobs, 1977 by permission of the publishers, IPC Business Press Ltd. ©1977.)

It is instructive to compare the different methods for estimating pyritic sulfur on a number of coals. In addition to comparing the method of Huffman and Huggins (1978) with the ASTM method, we attempt to use Montano's method by substituting his value for the recoil-free fraction (0.60) in Eq. (13) to calculate a value of the constant C and, where available, we use values for the weight percent of iron in coals so that we can calculate % S_{pyr} by Levinson and Jacobs' method. As the expression used in our analysis for peak area [Eq. (8), Section II,C)] differs from the approximate expression derived by Lang (1963), the comparison of our method with that of Montano is somewhat less rigorous than the comparison with the method of Levinson and Jacobs. Nevertheless, it should serve to illustrate the principal differences among the three Mössbauer techniques for determining pyritic sulfur. Table VIII is a comparison of the results.

Except for two of the samples, the Mössbauer values for pyritic sulfur estimated by the calibration curve (column H & H, Table VIII) are in good agreement with the ASTM estimates, and the small deviations appear to be in the direction of underestimation by the ASTM method, as expected. For two of the samples, the estimate by the ASTM method is much larger than the Mössbauer measurement, and it is probably no accident that these two samples are unusually low in pyrite and rich in other forms of iron. As the iron content is generally used to estimate the pyritic sulfur in the ASTM method, possible shortcomings of the ASTM method are (1) the HNO_3 treatment leached out significant amounts of iron from clay minerals, which was measured as pyritic iron; (2) the iron carbonate dissolved mostly in the HNO_3 treatment rather than in the HCl treatment. Of these possibilities, the former seems more probable.

Regarding the different Mössbauer methods, the disagreement between the values obtained by Montano's and Huffman and Huggins' methods is simply because of the difference in the value of the recoil-free fraction used for pyrite. The value obtained from the slope of the calibration curve is 0.48, whereas Montano obtained 0.6 by a different method, and hence the values obtained by Montano's method are uniformly 20% lower than from the calibration curve. Possibly, a more complete characterization of the temperature dependence in Montano's approach might yield a lower Debye temperature and recoil-free fraction that would give better agreement between the two methods. However, in the authors' opinion, the applicability of the Debye model for a complex crystal structure like pyrite is somewhat questionable.

The method outlined by Levinson and Jacobs (1977) (column L & J, Table VIII) shows good agreement with the calibration curve method,

TABLE VIII Comparison of Mössbauer Methods and ASTM Method for Estimating Pyritic Sulfur

Coal seam, state	% Fe_{pyr}	% Fe_{coal}[a]	X_{pyr}/d	Montano	L & J	H & H	ASTM	Reference
Herrin No. 6, Ill.	77.7	0.95	1.47	(0.46)[b]	0.84	0.58	0.61	Huffman and Huggins (1978)
Pittsburgh, Pa. (raw)	73.7	—	2.65	(0.83)	—	1.04	0.90	
Pratt, Ala.	59.0	0.63	0.90	(0.28)	0.42	0.35	0.33	
Bulli, Australia	4.5	0.33	0.03	(0.01)	0.02	0.01	0.13	This work
Pocahontas No. 3, W. Va. (washed)	3.7	—	0.05	(0.01)	—	0.02	0.21	
Pittsburgh, W. Va.				1.1		1.4[c]	1.33	Montano (1977)
Pocahontas No. 4, W. Va.				0.08		0.10[c]	0.06	
Redstone, W. Va.				1.2		1.5[c]	1.97	
Pittsburgh, Pa.	75.4	0.59	1.30	(0.41)	0.51	0.51	0.39	This work[d]
Herrin No. 6, Ill.	73.5	0.68	1.37	(0.43)	0.58	0.54	0.43	
Pratt, Ala.	45.0	0.44	0.52	(0.16)	0.22	0.20	0.17	
Pocahontas No. 3, W. Va.	25.2	0.42	0.33	(0.10)	0.12	0.13	0.07	
Gholson, Ala.	13.7	0.74	0.28	(0.09)	0.12	0.11	0.09	

[a] % Fe_{coal} = % Fe_{ash} · ash/10^2. See text for explanation of these and other symbols.

[b] Values in parentheses calculated using recoil-free fraction of 0.60 (Montano, 1977) and a value for C of 3.13 calculated according to Eq. (12).

[c] Values calculated using recoil-free fraction 0.48 derived from a value of C of 2.54 and Eq. (12).

[d] Aliquots for pyritic sulfur by Mössbauer technique and for chemical analysis of iron known to be taken from the same, homogenized sample of coal. Not so for other samples.

particularly for those samples for which considerable care was taken to ensure that the Mössbauer and chemical data were obtained on the same homogenized sample (bottom five coals in Table VIII).

Although the Mössbauer method gives a precise determination of the pyritic sulfur that may often be larger than the pyritic-sulfur value obtained by the ASTM standard method because of the problems associated with the different acid treatments (Levinson and Jacobs, 1977), it is likely that the ASTM estimate will closely correspond to the pyritic sulfur *that can be removed from coal.* This value is of most interest for many industrial purposes. For this reason, and also because of the length of time and specialized equipment needed to take Mössbauer spectra, it seems unlikely that the Mössbauer methods will supplant the ASTM method for industrial purposes. However, for research purposes and for investigating discrepant results obtained with ASTM methods (e.g., low-pyrite coals rich in other iron-bearing minerals), the Mössbauer methods should be utilized as an alternative and definitive technique for the determination of pyritic sulfur.

Finally, it should also be noted that Mössbauer spectroscopy could also be applied to determine the weight percentage of iron contained in other iron-bearing phases. For example, calibration curves similar to Fig. 9 could readily be determined for siderite, illite of a given iron content, $FeSO_4 \cdot H_2O$, jarosite, etc. It seems likely that such calibration data will become available in the future.

VI. MÖSSBAUER STUDIES OF COKE

One of the main uses of the bituminous class of coals is coke making. The commercial carbonization process consists of heating bituminous coals to temperatures of approximately 950° to 1000°C in an air-excluded oven for a period of about 16 hr. During such treatment, the mineral matter in coal undergoes transformations to phases stable at such high temperatures (mineral matter in bituminous coals has usually not been subjected to temperatures above 150°C during coal formation). Mössbauer studies of coke show how the iron-bearing minerals transform during carbonization; in addition, Mössbauer spectroscopy can be used to obtain quantitative estimates of the amount of sulfur in the coke in the form of iron sulfides.

A. Mineral Transformations during Carbonization

Mössbauer spectra of cokes made in experimental coke ovens (30-lb charges) or as residues in volatile-matter tests (volatile matter residue

or VMR samples) are shown in Figs. 11 and 12. The spectra of the
original coals are shown in Figs. 4 and 5; hence comparison between
the coals and cokes can be made. The principal contributions to the
spectra of cokes are identified as iron sulfides, iron metal, and paramagnetic phases such as iron-bearing glass or poorly crystalline silicate
or oxide phases. Minor contributions to the spectra include magnetic
iron oxides and an enigmatic iron phase giving rise to a single peak
(Huffman and Huggins, 1978). Herrin No. 6 coal from Illinois heated
to around 500°C under nitrogen for 48 hr (Smith *et al.*, 1978) exhibits
a spectrum very similar to that of the coke shown in Fig. 11.

Two distinct iron sulfides are recognized in Mössbauer spectra of
coke: stoichiometric FeS (troilite) and nonstoichiometric $Fe_{1-x}S$ (pyrrhotite, $x < 0.125$); the probable reason for the presence of two iron
sulfides, suggested on the basis of scanning-electron microscopic observations (Huffman and Huggins, 1978), is whether or not the iron

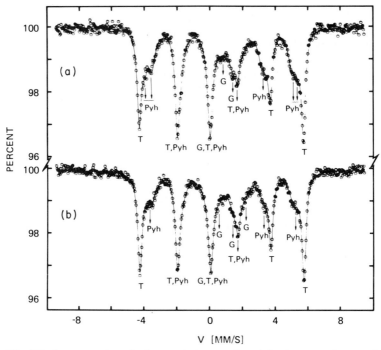

Fig. 11 Mössbauer spectra of cokes made in experimental coke ovens from coals from
Herrin No. 6 seam, Ill. (a) and Pittsburgh seam, Pa. (b) Mössbauer spectra of the coals
are shown in Fig. 4. The arrows labeled T, Pyh, and G indicate Mössbauer peaks due to
FeS (troilite), $Fe_{1-x}S$(pyrrhotite), and paramagnetic Fe^{2+}-bearing glass, respectively.

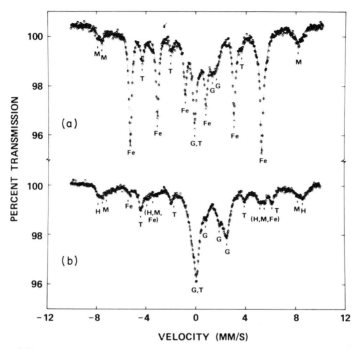

Fig. 12 Mössbauer spectra of volatile-matter residue from coal from Gholson seam, Ala. (a) and of coke (at 77 K) from Pocahontas No. 3 seam, W.Va (b). Mössbauer spectra of the coals are shown in Fig. 5. The arrows labeled T, H, M, Fe, and G indicate Mössbauer peaks due to FeS (troilite), Fe_2O_3 (hematite), Fe_3O_4 (magnetite), iron metal and paramagnetic Fe^{2+}-bearing glass, respectively.

sulfide particle is isolated from the coke-oven atmosphere. If the particle is isolated from the coke-oven atmosphere, the local atmosphere is likely to be rich in sulfur resulting from the decomposition of the pyrite crystal that gives rise to the iron sulfide particle. A sulfur-rich sulfide will be stable under such local conditions. Averaged Mössbauer parameters for the sulfides observed in coke and VMR samples are given in Table IX.

Also shown in Table IX are averaged Mössbauer parameters for the paramagnetic phases observed in coke. For the two quadrupole doublets, Q_1 and Q_2, the isomer-shift values are reasonably diagnostic of Fe^{2+} phases and their broadness, and low quadrupole-splitting values are suggestive of glass or poorly crystalline phases. As will be shown below, it is likely that these phases result from decomposed clays. However, exact identification of such phases does not appear possible on the basis of Mössbauer data alone. The other paramagnetic absorp-

TABLE IX Range and Average Values of Mössbauer Parameters Observed at Room Temperature for Iron Sulfide and Glass Plus Other Nonmagnetic Phases in Coke and Volatile Matter Residue Samples

Phase	Site	H(kG) (Range)	$\langle H \rangle$(kG) (Average)	δ(mm/s) (Range)	$\langle \delta \rangle$(mm/s) (Average)	ϵ(mm/s) (Range)	$\langle \epsilon \rangle$(mm/s) (Average)
FeS + Fe$_{(1-x)}$S	FeS + Fe$_{(1-x)}$S – 1	306.1-312.1	310.0	0.74- 0.76	0.749	-0.063--0.09	-0.074
	Fe$_{(1-x)}$S – 2	272.0-288.7	279.8	0.70- 0.81	0.75	-0.005--0.082	-0.01
	Fe$_{(1-x)}$S – 3	257.6-263.8	261.7	0.67- 0.71	0.69	0.01 - 0.02	0.017
Glass + ?	Q_1	—	—	0.82- 1.37	1.11	1.68 - 2.57	2.04
	Q_2	—	—	0.93- 1.05	1.01	0.62 - 1.06	0.88
	Single	—	—	-0.11--0.01	-0.055	—	—

tion, designated "single" in Table IX, has a small isomer shift, consistent with paramagnetic forms of iron metal, and could therefore be austenitic iron or superparamagnetic iron. However, other identifications, such as part of the absorption because of superparamagnetic iron sulfides, may also be possible. The remaining phases, iron metal and iron oxides, have the expected and diagnostic values for their Mössbauer parameters and do not appear unusual in any respect.

In Table X (Huffman and Huggins, 1978), a comparison is made of the distribution of iron between phases in coal and between phases in coke and VMR samples. For the most part, obvious correlations are apparent between pyritic iron in coal and sulfide iron in coke, between iron contained in siderite in coal and the sum of iron in hematite, magnetite and iron metal in coke, and between iron in clays in coal and iron in glass and other silicates in coke. These correlations suggest the following transformations:

$$\text{pyrite (FeS}_2) \rightarrow \text{FeS} + \text{Fe}_{1-x}\text{S}$$

$$\text{siderite (FeCO}_3) \rightarrow \text{Fe} \xrightarrow{\text{(during cooling)}} \text{Fe} + \text{Fe}_3\text{O}_4 + \text{Fe}_2\text{O}_3$$

$$\text{iron-bearing clays} \rightarrow \text{ferrous glass} + \text{other phases.}$$

The origin of the "single" peak in the spectra cannot be determined with any certainty as sampling uncertainties may be expected to give rise to differences between the iron distribution in coals and cokes of the order of the percentage absorption of this peak in most cases. At this point, it does not seem necessary to postulate any further reactions other than the three decomposition reactions listed above for the principal iron-bearing minerals in bituminous coals. A scanning-electron microscopy study of mineral matter in coal and its transformations during carbonization also shows that decomposition reactions dominate (Lee *et al.*, 1978).

One further point from Table X should be noted. The sub-bituminous coal from the Rosebud seam, Montana, contained all its iron in pyrite; however, the principal phases observed in the VMR sample prepared from this coal were iron metal and a phase that gave rise to a single peak with an isomer shift of -0.11 mm/s, close to that expected for austenitic Fe–C; the coal and VMR spectra are shown in Fig. 13. Apparently, the atmosphere created by carbonization of this sample is sufficiently rich in H_2 to effect the reduction of pyrite to iron metal. Whether this observation might be related to the difference in coking properties between this sub-bituminous coal and the bituminous coals or whether it is characteristic of low-rank coals remains to be tested.

TABLE X Comparison of the Percentages of Iron Contained in the Phases Present in a Number of Coal Samples with the Corresponding Percentages Observed in the Coke and VMR Samples Made From Those Coals

Coal seam, state	Pyrite	Illite	Siderite	Other	Troilite + Pyrrhotite	Glass + Other Silicates	Coke and VMR[a]			
							Non-magnetic, Single[b]	Metallic Fe	Hematite Magnetite	Hem + mag + met Fe
Pratt, Ala.	58.9	25.1	16.0		64.6	28.2	—	—	7.2	7.2
Eagle, W. Va.	74.0	17.4	8.6		70.3	12.7	—	9.9	7.1	17.0
Sewell, W. Va.	39.6	17.8	14.5	28.1 Magnetite	28.5	47.2	7.1	8.1	9.1	17.2
Pittsburgh, Pa.	86.9	13.1			80.5	19.5	—			
Herrin No. 6, Ill.	84.0	9.9		6.1 Jarosite	83.4	16.6				
Pocahontas No. 3, W. Va.	18.6	51.8[c]	29.6		18.4	49.0	6.9	25.7	32.6	
Clairton No. 1 Unit[a]	72.7	27.3			70.6	19.3	8.2			
Pittsburgh (washed), Pa.	85.0	15.0			79.5	11.7	—	3.4	5.4	8.8
Gholson, Ala.	16.2	5.2	78.6		7.4	15.4	10.5	55.6	11.1	66.7
Rosebud, Mont. (Sbb)	100.0				4.1	23.5	23.5	61.9	10.5	72.4

[a] The first seven samples are cokes and the last three are volatile matter residue (VMR) samples.

[b] Percentage of iron contributing to the single-peak near-zero velocity; listed separately only for samples where it accounted for more than 5% of the total iron.

[c] Contains a significant amount of chlorite.

[d] Coal blend and coke sample from U.S. Steel Clairton Works, Clairton, Pa.

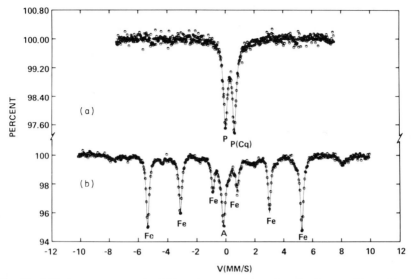

Fig. 13 Mössbauer spectra of subbituminous coal from Rosebud seam, Mont. (a) and its VMR (b). The labels P indicate Mössbauer peaks due to pyrite in the coal. The slight asymmetry is probably due to coquimbite (Cq). The labels Fe and A indicate Mössbauer peaks due to iron metal and austenitic iron in the VMR. Traces of magnetite and troilite are also present.

B. Quantitative Estimation of Sulfide Sulfur in Coke

The calibration curve method for the quantitative estimation of pyritic sulfur in coals can also be extended to cokes to obtain quantitative estimates of sulfide sulfur in coke (Huffman and Huggins, 1978). By using the previously discussed calibration data for pyritic sulfur in coal and noting that iron sulfides contain, on average, about 0.47 times the sulfur that pyrite contains per iron atom, it is estimated that the constant relating iron sulfide sulfur content to the total Mössbauer mass-absorption coefficient for iron sulfide in coke is approximately 0.47×0.394 [see Eq. (5), Section V,A,3]. That is,

$$\text{wt\% } S_{\text{sulfide}} \text{ (in coke)} \approx 0.185 \, [(X_{\text{pyrrh}} + X_{\text{tr}})/d], \qquad (16)$$

where X_{pyrrh} and X_{tr} are the effective Mössbauer absorption thicknesses for $Fe_{1-x}S$ and FeS, respectively, and d is in grams per square centimeters. This equation is strictly only semiquantitative because it is based on the pyrite calibration curve and therefore assumes that pyrite and iron sulfides have similar recoil-free fractions. Also, the multiplicative factor, 0.47, is only an average.

Comparison of iron sulfide sulfur contents determined by this method with those determined by a chemical procedure for several coke samples shows that the Mössbauer technique gives values that are quite reasonable but are systematically somewhat higher than those determined chemically (Huffman and Huggins, 1978). In view of the approximations noted above, it is premature to draw any conclusions regarding this difference. An experimentally determined iron sulfide calibration curve similar to that of Fig. 9 would clearly be desirable.

VII. MÖSSBAUER STUDIES OF ASH

The quantities of ash produced by coal combustion (between 50 and 100 million tons annually in the United States) make its disposal and potential uses serious environmental and economic factors in coal utilization. The characterization of the phases in coal ash may be needed to determine such factors. The phases found in coal ash are determined by the temperature of combustion, the length of time at high temperature, the cooling rate, and, particularly for iron phases, the redox nature of the atmosphere. Little work has been carried out by Mössbauer spectroscopy on coal ash, and we shall concentrate on two aspects only in this chapter: (1) transformations of iron-bearing minerals under ASTM test No. D3174 (ASTM, 1976) for ash content of coal and discussion of effects at higher temperatures; (2) low-temperature ash formed by radio-frequency stimulated oxidation of coal.

A. Mineral Transformations during the ASTM Test for Ash Content of Coal

In the ASTM test method, No. D3174, a 1 g sample of coal is slowly heated in air to a temperature of 700° to 750°C and held at this temperature until constant weight is achieved (see Volume I, Chapter 6, Section III,C). Mössbauer spectra of ash samples obtained in this manner are shown in Figs. 14 and 15. Three principal phases are identified, namely α-Fe_2O_3, paramagnetic Fe^{3+} phases (silicates, glass, or oxides), and $Mg_xFe_{3-x}O_4$ (magnesioferrite, $x < 1$). Typical values for the Mössbauer parameters for these phases in ash are shown in Table XI. The paramagnetic Fe^{3+} phases consist of at least two components. The component with the large quadrupole splitting is usually very broad and probably corresponds to Fe^{3+} in a glass. The other component has widths typical of Fe^{3+} crystalline phases. As shown below,

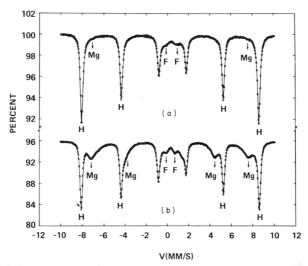

Fig. 14 Mössbauer spectra of ashes prepared from coals from Pittsburgh seam, Pa. (a) and from Gholson seam, Ala. (b) by heating in air to 750°C, for up to 24 hr. Arrows labeled H, Mg, and F indicate Mössbauer peaks due to α-Fe$_2$O$_3$ (hematite), Mg$_x$Fe$_{3-x}$O$_4$ (magnesioferrite) and paramagnetic ferric phases including glass, oxides, and silicate phases.

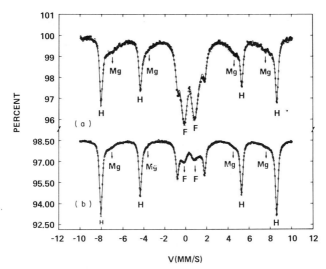

Fig. 15 Mössbauer spectra of ashes prepared from coals from Pocahontas No. 3 seam, W.Va. (a) and from Pratt seam, Ala (b). Conditions of ashing and symbols are given in caption to Fig. 14.

TABLE XI *Mössbauer Parameters for α-Fe$_2$O$_3$, MgFe$_2$O$_4$, and Fe^{3+} Paramagnetic Phases in Coal Ash (750°C, air)*

Coal seam, state	α-Fe$_2$O$_3$			MgFe$_2$O$_4{}^a$			Fe^{3+} Paramagnetic phases	
	δ	ϵ	H	δ	ϵ	H	δ	ϵ
Pocahontas No. 3, W. Va.	0.39	−0.10	514	0.42	0.02	453	0.35	0.91
							0.34	1.52
Pratt, Ala.	0.37	−0.10	515	0.37	0.00	484	0.33	0.88
				0.30	0.00	447	0.40	1.65
Eagle, W. Va.	0.34	−0.14	514	0.31	0.04	458	0.35	0.85
							0.36	1.44
Sewell, W. Va.	0.36	−0.11	516	0.39	0.01	460	0.37	0.77
							0.35	1.47
Gholson, Ala.	0.38	−0.09	517	0.40	0.01	482	0.32	0.90
				0.29	0.02	454		
				0.36	0.00	415		
Herrin No. 6, Ill.	0.38	−0.09	516	—	—	—	0.32	1.18
Pittsburgh, Pa.	0.38	−0.10	515	—	—	—	0.34	1.21

^{a} Very broad peaks. Can be fit by up to three magnetic subspectra in well-resolved spectra (e.g., Gholson). May also include minor calcium.

both components arise from clay minerals, but further identification has not been attempted.

A comparison of the percentages of iron in phases in ash with the percentages of iron in minerals in coal is shown in Table XII. A strong correlation between the percentage of Fe^{3+} contained in the paramagnetic phases in ash with the percentage of Fe^{2+} in clays is apparent. Correlations between other phases are not as obvious; however, it is

TABLE XII *Comparisons of Percentages of Iron in Minerals in Coal with Percentages of Iron in Phases in Ash*

Coal seam, state	Coal, % Fe in				Ash, % Fe in		
	Pyrite	Illite	Siderite	Other	α-Fe$_2$O$_3$	MgFe$_2$O$_4$	Glass
Pocahontas, No. 3, W. Va.	18.6	51.8	29.6		30.8	32.8	36.4
Pratt, Ala.	59.4	31.0	9.6		63.5	18.5	17.9
Eagle, W. Va.	74.0	17.4	8.6		61.2	18.1	20.7
Sewell, W. Va.	39.6	17.8	14.5	28.1 (Fe$_3$O$_4$)	59.2	24.5	16.3
Gholson, Ala.	13.7	6.0	80.3		52.3	41.5	6.3
Herrin No. 6, Ill.	84.0	9.9	—	6.1 (Jarosite)	87.2	—	12.8
Pittsburgh, Pa.	86.9	13.1	—		90.4	—	9.6

likely that part of the α-Fe_2O_3 in ash arises from pyrite and part from Fe^{2+} in siderite, whereas the magnesioferrite probably arises primarily from Fe^{2+} in ankerite. The transformations of iron-bearing minerals during the ASTM test method can be summarized as follows:

$$Pyrite\ (FeS_2) \rightarrow \alpha\text{-}Fe_2O_3,$$

$$Siderite\ (FeCO_3) \rightarrow \alpha\text{-}Fe_2O_3,$$

$$Ankerite\ [(Ca,Fe,Mg)CO_3] \rightarrow Mg_xFe_{3-x}O_4 + other\ phases,$$

$$Fe^{2+}\ in\ clays \rightarrow Fe^{3+}\ phases\ (silicates,\ glass,\ oxides).$$

Iron-rich chlorites in coal should also give rise to magnetic oxides of iron (Grim, 1968). Such transformations for the Pocahontas No. 3 coal, which is known to be chlorite rich (Lee *et al.*, 1978), would account for the sizable discrepancy between the percentage of Fe^{2+} in clays and of Fe^{3+} in paramagnetic phases in ash for this coal.

Most commercial combustion takes place at temperatures considerably higher than 750°C. As the temperature is increased, solid-state reactions between the different components in ash will occur, and the simple decomposition and oxidation reactions shown above will no longer apply. Eventually, at high enough temperatures, complete fusion will occur and the iron will be just one component in the molten slag. Depending on the quench rate and the composition, this slag will cool to a glass, or mixture of glass and crystals, or just crystals, and the Mössbauer spectra will reflect this difference. Figure 16 shows the spectra of completely molten slags quenched in about 10 s to ambient conditions. In one slag (Pratt seam, Alabama), a glass was obtained and the spectrum shows broad Fe^{2+} and Fe^{3+} peaks, typical of iron-bearing silicate glasses. In the other (Gholson seam, Alabama), the Mössbauer spectrum shows the presence of fine-grained (superparamagnetic) magnetite in the slag. These findings were confirmed in the optical microscope.

Much work can be done by Mössbauer spectroscopy in detailing the role of iron in determining high-temperature properties of slags and ash.

B. Mössbauer Studies of Low-Temperature Ash

The method of low-temperature ashing, as developed for coal by Gluskoter (1965), is a most convenient method for separating mineral matter from coal with the minimum of heating and of transformation of the mineral matter. Even though the temperature does not exceed 150°C during LTA, minor transformations, such as the formation of

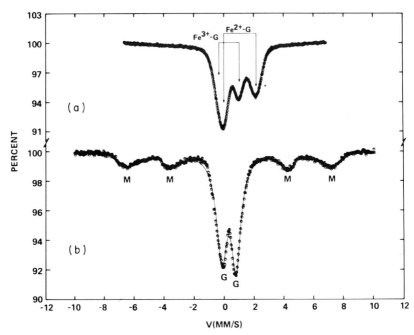

Fig. 16 Mössbauer spectra of rapidly quenched slags arising from mineral matter in coals from the Gholson (b) and Pratt seams, Ala. (a). The labels M and G indicate Mössbauer peaks due to magnetite and glass, respectively. For the Gholson slag (1325°C), the glass primarily contains ferric iron, whereas the Pratt slag (1540°C) contains both ferric and ferrous iron, as indicated.

sulfates and nitrates, have been suspected during this process. The formation of nitrates and sulfates during LTA is discussed, respectively, in Volume II, Chapter 20, Section IV,B,2, and Chapter 26, Section II. Montano (1977) documented minor changes in the appearance of the Mössbauer spectrum of LTA when compared with that of the coal, which were attributed to oxidation of both pyrite and ferrous sulfate to a ferric sulfate. A similar study in this laboratory on two Illinois coals, documented in Table XIII, agrees with his findings. In addition, differences between Mössbauer spectra of a ferrous-clay-rich coal and its LTA indicate that the iron in clay minerals can be partially oxidized to Fe^{3+} (Fig. 17). Spectra of a siderite-rich coal appeared to show no change after LTA, which indicates that $FeCO_3$ is much less susceptible to oxidation during LTA than pyrite or clays.

Although oxidation of the iron phases does appear to occur to a small extent during LTA, such findings do not appear to limit seriously the usefulness of LTA. However, for the estimation of relative amounts of

TABLE XIII *Comparison of Mössbauer Data on the Distribution of Iron between Minerals in Coals From Illinois and Their LTA[a]*

| | | % Fe in | |
Coal seam	Pyrite FeS$_2$	Szomolnokite FeSO$_4$·H$_2$O	Coquimbite Fe$_2$(SO$_4$)$_3$·9H$_2$O
Davis, C18105	92	8	0
LTA	80	14[b]	6
Herrin No. 6, C18903	90	10	0
LTA	87	tr	13[b]

[a] Samples supplied by H.J. Gluskoter, Illinois State Geological Survey.

[b] Also observed in x-ray diffraction patterns (H.J. Gluskoter, personal communication).

iron sulfates in coal and Fe^{2+}/Fe^{3+} ratios in clay minerals, experiments should be performed directly on the coal and not on the LTA.

VIII. OTHER APPLICATIONS OF MÖSSBAUER SPECTROSCOPY

A. Coal Cleanup Methods

A recent review by Jacobs *et al.* (1978) summarizes work by that group on coal cleanup procedures in which Mössbauer spectroscopy plays an integral part.

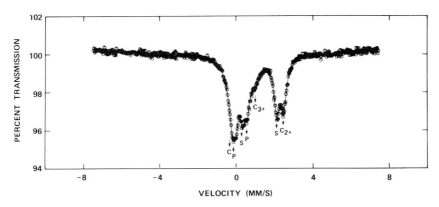

Fig. 17 Mössbauer spectrum of low-temperature ash of coal from Pocahontas No. 3 seam, W.Va. The spectrum of the coal is shown in Fig. 5. The differences between the two spectra suggest that oxidation of the Fe^{2+}-bearing clay minerals occurs during LTA. The arrows labeled, P, S, and C indicate Mössbauer peaks due to pyrite, siderite, and iron-bearing clays. The subscripts 2+ and 3+ indicate ferrous and ferric contributions to the clay absorption.

Mössbauer spectroscopy was used to document the efficiency of pyrite and clay removal from raw coal by float–sink methods using organic solvents to simulate washing in coal-preparation plants. For coal from the Upper Freeport seam, it was found that pyrite could be removed more efficiently than clays by such procedures. This result is not surprising because clay minerals have densities considerably closer to that of coal than pyrite, and the efficiency of removal can be expected to be a function of the density difference unless the particle-size distributions of the minerals are quite different.

Similar tests of the efficiency of high-gradient magnetic separation (HGMS) were also made by Mössbauer spectroscopy and showed that the removal of siderite ($FeCO_3$) was more efficient than removal of either pyrite or clay minerals. Detailed magnetic studies of this process have suggested that pyrrhotite formation may be responsible for the necessary enhancement of the magnetic properties of pyrite in order for pyrite to be removed efficiently by HGMS. High localized temperatures generated upon crushing of the coal were suggested to be responsible for converting a surface layer on pyrite particles to pyrrhotite.

B.   Studies of Coal-Conversion Processes

One further application of Mössbauer spectroscopy, examined by Jacobs *et al.* (1978), was to use the technique to investigate mineral transformations during coal-liquefaction processes. The objective was to find the procedure for liquefaction that also generated the most highly magnetic form of pyrrhotite ($Fe_{1-x}S$) for possible application of HGMS. Both Mössbauer and magnetic studies showed that most of the pyrite is converted during liquefaction of coal to pyrrhotite, which may be weakly or strongly ferrimagnetic, depending on the temperature-time pathways of the processes. By slow cooling from 225°C, a highly ferrimagnetic form of pyrrhotite, monoclinic Fe_7S_8, can be obtained, which is the preferred form for removal by HGMS.

IX.   CONCLUSIONS AND POSSIBLE FUTURE DIRECTIONS

In this chapter, much emphasis has been directed to the usefulness of Mössbauer spectroscopy for characterization of the iron-bearing phases in coal and coal derived solids. From such spectra, it is found that the principal iron-bearing minerals in coal are pyrite, ferrous clays (illite, chlorite), and carbonates (siderite, ankerite). Other phases such as sulfates, iron oxides, and oxyhydroxides may be found as a result of special conditions, the most important of which is weathering of coal upon atmospheric exposure. Mössbauer analysis of ash, coke, and coal-liquefaction products shows how the principal iron-bearing minerals transform under combustion, carbonization, and other coal-conversion processes. Additionally Mössbauer spectroscopy has been used to detail aspects of standard techniques, such as various ASTM tests and low-temperature ashing.

A number of problem-specific applications of Mössbauer spectroscopy were described in this chapter, including quantitative pyritic-sulfur determinative methods, coal cleanup and desulfurization applications, and environmental problems resulting from weathering reactions of pyrite. These applications suggest that Mössbauer spectroscopy can contribute significantly toward the solution of coal related problems.

To date, most of the activity in Mössbauer analysis of coal has been directed toward the characterization of phases in bituminous coals and coal related solids. However, future work will most likely be concentrated on applying Mössbauer spectroscopy as part of an integrated approach to problems arising from coal utilization. In addition to the areas listed above, other areas that should benefit from Mössbauer analysis are the investigation of ash and slag properties, paleoenvironmental and geological studies, and research on coal-conversion technologies.

APPENDIX: SUMMARY OF IRON DISTRIBUTION BETWEEN MINERALS FOR SOME UNITED STATES COAL SEAMS

Table A1 summarizes data obtained by Mössbauer spectroscopy on the distribution of iron between minerals for various United States coal seams. This table indicates the variation found within a seam (notably data for Pittsburgh seam, Pennsylvania and West Virginia and Herrin No. 6, Illinois) as well as the considerable variation found for United States coal seams.

ACKNOWLEDGMENTS

We are grateful for many useful suggestions arising from discussions with R.J. Gray (U.S. Steel Research), particularly with respect to the relationship of Mössbauer studies on coal to other areas of coal research. We are also grateful to G.R. Dunmyre (U.S. Steel Research) for his assistance with Mössbauer experimentation. We thank I.S. Jacobs and

TABLE A1 *Summary of Iron Distribution between Minerals for Some United States Coal Seams*

State, seam, county	Rank	Pyrite	% Fe in Clays	% Fe in Siderite	Other[a]	Reference
Pennsylvania						
Pittsburgh, Washington	HVB	87	10	3		This work
Pittsburgh, Washington	HVB	100				This work
Pittsburgh —	HVB	90	10			Lefelhocz et al. (1967)
Upper Freeport —	—	92	5	2		Jacobs et al. (1978)
Kittanning, Cambria	MVB	67	8		25(H, J)	Huffman and Huggins (1978)
West Virginia						
Pittsburgh —	(HVB)	100				Levinson and Jacobs (1977)
Pittsburgh, Barbour	(HVB)	92			8(Sz)	Montano (1977)
Redstone, Barbour	(HVB)	94			6(Sz)	Montano (1977)
Pocahontas No. 3, Wyoming	LVB	20	50	30		This work
Pocahontas No. 4, McDowell	(LVB)	45			55(Sz)	Montano (1977)
Pocahontas No. 4 —	LVB	47	53			Lefelhocz et al. (1967)
Eagle No. 2, Kanawha	HVB	70	17	13		Huffman and Huggins (1978)
Sewell, Nicholas	MVB	40	18	15	28(M)	Huffman and Huggins (1978)
Sewell, Fayette	MVB	50	24	26		Huffman and Huggins (1978)
Virginia						
Taggart, Wise	MVB	45	37	18		Huffman and Huggins (1978)
Jewell, Buchanan	MVB		100			Lefelhocz et al. (1967)
Kentucky						
High Splint —	HVB		100			Lefelhocz et al. (1967)
Harlan, Harlan	HVB	53	43	4		This work
Alabama						
Pratt, Jefferson	MVB	55	25	20		Huffman and Huggins (1978)
Black Creek, Walker	HVB	66	34			Huffman and Huggins (1978)
Black Creek, Carbon/ Winston	HVB	69	31			Huffman and Huggins (1978)
Gholson, Shelby	HVB	16	5	79		Huffman and Huggins (1978)
Mary Lee, Walker	LVB	12	51	33	4(J)	This work
Arkansas						
L. Hartshorne, Sebastian	LVB	24		76		Huffman and Huggins (1978)

TABLE A1 *(Continued)*

State, seam, county	Rank	Pyrite	% Fe in Clays	Siderite	Other[a]	Reference
Illinois						
Herrin No. 6, Franklin	HVB	99			1(Sz)	Huffman and Huggins (1978)
Herrin No. 6, Franklin	HVB	73	3		24(J)	This work
Herrin No. 6, Franklin	HVB	48			52(unk)	Smith *et al.* (1978)
Davis —	HVB	92			8(Sz)	This work
Colorado						
Somerset B, Gunnison	HVB	9	4	87		This work
Somerset C, Gunnison	HVB	31	11	58		This work
Montana						
Rosebud, Rosebud	SbB	99			1(Cq)	This work

[a] H—hematite; M—magnetite; J—jarosite; Sz—szomolnokite; Cq—coquimbite; unk—unknown.

L.M. Levinson (General Electric Research) for supplying us with a copy of their review paper in advance of publication. We are indebted to J.S. White, Jr., (U.S. National Museum) and H.J. Gluskoter (Illinois State Geological Survey) for supplying us with samples of iron sulfates.

REFERENCES

ASTM (1976). "1976 Annual Book of ASTM Standards. Part 26: Gaseous Fuels; Coal and Coke; Atmospheric Analysis." Am. Soc. Test. Mater., Philadelphia, Pennsylvania.
Baker, R. A. (1972). *Water Res.* **6,** 9–17.
Balme, B. E. (1956). *J. Inst. Fuel* **29,** 21–22.
Bancroft, G. M. (1973). "Mössbauer Spectroscopy. An Introduction for Inorganic Chemists and Geochemists." Wiley, New York.
Caric, S., Marinkov, L., and Slivka, J. (1975). *Phys. Status Solidi A* **13,** 263–268.
Coey, J. M. D. (1975). *Proc. Int. Conf. Moessbauer Spectrosc., Cracow* **2,** 333–354.
Daniels, J. M., and Rosencwaig, A. (1969). *J. Phys. Chem. Solids* **30,** 1561–1571.
Deer, W. A., Howie, R. A., and Zussman, J. (1962). "Rock-Forming Minerals," Vol. 3. Longmans, London.
Edwards, A. H., Jones, J. M., and Newcombe, W. (1964). *Fuel* **43,** 55–62.
Finkelman, R. B., and Stanton R. W. (1978). *Fuel* **57,** 763–768.
Gamayunov, N. I., Voitkovskii, Y. V., Yufik, Y. M., Cherkes, I. D., and Amusin, L. G. (1975). *Khim. Tverd. Topl.* **1,** 163–165.
Garrels, R. M., and Christ, C. L. (1965). "Solutions, Minerals, and Equilibria." Harper, New York.
Gluskoter, H. J. (1965). *Fuel* **44,** 285–291.

Gluskoter, H. J. (1975). *In* "Trace Elements in Fuel" (S. P. Babu, ed.), pp. 1–22. Am. Chem. Soc., Washington, D.C.

Gluskoter, H. J., Ruch, R. R., Miller, W. G., Cahill, R. A., Dreher, G. B., and Kuhn, J. K. (1977). *Ill. State Geol. Surv. Circ.* No. 499.

Goncharov, G. N., Ostanevich, Y. M., and Tomilov, S. B. (1970). *Izv. Akad. Nauk. SSSR, Ser. Geol.* **8,** 79.

Govaert, A., Dauwe, C., Plinke, P., De Grove, E., and De Sitter, J. (1976). *J. Phys. (Paris), (Colloq.)* **37**(C-6), 825–827.

Greenwood, N. N., and Gibb, T. C. (1971). "Mössbauer Spectroscopy." Chapman & Hall, London.

Greer, R. T. (1977). *In* "Scanning Electron Microscopy/1977" (O. Johari, ed.), Vol. 1, pp. 79–93. I.I.T. Res. Inst., Chicago, Illinois.

Grim, R. E. (1968). "Clay Mineralogy," 2nd Ed. McGraw-Hill, New York.

Hien, P. Z., and Shpinel, V. L. (1963). *Sov. Phys.—JETP* **17,** 268–273.

Hrynkiewicz, A. Z., Kubisz, J., and Kulgawczuk, D. S. (1965). *J. Inorg. Nucl. Chem.* **27,** 2513–2517.

Hucl, M., Janak, F., and Zapletal, K. (1975). *Int. Conf. Moessbauer Spectrosc., Proc., 5th, Bratislava, Czechoslovakia* **2,** 356–360.

Huffman, G. P. (1976). *Nucl. Instrum. Methods* **137,** 267–290.

Huffman, G. P., and Huggins, F. E. (1978). *Fuel* **57,** 592–604

Huffman, G. P., and Louat, N. (1967). *Bull. Am. Phys. Soc.* **12** (abstr.)

Huffman, G. P., Schwerer, F. C., Fisher, R. M., and Nagata, T. (1974). *Proc. Lunar Sci. Conf., 5th, Houston, Texas* **3,** 2779–2794.

Ingalls, R. (1964). *Phys. Rev. A* **133,** 787–795.

Jacobs, I. S., Levinson, L. M., and Hart, H. R., Jr. (1978). *J. Appl. Phys.* **49,** 1775–1780.

Janot, C., Chabanel, M., and Herzog, E. (1968). *Bull. Soc. Fr. Mineral. Cristallogr.* **91,** 166–171.

Kemezys, M., and Taylor, G. H. (1964). *J. Inst. Fuel* **37,** 389–397.

Klein, C., and Bricker, O. P. (1977). *Econ. Geol.* **72,** 1457–1470.

Lang, G. (1963). *Nucl. Instrum. Methods* **24,** 425–428.

Lee, R. J., Huggins, F. E., and Huffman, G. P. (1978). *In* "Scanning Electron Microscopy/ 1978" (O. Jahari, ed.), Vol. 1, pp. 561–568. SEM Inc., AMF O'Hare, Illinois.

Lefelhocz, J. F., Friedel, R. A., and Kohman, T. P. (1967). *Geochim. Cosmochim. Acta* **31,** 2261–2273.

Levinson, L. M., and Jacobs, I. S. (1977). *Fuel* **56,** 453–454.

Mackowsky, M.-T. (1968). *In* "Coal and Coal-Bearing Strata" (D. Murchison and L. S. Westoll, eds.), pp. 309–321. Elsevier, Amsterdam.

Mössbauer, R. L. (1958). *Z. Phys.* **151,** 124–143.

Montano, P. A. (1977). *Fuel* **56,** 397–400.

Montano, P. A., and Seehra, M. S. (1976). *Solid State Commun.* **20,** 897–898.

Morice, J. A., Rees, L. V. C., and Rickard, D. F. (1969). *J. Inorg. Nucl. Chem.* **31,** 3797–3802.

Palache, C., Berman, H., and Frondel, C. (1951). "Dana's System of Mineralogy," Vols. I and II. Wiley, New York.

Power, L. F., and Fine, H. A. (1976). *Miner. Sci. Eng.* **8,** 106–128.

Raj, D., Chandra, K., and Puri, S. P. (1968). *J. Phys. Soc. Jpn.* **24,** 39–41.

Reyes-Navarro, J., and Davis, A. (1976). "Pyrite in Coal: Its Forms and Distribution as Related to the Environments of Coal Deposition in Three Selected Coals from Western Pennsylvania," Spec. Res. Rep. SR-110. Coal Res. Sect., Pennsylvania State Univ., University Park.

Schafer, H. N. S. (1977). *Fuel* **56**, 45–46.

Schwartz, E. J., and Vaughan, D. J. (1972). *J. Geomagn. Geoelectr.* **24**, 441–458.

Smith, G. V., Lui, J.-H., Saporoschenko, M., and Shiley, R. (1978). *Fuel* **57**, 41–45.

Smyth, M. (1966), *Fuel*, **45**, pp. 221–231.

Spijkerman, J. J. (1971). *In* "Mössbauer Effect Methodology" (I. J. Gruverman, ed.), Vol. 7, pp. 85–96. Plenum, New York.

Stach, E., Mackowsky, M.-T., Teichmüller, M., Taylor, G. H., Chandra, D., and Teichmüller, R. (1975). "Coal Petrology." Borntraeger, Berlin.

Stevens, J. G., and Stevens, V. E. (1976). "Mössbauer Effect Data Index." Plenum, New York, and nine earlier volumes.

Stevens, J. G., Stevens, V. E., and Gettys, W. L., eds. (1978). *Moessbauer Eff. Ref. Data J.* **1**.

Temperley, A. A., and Lefevre, H. W. (1966). *J. Phys. Chem. Solids* **27**, 85–92.

Vaughan, D. J., and Burns, R. G. (1972). *Int. Geol. Congr., 24th, Sect. Rep.* **14**, 158–167.

Vaughan, D. J., and Ridout, M. S. (1970). *Solid State Commun.* **8**, 2165–2166.

Vertes, A., and Zsoldos, B. (1970). *Acta Chim. Acad. Sci. Hung.* **65**, 261–271.

Walker, J. C., Munley, F., and Loh, E. (1971). *Proc. Conf. Appl. Moessbauer Eff., Tihany, Hungary*, pp. 153–164.

Walker, L. R., Wertheim, G. K., and Jaccarino, V. (1961). *Phys. Rev. Lett.* **6**, 236–239.

Watson, R. E., and Freeman, A. J. (1961). *Phys. Rev.* **123**, 2027–2047.

Wertheim, G. K. (1964). "Mössbauer Effect: Principles and Applications." Academic Press, New York.

Zubovic, P. (1966). *In* "Coal Science" (R. F. Gould, ed.), pp. 221–246. Am. Chem. Soc., Washington, D.C.

Chapter 51

Fingerprinting Solid Coals Using Pulse and Multiple Pulse Nuclear Magnetic Resonance

B. C. Gerstein

AMES LABORATORY
U.S. DEPARTMENT OF ENERGY
AND DEPARTMENT OF CHEMISTRY
IOWA STATE UNIVERSITY
AMES, IOWA

I. INTRODUCTION

While nuclear magnetic resonance (NMR) has been an enormously useful tool for analyses of liquid species, use for characterization of fossil fuels in the solid state has been severely limited (Retcofsky and Friedel, 1973). For nuclei with the highest sensitivity to detection, e.g., ^{1}H, ^{19}F, and ^{31}P, homonuclear dipolar coupling (Slichter, 1963) in solids results in sufficient line broadening to obscure NMR chemical shifts that identify molecular environments. For low abundance, low gyromagnetic ratio nuclei in molecules containing a large portion of ^{1}H, e.g., ^{13}C in naphthalene, heteronuclear dipolar broadening of the ^{13}C by the

425

^{1}H is sufficient to mask chemical shifts of ^{13}C. In the absence of techniques to remove dipolar broadening, therefore, coal scientists utilizing NMR to infer identities of chemical entities in coals have had to resort to the use of relaxation times (Retcofsky and Friedel, 1968; Oth and Tschamler, 1963) and to average proton–proton distances inferred from the magnitude of the dipolar line broadening associated with these interactions (i.e., to second moment calculations). T_1's or longitudinal relaxation times in coals are a reflection of the content of effectively paramagnetic species present (e.g., dissolved O_2, or free radicals). T_2^*'s or transverse relaxation times are associated predominantly with dipolar interactions; heteronuclear, homonuclear, and free radical nuclear.† While second moments have been used to infer average proton-proton distances (the second moment associated with a pair of spins at a distance r_{ij} apart varies as the sixth power of the inverse of this distance), until recently (Gerstein et al., 1977a) the effect of free radical broadening an ^{1}H NMR spectra has not been known.

Even if the effect of dipolar broadening could be removed, however, solid state NMR spectra would still exhibit linewidths much larger than the total range of isotropic chemical shifts observable for any given nucleus. This phenomenon is a result of the fact that the chemical shift interaction is not really the scalar quantity observed in liquid state measurements. Rather, the shielding is dependent upon the direction of the external magnetic field with respect to the molecular coordinate system. This fact should be intuitively obvious from a consideration of the symmetry of the electronic environment about a proton in H_2O. A bit of reflection will convince one that the electronic symmetry of the proton environment is approximately C_{3v} (i.e., the proton resides in a sp^3 orbital and has an environment of roughly three other sp^3 orbitals directed symmetrically away from the oxygen center to which the proton is joined). Methane would be a case of exact C_{3v} electronic symmetry for the environment of a given proton. The group C_{3v} is axially symmetric, so the shielding is describable in terms of a "parallel" principal axis value, $\sigma_{33} = \sigma_{\parallel}$, and two "perpendicular" principal axis values, $\sigma_{11} = \sigma_{22} = \sigma_{\perp}$, of the effectively symmetric second rank chemical shift tensor. The total anistropy $(\sigma_{\parallel} - \sigma_{\perp})$ of this tensor in $H_2O(s)$ is 34 ppm (Ryan et al., 1977). This is to be compared to a *total range* of liquid state chemical shifts {that are the isotropic value [$\sigma_{iso} = \frac{1}{3}(\sigma_{11} + \sigma_{22} + \sigma_{33})$] of the chemical shift tensor} of 10 ppm. For many different chemical environments of any given nuclear species, therefore, the

---

† For a further discussion of relaxation times see Volume II, Chapter 23, Section II,B.

NMR spectrum of, e.g., ^1H in solid coals would be a hopeless, structureless jumble even in the absence of dipolar broadening.

The problem of chemical shift anisotropy broadening may be removed by the technique of "magic angle† spinning" at speeds generally (but not always) (Maricq and Waugh, 1977) greater than the chemical shift anisotropy. For protons, this means rotational speeds in excess of 2 kHz at fields of 1.4 T. The requirement becomes worse, of course, as the external field increases since the chemical shift is proportional to the field.

It is the purpose of the present chapter to discuss techniques that have been used for quantitative and qualitative analysis of solid coals via relaxation measurements, conventional magnetic resonance, and recently developed transient techniques in NMR (Mehring, 1976; Haeberlen, 1976).

II. PROTON NMR

A. Quantitative Analysis Using Pulse Techniques

In liquid state NMR, one is able, via comparison with a suitable standard, to infer concentrations of nuclei from relative areas under spectral peaks. Nuclear magnetic resonance spectra in solids are generally so broad that such analyses have not received general use. One can use the fact, however, that the initial value of a free induction decay (FID) in a pulse experiment is proportional to the total number of spins associated with the signal. With sufficiently short receiver recovery times (Adduci *et al.*, 1976) and pulse widths short compared to T_2^*, the effective transverse decay time (characteristically about 12 μsec in coals) initial amplitudes may be accurately determined. The time between data accumulations in pulse experiments is about 5 T_1 (Farrar and Becker, 1971). Proton T_1's in coals are dominated by the presence of effectively paramagnetic species, e.g., free radicals, and are in general of the order of 0.1 sec. Therefore, rapid data accumulation is possible. With care taken to assure that both sample and standard are in a uniform H_1 radio-frequency field, and that the zero of time is correctly determined, nondestructive analysis of protons in coals may easily be made (Gerstein and Pembleton, 1977) to an accuracy of 5%. Analysis times between sample reception and evaluation of proton concentration are less than 15 min and could certainly be less in an

---

† The magic angle is defined in Volume II, Chapter 23, Section II,C.

automated operation. Isotopic abundance of ^{13}C and ^{15}N in coals are sufficiently low, and T_1's are sufficiently large that use of single pulse techniques for quantitative analysis of these nuclei are unfortunately not practical (but see Section III,C).

B. Damping Constants of Protons in Coals under Transient Excitations T_1, T_2^*, and $T_{1\rho}$

Pulse techniques in NMR are superbly suited to a determination of relaxation parameters (Farrar and Becker, 1971). There are at least three damping constants that conceivably may be used to characterize protons in coals under NMR relaxation experiments. These are the longitudinal damping constant T_1, the effective transverse damping constant T_2^*, and the longitudinal damping constant in the rotating frame $T_{1\rho}$. T_1's of protons are of interest because they determine the time between data accumulations in experiments that depend upon proton longitudinal relaxation, e.g., simple one pulse experiments on ^1H alone and cross polarization experiments (see Section III). T_2^*'s could be used to distinguish the degree of rigidity of the solid state structure (Gerstein *et al.*, 1977a) as well as offering a measure of average interaction distances. $T_{1\rho}$'s are important in a determination of effective cross polarization periods used to detect isotopically rare nuclei (e.g., ^{13}C and ^{15}N) by polarization transfer from ^1H (see Section III). In Table I are listed some values of T_1, T_2^*, and $T_{1\rho}$ determined for some typical vitrain portions of Virginia and Iowa coals ranging in carbon content from 75 to 90%. Using the initial value of the decay associated with each portion of a multicomponent decay, it is sometimes possible to determine the fraction of nuclei associated with each decay. These fractions, when available, are also listed in Table I. Comparison of the values in Table I with model aliphatic, aromatic, and mixed aromatic–aliphatic compounds indicates that no distinction between aliphatic and aromatic protons may be inferred from T_2^* measurements alone.

C. Strong Homonuclear Decoupling

1. *Average Hamiltonian Theory*†

As mentioned in the introduction, NMR linewidths in solids are dominated by dipolar interactions. The purpose of the present section is to explain how spin systems may be manipulated to minimize these

† Haeberlen (1976).

TABLE I T_1, $T_2{}^*$, $T_{1\rho}$ of five coals

Coal	%C (MAF)	T_1 (msec)	$T_2{}^*$ (μsec)	$T_{1\rho}$ (μsec)[a,b]
Pocahontas No. 4	90.3	425	9.5	1354, 68%
				140, 32%
Powellton	85.1	412	17	1860, 77%
				84, 23%
Upper Mich	81	139	10, 96%	1255, 66%
			61, 4%	61, 34%
Star	78	99	6	2313, 81%
				61, 19%

[a] Values obtained on unheated, unpumped coals; all other values obtained on coals heated at 100°C for 8 hr and pumped to 10^{-7} torr.

[b] $T_{1\rho}$ measured at a locking field of 10^{-3} torr in order to match cross polarization conditions; with a locking field of 3.9×10^{-3} torr, e.g., Pocahontas No. 4 exhibits the values 2, 418, 77%, and 81, 23%.

interactions at specific observational "window times" in the experiment.

The observable in a pulse NMR experiment is a voltage that is proportional to the expectation value of the transverse magnetization at some phase in the xy plane set by the experimenter. For purposes of illustration, we choose to phase detect along the y axis of the rotating frame. The rotating frame, or interaction frame of the Zeeman Hamiltonian, is the frame of observation (via phase detection and resultant demodulation of the Larmor frequency) in an NMR experiment. Thus, the signal observed along the y axis as a result of a pulse NMR experiment will be proportional to $\langle I_y(t)\rangle$, the expectation value of the y component of angular momentum as a function of time. The NMR absorption spectrum is related to the time decay of $\langle I_y(t)\rangle$ by Fourier transformation, as illustrated in Fig. 1. An equivalent way of writing the familiar time dependent Schrödinger equation, which describes the time behavior of the state function ψ is via the density operator $\rho = \psi\psi^*$: (with $\hbar = 1$)

$$i\,(d\psi/dt) = \mathcal{H}\psi \rightleftarrows i\,(d\rho/dt) = [\mathcal{H}, \rho] \tag{1}$$

Expectation values of operators A are then given in terms of the density operator as

$$\langle A\rangle = \mathrm{Tr}(A\rho). \tag{2}$$

In particular, the expectation value of I_y is

$$\langle I_y(t)\rangle = \mathrm{Tr}\,\rho(t)I_y. \tag{3}$$

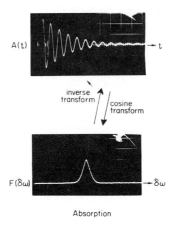

Fig. 1   Relation between time decay and NMR spectrum.

A specification of $\rho(t)$, the solution of the right-hand side of Eq. (1) will therefore specify the time decay observed as a result of a pulse experiment. The interactions governing the development of $\rho(t)$ are of two types; there are those controlled by the experimenter, $\mathscr{H}_{rf}(t)$ that is a result of radio-frequency pulses, and those internal to the system, $\mathscr{H}_{int}(t)$, such as dipolar and chemical shift interactions. We thus write

$$i\, d\rho/dt = [(\mathscr{H}_{rf} + \mathscr{H}_{int}), \rho]. \tag{4}$$

If the transformation

$$\rho(t) = U_{rf}^{-1}\tilde{\rho}(t)U_{rf} \tag{5}$$

is made, and U_{rf} is determined by $\mathscr{H}_{rf}$ via

$$i\, dU_{rf}/dt = \mathscr{H}_{rf}U_{rf}, \tag{6}$$

then it is found that in the new frame (the interaction frame of the rf) $\tilde{\rho}(t)$ is described by

$$i\, d\tilde{\rho}/dt = [\tilde{\mathscr{H}}_{int}, \tilde{\rho}], \tag{7}$$

where $\tilde{\mathscr{H}}_{int}(t)$ is controlled by $\mathscr{H}_{rf}$ through

$$\tilde{\mathscr{H}}_{int}(t) = U_{rf}^{-1}\mathscr{H}_{int}U_{rf}, \tag{8}$$

i.e., $\mathscr{H}_{rf}$ has been removed from the description of $\tilde{\rho}$, the density operator in the frame of $\mathscr{H}_{rf}$. In addition, U_{rf} is specified by the well known solution (Wilcox, 1967) to Eq. (6):

$$U_{rf}(t) = T \left\{ \exp -i \int_0^t \mathcal{H}_{rf}(t') \, dt' \right\}, \tag{9}$$

with T the Dyson time ordering operator.

Further, if the transformation

$$\tilde{\rho} = U_{int}^{-1} \tilde{\rho}(t) U_{int} \tag{10}$$

is made, and U_{int} is determined by $\tilde{\mathcal{H}}_{int}$, which in turn is controlled by $\mathcal{H}_{rf}$ through Eq. (8) via the relation

$$i \, dU_{int}/dt = \tilde{\mathcal{H}}_{int} U_{int}, \tag{11}$$

then the description of $\tilde{\rho}(t)$ becomes

$$i \, d\tilde{\rho}/dt = [0, \tilde{\rho}] = 0. \tag{12}$$

Therefore, in this frame, $\tilde{\rho}$ is time independent. In addition, the solution of Eq. (11) is generally given by the Magnus expression (Wilcox, 1967):

$$U_{int} = \exp\{-it_c[\bar{H}_{int}^{(0)} + \bar{H}_{int}^{(1)} + \cdots]\} \tag{13}$$

with t_c being a suitable chosen cycle time for the experiment, and

$$\bar{H}_{int}^{(0)} = \frac{1}{t_c} \int_0^{t_c} \tilde{\mathcal{H}}_{int}(t) \, dt$$

$$\bar{H}_{int}^{(1)} = \frac{-i}{2t_c} \int_0^{t_c} dt \int_0^t dt' \, [\tilde{\mathcal{H}}_{int}(t), \tilde{\mathcal{H}}_{int}(t')], \text{ etc.,} \tag{14}$$

with the strong convergence condition

$$\|\tilde{\mathcal{H}}_{int}\| t_c \ll 1. \tag{15}$$

Equations (5) and (9) yield

$$\rho(t) = U_{int} U_{rf} \tilde{\rho} U_{rf}^{-1} U_{int}^{-1}, \tag{16}$$

i.e., the time development of $\langle I_y \rangle$ in the frame of observation is related to $\tilde{\rho}$ via the products of U_{int} and U_{rf}. Clearly, if a time interval can be chosen such that both U_{rf} and U_{int} are unity, then *in such time intervals,* $\rho(t)$ appears unaffected by the Hamiltonians that otherwise strongly perturb the system, i.e., cause line broadening. In addition, if at multiples of cycle times $t = nt_c$ the condition $U_{rf} = 1$ is satisfied, then the density matrix formally develops in time at such times according to

$$\rho(0) = \exp\{-int_c[\bar{H}_{int}^{(0)} + \bar{H}_{int}^{(1)} + \cdots]\}. \tag{17}$$

By analogy with the solution of Eq. (1) for which $\mathcal{H}$ is a time independent interaction, $\mathcal{H}_0$,

$$\rho(t) = \exp\{-i\mathcal{H}_0 t\}\rho(0) \exp\{i\mathcal{H}_0 t\}, \tag{18}$$

Eq. (17) tells us that at times $t = nt_c$ the density operator appears to be acted on by an *average Hamiltonian* over the cycle times in question,

$$\mathcal{H}_{av} = \bar{H}^{(0)} + \bar{H}^{(1)} + \cdots. \tag{19}$$

U_{rf} may be made unity by a suitable choice of $\mathcal{H}_{rf}$ such that

$$\int_0^{t_c} \mathcal{H}_{rf}(t)\, dt = 0, \tag{20}$$

which causes all terms in the exponent of Eq. (9) to vanish. For example, a $\frac{1}{2}\pi$ rf pulse at time t' along x in the rotating frame, followed by a $\frac{1}{2}\pi$ pulse along $\bar{x}$ at some later time t'' will accomplish this purpose since in this case, for ideal delta function pulses,

$$\mathcal{H}_{rf}(t) = \tfrac{1}{2}\pi[I_x \delta(t - t') - I_x \delta(t - t'')]. \tag{21}$$

For times larger than t'' the integral of $\mathcal{H}_{rf}(t)$ vanishes. It does not take much imagination to realize that the integral of $\mathcal{H}_{rf}$ will vanish for any such "phase alternated" sequence of pulses. In particular for the "MREV-8" (Mansfield, 1971; Rhim *et al.*, 1973) sequence

$$[\tau, P_x, \tau, P_y, 2\tau, P_{\bar{y}}, \tau, P_{\bar{x}}, 2\tau, P_{\bar{x}}, \tau, P_y, 2\tau, P_{\bar{y}}, \tau, P_x, \tau], \tag{22}$$

where P_j is an ideal delta function $\pi/2$ pulse along the jth axis in the rotating frame,

$$\int_0^{t_c} dt\, \mathcal{H}_{rf}(t) \equiv 0.$$

This condition leads to $U_{rf} = 1$ at times $t_c = 12\tau$.

The value of U_{int} is in turn controlled by $\mathcal{H}_{rf}$ through Eq. (8). At the level of presentation appropriate here, it suffices to say that under the MREV-8 sequence, $\bar{H}_{int}^{(0)}$ and $\bar{H}_{int}^{(1)}$ vanish identically for $\mathcal{H}_{int}$ being the secular portion of the homonuclear dipolar interaction,

$$\mathcal{H}_D = \{\text{const}(1 - 3\cos^2\theta_{ij}/r_{ij}^3)\,(\mathbf{I}_i \cdot \mathbf{I}_j - 3I_{zi}I_{zj})\}. \tag{23}$$

In addition, cross terms between pulse imperfections and the homonuclear dipolar interaction vanish to zero order, so the effect of experimental artifact is minimized.

The chemical shift interaction, while not averaged to zero, is not unscathed, however. It is scaled by a factor roughly equal to $\sqrt{2}/3$, so that differences in chemical shifts are roughly halved in frequency under such a sequence.

As an example of the power of this sequence to remove homonuclear dipolar interactions, the results of a one pulse experiment on ¹H in solid trichloroacetic acid is compared with the decay observed in the 12τ window times of the MREV-8 experiment, as shown in Fig. 2.

2. Chemical Shift Anisotropies and Broadening because of Presence of Free Radicals, of ¹H Spectra in Coals

A typical ¹H spectrum in a vitrain portion of a coal is that shown for Upper Mich vitrain in Fig. 3. The spectrum shows a broad line, roughly 600 ppm wide, and a narrower component that sometimes disappears on heating at 100°C and evacuation. A spectrum taken under removal of homonuclear dipolar interactions via use of the MREV-8 sequence is shown in Fig. 4. While the width of the spectrum has been reduced to roughly 16 ppm, and there is a hint of structure at the peak (the peak width is roughly the proton aromatic–aliphatic separation), there is no detailed information available from which to infer chemical shifts.

There are at least two sources of broadening of ¹H spectra in coals under conditions of multiple pulse removal of homonuclear dipolar broadening. The first, mentioned earlier, is the chemical shift anisotropy of protons. The second is broadening because of the presence of

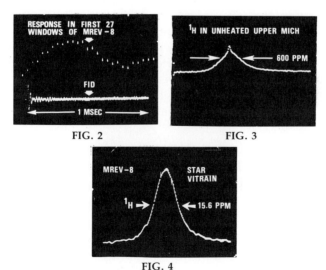

FIG. 2   FIG. 3

FIG. 4

Fig. 2 Comparison between decays of ¹H in solid CCl₃CO₂H under MREV-8 and single pulse experiment.

Fig. 3 ¹H NMR spectrum of Upper Mich vitrain.

Fig. 4 MREV-8 on Star Vitrain.

unpaired electron spins, or to dc field heterogenetics. As mentioned in the previous section, chemical shift anisotropies can be as large as 34 ppm for, e.g., hydroxyl protons. Evidence from infrared spectroscopy indicates a fair portion of these in coals (Schweighardt *et al.*, 1978, and references therein). The proton–unpaired electron spin interaction, however, could be this large or much larger since it is proportional to the product of the magnetic moments of the ^1H and the electron. The latter is 10^3 times larger than the ^1H moment, so free radical broadening could be 10^3 times as large as ^1H–^1H homonuclear dipolar broadening. An important question at this point is, "How large is the free radical–^1H broadening?" Both the chemical shift and the electron spin–nuclear spin interactions are proportional to the operator I_z. Under the MREV-8 sequence, with a preparation pulse along x, and observation along y in the rotating frame, an interaction equal to (const) $\times I_z$ is observed to be of magnitude $\sqrt{2/3}$ (const). In particular, if free radical broadening were large compared to chemical shift anisotropies in coals, this broadening would not be removed by MREV-8. The observation that the ^1H linewidth in coals is of order 16 ppm (corrected for the scaling factor), and that ^1H anisotropies can be larger than this by a factor of two indicates that a major source of the ^1H residual linewidth under the MREV-8 sequence is just proton shift anisotropy. This is an enormously useful piece of information since modulation of both "spin" and "real" space interactions under both rf multiple pulse techniques and physical "magic angle" spinning (*vide infra*) at speeds large compared to the linewidth to be narrowed is capable of removing both chemical shift *anisotropies and* ^1H nuclear–unpaired electron interactions. This is to say that in principle, these combined techniques are (but see Vega and Vaughan, 1978; Dybowski and Pembleton, 1979) capable of yielding liquid-like ^1H spectra!

D. Resolution of ^1H Spectra in Coals under Combined Rotation and Multiple Pulse Spectroscopy (CRAMPS)

In our previous discussion of the average Hamiltonian for the dipolar interaction under MREV-8, it was found that both $\bar{H}_D^{(0)}$ and $\bar{H}_D^{(1)}$ are zero for the secular portion of the homonuclear dipolar interaction. A tacit assumption in this treatment was that the "space" portion of $\mathcal{H}_D$ [cf. Eq. (22)], which is proportional to $(1 - 3 \cos^2 \theta_{12})$ for fixed internuclear distance r_{12} is constant in time. In fact, in a nonviscous liquid, molecular tumbling causes the time average of $\cos^2 \theta_{12}(t)$ to be exactly $\frac{1}{3}$, so $\langle \mathcal{H}_D \rangle = 0$ in such systems.

This fact gives us the hint that by proper manipulation of the spacial portion of $\mathcal{H}_\sigma$, the chemical shift Hamiltonian, chemical shift anisotropy broadening might be severely attenuated. This, in fact is true. If one combines an experiment in which the MREV-8 sequence is used to remove dipolar broadening and, in addition, the sample is rotated at an "appropriate" speed about an axis set at an "appropriate" angle to the magnetic field, both homonuclear dipolar broadening and chemical shift anisotropies can be severely attenuated. For purposes of the present discussion, "appropriate" rotation speeds are those for which rotational cycle times, $t_r = \omega_r^{-1}$, are large compared to multiple pulse cycle times ($t_c = 12\tau$ for the MREV-8), and the inverse of shift anisotropy. For example, for the coal spectrum shown in Fig. 4, the multiple pulse cycle time was 30 μsec. Chemical shift anisotropies of protons in hydrocarbons are about 30 ppm (roughly 2 kHz at fields of 1.4 T). Therefore, a rotation velocity of 2.5 kHz would satisfy both of the above conditions and is routinely achieved (Pembleton *et al.*, 1977).

A theoretical treatment of this combined experiment involves a description of both the dipolar and the chemical shift interactions when space and spin portions of the interactions are varying with time.

When the value of θ in the truncated homonuclear dipolar Hamiltonian is allowed to vary with time, $\mathcal{H}_D$ for a two spin interaction may be explicitly written as:

$$\mathcal{H}_D = (\gamma^2/r_{12}^3)b_{12}[\theta(t)][I_1I_2 - 3I_{z1}I_{z2}]. \tag{24}$$

Here (Pembleton, 1978), $b_{ij}(t) = A \cos \omega_r t - B \sin \omega_r t + C \cos 2\omega_r t - D \sin 2\omega_r t$. The A, B, C, and D terms are time independent trigonometric functions depending explicitly on the zero time whose values relate the molecular frame to the laboratory frame. These will change from one multiple pulse cycle to another, because the starting point of one multipulse cycle will not be the same as that for any other (unless the cycle time of the multiple pulse and physical rotation experiments is identical). At $\cos^2 \theta = \frac{1}{3}$, where θ is the angle between the rotation axis and dc field, and when the timing interval, τ, of the MREV-8 multiple pulse experiment (with cycle time $t_c = 12\tau$) is much less than the rotational cycle time, i.e., when $\tau/t_r \ll 1$, then the zero order average dipolar Hamiltonian becomes

$$\bar{H}_D^{(0)} = \gamma^2 r^{-3}(t_r/\tau)[(4A\tau/t_r) + (8C\tau/t_r)](X + Y + Z), \tag{25}$$

where

$$X = \mathbf{I}_1 \cdot \mathbf{I}_2 - 3I_{x1}I_{x2} \quad \text{(etc.)}. \tag{26}$$

But,

$$X + Y + Z \equiv 0, \tag{27}$$

so the zero order average Hamiltonian is zero at multiple pulse cycle times *regardless* of whether a common cycle time is chosen for both rotation and multiple pulse experiments!

To discuss the effect of "magic angle" spinning ($\cos^2 \theta = \frac{1}{3}$) upon chemical shift shielding parameters, it becomes useful to use irreducible spherical tensors (Mehring, 1976). The shielding Hamiltonian for a given nucleus has the form

$$\mathcal{H}_\sigma = \gamma \sum_{l=0,2} \sum_{m=-l}^{l} (-1)^m T_{l,m} \sigma_{l,-m}. \tag{28}$$

Here, $T_{l,m}$ describes the spin variables, and the $\sigma_{l,-m}$'s are the components of a second rank tensor. For nuclei such as ^{1}H and ^{13}C, the chemical shift tensor may be taken to be symmetric (Schneider, 1968). In this case, only the terms $l = 0, 2$ survive in Eq. (28). In the tensor principal axis system (PAS), only components of $\sigma_{l,-m}$ with $m = 0, \pm 2$ are nonzero. We denote these terms by $\rho_{l,m}$. The relation between the PAS values of $\underline{\sigma}$ (denoted by σ_{ii}) and the values of $\rho_{l,m}$ are

$$\rho_{00} = \tfrac{1}{3} \mathrm{Tr}\, \underline{\sigma}, \tag{29a}$$

$$\rho_{2,0} = \sqrt{3/2}(\sigma_{33} - \tfrac{1}{3} \mathrm{Tr}\, \underline{\sigma}) \equiv \sqrt{3/2}\, \delta, \tag{29b}$$

$$\rho_{2,\pm 2} = \tfrac{1}{2}(\sigma_{22} - \sigma_{11}) \equiv \tfrac{1}{2}\eta\, \delta. \tag{29c}$$

The relation between $\rho_{l,m}$ and the values of $\sigma_{l,-m}$ in the laboratory coordinate system (cs) is

$$\sigma_{l,-m}^{\mathrm{lab}} = \sum_{m'} D_{m,-m'}^{l} (\alpha, \beta, \gamma) \rho_{lm'}, \tag{30}$$

where (α, β, γ) are the Euler angles describing the rotations through which the laboratory cs is brought into coincidence with the PAS. Insertion of the appropriate values (Mehring, 1976) of $D_{m,m'}^{l}$ and $T_{l,m}$ from Eq. (30) and Eq. (29) into Eq. (28) yields

$$\mathcal{H}_\sigma = \omega_0 I_z[\tfrac{1}{3} \mathrm{Tr}\, \underline{\sigma} + \tfrac{1}{2}\delta(3 \cos^2 \beta - 1) + \tfrac{1}{2}\eta \sin^2 \beta \cos 2\gamma]. \tag{31}$$

Equation (31) is the expression for the normal powder pattern shielding Hamiltonian.

When the sample is now rotated about an axis at an angle θ from the static field H_0, with an angular speed ω_r, the result (Pembleton, 1978) is

$$\mathcal{H}_\sigma(t) = I_z[A + B e^{-i\omega_r t} + C e^{i\omega_r t} + D e^{-2i\omega_r t} + E e^{2i\omega_r t}] \tag{32}$$

where $A, \ldots, E$ are just trigonometric functions and contain information about the shift anisotropy. The effect of this Hamiltonian upon, e.g., the expectation value of the normalized magnetization observed along the y axis is directly calculated:

$$\langle I_y \rangle = T_r \rho(t) I_y / T_r I_y{}^2, \tag{33}$$

where

$$\rho(t) = U(t)\rho(0)U^{-1}(t), \tag{34}$$

and $U(t)$ is given by the Dyson expression (Wilcox, 1967)

$$U(t) = T \exp\left\{ -i \int_0^t \mathcal{H}(t)dt \right\}. \tag{35}$$

Use of the usual exponential operator formalism (Wilcox, 1967), i.e., of relations such as

$$\exp(il_z\theta)I_y \exp(-il_z\theta) = I_y \cos\theta + I_x \sin\theta \tag{36}$$

and of the angular momentum commutation relations (Slichter, 1963)

$$[I_l, I_m] = i\epsilon_{l,mn}I_n \tag{37}$$

yields, for example, for the contribution of the A and B terms in Eq. (32),

$$\langle I_y \rangle = \cos\omega_0 t(\tfrac{1}{3}\,\mathrm{Tr}\,\underline{\sigma} - \tfrac{1}{2}(3\cos^2\theta - 1)\tfrac{1}{2}\delta\{(3\cos^2\beta - 1)$$
$$+ \eta\sin^2\beta\cos2\gamma\} + \exp(-i\omega_r t)/\omega_r\{(\sigma_{33} - \tfrac{1}{3}\,\mathrm{Tr}\,\underline{\sigma})\tfrac{1}{2}\sin2\theta$$
$$\times (\sqrt{3}/2 D_{0,-1}(\Omega') + \tfrac{1}{2}\eta(D^{2(\Omega')}_{-2,-1} + D^{(\Omega')}_{2,-1}))\}). \tag{38}$$

Here, Ω' represents the Euler angles that bring the principal coordinate system into coincidence with the spinning coordinate system. Equation (38) now exactly determines the meaning of "appropriate" spinning speeds ω_r, and "appropriate" tilt angle θ_r. When

$$\omega_r > (\sigma_{33} - \tfrac{1}{2}\,\mathrm{Tr}\,\underline{\sigma}) \tag{39}$$

and when $\cos^2\theta = \tfrac{1}{3}$, then

$$\langle I_y \rangle = \cos\{\omega_0 t\,\tfrac{1}{3}\,\mathrm{Tr}\,\underline{\sigma}\}. \tag{40}$$

In the rotating frame of the Zeeman Hamiltonian, i.e., the frame in which the NMR signal is detected,

$$\mathcal{H}_\sigma = \tfrac{1}{3}(-\omega_0 I_z)\mathrm{Tr}\,\underline{\sigma} \tag{41}$$

the result for a nonviscous liquid. In addition, under the MREV-8 multiple pulse experiment, the average Hamiltonian of the operator I_z is $(I_x + I_z)$ so under combined MREV-8 and magic angle spinning, with phase detection along y in the rotating frame, the observed chemical shift is the isotropic value, scaled by a factor of $\sqrt{3}/2$, i.e.,

$$\sigma^{\text{obs}} = \sqrt{3}/2 \, \tfrac{1}{3} \, \mathrm{Tr} \, \underline{\sigma}. \tag{42}$$

Again, these results indicate that there is no need to syncronize clocking of the multiple pulse experiment and magic angle spinning period. The results of the CRAMPS experiment upon ^{19}F in polychlorotrifluoroethylene, ^{1}H in 4,4'-dimethylbenzophenone, and ^{1}H in a Virginia vitrain, Pocahontas No. 4, and an Iowa vitrain, Star, are shown in Figs. 5a–d. These results were obtained at a spinning speed of 2.5 kHz and a dc field of 1.4 T. The relatively large range of chemical shifts of ^{19}F allow a clear separation of the two aliphatic fluorines in polychlorotrifluoroethylene. On the other hand, only a clear separation of the aliphatic from the aromatic protons in 4,4'-dimethylbenzophenone is attained (Fig. 5b), and the enormous range of chemical shift dispersion of protons in coals yields an ambiguous signal indeed. At the top of

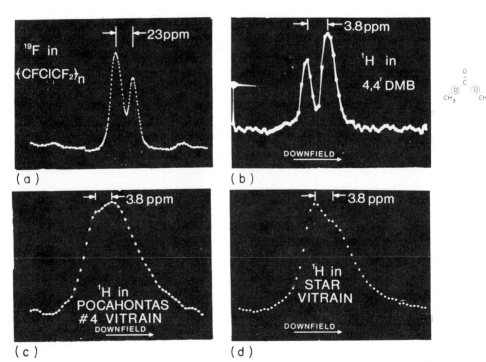

Fig. 5 (a) ^{19}F NMR of polychlorotrifluoroethylene under CRAMPS, (b) ^{1}H NMR of 4,4'-dimethylbenzophenone under CRAMPS, (c) ^{1}H NMR of vitrain under CRAMPS.

the spectra shown in Figs. 5c and 5d are two peaks roughly 4 ppm apart, suggestive of aliphatic and aromatic peak separation, but a unique deconvolution is not easily attained under conditions of the experiment. One method of further separation would be operation at higher fields with appropriately increased spinning rates.

III. CARBON NMR

A. Signal Enhancement by Cross Polarization

Using ideas suggested in 1962 (Hartman and Hahn, 1962), Pines *et al*. (1973) published a seminal paper that has been of enormous value to those wishing to determine natural abundance ^{13}C NMR in hydrocarbons. The idea was immediately adopted (Schaefer *et al*., 1977) and applied to glassy polymers, with the addition of magic angle spinning. Subsequent work on ^{13}C in coals was produced by VanderHart and Retcofsky (1976) (see Volume II, Chapter 24, Section II,C) by Bartuska *et al*. (1977), and by workers in the author's group (*vide infra*).

The cross polarization technique developed by Pines *et al*. (1973) may be schematically represented by the situation depicted in Fig. 6a and b. Figure 6a depicts the idea that a large (high natural isotopic

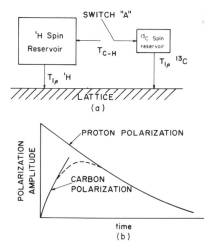

Fig. 6 (a) Spin temperature depiction of cross-polarization experiment. (b) Amplitudes of ^{1}H and ^{13}C magnetization as a function of time; ^{1}H magnetization decays as the proton $T_{1\rho}$. ^{13}C magnetization initially decreases exponentially as T_{C-H}, and then decays with the proton $T_{1\rho}$. $M(t) \cong M_0 \exp[-(t/T_{1\rho})]$.

abundance) proton spin system is cooled (polarized in a dc field, flipped perpendicular to this initial polarization, and locked into this flipped position by an rf locking field). Before application of the transverse locking field, the proton magnetization can decay back to its original longitudinal polarization with time constant T_1 via spin-lattice energy interchange. During application of the transverse locking field, the proton magnetization along this field decays with time constant $T_{1\rho}$ or T_1 in the rotating frame. Application of simultaneous rf irradiation of both ^{13}C and ^1H with rf fields of intensities $H_1(^{13}C)$ and $H_1(^1H)$ satisfying the Hartman–Hahn condition,

$$\gamma(^{13}C)H_1(^{13}C) = \gamma(^1H)H_1(^1H) \tag{43}$$

or

$$\omega_1(^{13}C) = \omega_1(^1H), \tag{44}$$

results in a common oscillating component of the longitudinal magnetization for both nuclear species. Under these conditions, the large, cold ^1H reservoir can cool the small, hotter [less polarized because of lower gryomagnetic ratio $\gamma(^{13}C)$] carbon-13 reservoir with time constant T_{C-H}. The maximum enhancement in ^{13}C polarization thus achieved is the ratio of $\gamma(^1H)/\gamma(^{13}C)$, or 4. As Fig. 6b indicates, however, maximum effectiveness is achieved when $T_{C-H} \ll T_{1\rho}$, and common polarization enhancements are roughly two. The enormous enhancement of signal-to-noise ratio (S/N) in such experiments lies in the fact that the cross polarization may be repeated every five proton T_1, rather than every five ^{13}C T_1. Since proton T_1's in coals are typically tens of milliseconds, whereas ^{13}C T_1's are of order of tens of seconds and the S/N in a time averaging experiment increases as the square root of the number of scans, signal averaging times for given S/N may be decreased by as much as $\times 1000$! The actual experiment consists of a three-step process: (1) flip the polarized ^1H spins from the dc field direction (z) to a transverse direction (e.g., y) in the rotating frame; (2) spin lock the ^1H spins along y, and simultaneously turn on a steady ^{13}C rf resonant field. Typical values of $H_1(^{13}C)$ and $H_1(^1H)$ are 4×10^{-3} and 10^{-3} T, respectively. This condition is maintained for a time of order of T_{C-H} (about 2 msec in practice); (3) after the cross polarization period, the ^{13}C rf field is removed, and the transverse ^1H spin locking rf field is maintained while the response of the ^{13}C spin system is monitored. At the level of presentation appropriate here, it suffices to say that maintenance of a 10^{-3} T (45 kHz) proton rf field is sufficient to decouple the $^{13}C-^1H$ dipolar interaction, which is of order of 30 kHz. The ^{13}C response after turn off of the ^{13}C rf field is therefore characterized by predominantly chemical shift Hamiltonians, and Fourier transform of such a decay leads to the anisotropic ^{13}C chemical shift spectrum. The result of 100 scans at 3 sec between scans on 0.4 g of natural abundance ^{13}C in adamantane is shown in Fig. 7. Rapid molecular tumbling at room temperature in the almost spherical adamantane molecule in the solid state results in considerable averaging of the ^{13}C chemical shift tensor. Peaks associated with both methylene and bridgehead carbons are therefore quite resolved.

The situation for ^{13}C in vitrain portions of coals, however, is quite different. For example, see Volume II, Chapter 24, Fig. 4.

B. Resolution under Strong Heteronuclear Decoupling

As evidenced by short proton T_2^{*}'s (see Section I) vitrains are quite rigid, and the full anisotropy of ^{13}C chemical shift tensors under cross polarization and heteronuclear decoupling appears as shown in Fig. 8.

Assuming an aromatic ^{13}C anisotropic spectrum as inferred from the spectrum of anthracite, and an anisotropic ^{13}C aliphatic spectrum inferred by the difference between lower carbon content coals and anthracite, VanderHart and Retcofsky (1976) were able to infer ^{13}C aromaticities in a number of vitrains. Values of these anisotropies depend upon the method chosen to deconvolute the spectra, however. The use of magic angle spinning to remove anisotropies, as discussed in Section II,D, is capable of removing the ^{13}C chemical shift anisotropies sufficiently to allow full resolution of the aliphatic and aromatic carbons, as shown by Schaefer *et al.* (1977).

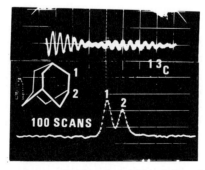

Fig. 7 NMR Spectrum of naturally abundant ^{13}C in 0.4 g of adamantane. Data accumulation time = $2\frac{1}{2}$ min.

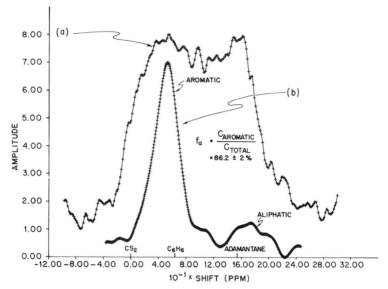

Fig. 8 ^{13}C spectrum of coal utilizing combined cross-polarization, and strong heter-onuclear decoupling (Pocahontas No. 4, Vitrain); (a) without magic angle spinning, 8000 scans, 0.8 g; t_{cp} = 1.5 msec, H_1 = 9.36 G: (b) With magic angle spinning: (3 kH$_2$) 20,000 scans, 0.25 g; t_{cp} = 1.5 msec, H_1 = 10.4, 41.6 G. Al = aliphatic carbon; AR = aromatic carbon.

C.   Resolution under Combined Strong Heteronuclear Decoupling and Magic Angle Spinning. The Question of ^{13}C Aromaticity

Schaefer *et al.* (1977) have published an extensive study of the ap-plication of combined cross polarization, strong heteronuclear decou-pling, and magic angle spinning to resolution of chemically shifted ^{13}C in glassy polymers. Application of these techniques to a sample of whole coal indicated that at minimum, aliphatic and aromatic carbons could be resolved in such samples. An example of the ^{13}C spectrum obtained using these combined techniques is shown in Fig. 8. To insure that data thus obtained are free from artifacts, a number of points must be considered. The first is that the relative effectiveness of the cross polarization experiment be the same for all carbons de-tected. To this end, it is important that the H_1 field be as uniform as possible over the entire sample. As indicated in Section III,A, the importance of the relative cross polarization times, T_{C-H}, for each type of carbon, in relation to the decay of the proton magnetization with damping constant $T_{1\rho}$ must be kept in mind. Ideally one would determine the shape of the ^{13}C spectrum as a function of cross polarization time and arrange this contact time to yield a spectrum invariant to increasing contact time. To enhance sensitivity, however, the minimum carbon–proton contact time in accord with this boundary condition would be used, as the total carbon magnetization decreases with the proton $T_{1\rho}$ at long contact times (cf. Fig. 6). Since this contact time depends upon the H–C dipolar interaction, which is proportional to r^3_{C-H}, it is conceivable that carbons in diamond-like structures in coal would not be seen at all. A comparison of a quantitative analysis for carbon utilizing combustion and spin counting (Gerstein and Pembleton, 1977), e.g., by determination of initial amplitudes of decay, would be helpful in shedding light on possible problems of this nature. That magic angle spinning does not adversely affect T_{C-H} for different types of carbons is also a consideration. A comparison of second moments obtained with and without spinning is helpful in settling this question, as is a comparison of ^{13}C spins counted with and without spinning. Spinning sidebands can seriously distort the spectrum if, e.g., the sideband of the aromatic carbons fell under the aliphatic peak. Again, the second moment of the spinning relative to the nonspinning spectrum should help to determine if this artifact is present. The use of combined cross polarization, strong heteronuclear decoupling, and magic angle spinning have yielded identical values for the carbon aromaticities obtained for Pocahontas No. 4 vitrain in the author's laboratory and by VanderHart (1978) and Bartuska and Maciel (1978). It now appears that at minimum, chemical processes on solid coals may be qualitatively monitored using ^{13}C NMR. Whether or not the results are quantitatively remains a problem in some instances (Volume II, Chapter 24, Section III,C).

IV. THE FUTURE

Almost every atomic specie one could conceive would be important in determining if coal structure and chemistry have an NMR active nuclear isotope. In addition, the nuclear magnetic resonance phenomenon is sensitive to a broad range of interactions. The use of NMR to fingerprint solid coals (and liquid coal products) would therefore appear to be limited only by the imagination. Some future applications that immediately come to mind are: (a) probing the connective (primary) structure in coals, as affected by molecules such as benzene, ammonia, and tetralin via relaxation times that sample motion in the zero frequency (T_2^*) to MHz (T_1) range; (b) probing chemical environments of heteroatoms such as oxygen via selective labeling with a high abundance, NMR active nucleus, e.g., selective labeling of hydroxyl groups by trimethyl silylation, and use of cross polarization, strong heteronuclear decoupling, and magic angle spinning to determine ^{29}Si chemical shifts; (c) direct detection of chemical environments of heteroatoms such as nitrogen via techniques mentioned in (b); (d) use of ^{1}H polarization depletion techniques (Brown, 1978) to detect NMR of heteroatoms such as sulfur, oxygen, and nitrogen as a means of probing chemical environments of these species.

It is clear that the use of NMR to probe the solid state and to probe unusual situations in liquids, e.g., quadrupolar broadened lines (Retcofsky and Friedel, 1972), is in a period of exciting development. Because of their heterogeneity, their broad range of possible chemical structures, and their inherent importance to the world's energy and chemical reserves, it would appear that coals offer one of the most exciting test systems for applications of these developments.

ACKNOWLEDGMENT

The portions of experimental results herein reported that were obtained by the author's research group were supported by the U.S. Department of Energy, Office of Basic Energy Science, Molecular Science Division.

REFERENCES

Adduci, D. J., Hornung, P. A., and Torgeson, D. R. (1976). *Rev. Sci. Instrum.* **47,** 1503–1505.

Bartuska, V. J., and Maciel, G. E. (1978). Personal communication.

Bartuska, V. J., Maciel, G. E., Schaefer, J., and Stejskai, E. O. (1977). *Fuel* **56,** 354–358.

Brown, T. L. (1978). *Exp. Nucl. Magn. Reson. Spectrosc. Conf., 19th, Blacksburgh, Va.*

Dybowski, C., and Pembleton, R. G. (1979). *J. Chem. Phys.* **70,** 1962–1966.

Farrar, T. C., and Becker, E. D. (1971). "Pulse and Fourier Transform NMR," Sect. 5.4. Academic Press, New York.

Gerstein, B. C., and Pembleton, R. G. (1977). *Anal. Chem.* **49,** 75–77.

Gerstein, B. C., Chow, C., Pembleton, R. G., and Wilson, R. C. (1977a). *J. Phys. Chem.* **81,** 565–570.

Gerstein, B. C., Pembleton, R. G., Wilson, R. C., and Ryan, C. M. (1977b). *J. Chem. Phys.* **66,** 361–362.

Haeberlen, U. (1976). "High Resolution NMR in Solids; Selective Averaging." Academic Press, New York.

Hartmann, S. R., and Hahn, E. L. (1962). *Phys. Rev.* **128,** 2042–2053.

Mansfield, P. (1971). *J. Phys. C* **4,** 1444–1452.

Maricq, M., and Waugh, J. S. (1977). *Chem. Phys. Lett.* **47,** 327–329.

Mehring, M. (1976). "High Resolution NMR Spectroscopy in Solids." Springer-Verlag, Berlin and New York.

Oth, J. F. M., and Tschamler, T. (1963). *Fuel* **42,** 467–478.

Pembleton, R. G. (1978). Ph.D. Thesis, Iowa State Univ., Ames.

Pembleton, R. G., Ryan, L. M., and Gerstein, B. C. (1977). *Rev. Sci. Instrum.* **48,** 1286.

Pines, A., Gibby, M. G., and Waugh, J. S. (1973). *J. Chem. Phys.* **59,** 569-595.

Retcofsky, H. L., and Friedel, R. A. (1968). *Fuel* **47,** 391-395.

Retcofsky, H. L., and Friedel, R. A. (1972). *J. Am. Chem. Soc.* **94,** 6579-6584.

Retcofsky, H. L., and Friedel, R. A. (1973). *J. Phys. Chem.* **77,** 68-71.

Rhim, W. K., Elleman, D. D., and Vaughan, R. W. (1973). *J. Chem. Phys.* **58,** 1772-1773.

Ryan, L. M., Wilson, R. C., and Gerstein, B. C. (1977). *Chem. Phys. Lett.* **52,** 341-344.

Schaefer, J., Stejskal, E. O., and Buchdahl, R. O. (1977). *Macromolecules* **10,** 384-405.

Schneider, R. F. (1968). *J. Chem. Phys.* **48,** 4905-4909.

Schweighardt, F. T., Retcofsky, H. L., Friedman, S., and Hough, M. (1978). *Anal. Chem.* **50,** 368-371.

Slichter, C. P. (1963). "Principles of Magnetic Resonance," Chap. 3. Harper, New York.

VanderHart, D. L. (1978). Personal communication.

VanderHart, D. L., and Retcofsky, H. C. (1976). *Fuel* **55,** 202-204.

Vega, A., and Vaughan, R. W. (1978). *J. Chem. Phys.* **68,** 1958-1966.

Wilcox, R. M. (1967). *J. Math. Phys.* **8,** 962-974.

Chapter 52

Differential Thermal Analysis of Coal Minerals

S. St. J. Warne
DEPARTMENT OF GEOLOGY
THE UNIVERSITY OF NEWCASTLE
NEW SOUTH WALES, AUSTRALIA

I. INTRODUCTION

Differential thermal analysis (DTA) under controlled atmosphere conditions provides a method of mineral identification as opposed to chemical constituent analysis.

The technique has the marked advantage of being readily applicable to whole coal samples that do not have to be pretreated in any way to remove organic matter or concentrate the mineral fractions. In this way none of the mineral grains present, no matter how small, are lost or

447

altered in any way, and all contribute to the resultant DTA curves that therefore represent all of the adventitious mineral matter present.

The presence and occurrence of minerals in coal as groups such as carbonates or clays or as specific minerals such as siderite ($FeCO_3$) or kaolinite [$Al_2Si_2O_5(OH)_4$] are of considerable technological and economic importance.

For example, the decomposition temperatures and products of the carbonate group of minerals in general and their interaction with other components present can significantly influence coal combustion reactions, (Estep *et al.*, 1968a), sulfur retention (Ode and Gibson, 1962; Whittingham, 1954), and variations in ash yield and fusion temperatures (Sprunk and O'Donnell, 1942; Stach *et al.*, 1975). Specific minerals such as kaolinite and siderite raise and lower the ash fusion points, respectively, while pyrite (FeS_2) results in excessive undesirable sulfur contents in coke (Stach *et al.*, 1975).

Furthermore, the physical forms in which all these minerals occur are not the same (e.g., ranging from bands, nodules, relic cell lumen, and cleat fillings to idiomorphic crystals), and the mineral suite may vary within a seam and from one seam to another. Thus, the actual minerals present influence the choice of method and the success with which they may be removed from coals by coal washing and preparation processes. (Warne, 1977a).

Further details of the minerals found in coal are included elsewhere (e.g., Volume II, Chapters 26 and 27).

II.   DIFFERENTIAL THERMAL ANALYSIS (DTA)

Differential thermal analysis (DTA) has been defined by the International Confederation for Thermal Analysis (Lombardi, 1977) as "A technique in which *the temperature difference between a substance and a reference material* is measured as a function of temperature while the substance *and the reference are* subjected to a controlled temperature programme." For a further definition of DTA, see Volume II, Chapter 37, Section II,A.

A.   Method

Detailed descriptions of the basic method, effects of variables, and the equipment used to obtain reproducible diagnostic DTA curves of minerals and compounds have been published by Bayliss and Warne

(1962), Garn (1965), Mackenzie (1970), Smykatz-Kloss (1974), and Wendlandt (1974).

Very briefly, the technique involves the simultaneous heating (or cooling) at a constant rate of two samples of equal mass under identical conditions, e.g., actual heating rate, furnace atmosphere, pressure, composition and mobility (static or flowing), sample container type, composition, shape and size, sample size, grain size and sizing, degree of packing, and the type and composition of the thermocouples used. Of the two samples, one is an inert reference material, such as calcined alumina (Al_2O_3) that undergoes no modifications at all during heating. It merely heats up in response to the progressively increasing temperature of the furnace whose temperature it represents. The other is the "unknown" or sample under test.

As these two samples are heated at a constant rate, their individual temperatures are constantly measured electronically by thermocouples situated ideally at their centers. From these signals their temperature difference (ΔT) is continuously plotted against the true temperature of the furnace (T) as measured in the inert reference sample. The result is an automatic print out on an X-Y recorder that plots T on the X (abscissa) axis (increasing from left to right) against ΔT on the Y (ordinate) axis to give a DTA curve of the sample under test.

Any reactions that may occur in the "unknown" sample, e.g., dehydration, decomposition, reaction between components, oxidation, reduction, melting, solidification, crystallographic inversions, and magnetic transformations, all either take in heat (endothermic) or give it out (exothermic). Such reactions are recorded as endothermic or exothermic peaks on DTA curves, the size, shape, temperature, and number of which provide a diagnostic method for mineral identification.

By convention DTA curves are oriented for reference with $+ve$ ΔT values (exothermic) and $-ve$ ΔT values (endothermic) plotted in the directions of the top and bottom of the X-Y record, respectively.

With the publication of the "Scifax" punched card index for DTA data for inorganic and organic compounds (Mackenzie, 1962, 1964) and the multivolume "Atlas of Thermoanalytical Curves" by Liptay (1971–1976) the application of DTA to diagnostic mineralogy is well catered for with respect to reference data. This has resulted in the wide acceptance and routine use of this technique.

It is vitally important that the internationally recognized nomenclature should be adhered to for the detailed description of all aspects of

DTA. This nomenclature is readily available [see Mackenzie *et al.* (1975) and Lombardi (1977)] and represents the rulings of the International Confederation for Thermal Analysis.

B. Special Equipment Aspects

Standard commercially available equipment is suitable, provided provision for adequate furnace atmosphere control has been made and reaction between coal constituents, particularly the sulfides (FeS_2), and thermocouples and certain metallic sample holders does not occur.

1. Controlled Atmosphere DTA

Furnace atmosphere conditions should be classified as primarily static or dynamic (flowing) and by the composition of the gas used, as the difference may exert considerable influence on the DTA peaks that result. A number of different atmosphere *types* may be employed: vacuum, static air, static gas, gas flow over the sample, gas flow through the sample, self-generated, and the employment of positive gas pressures both in the static and dynamic modes.

The furnace gas *composition* may also be varied to promote or inhibit certain reactions, such as oxidation, by using oxygen or nitrogen, respectively. This will enlarge or remove oxidation peaks from the DTA curves of minerals prone to this reaction, thus aiding in the identification of such peaks and solving many problems of peak superposition on the DTA curves of minerals that occur singly (Warne, 1976b) or in mixtures (Bayliss and Warne, 1972).

It should be clearly noted that although oxidation reactions are effectively suppressed, not all "inert" gases affect the resultant DTA curves in a uniform manner as far as other reactions are concerned. The case of carbon dioxide in connection with carbonate minerals in coal is discussed in detail below (e.g., Section V,B).

In addition, the marked decrease of DTA peak size with increasing gas thermal conductivity, which is particularly noticeable in the case of helium as compared to nitrogen, has been described by Warne (1978). This leads to considerable variation in detection limits and content evaluations, dependent upon the actual composition of the purge gas used. For example, the thermal conductivity of the noble gases have a large range from several times less to several times greater than nitrogen, i.e., krypton, argon, and helium have values approximately $\frac{1}{3}$, $\frac{2}{3}$, and 6 times that of nitrogen.

In the present case, where the main purpose is to exclude the active gas oxygen and remove the self-generated coal gas, the method of dynamic gas flow (of an inert gas) over the coal sample has proved very suitable and simple to achieve (Warne, 1961, 1965). It will also alter, but in a predictable way, the DTA curve configurations of certain minerals (Warne, 1976b, 1977b,c).

Ideally, the method of gas flow through the sample could be expected to give the best results, but this has the disadvantage of needing much more sophisticated equipment and may assist in the physical loss of material from the "unknown" sample. In the case of coal, this factor would be compounded with the rapid evolution of coal gas in the middle temperature range.

2. Thermocouple and Sample Holder Reaction

Some minerals are notorious for their ability to combine with, attack, or corrode the metallic materials from which thermocouples and sample holders are commonly made. In so doing they may cause spurious thermal effects because of thermocouple attack and failure and alteration of the size of the sample "well" and the composition of its walls (McLaughlin, 1957; Kopp and Kerr, 1958; Smykatz-Kloss, 1966; Cole and Crook, 1966).

In coal, the only minerals that react in this way are the sulfides, pyrite, and marcasite. Both have the same chemical composition (FeS_2) but belong to different crystal systems. Pyrite is the most common form in coal.

Many methods have been used to combat these adverse effects; inert atmospheres (Hiller and Probsthain, 1956; Warne, 1965), vacuum (Whitehead and Breger, 1950; Levy, 1958), marked dilution with alumina (Sabatier, 1956) or coal (Warne, 1965), ceramic or alumina (Al_2O_3) thermocouple protection (Kopp and Kerr, 1957; Pickering, 1963; Smykatz-Kloss, 1974), special sample holders (Dunne and Kerr, 1960, 1961; Bollin, 1961), and various types of evacuated sealed glass vials in which the thermocouples are not in contact with the samples but protrude into them protected within a glass "dimple" or sleeve (Bollin et al., 1960; Bollin, 1970).

Of these, the simplest and probably the most satisfactory method appears to be a combination of a dynamic inert furnace atmosphere coupled with ceramic sample holders and thermocouples covered with an inert protective material. Thermocouple attack may also be minimized by working only with samples of relatively low sulfide content, i.e., in natural mixtures or by artificial dilution with inert powdered alumina.

Alternatively, platinum sample holders and platinum/platinum–rhodium thermocouples have been recommended by Bollin (1970), but Smykatz-Kloss (1974) states that platinum becomes alloyed with sulfur to form cooperite (PtS).

3. Sample Holder Size

The amount of powdered material required for the unknown sample, in this case, controls the size of the sample holders required.

For mineral identification purposes alone, any sized sample holder recommended by the manufacturers of modern equipment will give suitable and reproducible results. If specific minerals are to be selected out or concentrated, small samples are often easier to obtain. For these, small sample holders give the best DTA results.

If, however, "quantitative" evaluations and comparisons of the mineral contents are to be made, the question of sample representivity arises. From this viewpoint the larger the sample examined at a given grain size, the greater is its theoretical representivity, assuming it has been prepared correctly.

C. Sample Preparation and Analysis Conditions

Coal samples, taking into consideration the demands of representivity, have been found to give good DTA results when reduced to a grain size of -150 mesh(B.S. sieve), heated at a uniform rate of $15°C/min$, under dynamic furnace atmosphere conditions of nitrogen or carbon dioxide,† using flow rates of 2 liters/min for the equipment described for the definitive work published in this field by Warne (1965, 1970, 1975, 1976a, 1977a,b, 1979a). Additional work on the effects of variable atmosphere DTA on the detection limits of anhydrous carbonates, using the most modern equipment, has required gas flow rates of only 100–200 mliter/min (Warne, 1977c, 1978).

III. CHARACTERISTIC DTA CURVES OF MINERALS FOUND IN COAL

A considerable number of different mineral species have been recorded in coal, a detailed list of which appears in Stach et al. (1975). To this should be added, for completeness, witherite ($BaCO_3$) and millerite (NiS), which have been recorded in small amounts by Bethell (1963),

---

† For example, these gases flowed over and not through the samples.

Gibson and Selvig (1944), and Lawrence *et al.* (1960). For a list of 27 common minerals found in coals, see Volume II, Chapter 26, Table I.

Unless otherwise stated, the DTA curves described hereafter were determined under dynamic nitrogen furnace atmosphere conditions with the gas flow over the samples.

However, it is only the minerals listed in Sections III,A–D, that occur in any appreciable quantities in coal and are therefore suitable for identification by DTA.

A. Carbonates—Calcite, Magnesite, Dolomite, Ferroan Dolomite, and Ankerite

The carbonate minerals that commonly occur in coal may be divided conveniently into those whose DTA curves exhibit one major peak (calcite, magnesite, and siderite) and those which show more than one (dolomite, ferroan dolomite, and ankerite).

All these carbonates are most suitable for detection by DTA.

1. Calcite, Magnesite, and Siderite

The DTA curves of calcite ($CaCO_3$), magnesite ($MgCO_3$), and siderite ($FeCO_3$), when determined under inert conditions, are each composed essentially of a single large endothermic peak. These occur with peak temperatures that range between approximately 800°–980°C, 580°–690°C, and 540°–620°C, dependent on the proportion of the individual carbonate present in the sample. Compare Fig. 1 curves 1, 3, 5, 7, 10, and 11; Fig. 2 curves 1, 3, 5, 7, 9, and 11; and Fig. 3 curves 1, 3, 5, 8, 9, and 11.

The three peaks in question are caused by the decomposition reactions

$$CaCO_3 \rightarrow CaO + CO_2 \uparrow$$

$$MgCO_3 \rightarrow MgO + CO_2 \uparrow$$

$$FeCO_3 \rightarrow FeO + CO_2 \uparrow$$

The effects of determining the DTA curves of these three minerals under conditions of flowing *carbon dioxide* have been described in detail by Warne (1976b, 1977a,c, 1978) and Warne and Mitchell (1979b).

Briefly it may be seen that in each case

(a) the peaks become much more sharply attenuated with a commensurate increase in peak height;

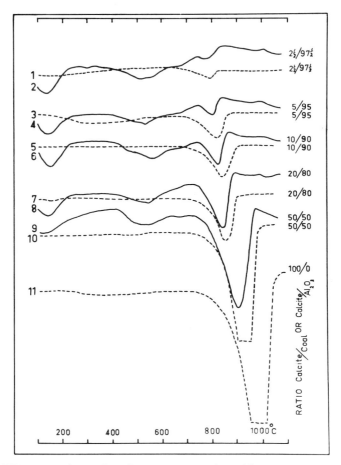

Fig. 1 DTA curves obtained, in flowing nitrogen, from dilution sequences of artificial mixtures of calcite with Al_2O_3 (- - -) or the "reference" coal (———). (After Warne, 1965.)

(b) the individual peaks, in toto, move *up scale* to occur at higher temperatures;

(c) the detection limits are, as a result, considerably increased.

2. Dolomite, Ferroan Dolomite, and Ankerite

A continuous mineralogical series exists (i.e., dolomite–ferroan dolomite–ankerite), such that in the dolomite [$CaMg(CO_3)_2$] lattice Fe may replace Mg in all proportions toward the end member $CaFe(CO_3)_2$.

No natural occurrences of this end member are however known, and
this isomorphous series appears to end at about 70% CaFe(CO$_3$)$_2$ according to Hey (1955).

Under inert conditions the DTA curves of the end members, dolomite
and ankerite, may be recognized by their typical two and three endothermic peaked configurations, respectively (compare curve 9, Fig.
4 with curve 9, Fig.. 5). These peaks occur in the upper (750°–950°C)
temperature bracket, with the three peaks of ankerite spread over a
wider temperature range than the two peaks of dolomite.

Under these conditions, the range of compositions from Fe rich

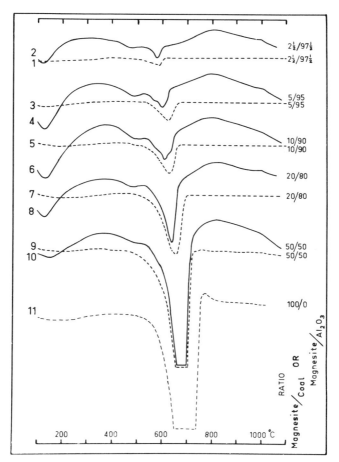

Fig. 2 DTA curves obtained, in flowing nitrogen, from dilution sequences of artificial
mixtures of magnesite with Al$_2$O$_3$ (- - -) or the "reference" coal (———).

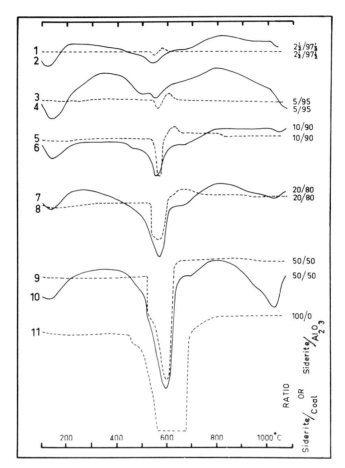

Fig. 3 DTA curves obtained, in flowing nitrogen, from dilution sequences of artificial mixtures of siderite with Al_2O_3 (- - -) or the "reference" coal (———). (After Warne, 1965.)

dolomites to Fe poor ankerites, which constitute the ferroan dolomite members, cannot be recognized.

Furthermore, with the progressive dilution (decrease in carbonate mineral content in natural mixtures) of either of these two minerals, their characteristic double- and triple-peaked curves become progressively less and less well differentiated and eventually "fuse" or "coalesce" (Warne, 1965), compare curves 11, 9, 7, 5, 3, and 2, Fig. 4 and curves 11, 9, 8, 5, 3, and 1, Fig. 5.

For both minerals at contents at or below 10% (Warne, 1975), the resultant single relatively broad endothermic peaks are similar to each other and occur with consistantly different, but close, peak temperatures.

Under conditions of flowing carbon dioxide, however, these two minerals produce quite different DTA curves (Warne, 1975).

(a)   With the exception of the lowest temperature endothermic peak for each, the other peaks of dolomite and ankerite suffer similar modifications to those described above (see Section III, 1, a–c).

(b)   In both cases the initial endothermic peak becomes displaced

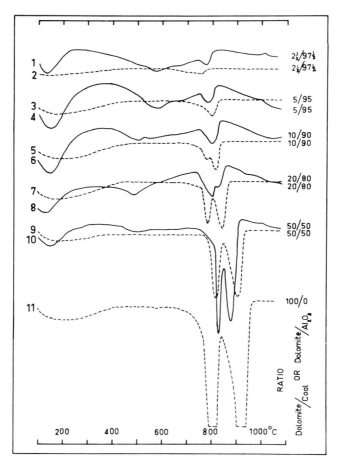

Fig. 4   DTA curves obtained, in flowing nitrogen, from dilution sequences of artificial mixtures of dolomite with Al_2O_3 (– – –) or the "reference" coal (———). (After Warne, 1965.)

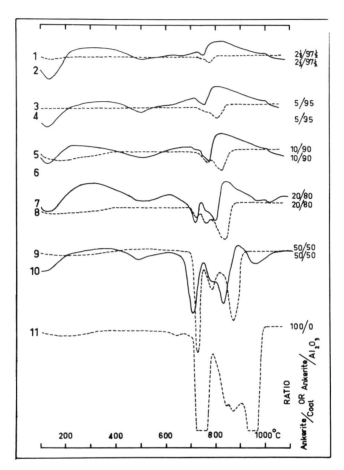

Fig. 5 DTA curves obtained, in flowing nitrogen, from dilution sequences of artificial mixtures of ankerite with Al_2O_3 (- - -) or the "reference" coal (———). (After Warne, 1965.)

down scale to occur with lower peak temperatures and modified peak shape.

(c)   As a result, the typical two and three endothermic peaked configurations showed marked increases in separation and definition.

(d)   Thus, identification accuracy and detection limits are greatly improved.

For example, compare curve 3 with 4 and 5 with 6, Fig. 12.

Furthermore, it has been established (Kulp *et al.*, 1951) that the size of the middle endothermic peak of ankerite is dependent on the iron content. However, in nitrogen (or air) because of the phenomena of peak coalescence this peak is not recorded under conditions of low ankerite sample content or when the actual iron content of the ankerite is low.

In this way members of this mineralogical series that have low iron contents, i.e., the ferroan dolomites, will be missed and classified simply as dolomite, not only by DTA, but by x-ray diffraction (XRD) and routine microscopy. They will, however, not be missed by differential staining (Warne, 1962, Evamy, 1963) or infrared spectroscopy (Farmer and Warne, 1978).

However, by using a furnace atmosphere of flowing carbon dioxide, the greatly improved peak resolution and separation results in the recognition of this middle peak right down to the limits of detection (Warne, 1975).

Thus, in addition to the identification of the end members, dolomite and ankerite, the ferroan dolomites may be recognized and their degree of iron substitution evaluated.

B. Clay Minerals—Kaolinite, Montmorillonite, and Illite

Of the three clay minerals commonly found occurring in coal, kaolinite is the most easily recognized by DTA and has the best detection limits, while illite gives the least satisfactory results.

Furthermore, determinations in flowing carbon dioxide have virtually no additional effects on the resultant DTA curves of these three minerals, as the reactions involved are not influenced by variations in the partial pressure of this gas.

1. *Kaolinite*

The DTA curve of kaolinite $Al_2SiO_5(OH)_4$ is composed of two large peaks, one endothermic and one exothermic, that occur in the temperature ranges 550–620°C and 990°–1010°C, respectively, dependent upon sample content (compare curves 2, 4, 5, 7, and 9, Fig. 6).

These two peaks are caused by dehydroxylation (endothermic), Mackenzie (1970), and crystallization of a spinel phase (exothermic), Smykatz-Kloss (1974). For discussion see Mackenzie (1972).

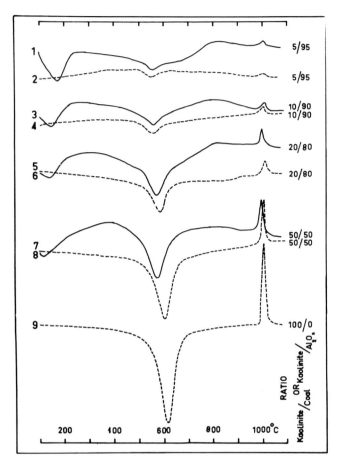

Fig. 6 DTA curves obtained, in flowing nitrogen, from dilution sequences of artificial mixtures of kaolinite with Al_2O_3 (- - -) or the "reference" coal (———). (After Warne, 1965.)

2. Montmorillonite

The DTA curve of montmorillonite[†] is composed of two endothermic peaks followed by a sigmoidal feature. The initial large low-tempera-ture peak is followed by a medium one at 700° to 730°C and the S-shaped endothermic–exothermic complex between 850° and 1020°C (compare curves 2, 4, 5, and 7, Fig. 7).

[†] $R_{0.33} \cdot (Al,Mg)_2 \, Si_4O_{10}(OH)_2 \cdot nH_2O$, where R˙ includes one or more of the cations Na^+, K^+, Mg^{2+}, and Ca^{2+}, (Hey, 1955).

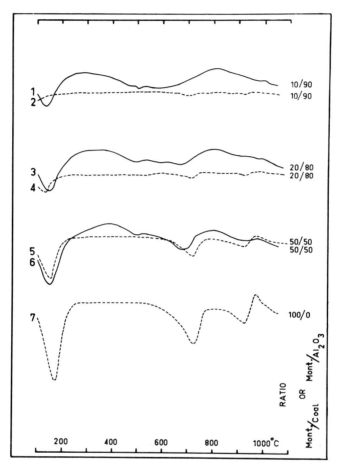

Fig. 7 DTA curves obtained, in flowing nitrogen, from dilution sequences of artificial mixtures of montmorillonite with Al_2O_3 (- - -) or the "reference" coal (———). (After Warne, 1965.)

These peaks are attributed to loss of absorbed water, dehydroxylation, and structural changes, respectively (Mackenzie, 1970).

3. Illite

Illite,‡ exhibits the least distinctive DTA curve of the three clay minerals, with considerably smaller peaks. The curve is characterized by a low-temperature endothermic peak at 100°–150°C because of water

‡ $(H_3O,K)_4Al_8(Si,Al)_{16}O_{40}(OH)_8$, with K about 2–3.

loss, followed by a dehydroxylation endotherm between 500° and 550°C, and finally an S-shaped endothermic-exothermic feature in the 900°-1000°C region. The latter is considered to represent formation of a spinel and is affected by composition according to Muñoz Taboadela and Aleixandre Ferrandis (1957) (compare curves 2, 3, and 6, Fig. 8).

C. Silica Minerals—Quartz, Chalcedony, Agate, and Opal

Warne (1970) has drawn attention to the various forms in which silicon dioxide minerals may be found in coal. These are quartz, chalcedony, agate, and probably the "amorphous" form, opal.

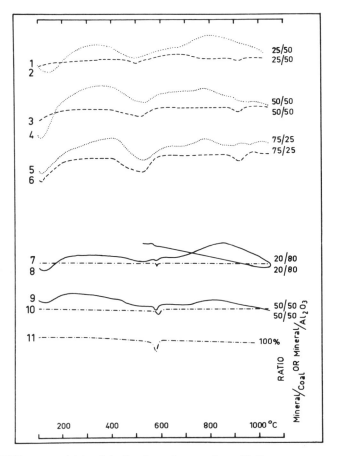

Fig. 8 DTA curves obtained, in flowing nitrogen, from dilution sequences of artificial mixtures of (a) illite with Al_2O_3(- - -) or the "reference" coal (···) and (b) quartz with Al_2O_3 (-·-·-) or the "reference" coal (———). (After Warne, 1965.)

Chalcedony and agate are mineralogically identical, being composed dominantly of cryptocrystalline quartz with subsidiary amounts of hydrated silica (opal). Agate, however, is banded and chalcedony is not. The remaining form, opal, is composed of "amorphous" hydrous silica $SiO_2 n H_2O$.

These forms of silica may occur in coal as detrital grains, cell lumen, and other "void" infillings, veins, lenses, nodules, or actual replacement structures.

The polymorphism of the SiO_2 minerals—quartz, cristobalite, and tridymite is well known, but in the present case it is only the quartz type minerals, listed above, that are of concern.

1. Quartz

Quartz on heating does not decompose, but within the normal DTA temperature range (ambient to 1100°C) suffers a relatively minor internal crystallographic rearrangement from the α to the β form. This inversion is reversible and takes place rapidly to produce a relatively small but sharp endothermic peak.

The DTA curve of quartz therefore shows a small sharp endothermic peak that usually occurs close to 573°C on the heating curve, because of the reaction

$$\alpha \text{ quartz} \rightarrow \beta \text{ quartz}.$$

Because of the reversibility of this reaction a comparable exothermic peak will be recorded on the cooling curve, if this is determined (see curves 1 and 2, Fig. 9). The value of recording such cooling curve peaks is discussed below.

As crystallographic inversions do not involve any chemical reactions or weight changes, they are not affected by changes in furnace atmosphere composition or many of the other variables that may distort and modify the DTA peaks produced by other types of reactions (Bayliss and Warne, 1962).

Thus, this peak temperature determined by a number of workers, with different DTA equipment, under often dissimilar "normal" operating conditions is in close agreement with the usual figure of 573°C.

Because of this peak temperature constancy, quartz finds use as a convenient internal temperature reference (Zuberi and Kopp, 1976), particularly on cooling curves where peak superposition is unlikely as most reactions are not reversible (Warne, 1970). However, some instances of low inversion temperatures have been recorded (Fields, 1952). It appears wise, therefore, to check any quartz specimen material to be used in this way against another, known to invert at the usual temperature.

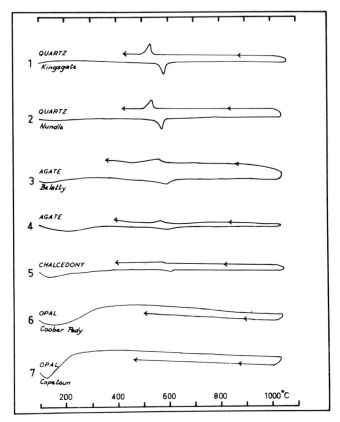

Fig. 9 DTA curves obtained, in static air, from artificial mixtures of samples of natural quartz, agate, chalcedony and opal each diluted with Al_2O_3 in the ratio 1:1. (After Warne, 1970.)

2. *Chalcedony and Agate*

The DTA curves of chalcedony and agate are almost identical to those of quartz with the exception that their α–β inversion peaks occurred at a somewhat higher temperature and were considerably smaller and less clearly defined (Warne, 1970) (see curves 3, 4, and 5 and compare with curves 1 and 2, Fig. 9).

3. *Opal*

In marked contrast are DTA curves of opal that are featureless except for an initial broad endotherm because of low-temperature water loss (see curves 6 and 7, Fig. 9).

D. Sulfides—Pyrite and Marcasite

The sulfide minerals pyrite and marcasite are cubic and orthorhombic dimorphs of FeS_2 (Deer *et al.*, 1962). Under inert conditions, at the concentrations found in coal, each dimorph is represented by virtually indistinguishable single endothermic DTA peaks (Warne, 1965). These apparently represent the same decomposition reaction, although no evidence for a crystallographic inversion reaction has been found on the curve of marcasite (Warne, 1965; Todor, 1976). Perhaps this is because the change is relatively slow (Winchell, 1948) or the inversion occurs at only a little lower temperature (Hiller and Probsthain, 1956) than the decomposition reaction that would by superposition mask it.

As pyrite and marcasite are indistinguishable by DTA they will be further referred to as "pyrite," for which read pyrite or marcasite.

For the following reasons, the determination of the true DTA curve of "pyrite" has proved to be a difficult task and is not as yet fully resolved for all furnace atmosphere and other conditions.

(a) The intense affinity shown for oxygen. Thus, at the decomposition temperature any oxygen present or which becomes available, immediately causes very rapid oxidation that often shows as sharp exothermic peaks because of successive bursts of oxidation (McLaughlin, 1957).

(b) Thermocouple attack or corrosion.

(c) Reactions with metallic sample holders.

For details of these, see above (e.g., Section II,B,1 and 2).

Probably because of a combination of the first two of these factors, the control sequences of "pyrite" (diluted with Al_2O_3) as determined by Warne (1965) show some irregularities (see Warne, 1965, Figs. 9 and 10).

However, in the complementary sequence of "pyrite"/coal mixtures, as a result of the combination of increased thermocouple protection because of decreasing "pyrite" content (dilution) and the considerable evolution of coal gas before, during, and after the "pyrite" decomposition the results were very acceptable.

Under these stringent inert conditions the thermal effects of "pyrite" showed up as a single endothermic decomposition peak, modifying the coal curve, with a peak temperature at approximately 580°C (compare curves 2 and 7 with curve 5, Fig. 10).

Again, because of the type of reaction involved, this diagnostic endothermic peak of "pyrite" is not affected by determinations in carbon dioxide instead of nitrogen.

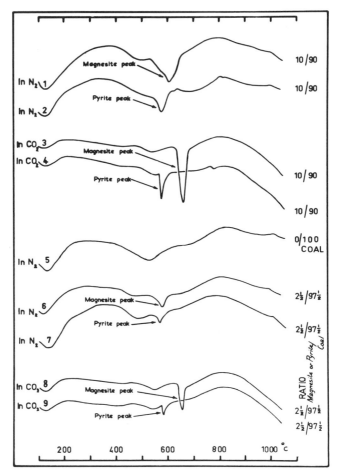

Fig. 10 DTA curves obtained, in flowing nitrogen or carbon dioxide from artificial mixtures of the "reference" coal with various proportions of magnesite or pyrite. (After Warne, 1979a.)

IV. DIFFERENTIAL THERMAL ANALYSIS OF COAL

The DTA of coal and the applications and interpretations of this data have been discussed in detail in Volume II, Chapter 37.

In the present case the object is to suppress preferentially and not to promote or clarify the coal DTA peaks so that the smaller peaks produced by the lesser amounts of inorganic mineral matter present may be recorded, identified, and evaluated.

The DTA curves of coal produced previously (e.g., Klimov, 1953;

Glass, 1954, 1955; Clegg, 1955), were often complex and variable. However, by using a technique of flowing nitrogen (Warne, 1965) the DTA curve of coal was reduced to a relatively featureless curve (see Fig. 11), which no longer swamped the mineral caused peaks, thus allowing for their identification.

V.   DIFFERENTIAL THERMAL ANALYSIS OF COAL/ MINERAL MIXTURES

In any investigation of the application of DTA to the detection and evaluation of the minerals found in coal samples, four points must be considered in connection with the actual coal and DTA unit used.

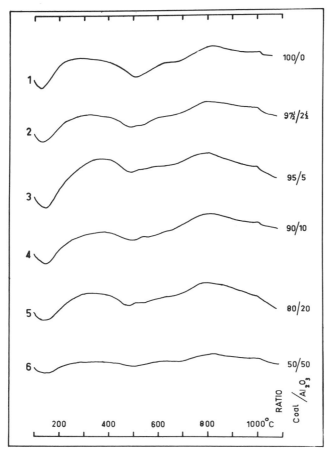

Fig. 11   DTA curves obtained, in flowing nitrogen, from a dilution sequence of the "reference" coal with Al_2O_3 (———). (After Warne, 1965.)

(a)   The effects on the DTA curve configuration of coal as the coal content decreases.

(b)   The effects of decreasing content on the DTA curve of each of the minerals concerned.

(c)   The actual modifications to the coal curve, caused by the presence of various known contents of each mineral.

(d)   The minimum detection limits of each mineral when in mixtures with coal.

These aspects are best dealt with by the production of two sets of reference DTA curves for each mineral, obtained from artificial mixtures of various proportions of each mineral (1) with Al_2O_3 and (2) the coal in question,† both determined under identical furnace atmosphere gas conditions.

A.   In Dynamic Furnace Atmosphere Conditions of Nitrogen

From such sets of "dilution" curves it can be clearly seen that all the minerals calcite, magnesite, siderite, dolomite, ankerite, kaolinite, montmorillonite, illite, quartz, chalcedony, and agate, but *not* opal, cause easily recognizable and characteristic modifications to the coal DTA curve (see Figs. 1-11).

Even with the equipment used by Warne (1965) to determine these curves, the detection limits are in the order of 0.5% for pyrite or marcasite, 1% for calcite, magnesite, dolomite, and ankerite, 2% for siderite and kaolinite, 15% for montmorillonite and quartz, and perhaps as high as 30% for illite.

Further work using modern commercially available equipment with much higher sensitivity has shown that the detection limits may be considerably improved, e.g., in the case of quartz by a factor of at least three (Warne, 1970).†

In the same paper, attention has been drawn to the much greater intensity of the $\alpha-\beta$ peak of quartz, compared to chalcedony or agate (in the order of 50-60%), while no such peak occurs at all for opal.

This author warns further, that the content evaluation of "quartz" in samples by DTA and XRD will produce different results when silica is present, not as quartz but as chalcedony or agate. Even if based on XRD data alone, the contents of chalcedony or agate will appear some-

† Extends to brown coals, Aleksandrov and Kamneva (1976) and Majumdar and Mitra (1974).

† Reduced further to <1% by Rowse and Jepson (1972).

what low because of the opal content, which, according to Deer *et al.* (1962), may be about 10%. Furthermore, if the silica content, as detected chemically, is actually in the form of opal, it will not be detected at all by either DTA or XRD.†

B. In Dynamic Furnace Atmosphere Conditions of Carbon Dioxide

From the DTA curves determined in nitrogen it was concluded that all the minerals could be clearly identified when in mixtures with coal (e.g., Figs. 1–11).

However, the three minerals calcite, dolomite, and ankerite present a special case.

When the contents of these minerals in coal samples are high, their identification is straightforward from their one-, two-, and three-peaked DTA curves, respectively (compare curve 9, Fig. 1, curve 10, Fig. 4, and curve 10, Fig. 5, and curves 2, 4, and 6, Fig. 12). However, because of peak coalescence, contents of 10% or less cause curve modifications consisting in each case of a single endothermic peak of similar appearance. Their peak temperatures, although close, are sufficiently different to be of diagnostic value (compare curves 2, 3, and 4, Fig. 13), although some concern might be felt in relation to their accurate differentiation and identification.

This matter has been alleviated because of the marked effects (see Section III,A,1 and 2) that determinations in carbon dioxide have on the DTA curves of these three minerals (Warne 1975; see also Fig. 12). Of particular importance are

(a)   the preservation of the diagnostic one-, two-, and three-peaked curve configurations without coalescence right to the limits of detection,

(b)   the improved detection limits (compare curve 5 to curve 10, Fig. 13).

Further work in carbon dioxide, using the most modern equipment, has confirmed the much improved detection limits and established the detection down to at least 0.25% contents of the individual common anhydrous carbonate minerals. (Warne, 1977c; Warne and Mitchell, 1979b).

The DTA curves of the other mineral groups, clays, sulfides, and silica minerals, are not affected by determinations in carbon dioxide.

† Further supporting evidence has been put forward by Rowse and Jepson (1972).

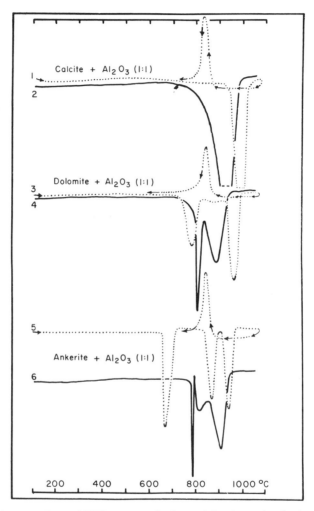

Fig. 12 A comparison of DTA curves of calcite, dolomite and ankerite, diluted 1:1 with Al_2O_3 and obtained in static air (——) or flowing carbon dioxide (···) furnace atmospheres. (After Warne, 1975.)

VI. MINERAL COMPONENT INTERACTION AND PEAK SUPERPOSITION

The possibilities of mineral constituent interactions and the effects of DTA peak superpositions have been investigated in detail and the results presented in Figs. 11–14 of Warne (1965). For the fullest understanding, these curves should be viewed in conjunction with the reading of this section.

The DTA curves of artificial mixtures of carbonates showed no sign of interaction as the effects of all mineral peaks could be observed. Some peak superposition did occur that resulted in peak enlargement, which being atypical was of diagnostic value. These curves also indicate that

(a)   dolomite and ankerite can be distinguished from mixtures containing the same proportions of magnesite + calcite and siderite + magnesite + calcite;

(b)   in mixtures together, the lowest temperature peak of dolomite

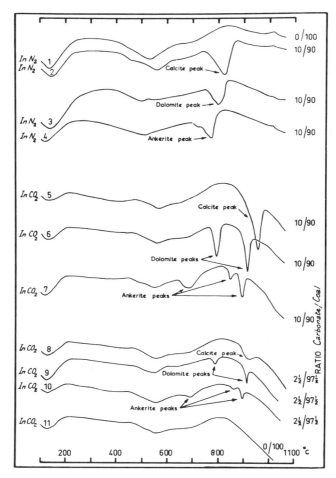

Fig. 13 DTA curves obtained, in flowing nitrogen or carbon dioxide, from artificial mixtures of the "reference" coal and calcite, dolomite and ankerite in various proportions, to illustrate the improved identification and detection which results in carbon dioxide. (After Warne, 1975.)

and ankerite is recorded separately, but the highest temperature peak of each is superimposed to give an abnormally large composite endotherm;

(c) the presence of calcite with dolomite or ankerite is indicated by an abnormal enlargement of their highest temperature peak because of superposition of the calcite peak;

Curves of the three clay minerals in mixtures with siderite (because of the likelihood of peak superposition) and ankerite (because of the release of oxides of Fe, Mg, and Ca) showed further interesting points.

(a) Despite first peak superposition of siderite and illite, the latter is detected by the presence of its highest temperature peak and siderite by the abnormal enlargement of the "600°C" composite endotherm. In flowing carbon dioxide an up scale modification of this peak is indicative of siderite.

(b) Siderite can only be detected within the content limits of 10–40% in mixtures with kaolinite, because of the close, but not perfect, superposition of their "600°C" peaks. Siderite of poorer crystallinity,† will decompose somewhat earlier, which aids peak separation and thus increases its detection limit with kaolinite (Bayliss and Warne, 1972). This problem can, however, be solved by double DTA (e.g., Section VII,A).

(c) Although in high contents, the ankerite peaks mask the presence of montmorillonite (except for its initial low-temperature water loss peak), this problem ceases to exist because of peak coalescence and size reduction as the proportion of ankerite falls to contents typically found in coal.

(d) Peak superposition is unimportant for *dolomite*/montmorillonite mixtures.

The presence of pyrite presents several problems.

(a) A major superposition of the "600°C" endotherms of pyrite and kaolinite occurs with the much smaller pyrite peak being obscured and its presence missed because the small enlargement of the resultant peak is overlooked. The solution lies again with double DTA (e.g., Section VII,A). Conversely the presence of kaolinite is always indicated by its "1000°C" exotherm.

(b) In nitrogen, at the concentrations likely in coal, i.e., <10%, the single endotherm of the following minerals occurs with increasing peak temperatures in the order, siderite, pyrite, and magnesite over a range of approximately 60°C. The peak temperature of siderite is sufficiently

† This is usual for siderite of sedimentary as opposed to hydrothermal origin.

lower than that of pyrite for its identification to be assured, while that of magnesite, although higher than that of pyrite, is perhaps sufficiently close to cause concern regarding its accurate identification.

Clarification is obtained by running a duplicate sample in carbon dioxide that preferentially moves the magnesite peak up scale some 80°C, while the pyrite peak remains unaffected. Thus, in addition to this diagnostic up scale shift, the greatly increased peak temperature difference allows for the accurate identification of these two minerals (Warne, 1979a).†

(c)   In the presence of pyrite, the DTA curves of calcite, dolomite, and ankerite become modified by a considerable size reduction of their highest temperature peaks. This is always followed by a small exothermic fluctuation. Modification intensities increase with larger pyrite contents and appear because of $Fe_2O_3 \cdot CaCO_3$ formation (for discussion, see Warne, 1965).

Reactions between the silica minerals and carbonates to form wollastonite $CaSiO_3$, forsterite Mg_2SiO_4, and fayalite Fe_2SiO_4 must be considered. However, no DTA evidence for such reactions was found in natural coal/mineral mixtures. Warne (1965) further investigated this from artificial mixtures of powdered calcite, magnesite, and siderite with quartz, agate, opal, and fine chemically precipitated SiO_2 in the ratio of 1:1.‡ Again, *no DTA reactions* showed except for the calcite/ precipitated silica mixture, where a diminished calcite peak was followed by an additional small exotherm attributed to wollastonite formation.

For coal, it was concluded that no such reactions occurred because of the "coarse" grain size used and the dilution effects of the coal.

VII.   ADDITIONAL ASPECTS OF SAMPLE PREPARATION AND PRETREATMENT

A.   Sample Preparation—Double Differential Thermal Analysis

In this technique a reactive "unknown" and a reactive reference are compared as an aid to peak identification or to remove the peak of one mineral superimposed on that of another.

† In the same way a siderite peak may be confirmed by its upscale movement in carbon dioxide.

‡ Finer than usual crushing, i.e., to −200 mesh (British Standard Sieve) was used to facilitate reaction.

This method involves adding to the reference an equal amount of one of the minerals in the test sample. Thus the peaks of this mineral will be nullified and disappear from the DTA curve to leave all other peaks free from interference. Also used as an aid to content evaluations, the known content added being the same as in the "unknown" sample when the peaks of a given mineral completely disappear from the resultant DTA curve (Garn, 1965).

B. Sample Pretreatment—Oxidation

Because of their general low mineral content, several methods exist that remove the coal to leave a mineral concentrate. Such methods *must not* alter or attack the minerals in any way.

Chemical oxidation with hydrogen peroxide is useful, but may result in the formation of oxalates, oxalatoaluminates or oxalatoferrates, each of which have their own DTA peaks and not all the organic matter is necessarily destroyed [Mackenzie and Mitchell (1972)]. Nalwalk *et al.* (1974) and Ward (1974) have successfully applied this technique to free the minerals from coal.

The development of commercial equipment for producing radio-frequency excited gases at low pressures (Gleit and Holland, 1962; Gleit, 1963) led to a new method for low-temperature (150°–200°C) oxidation of organic matter, when used with oxygen, Gluskoter (1965) and O'Gorman and Walker (1971).

This technique has been applied with success to the removal of the coal components in mineralogical studies, e.g., by Estep *et al.* (1968b) prior to infrared spectroscopy, Rao and Gluskoter (1973) and O'Gorman and Walker (1971, 1973) prior to XRD, IR, *DTA,* and other thermal analysis determinations. The latter authors have also extended the range of their thermal studies to 1400°C.†

Further developments by Frazer and Belcher (1973) led to its establishment as a routine laboratory method for the quantitative determination of the mineral matter content of coals under standardized conditions.‡ Under these, all minerals were stable including siderite, pyrite, kaolinite, and montmorillonite, and although gypsum dehydrated to the hemihydrate, it is reconverted by overnight exposure to saturated water vapor at room temperature. Detailed discussions of low-temperature ashing and some problems presented by it are given in Volume II, Chapter 20, Section IV,B,2, and Chapter 26, Section II.

---

† The usual maximum temperature for DTA is 1050°–1100°C.
‡ Particularly important is the maximum operating temperature of ≈150°C.

VIII.   CONCLUSIONS

Flowing atmosphere DTA (nitrogen) offers a valuable method for the identification and content evaluation of minerals in coal, particularly as whole coal samples may be used. Detection limits differ between minerals but for most are very satisfactory.

Determinations in flowing carbon dioxide increase the detection limits of carbonates to such a degree (approx. 0.25%) that DTA presents a prime method for their identification, which is further assisted by their preferential up scale movements.

Problems of peak superposition may be solved by determinations in flowing carbon dioxide or double DTA, while unsuitably low mineral contents can be satisfactorily upgraded by radio-frequency oxidation before the determination of DTA curves.

REFERENCES

Aleksandrov, I. V., and Kamneva, A. I. (1976). *Khim. Tverd. Topl.* **2,** 90–94.
Bayliss, P., and Warne, S. St. J. (1962). *Am. Mineral.* **47,** 775–778.
Bayliss, P., and Warne, S. St. J. (1972). *Am. Mineral.* **57,** 960–966.
Bethell, F. V. (1963). *J. Inst. Fuel* **36,** 478–492.
Bollin, E. M. (1961). Ph.D. Thesis, Columbia Univ., New York.
Bollin, E. M. (1970). *In* "Differential Thermal Analysis, Vol. I: Fundamental Aspects" (R. C. Mackenzie, ed.), pp. 193–236. Academic Press, New York.
Bollin, E. M., Dunne, J. A., and Kerr, P. F. (1960). *Science* **131,** 661–662.
Clegg, K. E. (1955). *Ill. State Geol. Surv., Rep. Invest.* No. 190.
Cole, W. F., and Crook, D. N. (1966). *Am. Mineral.* **51,** 499–502.
Deer, W. A., Howie, R. A., and Zussman, J. (1962). "Rock Forming Minerals. Vol. 5: Silicates." Longmans, London.
Dunne, J. A., and Kerr, P. F. (1960). *Am. Mineral.* **45,** 881–883.
Dunne, J. A., and Kerr, P. F. (1961). *Am. Mineral.* **46,** 1–11.
Estep, P. A., Kovach, J. J., Hiser, A. L., and Karr, C., Jr. (1968a). *In* "Spectrometry of Fuels" (R. A. Friedel, ed.), pp. 228–247. Plenum, New York.
Estep, P. A., Kovach, J. J., and Karr, C., Jr. (1968b). *Anal. Chem.* **40,** 358–363.
Evamy, B. D. (1963). *Sedimentology* **2,** 164–170.
Farmer, V. C., and Warne, S. St. J. (1978). *Am. Mineral.* **63,** 779–781.
Fields, M. (1952). *Nature (London)* **170,** 366–367.
Frazer, F. W., and Belcher, C. B. (1973). *Fuel* **52,** 41–46.
Garn, P. D. (1965). "Thermoanalytical Methods of Investigation." Academic Press, New York.
Gibson, F. H., and Selvig, W. A. (1944). *U.S. Bur. Mines, Tech. Pap.* No. 669.
Glass, H. D. (1954). *Econ. Geol.* **49,** 294–309.
Glass, H. D. (1955). *Fuel* **34,** 253–268.
Gleit, C. E. (1963). *Am. J. Med. Electron.* **2,** 112.
Gleit, C. E., and Holland, W. D. (1962). *Anal. Chem.* **34,** 1454–1457.
Gluskoter, H. J. (1965). *Fuel* **44,** 285–291.
Hey, M. H. (1955). "An Index of Mineral Species and Varieties." Br. Mus., London.
Hiller, J. E., and Probsthain, K. (1956). *Geologie* **5,** 607–616.

Klimov, B. K. (1953). "Khimia i Genesis Tverdych Gorjučich Iskopaemych," p. 235. Izd. Akad. Nauk SSSR, Moscow.

Kopp, O. C., and Kerr, P. F. (1957). Am. Mineral. 42, 445–454.

Kopp, O. C., and Kerr, P. F. (1958). Am. Mineral. 43, 1079–1097.

Kulp, J. L., Kent, P., and Kerr, P. F. (1951). Am. Mineral. 36, 643–670.

Lawrence, L. J., Warne, S. St. J., and Booker, M. (1960). Aust. J. Sci. 23, 87–88.

Levy, C. (1958). Bull. Soc. Fr. Mineral. Cristallogr. 81, 29–34.

Liptay, G., ed. (1971–1976). "Atlas of Thermoanalytical Curves," Vols. 1–5. Heyden, London.

Lombardi, G. (1977). "For Better Thermal Analysis." Inst. Min. Pet. Uni., Rome.

Mackenzie, R. C., comp. (1962). "'Scifax' Differential Thermal Analysis Data Index." Cleaver-Hume, London.

Mackenzie, R. C., comp. (1964). "'Scifax' Differential Thermal Analysis Data Index," First Supplement. Macmillan, New York.

Mackenzie, R. C., ed. (1970). "Differential Thermal Analysis. Vol. 1: Fundamental Aspects." Academic Press, New York.

Mackenzie, R. C., ed. (1972). "Differential Thermal Analysis. Vol. 2: Applications." Academic Press, New York.

Mackenzie, R. C., and Mitchell, B. D. (1972). In "Differential Thermal Analysis. Vol 2: Applications" (R. C. Mackenzie, ed.), pp. 267–297. Academic Press, New York.

Mackenzie, R. C., Keattch, C. J., Daniels, T., Dollimore, D., Forrester, J. A., Redfern, J. P., and Sharp, J. H. (1975). J. Therm. Anal. 8, 197–199.

McLaughlin, R. J. W. (1957). In "The Differential Thermal Investigation of Clays" (R. C. Mackenzie, ed.), pp. 364–379. Mineral. Soc., London.

Majumdar, S. K., and Mitra, B. (1974). J. Mines, Met. Fuels 22, 202–208.

Muñoz Taboadela, M., and Aleixandre Ferrandis, V. (1957). In "The Differential Thermal Investigation of Clays" (R. C. Mackenzie, ed.), pp. 165–190. Mineral. Soc., London.

Nalwalk, A. J., Friedel, R. A., and Queiser, J. A. (1974). Energy Sources 1, 179–187.

Ode, W. H., and Gibson, F. H. (1962). U.S. Bur. Mines, Rep. Invest. No. 5931.

O'Gorman, J. V., and Walker, P. L. (1971). Fuel 50, 135–151.

O'Gorman, J. V., and Walker, P. L. (1973). Fuel 53, 71–79.

Pickering, R. J. (1963). Am. Mineral. 48, 1383–1388.

Rao, C. P., and Gluskoter, H. J. (1973). Ill. State Geol. Surv., Circ. No. 476.

Rowse, J. B., and Jepson, W. B. (1972). J. Therm. Anal. 4, 169–175.

Sabatier, G. (1956). Bull. Soc. Fr. Mineral. Cristallogr. 79, 172–174.

Smykatz-Kloss, W. (1966). Contrib. Mineral. Petrol. 13, 207–231.

Smykatz-Kloss, W. (1974). "Differential Thermal Analysis—Application and Results in Mineralogy." Springer-Verlag, Berlin and New York.

Sprunk, G. C., and O'Donnell, H. J. O. (1942). U.S. Bur. Mines, Tech. Pap. No. 648.

Stach, E., Mackowsky, M.-Th., Teichmüller, M., Taylor, G. H., Chandra, D., and Teichmüller, R. (1975). "Stach's Textbook of Coal Petrology," 2nd Ed. Borntraeger, Berlin.

Todor, D. N. (1976). "Thermal Analysis of Minerals," Abacus Press, Tunbridge Wells, England.

Ward, C. R. (1974). Fuel 53, 220–221.

Warne, S. St. J. (1961). Bull. Soc. Fr. Mineral. Cristallogr. 84, 234–237.

Warne, S. St. J. (1962). J. Sediment. Petrol. 32, 29–39.

Warne, S. St. J. (1965). J. Inst. Fuel 38, 207–217.

Warne, S. St. J. (1970). J. Inst. Fuel 43, 240–242.

Warne, S. St. J. (1975). J. Inst. Fuel 48, 142–145.

Warne, S. St. J. (1976a). *Proc. Eur. Symp. Therm. Anal., 1st, Salford, Eng.* pp. 359–360.
Warne, S. St. J. (1976b). *Chem. Erde* **35,** 251–255.
Warne, S. St. J. (1977a). *Proc. Int. Confed. Therm. Anal. Conf., 5th, Kyoto,* pp. 460–463.
Warne, S. St. J. (1977b). *Netsusokutei (J. Calor. Therm. Anal. Jpn.)* **4,** 105–106.
Warne, S. St. J. (1977c). *Nature (London)* **269,** 678.
Warne, S. St. J. (1978). *J. Therm. Anal.* **14,** 325–330.
Warne, S. St. J. (1979a). *J. Inst. Fuel* **52,** 21–22.
Warne, S. St. J., and Mitchell, B. D. (1979b). *J. Soil. Sci.* **30,** 111–116.
Wendlandt, W. W. (1974). "Thermal Methods of Analysis," 2nd Ed. Wiley, New York.
Whitehead, W. L., and Breger, I. A. (1950). *Science* **111,** 279–281.
Whittingham, G. (1954). *Br. Coal Util. Res. Assoc., Mon. Bull.* **18,** 581–590.
Winchell, A. N. (1948). "Elements of Optical Mineralogy." Wiley, New York.
Zuberi, Z. H., and Kopp, O. C. (1976). *Am. Mineral.* **61,** 281–286.

Part XII

BY-PRODUCT UTILIZATION AND MISCELLANEOUS PROBLEMS

Chapter 53

Prediction of Ash Melting Behavior from Coal Ash Composition

Karl S. Vorres
INSTITUTE OF GAS TECHNOLOGY
CHICAGO, ILLINOIS

I. INTRODUCTION

The inorganic residue or ash associated with coal usually passes through a coal utilization process with the fuel or is separated during a combustion step and is largely removed. The ash experiences a range of high temperatures and may, depending on the ash properties, result in troublesome deposits in different parts of the coal utilization or conversion equipment. The design of this equipment will involve consideration of the melting or softening properties of the coal ash through the temperature range of the process (Babcock and Wilcox, 1978). Since coal ash composition varies considerably, and the properties of the ash vary with composition, the melting properties also vary considerably. Experimental determinations of melting, and especially of viscosity, are time consuming and expensive. Some method of reasonably accurate prediction of melting and viscosity behavior would be of value to engineers in the design of coal-consuming equipment.

481

II. MELTING AND SOFTENING BEHAVIOR

Coal ash is a mixture of many materials and would not be expected to exhibit a sharp melting point that is characteristic of a pure compound. The melting behavior of coal ash is characterized by several temperatures, such as initial deformation, softening, hemispherical, and fluid that refer to the progressive changes in a conical sample as it undergoes heating in a furnace (ASTM, 1977). Fusibility of ash and the critical temperatures are discussed in Volume I, Chapter 6, Section XI.

The viscosity of the molten ash is also useful in characterization. The changes in viscosity with temperature vary over a wide range (Corey, 1964). Newtonian and non-Newtonian behavior have been observed. Changes in viscosity also are effected by variations in the gaseous environment in contact with the melt. This is especially noticeable when polyvalent ions, such as iron, are present in substantial concentrations (Babcock and Wilcox, 1978). Further, variation in viscosity with time has been observed with some melts.

The plastic range between the solid and freely flowing liquid state is of special concern since this will affect the accumulation of sticky deposits on surfaces within the coal processing equipment.

A conceptual framework is needed to facilitate the understanding of the observed phenomena in a consistent manner and to provide the basis for prediction of behavior of different compositions through a range of temperatures and gaseous environments, and over periods of time as well. The purpose of this chapter is to suggest the use of a concept involving the structural chemistry of the species that make up the melt so that the concept can be applied where appropriate.

III. ACIDS AND BASES

The melting and softening behavior of coal ash has been described in terms of melting, viscosity, fouling, and slagging. The latter terms refer to the accumulation of ash material on tubes or on walls of boilers. Correlations of the behavior of coal ash materials with regard to these aspects of melting behavior have been made with a number of ash composition parameters, including silica ratio, dolomite percentage, ferric percentage, sodium content, base content, and base-to-acid ratio (Sage and McIlroy, 1960; Rees, 1964; Winegartner, 1974; Bryers and Taylor, 1975; Winegartner and Rhodes, 1975). The ability to develop a correlation is useful from a design standpoint but should be extended to an understanding of the underlying phenomenon to provide reliable predictions.

The concept of acids and bases should prove useful in this regard. In earlier correlations, acids were defined as oxides of Al, Si, and Ti, while bases were defined as oxides of Na, K, Ca, Mg, and Fe (Winegartner, 1974). Other species were not defined, nor were oxidation states involved in the definition, other than to refer to the form of the oxide commonly reported in analysis.

Behavior as an acid or base is related to the structural inorganic chemical characteristics of the ions. The inorganic cations may be ranked according to their size or ionic radii for their usual valences. In general, ionic radii decrease with increasing charge and also decrease with increasing atomic weight for a given valence.

Values for the cation radii selected from the set given by Ahrens (1952) for the metals indicated above are listed below in angstrom units

Si^{4+}	0.42	Fe^{2+}	0.74
Al^{3+}	0.51	Na^+	0.94
Fe^{3+}	0.64	Ca^{2+}	0.99
Mg^{2+}	0.67	K^+	1.33
Ti^{4+}	0.68		

Two values are given for iron because each of the oxidation states is important, and to begin to indicate reasons for the special behavior of iron compounds in different gaseous environments.

In coal ash the predominant anion is the oxide ion. The radius of 1.40 Å is usually assigned to that ion, which indicates that it acts like a larger sphere than any of the indicated cations. In ionic crystals there tends to be a closely packed array of the electronegative anions. The cations then fit in the interstices between the anions in a repetitive array. The cations are surrounded by a specific number of anions, depending on the type of interstice they occupy. The coordination number, or number of adjacent anions, is determined by the ratio of the radii for the two oppositely charged species. The radius ratio of 0.225–0.414 results in a coordination number of four and involves the shape of a tetrahedron for the four oxide ions. Radius ratios of 0.414–0.732 give a coordination number of six and produce an octahedron of oxide ions. A coordination number of eight is involved for ions with somewhat larger radius ratios.

Referring to the list of cation radii and using the value of 1.40 Å for the oxide ion, a set of radius ratios can be calculated. From this set, tetrahedral coordination would be expected for Si^{4+} and Al^{3+}. The border between tetrahedral and octahedral coordination on the basis of geometry would be expected at 0.58 Å. The cations Fe^{3+}, Mg^{2+}, Ti^{4+},

Fe^{2+}, Na^+, and Ca^{2+} have values in the range for octahedral configuration. The border between octahedral and cubic coordination comes at 1.03 Å and, K^+ would be expected to exhibit a coordination number of eight.

IV. IONIC POTENTIAL

Another structural concept is that of the ionic potential. This potential is defined as the quotient of the valence and ionic radius for a given cation. This parameter indicates something about the relative ability of a cation to compete with others to coordinate anions around it. A high value of the ionic potential indicates a capacity to compete effectively with other cations for oxide ions to form some coordinated complex. This would be expected to be more significant in a mobile liquid state than in the fixed solid state. The values calculated from the cation radii indicated earlier and the oxidation states of species of interest are listed below:

Si^{4+}	9.5	Fe^{2+}	2.7
Al^{3+}	5.9	Ca^{2+}	2.0
Ti^{4+}	5.9	Na^+	1.1
Fe^{3+}	4.7	K^+	0.75
Mg^{2+}	3.0		

The highest values belong to the acid group of Si, Al, and Ti, while the lower values are associated with bases. This association leads to the suggestion that the ionic potential may be the physical characteristic that can be useful in anticipating and quantifying acid and base behavior. This parameter should also be useful in further efforts to correlate chemical composition with melting and softening behavior and viscosity.

The ionic potential is an indication of a cation's capacity to attract anions to form a complex ion of the general formula MO_x^{n-}, such as SiO_4^{4-}. The formation of these ions is also dependent on the availability of anions, and specifically in coal ash systems, oxide ions. A compound such as SiO_2 cannot provide all of the four oxide ions needed by the Si for tetrahedral coordination in a liquid unless there was a mechanism for sharing as there is in a solid. The sharing of oxide ions between different Si ions in the liquid state is apparently possible to form polymer groupings. Individual V-shaped SiO_2 groups may provide for chains, sheets, and other arrangements. Long chains would require additional oxide ions for termination. Oxides of cations with lower ionic potential than Si could provide the oxide ions. As more oxide ions became available, chains could terminate more frequently or the polymer average size would become smaller. The oxide ions would be most available from the oxides of cations with the lowest ionic potentials such as the alkali metals. These oxides would be expected to dissociate into hard sphere ions. The oxide ions would then be attracted to and coordinated by the cations of highest ionic potential.

The role of acids in coal ash melts, then, would be that of polymer formers. The greatest tendency to form polymers would be observed for ions with the largest ionic potential. The bases serve as oxide ion donors. The oxide ions would be attracted by ions of the highest ionic potential in the system to break up polymers and reduce viscosity. The remaining cations of low ionic potential would remain as unattached hard spheres and could facilitate slippage between polymer aggregates.

Therefore, in mixed oxide systems, such as coal ash, the available oxide ions from cations of low ionic potential would tend to reduce the size of polymeric groups associated with cations of high ionic potential. Addition of material that would increase the available oxide ion concentration would decrease the viscosity of a melt consisting primarily of Si and Al that is characteristic of eastern coal ash compositions.

V.   THE BEHAVIOR OF IRON

The ferrous and ferric ions have significantly different ionic potentials. Ferric iron has a value between those of the acid group and of the base group. The special behavior and importance of iron appears to be because of the different behavior associated with each valence state and the ability of iron to transfer electrons to alternate between these valence states and act as either an acid or base. This may be another form of amphoteric behavior that has been observed in other species. Considering the range of ionic potentials in the above listing, the ferric iron may be considered a weak acid and the ferrous ion a base. In combustion the iron in coal ash, as in boilers, will exist as a mixture of the two states.[†] A small part, frequently around 20%, is ferric iron. The remainder is ferrous iron, although under some extreme conditions elemental iron may form. The preponderance of the iron in the ferrous state indicates the appropriateness of classifying iron oxides (assuming they must be classed as one type) with the bases, even though they are identified as Fe_2O_3.

The gaseous environment is significant in determining the role of

[†] As an example, Figs. 1 and 4 in Chapter 42 show crystals of magnetite and hematite, respectively, from Lurgi ash.

iron in the coal ash systems. Since iron is an important component in eastern coal ash, usually following Si and Al in abundance, a change in the iron also affects the whole system. Studies have shown a marked reduction of the viscosity of melts in going from oxidizing to reducing conditions that do significantly alter the proportion of ferric and ferrous ions. This change in properties would be consistent with a complex ion forming tendency for the ferric ion with its higher ionic potential and an oxide ion donor role for the ferrous ion with its lower ionic potential. The reduction of viscosity in a reducing environment is because of two factors: (1) reduced concentration of polymer-forming ferric ions and (2) increased oxide ion concentration to terminate polymer groups around cations with high ionic potential. The oxide ions would be released by the ferrous ions.

VI.   CORRELATION INDICATING BASE BEHAVIOR

Winegartner and Rhodes (1975) correlated ash fusion temperatures with coal ash composition using equations involving the base-to-acid ratio in terms of the individual metal oxide species. They indicated that the coefficients in front of the bases CaO, MgO, K_2O, and Na_2O should each be one when the ash composition was given in *mole* percentages rather than weight percentages as had customarily been done. The implication of this finding is that these bases are equally effective as oxide ion donors. It is worth emphasizing that while the number of ions or cations may vary in these oxides, the number of oxide ions remains constant. They also indicated that Fe_2O_3 should really be expressed as FeO. This treatment is consistent with characterizing FeO as a base and indicating the role of the oxide ion donor for FeO.

VII.   CONCEPTUAL STRUCTURAL CONSIDERATION

The melting process involves the utilization of thermal energy to disrupt the crystal lattice of a solid. The lattice may be strained by insertion of impurities with the result that the thermal requirement for lattice disruption is reduced. Addition of a different metal oxide to an existing oxide lattice would extend the oxide lattice, while the cations would tend to distribute themselves on sites with an appropriate co-ordination number. The further implication of Winegartner and Rhodes' work is that the number of oxide ions is most significant and that the nature of the base cations is not significant in affecting the softening temperatures.

VIII.   SUMMARY

The ionic potential (valence divided by ionic radius) may be used (1) for classification of metals into acids and bases, (2) for ranking of strength in that classification, and (3) for prediction of coal ash melting behavior and viscosity from the chemical composition. The role of an acid is that of a complex ion former with anions provided by bases that in coal ash systems are oxide ion donors. For systems containing acids and limited oxide ion concentrations, polymers form in the melts. The addition of bases would reduce polymer size and would decrease viscosity. The gaseous environment may alter the relative concentration of ferric and ferrous ions, and change acid to base ions, and vice versa. Bases are reported to be equally effective oxide ion donors. The effects of base addition are proportional to the amount of oxide ion provided to the system.

REFERENCES

Ahrens, L. H. (1952). *Geochim. Cosmochim. Acta* **2,** 155.
ASTM (1977). "1977 Annual Book of ASTM Standards. Part 26: Gaseous Fuels; Coal and Coke; Atmospheric Analysis," pp. 266–271. ASTM, Philadelphia, Pennsylvania.
Babcock and Wilcox (1978). "Steam—Its Generation and Use," Chap. 15. Babcock and Wilcox, New York.
Bryers, R. W., and Taylor, T. E. (1975). *Am. Soc. Mech. Eng., Pap.* 75-WA/CD-3.
Corey, R. C. (1964). *U.S. Bur. Mines, Bull.* No. 618.
Rees, O. W. (1964). *Ill. State Geol. Surv., Circ.* No. 365.
Sage, W. L., and McIlroy, J. B. (1960). *J. Eng. Power* **82,** 145, 155.
Winegartner, E. C., ed. (1974). "Coal Fouling and Slagging Parameters," ASME Res. Comm. Corros. Deposits Combust. Gases. Am. Soc. Mech. Eng., New York.
Winegartner, E. C., and Rhodes, B. T. (1975). *J. Eng. Power* **97,** 395–406.

Chapter 54

Size Dependence of the Physical and Chemical Properties of Coal Fly Ash

G. L. Fisher

D. F. S. Natusch

RADIOBIOLOGY LABORATORY
UNIVERSITY OF CALIFORNIA
DAVIS, CALIFORNIA

DEPARTMENT OF CHEMISTRY
COLORADO STATE UNIVERSITY
FORT COLLINS, COLORADO

I. INTRODUCTION

In order to assess the environmental significance and potential health hazards associated with exposure to environmental pollutants, detailed studies of physical and chemical properties are required. It is these properties that determine the route and biological consequences of exposure. The aerodynamic behavior of aerosols released during coal

489

combustion will determine the potential for atmospheric transport and subsequent human exposure. Large particles (>10 μm) escaping the power plant's control technology will fall out near the plant, which may ultimately result in general population exposure by ingestion of agricultural products or water. Thus, exposure of agricultural products by soil or foliar deposition or contamination of water sources in the environs of the power plant will reflect, for the most part, the chemical composition of the larger particles. Long-range transport and general population exposure will be associated with the more stable aerosols. These fine particles (<10 μm) are of special interest because they are less efficiently collected by existing control technologies, have a relatively long atmospheric residence time, and upon inhalation, are efficiently deposited and slowly removed from the pulmonary region of the respiratory tract.

In a review of particulate abatement technologies, Vandegrift *et al.* (1973) described collection efficiency as a function of particle size for a variety of control technologies including electrostatic precipitators, fabric filters, wet scrubbers, and cyclones. Average collection efficiencies for a medium-efficiency electrostatic precipitator (ESP) were 90, 70, and 35% for 1.0, 0.1, and 0.01 μm particles, respectively. Interestingly, the Venturi wet scrubber (VWS) was more efficient (99.5%) for 1.0 μm particles and less efficient ($<1\%$) for 0.01 μm particles. A crossover in the ESP- and VWS-efficiency curves was observed at 0.35 μm.

Respiratory tract deposition of inhaled particles is determined by the physics and chemistry of aerosols, the anatomy of the respiratory tract, and the airflow patterns in the lung airways (Yeh *et al.*, 1976).† The most important physical factors affecting lung deposition of inhaled particles are the aerodynamic properties of the aerosol and the chemical reactivity in the airways. Lung deposition is generally described in terms of fractional particulate deposition by mass or number in the three major regions of the respiratory tract: the nasopharyngeal, tracheobronchial, and pulmonary regions (Task Group on Lung Dynamics, 1966). The nasopharyngeal region is composed of the nose and throat, extending to the larynx; the tracheobronchial region consists of the trachea and bronchial tree, including the terminal bronchioles; and the pulmonary region consists of the respiratory bronchioles and the alveolar structures. Particles greater than 10 μm are effectively collected in the nasopharyngeal region; tracheobronchial and pulmonary deposition generally increase with decreasing particle size. Fractional deposition in the pulmonary region ranges from 30 to 60% of the inhaled

† Coal workers' pneumoconiosis and analysis of miners' lung tissue are discussed in Volume II, Chapter 28.

aerosol for particles ranging in size from 1.0 to 0.01 μm (Task Group on Lung Dynamics, 1966). Similarly, tracheobronchial deposition ranges from 5 to 30% for inhaled aerosols from 1.0 to 0.01 μm, respectively. Respiratory tract deposition profiles have been calculated for iron, lead, and benzo(a)pyrene in urban aerosols (Natusch and Wallace, 1974). The hygroscopicity or reactivity of an aerosol in the airways may dramatically alter the particle size and hence the regional deposition. Parks *et al.* (1977) have shown that, upon inhalation, ammonium sulfate aerosols with initial aerodynamic diameters of 0.8 μm at 8% relative humidity may rapidly grow to 2.3 μm in the water vapor saturated atmosphere of the respiratory tract. The rapid growth of the aerosols resulted in deposition predominantly in the nasopharyngeal region and lower than expected deposition in the tracheobronchial and pulmonary regions.

The rate of clearance of deposited particulate matter from the respiratory tract will be determined, in part, by the chemical behavior in the lung's unique microenvironment in the vicinity of the particle. Hygroscopic particles deposited in the respiratory tract will be rapidly cleared by dissolution and subsequent passage into the bloodstream for ultimate exposure of internal organs. Less soluble particles deposited on the mucocilliary escalator of the tracheobronchial region and on the ciliated epithelium of the nasopharyngeal region will be rapidly cleared with half-times on the order of one day and a few minutes, respectively (Task Group on Lung Dynamics, 1966). Relatively insoluble particles deposited in the pulmonary region will be phagocytized by the pulmonary alveolar macrophages (PAM). These particles will be slowly removed by either dissolution within PAM or transport within PAM to the mucocilliary escalator. The biological half-time of material in the pulmonary region is very much a function of particulate chemical composition; half-times of hundreds of days have been reported for insoluble particles.

It should be emphasized, however, that dissolution of surface-associated chemical components need not be a requisite for their interaction with the biological system. For example, inhaled particles may be phagocytized by macrophages where direct particle surface-cell interaction will take place. A reasonable comparison of "insoluble" particle interaction may be made with asbestos.

In this chapter, the size dependence of physical and chemical properties of coal fly ash is reviewed. Because the size dependence of many of the chemical properties results from surface-associated chemical phenomena, a detailed description of surface analysis is provided. An understanding of the bioenvironmental significance of ambient fly ash requires a detailed understanding of its chemical reactivity and biological interactions with fly ash surfaces. This chapter reproduces the material found in a DOE report published through NTIS (Fisher and Natusch, 1979).

II. MORPHOLOGY AND FORMATION OF COAL FLY ASH

A. Morphological Analysis

Morphological studies by light and electron microscopy have described the heterogeneity and structural complexity of coal fly ash. Based on morphological appearance, much can be inferred concerning origin, formation, and chemical composition. McCrone and Delly (1973) indicate that particulate matter derived from combustion products is readily identified under the light microscope. The fused glassy spheres in coal fly ash are the result of exposure to boiler temperatures >1200°C. Aside from the water-white glassy spheres, McCrone and Delly (1973) also describe the presence of opaque "magnetite" spheres and spheres containing trapped gas bubbles.

Light microscopy has been used to define 11 major morphological classes of coal fly ash particles (Fig. 1) in stack-collected, size-fractionated material (Fisher et al., 1978). The characteristics employed in morphological characterization were particle shape and degree of opacity. The 11 classes include (a) amorphous, nonopaque particles, (b) amorphous, opaque particles, (c) amorphous, mixed opaque and nonopaque particles, (d) rounded, vesicular, nonopaque particles, (e) rounded, vesicular, mixed opaque and nonopaque particles, (f) angular, lacy, opaque particles (g) cenospheres (hollow spheres), (h) plerospheres (sphere filled with other spheres), (i) nonopaque, solid spheres, (j) opaque spheres, and (k) spheres with either surface or internal crystals. A morphogenesis scheme (Fig. 2) has been developed relating the 11 morphological classes to extent and duration of exposure to combustion zone temperatures and probable matrix composition. Opaque amorphous particles and angular, lacy, opaque particles were tentatively classified as unoxidized carbonaceous material or iron oxides (Fisher et al., 1978). Subsequent SEM-x-ray analysis (Fisher et al., 1979a) indicated that these opaque particles were composed of low atomic number matrices. Furthermore, calculation of the effective atomic number of class b particles based upon Bremstrahlung production indicated that this class is predominantly composed of elemental carbon (Fisher

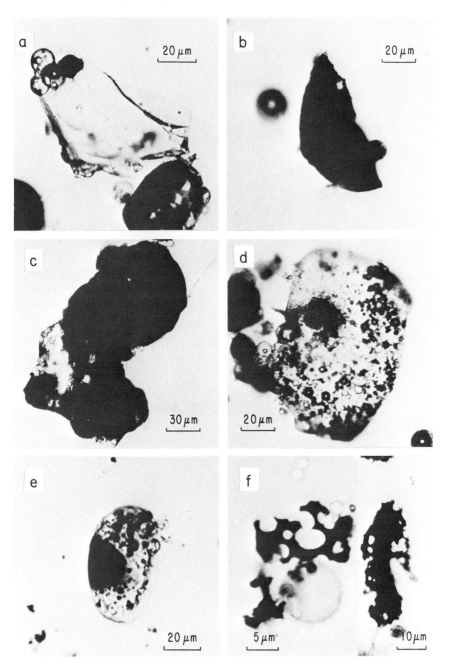

Fig. 1

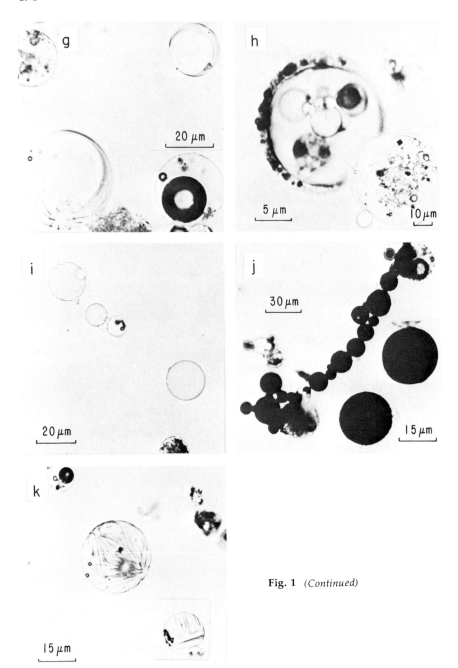

Fig. 1 (Continued)

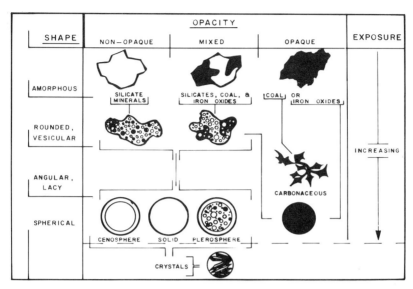

Fig. 2 Fly ash morphogenesis scheme illustrating probable relationship of opacity to particle composition, and relationship of particle shape to exposure in combustion chamber. (From Fisher *et al.*, 1978, p. 450.)

et al., 1979b). The opaque spheres (class j) appear to be predominantly magnetite and may be identified by (1) magnetic separation or passing a magnet near a liquid mount of the sample under a microscope and (2) by observation of small clusters of these particles. The amorphous and rounded-vesicular, nonopaque particles (classes a and d) appear to be aluminosilicate particles. Rounding and vesicularity reflect increased exposure to boiler conditions. Further heating of these particles will give rise to nonopaque spheres that are either solid, hollow, or packed with other particles. Similarly, the mixed opaque, nonopaque, amorphous, or rounded classes will give rise to spherical particles upon increased exposure to combustion conditions in the boiler. The nonopaque, solid spheres ranged in color from water white to yellow to orange and deep red. Analysis of single particles in this class by SEM-

Fig. 1 Light photomicrograph demonstrating the 11 major morphological classes of coal fly ash: (a) amorphous, nonopaque particles; (b) amorphous, opaque particles; (c) amorphous, mixed opaque and nonopaque particles; (d) rounded, vesicular, nonopaque particles; (e) rounded, vesicular, mixed opaque and nonopaque particles; (f) angular, lacy, opaque particles; (g) cenospheres; (h) plerospheres; (i) nonopaque, solid spheres; (j) opaque spheres and (k) spheres with either surface or internal crystals. (From Fisher *et al.*, 1978, p. 449.)

x-ray techniques indicated that the variation in color was associated with iron content (Fisher *et al.*, 1979b). Cenosphere and plerosphere formation will be discussed in detail in the following sections. Crystals within glassy spheres (as determined by light microscopy) are probably formed by heterogeneous nucleation at the surface of the molten silicate droplet (Fisher *et al.*, 1979a). In this regard, Gibbon (1978) has demonstrated the presence of mullite crystals within and on the surface of fly ash particles (Fig. 3). Crystal formation within glassy spheres was demonstrated by transmission electron microscopy (TEM) of hydrofluoric acid-etched replicas. In this process the original glassy material is dissolved, but the insoluble mullite remains. Mullite structure was confirmed by electron-diffraction analysis.

Fisher *et al.* (1978) have quantified the relative abundances of the 11 light-microscopically defined morphological classes in four size-classified, stack-collected fly ash fractions (McFarland *et al.*, 1977). The four fractions had volume median diameters (VMD's) of 2.2, 3.2, 6.3,

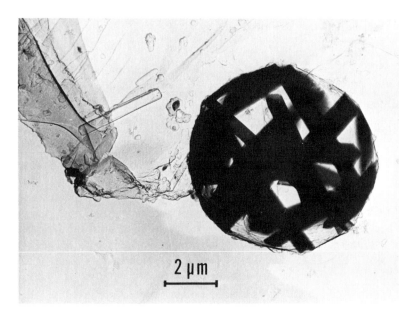

2 µm

Fig. 3 Transmission electron micrograph of a replica of a fly ash sphere showing abundant mullite needles. The association of crystals within sphere is retained by the replica; the original glassy material is dissolved during the replication process, but mullite is insoluble in HF. (Photos courtesy of G. A. Waits, D. S. McKay, and D. L. Gibbon, Lyndon B. Johnson Space Center, Houston, Texas.)

and 20 μm with associated geometric standard deviations (σ_g) of approximately 1.8 for all fractions. The data in Table I demonstrate that the relative abundances of all particle classes are size dependent. In particular, only the nonopaque solid spheres increased in abundance with decreasing particle size; all other morphological classes appeared to increase in frequency with increasing particle size. Amorphous and vesicular particles (classes a, b, c, d, e, and g) predominated in the coarsest fraction (66% by number), while solid, nonopaque spheres predominated in the finest fraction (87% by number).

B. Morphogenesis

1. Cenosphere Formation

The mechanism of formation of cenospheres, i.e., hollow spheres, has been the subject of a number of reports. Raask (1966) demonstrated that sphere formation may result from melting of mineral inclusions in coal on a nonwetting surface, namely carbon. He also demonstrated that gas generation inside the molten droplet resulted in cenosphere formation. He reported two stages of gas evolution. In the first stage, directly after melting coal-ash slag, SO_2 and N_2 were released. The SO_2

TABLE I *Relative Abundance (%) of Morphologic Particle Classes in Four Fly Ash Fractions*[a]

	Fraction			
Particle class	VMD[b] = 20 μm	VMD = 6.3 μm	VMD = 3.2 μm	VMD = 2.2 μm
(A) Amorphous, nonopaque	7.25	2.13	0.79	0.33
(B) Amorphous, opaque	0.42	0.18	—	—
(C) Amorphous, mixed opaque and nonopaque	0.77	0.09	—	—
(D) Rounded, vesicular, nonopaque	12.39	6.67	2.91	2.99
(E) Rounded, vesicular, mixed opaque and nonopaque	2.27	0.24	—	0.03
(F) Angular, lacy, opaque	1.34	0.57	0.27	0.33
(G) Nonopaque, cenosphere	41.11	26.22	13.20	7.91
(H) Nonopaque, plerosphere	0.51	0.21	—	—
(I) Nonopaque, solid sphere	25.58	56.01	79.16	86.99
(J) Opaque sphere	1.56	0.90	0.33	0.24
(K) Nonopaque sphere with crystals	6.80	6.79	3.18	0.95

[a] From Fisher *et al.* (1978).
[b] Volume median diameter.

was thought to result from sulfate decomposition and N_2 from air trapped in the melt. Further heating resulted in CO evolution that was catalyzed by addition of iron or iron oxide to the melt. The author hypothesized that iron carbide was formed at the slag–carbon interface and then reacted with silica resulting in CO evolution:

$$2Fe_3C + SiO_2 \rightarrow Fe_3Si + 3Fe + 2CO. \qquad (1)$$

In a subsequent report, Raask (1968) describes the physical and chemical properties of cenospheres in pulverized fuel ash collected by the electrostatic precipitators at 10 power plants.† The analysis of major elements indicated that the mass of the cenospheres consisted of 75–90% aluminosilicate, 7–10% iron oxide, and 0.2–0.6% calcium oxide. The mass median diameter of the sieved cenospheres from four power plants ranged from 80 to 110 μm. Raask (1968) analyzed the gas content of the cenospheres after breaking the particles in a hydrogen atmosphere. Approximately 0.2 atm (20°C) of gas composed of CO_2 and N_2 was calculated to be present in each of the four ashes studied. In contrast to his previous work (Raask, 1966), no detectable CO was present. Raask suggested that the source of the CO_2 was the oxidation of carbon by iron oxide:

$$2Fe_2O_3 + C \rightarrow 4FeO + CO_2. \qquad (2)$$

This hypothesis was supported by the observation of a higher $FeO:Fe_2O_3$ ratio in the cenospheres than in the denser ash. He also speculated that cenosphere nitrogen may result from decomposition of silicon nitride:

$$Si_3N_4 + 6Fe_2O_3 \rightarrow 3SiO_2 + 12FeO + 2N_2. \qquad (3)$$

It is also possible that the observed CO_2 evolution was due to carbonate mineral decomposition. Assuming a diameter of average volume of 100 μm, a density of 0.5 g/cm³ and 0.5% CaO, only 20% of the calcium need be associated with carbonate mineral to provide sufficient CO_2. In this regard, Fisher et al. (1976) have postulated that CO_2 released by crushing fly ash under vacuum (after thorough degassing) was the result of carbonate mineral decomposition, which occurred during coal combustion. In those studies, CO_2 and H_2O were thought to be due to clay mineral decomposition. In particular, based on the stoichiometry of major elements, Fisher et al. (1976) suggested that the major clay mineral in the parent coal (western United States) was kaolinite. In a detailed study of the transformation of mineral matter

† Thermal analysis of fly ash cenospheres is given in Volume II, Chapter 37, Section V,B,3.

in pulverized coal, Sarofim *et al.* (1977) demonstrated that the three major inorganic components in a bituminous coal and lignite were kaolinite, a mixture of calcium carbonate and sulfate, and pyrites. These authors estimated that the mineral matter in the bituminous coal was 50% kaolinite, 40% pyrite, and 10% calcium sulfate and carbonate, and that in lignite it was 50% kaolinite, 10% pyrite, and 40% calcium sulfate and carbonate. The mass median diameter of the mineral matter was 2 μm for both coals. The optimal temperature for cenosphere formation based on ash density was demonstrated to be 1500° K. This optimum was rationalized by calculating the time for sphere formation. At higher temperatures gas evolution is too rapid and gas will escape from the molten ash; while at lower temperatures sphere formation is too slow relative to the duration of the molten state within the furnaces. Padia *et al.* (1976) have summarized the principal reactions that take place in mineral matter during coal combustion:

$$Al_2Si_2O_5(OH)_4 \rightarrow Al_2O_3 \cdot 2SiO_2 + 2H_2O, \tag{4}$$

$$\downarrow$$

$$Al_2O_3 + 2SiO_2, \tag{4a}$$

$$2FeS_2 + 5.5O_2 \rightarrow Fe_2O_3 + 4SO_2, \tag{5}$$

$$CaSO_4 \rightarrow CaO + SO_3, \tag{6a}$$

$$MgSO_4 \rightarrow MgO + SO_3, \tag{6b}$$

$$Fe_2(SO_4)_3 \rightarrow Fe_2O_3 + 3SO_3, \tag{6c}$$

$$CaCO_3 \rightarrow CaO + CO_2, \tag{7a}$$

$$CaMg(CO_3)_2 \rightarrow CaO + MgO + 2CO_2. \tag{7b}$$

These reactions, Eqs. (4)–(7), all generate gaseous decomposition or oxidation products. Kaolinite decomposition [Eq. (4)], pyrite oxidation [Eq. (5)], calcium and magnesium sulfate decomposition [Eqs. (6a)–(6b)], and calcium carbonate [Eq. (7a)] and dolomite decomposition [Eq. (7b)], may all occur at 1000°C or less, and thus may readily provide gas pressure for cenosphere formation.

2. Plerosphere Formation

Light and electron microscopic studies have identified a morphological class of spherical particles containing encapsulated smaller spheres (Matthews and Kemp, 1971; Natusch *et al.*, 1975; Fisher *et al.*, 1976) (Fig. 4). These encapsulating spheres or plerospheres (Fisher *et al.*, 1976) are similar to cenospheres in that they are composed of an aluminosilicate shell but are filled with individual particles rather than

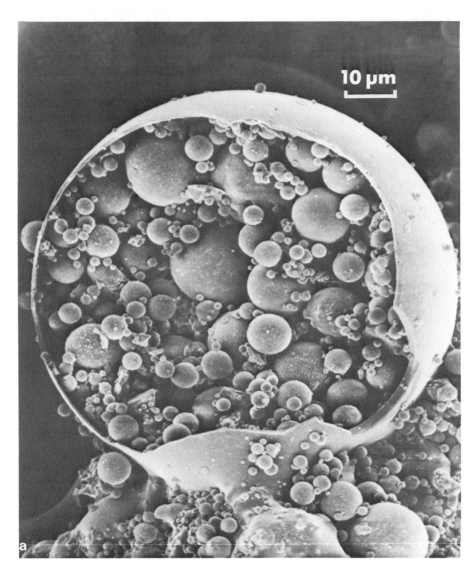

Fig. 4 (a) Spheres-within-sphere structure of a fly ash agglomerate (plerosphere) collected from the electrostatic precipitator of an operating coal-fired electric plant. (From Fisher *et al.*, 1976, p. 534.) (b) Crushed plerosphere revealing encapsulated spheres. (Photo courtesy of D. F. S. Natusch.) (c) Plerosphere containing amorphous core of parent material. Smaller spheres appear to be boiling out of the core. (From Fisher *et al.*, 1976, p. 534.)

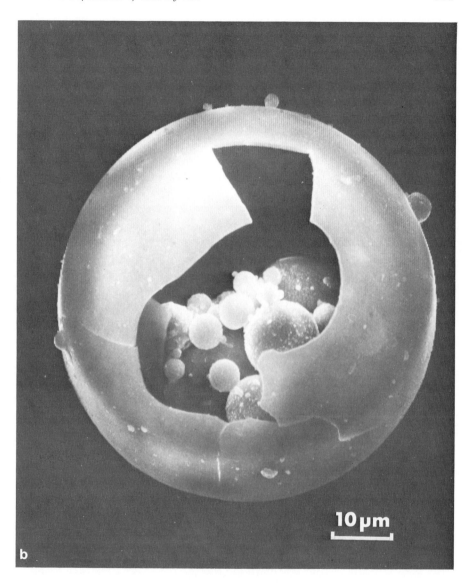

Fig. 4 *(Continued)*

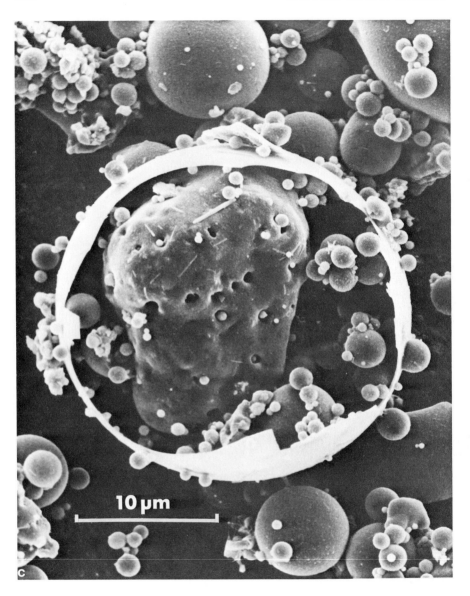

Fig. 4   (*Continued*)

gas (Fig. 4a–4b). Matthews and Kemp (1971) and Natusch *et al.* (1975) have established that plerosphere formation is truly the result of encapsulation during particle formation rather than filling of a ruptured cenosphere. For these studies, either the electron beam of a scanning electron microscope (Matthews and Kemp, 1971) or an argon–ion milling apparatus (Natusch *et al.*, 1975) were used to etch through the individual particle surfaces. Subsequent examination of the etched particles indicated the presence of numerous smaller particles within the plerosphere, thus confirming that encapsulation occurred during particle formation.

Plerosphere formation is hypothesized to result from a process similar to cenosphere formation (Fisher *et al.*, 1976). As the aluminosilicate particle is progressively heated, a molten surface layer develops around the solid core (Fig. 4c). Mineral decomposition, with evolution of CO_2 and H_2O, then results in formation of a bubble around the core, which remains attached to the molten shell. Further heating leads to additional gas formation causing the core to boil away from the shell. This process may result in concomitant formation of fine particles. The process can be repeated until the plerosphere is full of other plerospheres or solid particles or until the particle freezes. The characteristic time for formation of a 50-μm-diam plerosphere was calculated to be about 1 msec. Similarly, Raask (1968) calculated the formation time of a 50-μm cenosphere to be about 0.3 msec.

3. Surface Crystal Formation

Surface crystals (Fig. 5) identified by SEM have been explained by reaction of sulfuric acid with metal oxides (Fisher *et al.*, 1976). This crystal formation process is relatively slow compared to the time required for particle formation. Fisher *et al.* (1976) have hypothesized that surface crystal formation results from SO_2 hydration and subsequent oxidation on fly ash surfaces to form H_2SO_4, which then reacts with metal oxides, predominantly CaO, or with ambient NH_3 to form either $CaSO_4$ or $(NH_4)_2SO_4$. Such a mechanism could also result in formation of soluble compounds from relatively insoluble oxides, e.g., conversion to PbO to $PbSO_4$. Fine particulate matter has also been observed by SEM on fly ash surfaces by a number of investigators (Cheng *et al.*, 1976; Fisher *et al.*, 1976, 1978; Matthews and Kemp, 1971; Natusch *et al.*, 1975; Sarofim *et al.*, 1977; Small, 1976). Small (1976) has identified four surface morphologies based on SEM analysis (Fig. 6). Spheres with smooth surfaces comprised the most commonly encountered particle morphology (Fig. 6a). Spheres with small surface

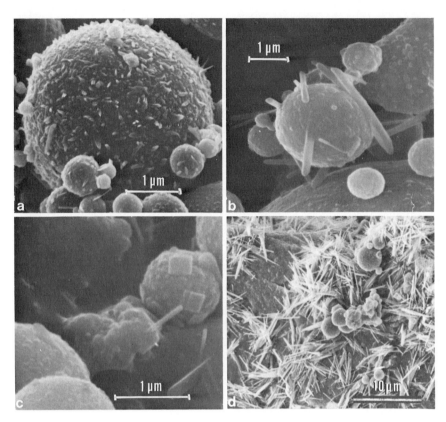

Fig. 5 Scanning electron micrographs of fly ash collected from electrostatic precipitator hopper, indicating (a–c) the variety of observed crystal habits of fine fly ash particles. (From Fisher *et al.*, 1978, p. 444.) (d) Larger crystals analyzed by electron microprobe analysis were tentatively identified as calcium sulfate, probably, anhydrite. (From Fisher *et al.*, 1976, p. 534.)

particles (Fig. 6c) or relatively large surface-associated droplets, resulting from condensation and solidification (Fig. 6b), were also observed. Elemental analysis of these two particle classes indicated the surface was predominantly Si and Al and the underlying particle was mainly Fe. A fourth class of spherical particles with high Fe concentrations was found to display an unusual pattern of coarse surfaces (Fig. 6e). Sarofim *et al.* (1977) reported the presence of submicron silica particles on laboratory-generated fly ash. As an extension of the vaporization-condensation mechanism of Davison *et al.* (1974) for trace elements, Sarofim *et al.* (1977) suggest that silica deposition resulted from for-

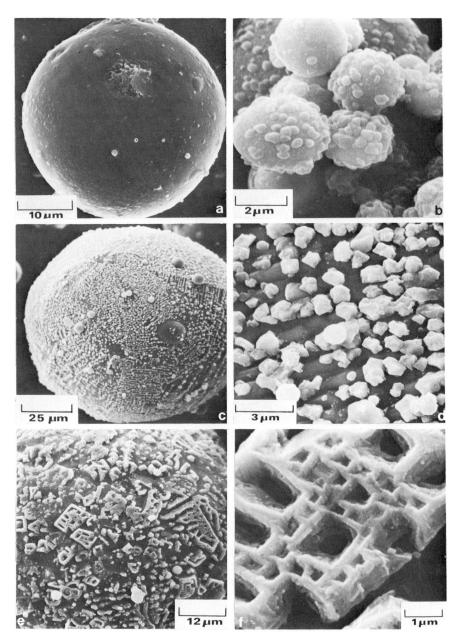

Fig. 6 Scanning-electron micrographs of fly ash indicating four surface morphologies: (a) spheres with smooth surfaces, (b) spheres with relatively large surface droplets, (c,d) spheres with small surface particles, and (e,f) an unusual pattern ("alphabet soup") of surface features. (From Small, 1976.)

505

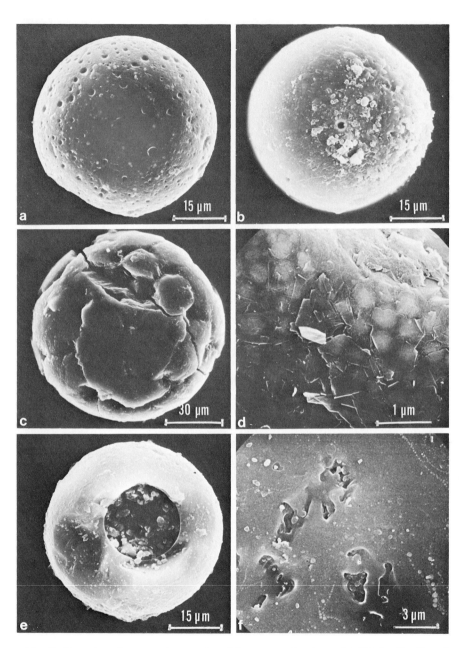

Fig. 7 Scanning-electron micrographs of lunar samples from the US Apollo 17 and the Russian Luna missions: (a) smooth vesicular sphere, (b) sphere with aggregate fine particles, (c) knobby sphere, (d) surface detail indicating submicron, lath-shaped crystals, and (e and f) spheres-within-sphere (plerosphere) structure. (Photos courtesy of G. A. Waits, D. S. McKay, and D. L. Gibbon, Lyndon B. Johnson Space Center, Houston, Texas.)

506

mation of fine silica particles that agglomerate on fly ash surfaces. The formation of submicron silica particles was thought to be because of nucleation of SiO resulting from reaction of SiO_2 with carbon.

4. Spherule Formation from Natural Processes

It is interesting to note the natural (nonanthropogenic) occurrence of glassy spheres with morphological appearance similar to fly ash. Glassy spherules have been reported to be present on shatter cone surfaces (Gay, 1976). These spherules were presumed to be the result of meteorite impact. Similarly, glassy spheres have been identified in lunar dust samples. Nagy *et al.* (1970) reported the presence of fine glass beads that were either spherical, nearly spherical, or dumbbell shaped in Apollo II lunar samples. Some broken glass beads were hollow and vesicular, similar to cenospheres; particle surfaces generally were coated with fine particulate matter. CH_4, CO, and CO_2 were found to be entrapped in the glass beads. Carter and MacGregor (1970) reported that glass spheres ranged in color from colorless through green, brown, wine-red, to opaque. In discussing the formation of the lunar spherules, Gibbon (1978) pointed out that the morphology is related to either impaction or volcanism. The loose lunar soil is primarily composed of spherical particles, indicating a high-temperature origin. Waits *et al.* (1978) have performed detailed morphological studies of lunar samples from the Apollo 17 and the Russian Luna missions (Fig. 7). Smooth vesicular spheres (Fig. 7a) and spheres with internal vesicles have been identified. Fine particulate matter on sphere surfaces (Fig. 7b) is thought to reflect the in-flight aggregation of fine particles on molten sphere surfaces. Knobby spheres (Fig. 7c) that are predominantly crystalline are thought to be feldspar. Detailed surface analysis of some lunar spherules indicate submicron crystals with lath-shaped habit (Fig. 7d). Plerospheres have also been observed (Figs. 7e and 7f), although it is not known whether particles in the plerospheres of Figs. 7e and f filled a fracture after vesicle formation or were formed inside the sphere. Thus it appears that studies of coal fly ash formation and of lunar spherule formation are complementary and should provide mutual support in understanding the physical and chemical processes involved in the high-temperature morphogenesis of particulate matter.

III. PHYSICAL PROPERTIES OF COAL FLY ASH: PARTICLE SIZE DEPENDENCE

In addition to particle morphology, a number of other physical properties have been investigated in attempts to elucidate the formation and behavioral characteristics of fly ash. Such properties include the mass distributions of particle size and the particle density, specific surface area, electrical resistivity, and ferromagnetic susceptibility. Unfortunately, the available data are sparse and apply to fly ash collected after a control device or from the device itself. Consequently, it is not presently possible to relate physical properties to parameters such as plant operating conditions or the type of coal from which the fly ash was derived. Nevertheless, it is qualitatively apparent that the physical properties of fly ash depend upon both of these parameters.

A. Mass Distribution

Measurements of the distribution of fly ash particle mass with size are of two distinct types. The first involves determination of the aerodynamic particle size distribution, which normally involves isokinetic collection of fly ash directly from a stack gas stream.† Several sampling devices are available, but the most common involve the principle of inertial impaction and enable collection and size classification of fly ash *in situ*. The principles and methodology of inertial sampling have recently been reviewed by Raabe (1976), Newton *et al.* (1977) and Natusch *et al.* (1978). Alternatively, bulk fly ash, such as might be collected from an electrostatic precipitator or bag house, can be differentiated into physical size fractions by mechanical sieving. In either case, it is necessary to separate the various size fractions prior to weighing or microscopic counting for size estimation.

These two types of measurement are quite distinct. Aerodynamic size determination enables prediction of the atmospheric and airstream behavior of each size fraction, such as is important for establishing particulate collection efficiency, atmospheric residence, and inhalation characteristics (White, 1963; Butcher and Charlson, 1972; Natusch and Wallace, 1974; Yeh *et al.*, 1976). On the other hand, physical size determination provides a straightforward measurement of the physical dimensions of the particles and can be directly related to particle number. Interconversion between aerodynamic and physical sizes can be accomplished using the relationship (Koltrappa and Light, 1972):

$$D_{ae}^2 = D_r^2 C(D_r) \rho_r / C(D_{ae}) \rho_{ae} \tag{8}$$

D_{ae} is the aerodynamic particle diameter, D_r is the physical particle diameter, ρ_r is the particle density, ρ_{ae} is 1 g/cm^3 by definition, and $C(D_{ae})$ and $C(D_r)$ are the Cunningham slip correction factors for the

† Isokinetic sampling of particulates from a stack gas stream is described in Volume II, Chapter 36, Section II,B.

diameters D_{ae} and D_r. Raabe (1976) has compared the commonly used conventions for description of the aerodynamic size of respirable aerosols, including a detailed description of the slip correction and dynamic shape factors. Aerodynamic size distributions are often presented as log–normal probability functions of the grouped mass or number data derived from inertial impaction instrumentation. The general mathematical approach to fitting size distributions to aerosol data has recently been described by Raabe (1978).

Despite the comparative simplicity of determining both the aerodynamic and physical size distributions of fly ash mass, the number of available measurements of fly ash prior to collection by control equipment is relatively sparse. It has been established, however (Southern Research Institute, 1975), that the size and morphology of fly ash depend not only on the nature of the mineral inclusions in coal, as discussed in the previous section, but also on the manner in which the coal is burned. This latter dependence is illustrated in Fig. 8 for fly ash derived from coal burned in a chain grate stoking unit, a pulverized coal fed unit, and a cyclone fired unit (Southern Research Institute, 1975). In each case the fly ash was sampled upstream from control equipment so it is representative of that generated by combustion.

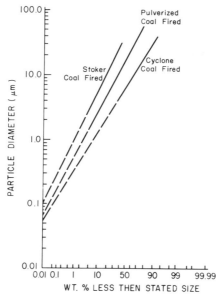

Fig. 8 Size distributions for boiler particulate emissions from coal combustion in a chain grate stoker, a pulverized coal fed unit and a cyclone fired unit. (From Southern Research Institute, 1975.)

It is apparent from Fig. 8 that the fly ash mass approximates a log-normal distribution over the aerodynamic size range considered. Furthermore, although the geometric standard deviations of the three distributions are similar, their mass median diameters differ considerably (i.e., Cyclone ≃6 μm; Pulverized ≃18 μm; Stoker ≃42 μm). These observations are in accord with the general principles of particle formation outlined in the previous section.

From a practical standpoint, one is primarily interested in the aerodynamic size distribution of the fly ash that is actually emitted from a coal-fired power plant. This is, of course, largely determined by the collection efficiency of the particle control equipment. Specifically, the size distribution of the emitted fly ash is determined by the product of the functions describing the size dependence of fly ash mass entering a control device and describing the dependence of collection efficiency of this device on particle size. Examples of the aerodynamic size distribution of fly ash mass emitted from a coal-fired power plant equipped with different control devices are presented in Fig. 9.

B. Density and Magnetic Distributions

Determination of the density of coal fly ash as a function of particle size is largely of interest in obtaining an understanding of aerodynamic behavior and of the factors responsible for the intrinsic heterogeneity of coal fly ash. Thus, determination of the densities of different fly ashes, and subfractions thereof, provides a means of interconverting aerodynamic and physical sizes according to Eq. (8). In addition, some differentiation between distinct morphological and compositional characteristics can be achieved. For example, cenospheres can readily be

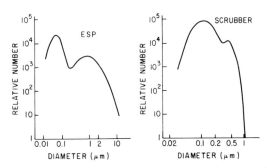

Fig. 9 Size distributions for particulate emissions from similar production units with either electrostatic precipitator (ESP) or Venturi wet scrubber at the same power plant. (Modified from Ondov et al., 1976.)

distinguished from solid particles on the basis of density as can predominantly carbonaceous particles from aluminosilicates.

Determination of fly ash particle density is most simply achieved by means of the traditional "float–sink" method that employs a series of liquids of different densities to separate particles of greater and lesser density than the liquid (Ruch *et al.*, 1974; Olsen and Skogerboe, 1975).[†] Alternatively, separation can be achieved by placing the particles in a liquid in which a density gradient has been established.

While determination of particle density is of considerable interest in its own right, more definitive insights are obtained if density separations are carried out in conjunction with sequential size separations and with differentiation between ferromagnetic and nonferromagnetic particles. Such a three-dimensional fractionation scheme has been presented by Natusch *et al.* (1975), and resulting mass distributions are presented in Tables II and III for fly ashes derived from typical midwestern United States bituminous and western sub-bituminous coals. These data were obtained by separation of bulk fly ash into several physical size fractions by sieving. Each size fraction was then subdivided into a number of density fractions that were, in turn, separated into magnetic and nonmagnetic fractions according to whether the particles adhered to a magnet or not. The designation of magnetic and nonmagnetic is entirely operational in nature.

A number of characteristics of coal fly ash can be distinguished from the data presented in Tables II and III. It is apparent that both fly ashes are compositionally extremely heterogeneous, although there are very considerable differences between the mass distributions for these two fly ashes. As discussed in the previous section, much of the variation in densities observed is attributable to morphological rather than compositional characteristics. This is rather well illustrated by the data in Fig. 10, where density distributions have been determined as a function of particle size both before and after crushing the fly ash. The observed shift to higher density on crushing indicates the presence of vesicular particles and cenospheres in the larger size fractions, as discussed previously.

Interestingly, determination of the x-ray powder diffraction[‡] patterns of each of the subfractions presented in Tables II and III reveals no convincing differences in matrix composition that depend upon either size or density (Natusch *et al.*, 1975). This finding further supports the contention that the density distributions in fly ash are largely determined by morphology rather than by composition. There are, however,

† Float–sink separations are described in Volume I, Chapter 15, Section II,A and B.
‡ X-ray diffraction of ash is described in Chapter 42, Section III.

TABLE II Mass distribution of size-classified, magnetic and nonmagnetic fractions of a midwestern bituminous coal fly ash (%)[a]

| Size | Density (g/cm³) | | | | | | | | | | | |
| | Nonmagnetic | | | | | | Magnetic | | | | | |
(μm)	<1.6	1.6-2.0	2.0-2.3	2.3-2.7	2.7-3.0	>3.0	<2.1	2.1-2.5	2.5-2.9	2.9-3.4	3.4-3.6	>3.6
<20	b	b	0.2	28.0	b	b	b	0.6	0.4	1.3	14.9	0.5
20-60	1.4	1.3	12.1	12.9	0.1	b	0.2	0.6	1.8	11.5	3.1	0.1
60-90	0.7	1.0	0.6	1.1	0.6	0.1	0.5	0.8	1.0	0.2	b	b
>90	0.1	0.1	0.1	0.6	0.5	0.2	0.1	0.2	0.3	0.2	0.1	b

[a] From Natusch (1978c), unpublished results.
[b] Less than 0.05%.

TABLE III Mass distribution of size-classified, magnetic and nonmagnetic fractions of a western sub-bituminous coal fly ash (%)[a]

| Size | Density (g/cm³) | | | | | | | | | | | |
| | Nonmagnetic | | | | | | Magnetic | | | | | |
(μm)	<1.6	1.6-2.0	2.0-2.3	2.3-2.7	2.7-3.0	>3.0	<1.6	1.6-2.0	2.0-2.3	2.3-2.7	2.7-3.0	>3.0
<20	b	b	b	0.7	b	b	b	b	b	b	b	b
20-44	0.2	0.4	0.5	21.3	0.3	0.2	b	b	b	1.0	b	b
44-74	0.5	0.8	1.0	45.6	0.6	0.5	0.1	0.1	0.2	6.8	0.1	0.1
>74	0.2	0.3	0.4	16.1	0.2	0.2	b	b	b	1.5	b	b

[a] From Natusch (1978c), unpublished results.
[b] Less than 0.05%.

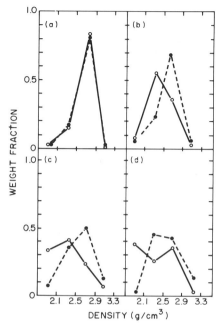

Fig. 10 The effect of crushing on the mass distribution of size-classified fly ash fractions. The shift to higher densities indicates the presence of hollow or vesicular particles. (a) Particle diameter < 20 μm, (b) particle diameter 20-44 μm, (c) particle diameter 44-74 μm, (d) particle diameter >74 μm. - - -, crushed weight fraction and —— uncrushed weight fraction. (Figure courtesy of D. F. S. Natusch.)

very distinct differences between the amount of magnetite (Fe_3O_4) present in the magnetic and nonmagnetic fractions (Fig. 11). This suggests that magnetite is primarily responsible for the ferromagnetic susceptibility of coal fly ash.

C. Electrical Resistivity Distribution

The electrical resistivity of coal fly ash is an important physical property from the standpoint of control. Thus, it has been established (Bickelhaupt, 1974, 1975) that the collection efficiency of electrostatic precipitators increases with decreasing fly ash resistivity. Bickelhaupt (1974, 1975) has further shown that both the surface and volume resistivities of fly ash, at precipitator operating temperatures, are inversely proportional to the specific concentrations of alkali metals, which are thought to act as charge carriers. These studies have shown that considerable differences in electrical resistivity occur between different fly ashes, and correlations are observed between fly ash resistivity and

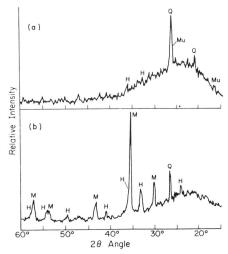

Fig. 11 X-ray powder diffraction patterns demonstrating the compositional differences between magnetic (a) and nonmagnetic (b) fly ash fractions. H: hemalite, Q: quartz, and Mu: mullite. (Figure courtesy of D. F. S. Natusch.)

alkali metal content, but no measurements have been made relating resistivity directly to particle size.

Some insight into the dependence of resistivity on particle size can be obtained by considering the data presented in Table IV. This table lists concentrations of potassium measured in fly ash that has been fractionated sequentially according to size, density, and ferromagne-

TABLE IV *Concentration (%) of Potassium in Fly Ash Separated Sequentially by Size, Density, and Ferromagnetism[a]*

Particle type	Particle size (μm)	Density (g/cm³)			
		<2.1	2.1–2.5	2.5–2.9	>2.9
Nonmagnetic	<20	2.69	2.34	2.22	1.73
	20–44	2.63	2.28	1.33	1.09
	44–74	1.89	1.63	1.05	0.45
	>74	1.79	1.48	1.06	0.13
Magnetic	<20	—[b]	—	0.76	0.70
	20–44	—	1.92	1.48	0.73
	44–74	1.78	1.60	1.27	0.85
	>74	1.37	1.62	1.49	0.83

[a] From Natusch *et al.* (1975).
[b] No meaningful data.

tism as described previously. It can be seen that in the nonmagnetic fractions (that account for 64% of this fly ash) there is a pronounced increase in the concentration of potassium (and also of sodium), both with decreasing particle size and with decreasing density. This suggests that, for this fly ash sample, resistivity decreases with particle size and with density. Similar, though less pronounced, density dependencies are observed in the magnetic fractions, but size dependencies, if any, are obscure. Since both decreasing density and decreasing physical size contribute to decreasing aerodynamic size, it is apparent that the efficiency of electrostatic precipitation per unit mass of these size-classified fly ashes increases with decreasing aerodynamic particle size. This is an extremely desirable characteristic. It should be pointed out, however, that these studies require extension to respirable particle sizes.

D. Surface Area Distribution

The specific surface area of fly ash particles is an important parameter in determining a number of the behavioral characteristics of coal fly ash. It is the surface area of a particle that determines the number of electrostatic charges that can be placed on that particle in an electrostatic precipitator (White, 1963; Bickelhaupt, 1974, 1975); it is the surface area of a particle that determines the extent of condensation or adsorption of species from the gas phase (Davison *et al.*, 1974; Natusch and Tomkins, 1977); and it is the surface area of fly ash that determines the rate and extent of its aqueous leaching (Natusch *et al.*, 1975; Matusiewicz and Natusch, 1979).

To a reasonable approximation, one would expect the specific surface area (square meters per gram) of fly ash to increase linearly with decreasing particle diameter since the particles are predominantly spherical. Similar trends would also be expected for nonspherical particles having similar shape factors (Butcher and Charlson, 1972). In fact, the expected trend is observed; however, two important points are noted. The surface areas that are measured for spherical fly ash particles are considerably greater than those calculated from measured particle diameters. Even taking into account the assumptions inherent in surface area measurements, it appears that coal fly ash has a significant "internal" surface area. This is probably in the form of pores or cracks or a porous surface layer, although, as previously described, surface crystal formation may contribute significantly to the measured surface area. However, several fly ashes show no significant dependence of surface area on particle diameter (Table V), especially for small particles. These

TABLE V *Comparison of Measured and Calculated Specific Surface Areas of Size-Classified Fly Ash Fractions*[a]

Physical size (μm)	Measured (m²/g)	Calculated (cm²/g)
<45	2.02	>267
45–63	3.55	191–267
63–90	2.55	133–191
90–125	2.43	96–133
125–180	1.20	67–96
>180	3.11	<67

[a] From Kim and Natusch (1978). Unpublished results.

data indicate the existence of substantial internal surface area that is effectively proportional to particle volume rather than external surface area. In this regard, it has recently been suggested (Natusch, 1978a) that collisionally efficient condensation processes may result in deposition of material from the gas phase predominantly onto the external particle surface, whereas much less efficient adsorption processes (Natusch and Tomkins, 1977) can deposit gases and vapors on both the internal and external surfaces of a particle.

IV.   ELEMENTAL COMPOSITION OF COAL FLY ASH: PARTICLE SIZE DEPENDENCE

Studies of the size dependence of the elemental concentration of fly ash can be classified into two categories. The first category consists of those studies that relate the elemental concentration to the particle size of size-classified material. For these studies, sufficient mass of size-classified material is collected to allow gravimetric determination prior to elemental analysis. The second category are the many studies that have employed inertial cascade impactor systems for aerodynamic size classification. Aerosol sampling is performed isokinetically to avoid anomalous alteration of the particle size distribution. Because impactor stages are often coated with sticky adhesive to prevent particle bounce off and reentrainment effects and because only small masses of material may be collected on the stages, accurate gravimetric determination of sample mass is difficult. To obviate this complication, specific elemental masses of deposited particles on each stage are often ratioed to the mass of an element that does not demonstrate a marked concentration dependence with particle size. In this regard, Ondov *et al.* (1977a) have analyzed four size-classified, stack-collected fly ash samples ranging in particle size (VMD) from 2.2 to 20μm (McFarland *et al.*, 1977). The elements Al, Si, Ca, K, Ce, La, Rb, Nd, Hf, Sm, and Cs varied in concentration by less than 20% among all fractions and should therefore be suitable for mass estimation. A second approach that has been used in the analysis of impactor data, reports the size distribution for the mass of each element analyzed, thus avoiding the compounded errors in data derived from elemental ratios. Impactor studies also report elemental concentrations in terms of mass per unit volume of aerosol sampled. Thus, because of the limitation of gravimetric determination, results from impactor studies are often reported as ratios of elemental masses or mass-to-volume ratios rather than specific concentrations.

Many studies employ the enrichment factor (EF) of Gordon and Zoller (1973). The EF is defined as the ratio of an elemental concentration in the fly ash sample to the elemental concentration in the coal. To provide normalization relative to total mineral content, EF's are often calculated from the ratios of specific elemental contents in the fly ash samples and coal, respectively, to those of mineral matrix elements in the fly ash samples and coal, respectively. Thus, the EF may be calculated from

$$EF = ([X]_s/[M]_s)/([X]_c/[M]_c), \tag{9}$$

where $[X]_s$ and $[X]_c$ represent the mass of element X in the sample and coal, respectively, and $[M]_s$ and $[M]_c$ represent the content of the matrix element in the sample and coal, respectively. A number of "matrix" elements have been used in the EF calculation: Al (Gordon *et al.*, 1974), Fe (Ragaini and Ondov, 1977), Sc (Ondov *et al.*, 1977c), Ce (Coles *et al.*, 1979), and ^{40}K (Coles *et al.*, 1978). In the following section, studies of fly ash analyses using gravimetrically determined masses will be discussed separately from studies employing smaller masses.

A number of analytical techniques have been employed in the determination of the elemental composition of coal fly ash. For complete analysis of the major, minor, and trace elements, a combination of analytical techniques is usually employed. The physical and chemical heterogeneity in terms of particle size, chemical distribution within and among individual particles, and the fused aluminosilicate matrix provide a unique combination of difficulties for the analyst. The techniques employed for elemental analysis may be divided into two categories: (1) single element techniques that generally require matrix dissolution and (2) multielement techniques that generally are performed on the undissolved ash. A detailed and extensive review of the elemental analysis of particulate matter has recently been published by

Natusch et al. (1978). See also Chapters 11 through 14 in Volume I, and Chapter 46 in this volume.

A. Studies of Specific Concentrations

Davison et al. (1974) published the first detailed elemental analysis of coal fly ash as a function of particle size. The ash was collected from a power plant using southern Indiana coal. Two types of fly ash samples were analyzed: (1) fly ash collected by the plant's cyclonic precipitator and (2) stack-collected material. The precipitator ash was size separated by sieving the larger particles and aerodynamically separating the remaining mass. The stack-collected fly ash was aerodynamically classified using an Anderson impactor. These authors presented the elemental concentrations in three categories based on the degree of concentration dependence on particle size. The elements showing "pronounced" concentration trends of increased concentration with decreasing particle size were Pb, Tl, Sb, Cd, Se, As, Ni, Cr, Zn, and S. Elements classified as showing limited concentration trends were Fe, Mn, V, Si, Mg, C, Be, and Al. Iron concentrations decreased with particle size for the precipitator ash, while no trend was observed in the stack-collected samples. The elements described as showing no concentration trends were Bi, Sn, Cu, Co, Ti, Ca, and K. The mechanism of concentration enhancement has been postulated to be volatilization of the element (or compound) at combustion temperatures (1400°–1600°C) followed by condensation on particle surfaces (Natusch et al., 1974; Davison et al., 1974). Thus, fine particles with their large ratio of surface area to mass will preferentially concentrate volatile inorganic species. In particular, those elements displaying the greatest concentration dependence with particle size generally are associated with elemental forms that boil or sublime at coal combustion temperatures.

Fisher et al. (1977) and Fisher and Chrisp (1978) have described the size dependence of the elemental concentrations in coal fly ash collected from the stack of a power plant burning low-sulfur, high-ash, western United States coal. The fly ash was size classified in situ, downstream from the ESP, using a specially designed instrument employing two cyclone separators in series followed by a 25 jet centripeter (McFarland et al., 1977). Elements were classified into two categories: elemental concentrations (1) dependent on particle size and (2) independent of particle size. Concentration dependence with particle size was determined qualitatively with the criterion that constant concentration trends beyond experimental uncertainty were observed for each of the four fractions analyzed. In order of decreasing dependence on particle size, the elements Cd, Zn, As, Sb, W, Mo, Ga, Pb, V, U, Cr, Ba, Cu, Be, and Mn displayed increased concentration with decreasing particle size. Silicon was the only element to decrease in concentration with decreasing particle size.

The elements not displaying clear-cut concentration dependence on particle size for all fractions analyzed were Al, Fe, Ca, Na, K, Ti, Mg, Sr, Ce, La, Rb, Nd, Th, Ni, Sc, Hf, Co, Sm, Dy, Yb, Cs, Ta, Eu, and Tb. Of these elements, Na, Sr, Ni, and Co displayed marked enhancement in the finest fraction relative to the coarsest fraction. Coles *et al.* (1979) have described the elemental behavior in the four size-classified fractions in terms of elemental enrichment factors relative to the parent coal.

The elements were grouped into three classes: group I elements displayed little or no enrichment in fine particles and were lithophilic; group II elements displayed marked enrichment and were chalcophilic (sulfur associated); and group III consisted of elements with behaviors intermediate to groups I and II. Group I elements included Al, Ca, Cs, Fe, Hf, K, Mg, Mn, Na, Rb, Sc, Ta, Th, Ti, Ce, Dy, Eu, La, Nd, Sm, Tb, and Yb; group II elements were As, Cd, Ga, Mo, Pb, Sb, Se, W, and Zn; and group III consisted of Ba, Be, Co, Cr, Cu, Ni, Sr, U, and V.

In a separate report, Coles *et al.* (1978) described enrichment factors for [228]Th, [228]Ra, ^{210}Pb, [226]Ra, [238]U, and [235]U relative to [40]K in the four size-classified fractions of stack fly ash. Although the EF's for all radionuclides appeared to increase with decreasing particle size, ^{210}Pb, the most volatile radionuclide, showed the greatest size dependence. The authors proposed that U is present as either a carbonate [$Na_2UO_2(CO_3)_2$ or $Na_4UO_2(CO_3)_3$] that upon heating in an oxidative atmosphere may give rise to either volatile UO_3 from oxidation of uranite (UO_2) or the silicate-soluble, nonvolatile mineral, coffinite [$U(SiO_4)_{1-x}(OH)_{4x}$]. Thus, U behavior would be expected to display an intermediate behavior depending on the relative concentrations of uranite and coffinite. The behavior of Th was rationalized to be due to coexistence in submicron zircon grains in the coal. The authors suggested that [226]Ra enrichment may have been due to [238]U, while no explanation of [228]Ra enrichment was presented.

Campbell *et al.* (1978) have studied the elemental distribution of size-classified ESP-collected coal fly ash from a western United States power plant. Reaerosolized ESP fly ash was separated into nine size fractions ranging in size (VMD) from 0.5 to 50 μm. The authors describe fine particle enhancement for elements "volatilized during combustion," i.e., As, Co, Cr, Ga, Pb, Se, and Zn. Their data also demonstrate that K, Al, Mn, Mg, Na, Ba, S, Ni, V, Cu, Cs, Rb, Sb, Br, Mo, and Sn display an inverse concentration dependence on particle size. Silicon and possibly Zr were reported to increase in concentration with increasing particle size. The concentrations of Ca and Sr demonstrated a maximum at approximately 5 μm. A similar concentration pattern was reported for Ce, Eu, and Yb.

These studies are in basic agreement with the hypothesis of Natusch et al. (1974) in that the most volatile elements (or their oxides), Cd, Zn, Se, As, Sb, W, Mo, Ga, Pb, and V, displayed the greatest size dependence. Furthermore, the least volatile elements did not display a strong particle size dependence. With regard to enhancement of Ba and U, Coles et al. (1979) postulated that Ba may form the volatile species $Ba(OH)_2$ and U may be volatilized in part as UO_3. Fisher et al. (1977) have proposed that the presence of Cr in the organic fraction of coal, Mn and Sr as carbonate minerals, and Cu as sulfides, may explain the behavior of these relatively refractory elements. Campbell et al. (1978) speculated that the concentration profiles exhibiting maximum particle sizes of approximately 5 μm for Ca, Sr, and the rare earth elements were because of the presence of these elements in apatite.

B. Studies of Relative Concentrations

Most studies of the chemical properties of size-classified fly ash have employed cascade impactors for stack or plume sampling. Zoller et al. (1974) reported enrichment factors relative to Al for stack-collected fly ash. The ash studied was collected downstream from the ESP at a power plant burning pulverized coal containing 10% ash and 1% S. In agreement with the previously described studies, enhancement of the volatile elements, Sb, Se, As, Pb, Zn, Ni, and I, was observed in the stack fly ash relative to their concentrations in the coal. Bromine was depleted in the stack ash relative to the coal. The authors point out that the EF's for Se, I, and Br are underestimates because portions of these elements were probably in the vapor phase. Elements not displaying enrichments included Ti, Sc, Th, Ta, Na, K, Rb, Mg, Sr, Ca, Ba, V, Cr, Mn, Fe, Co, and six rare earth elements. It should be pointed out that although the stack sample was not size classified, a relatively fine particle distribution (i.e., MMD 5–10 μm) may be presumed for this post-ESP material. In a subsequent report (Gladney et al., 1976), the research team described the size dependency of the EF's in the stack fly ash.

Three patterns of elemental behavior were described. The elements

Na, K, Rb, Mg, Ca, Sr, Ba, Sc, Ti, V, Mn, Co, Zr, Th, Hf, Ta, and all rare earths except Ce displayed an EF distribution that was not size dependent. Interestingly, the authors also report that the relatively volatile elements, Cr, Zn, Ni, and Ga, also exhibited little size dependence. A definite increase in EF of fine particles was observed for Pb, As, and Sb. The volatile elements, Se, Br, I, and, to a lesser extent, Hg, displayed bimodal activity. An enrichment minimum was observed from 0.7 to 5.0 μm. Iron and Ce displayed EF's that decreased with decreasing particle size.

Klein *et al.* (1975) described the pathways of 37 trace elements through a cyclone-fed power plant burning coal of 3% S and 11% ash. Concentration ratios for ESP outlet versus inlet ash indicated enhancement of As, Cd, Cr, Pb, Sb, Se, V, and Zn in the finer fly ash fraction. The authors point out that the ESP efficiency was 96.5% during their first sampling trip, as compared to 99.5% during their second sampling trip. Interestingly, the removal of the major elements was more complete during the second trip, although no change in capture efficiency was observed for Cd, Pd, and Zn because of association with fine particles. The authors estimate that 60–90% of the Hg was released from the stack as a vapor. In a subsequent study, Andren and Klein (1975) presented extensive data on the mass balance and chemical form of selenium emissions from the same power plant. The authors concluded that 68% of the Se was incorporated into fly ash. Based on an ESP efficiency of 99.6%, the authors also concluded that 93% of the Se released to the environment is in the vapor phase. The oxidation state of Se was determined to be Se^0 based upon inefficient extraction in HCl and complete elemental extraction in Br/Br^--redox buffer, 16M HNO_3, 18M H_2SO_4, or 1:1 $HNO_3:HClO_4$.

Mercury emissions from coal-fired power plants have been described in detail. Billings and Matson (1972) and Billings *et al.* (1973) studied mercury emission from a power plant burning low sulfur (<1%), high ash (21%) pulverized coal. The authors concluded that 90% of the Hg was released from the stack as a vapor and that fly ash particles represented less than 1% of the Hg emissions. The annual release of Hg from all coal-fired United States power plants was estimated to be 10^3 metric tons in 1971. Similarly, Diehl *et al.* (1972) studied Hg emissions from a 100-g/hr pulverized coal combustor and a 500-lb/hr pulverized coal combustor. Although these authors experienced difficulties in their collection of Hg from the flue gas, 35 and 60% of the total Hg was found in the fly ashes generated from combustion of coals having ash contents of 21.6 and 6.9%, respectively, and sulfur contents of 5.2 and 1.2%, respectively. Subsequent studies in the larger combustor using coal with 10.1% ash and 2.1% S, resulted in fly ash containing 12% of the total Hg. The authors present calculations for two Illinois power plants, indicating that the Hg content of ash contained within the plants accounted for 7 and 19% of the total Hg in the coal. Thus, in agreement with Billings' work, most of the Hg in coal is volatilized and released as a vapor to the atmosphere. Similarly, Kalb (1975) has reported that the major portion of Hg in coal is volatilized during combustion and released to the atmosphere. Approximately 10% of the volatilized Hg was found to be adsorbed onto fly ash; organomercury compounds were not observed. The author points out that Hg emissions could be reduced by coal cleaning, which results in removal of higher density minerals, including pyrite that is relatively high in Hg contents.

In a review of trace element studies related to low sulfur, high ash coal combustion in Four Corners, New Mexico, Wangen and Wienke (1976) described enrichment factors for electrostatic precipitator ash relative to bottom ash. Enhancement in the precipitator ash was observed for the following elements in order of decreasing magnitude: Se, As, F, Sb, Zn, Tl, Hg, Mo, Ga, B, Pb, V, and Cr. Enrichment factors near unity were observed for the other 22 elements studied.

Kaakinen *et al*. (1975) studied the behavior of 17 elements in the inlets and outlets of a power plant burning pulverized coal containing 0.6% S and 6% ash. Although particle size was not reported, the author described the specific surface area of his samples. The surface areas measured by nitrogen adsorption for the bottom ash, mechanical collector hopper ash, electrostatic-precipitator hopper ash, and electrostatic-precipitator-outlet fly ash were 0.38, 1.27, 3.06, and 4.76 m^2/g, respectively. Enhancement in trace element concentration relative to Al was observed for Pb, Mo, As, Zn, Sb, and Cu.

The magnitude of the EF's correlated with relative distance of each outlet downstream from the boiler and the specific surface area of the ashes. The authors point out that As enrichment depends on the Ca content of the coal; As_2O_3 is associated with low Ca coals while As_2O_5 is associated with high Ca. Zirconium was the only element displaying a decrease in fly ash downstream from the mechanical collector. The decrease was thought to be because of the occurrence of Zr as zircon, a relatively high density mineral that may be more efficiently captured by the mechanical collector. Contrary to this observation, little or no enrichment was reported for Nb, Sr, Fe, Rb, and Y.

Ondov *et al*. (1977b,c) have performed extensive analyses of element enrichments in fly ash as a function of particle size. In the study of two large western power plants burning high ash, low sulfur, pulverized coal, Ondov *et al.* (1977c) reported considerable enrichment of W, U, Ba, Zn, V, In, Ga, Br, As, Se, Sb, and Mo in fine particles for the plant with an ESP rated at 99.5% efficiency. In the second plant, with a 97% efficient ESP, EF distribution tended to be bimodal for these elements, with a broad maximum of 2–10 μm. The authors also point out that Br, Se, Cr, Mn, Ta, Co, and Zn displayed enrichment in both the fine and the large particles, i.e., an EF minimum was observed from approximately 1 to 8 μm. The authors indicate that the biphasic distributions may be the result of artifacts in collection because the larger particles will be collected on the first impactor stages, through which vapor containing volatile elements is initially drawn.

Fisher *et al.* (1979d) also reported data supporting bimodal elemental distributions. Filtration studies with neutron-activated, stack-collected fly ash (VMD = 2.2 μm; σ_g = 1.8) were performed by dispersing ash samples in buffer at pH 7.4 and filtering through a membrane with pore size of 5, 2, 0.8, 0.4, 0.2, 0.1, 0.05, or 0.03 μm. The elements were classified into four groups based on their behavior: (1) Na, Ca, Co, Se, Mo, and Ba were partially soluble and did not display filtrate concentrations that were pore-size dependent; (2) Sb, As, Zn, W, Cr, and U displayed a pattern of filtrate concentrations that appeared to be bimodal; (3) K, Si, Fe, Ce, Sm, Eu, and Th were only detected in filtrates from membranes >2 μm in pore size; and (4) Zr, Cs, Nd, Rb, Tb, Yb, Hf, and Ta were not detected in the filtrates. For those elements displaying bimodal behavior, a relatively large increase in concentration was observed in filtrates derived from the 0.4 μm membrane. The concentration profile remained constant thereafter. These data suggest a concentration maximum for Sb, As, Zn, W, Cr, and U in fine particles less than 0.4 μm in diameter.

Ondov *et al.* (1977c) have compared enrichment factors for the two power plants to those published by Klein *et al.* (1975), Kaakinen *et al.* (1975), and Gladney *et al.* (1976). The comparison (Table VI) of EF's for elements in stack-collected fly ash indicates relatively good agreement between studies of different power plants with ESP control systems employing a wide variety of coals. In light of the uncertainties, only Mo, Se, and Mn showed significant differences between plants. The volatile elements Sb, As, and Pb were clearly enhanced in samples from all power plants; Zn, Se, Cr, and V were enhanced in stack ash from those plants with the most efficient ESP's, i.e., those plants presumably releasing the finest ash. Bromine was the only element displaying a significant fractional EF. Ondov *et al.* (1977c) point out that the EF's for stack ash collected from a unit with a venturi wet scrubber (VWS) are generally much higher than those for plants with ESP's. The

TABLE VI *Enrichment Factors for Elements in Stack Fly Ash from Coal-Fired Power Plants[a]*

	Western U.S. plant A[b]	Western U.S. plant B (ESP)[c]	Allen Steam plant[d]	Chalk point[e]	Valmont[f]	Western U.S. plant B (VWS)[c]
Sb	7.0	5.3	6.7	4.0	—	120
Cd	—	6.0	—	—	—	—
W	—	4.9	—	—	—	70
As	6.6	7.9	6	6.3	—	100
In	5.5	3.7	—	—	—	20
Zn	4.3	4.3	7.8	1.5	2.5	19
Pb	—	3.8	8.1	3.7	3.1	—
Ga	4.3	3.0	—	1.2	—	—
U	3.3	2.5	—	—	—	13.5
Se	3.0	5.3	5.5	5.7	1.7	400
Ba	2.5	2.7	0.7	0.92	—	13
Cr	2.5	2.6	3.0	1.1	—	100
Co	2.3	1.7	1.4	1.0	—	4.3
V	2.0	2.5	2.5	0.75	—	21
Mo	1.8	3.5	—	—	3.0	43
Mg	1.1	0.8	0.54	—	—	2.7
Fe	1.1	0.90	0.84	0.83	1.0	2.0
Na	1.0	1.1	0.99	—	—	3.2
Sc	1.0	1.0	1.0	1.0	—	1.0
K	1.0	0.7	0.95	0.83	—	0.86
Th	0.95	0.90	0.76	—	—	0.89
Al	0.86	0.75	0.44	0.83	0.94	1.3
Ca	0.76	0.89	—	0.92	—	7.6
Mn	0.68	1.1	0.78	—	—	21
Be	—	0.6	—	0.64	—	—
Br	0.2	0.1	—	0.17	—	57

[a] Modified from Ondov *et al.* (1977c).

[b] Plant A employed an ESP with removal efficiency of 99.6% (Ondov *et al.*, 1977c).

[c] Plant B employed an ESP with efficiency of 97% on one unit and a venturi wet scrubber (VWS) on a second unit (Ondov *et al.*, 1979).

[d] Employed an ESP with 99.5% efficiency (Klein *et al.*, 1975).

[e] Employed an ESP with 75% efficiency (Gladney *et al.*, 1976).

[f] Employed a mechanical collector and an ESP with 91% efficiency (Kaakinen *et al.*, 1975).

authors attribute these findings, in part, to the high efficiency (>99%) of removal of particles >2 μm and the low efficiency (40%) of removal of particles <2 μm by the VWS. In another study, Ondov *et al.* (1979) indicated that the ratio of VWS-to-ESP fractional emissions of submicron, supermicron, and total suspended particles were 1:6, 11:1, and 10:1, respectively. They also proposed that corrosion may enhance VWS emissions of Cr, Co, Cu, and Zn. Thus, although the VWS may have a higher removal efficiency of total suspended particulate matter, the ESP may more efficiently remove respirable particles.

Ondov *et al.* (1977b) have reported enrichment factors for plume samples collected from a power plant with five generating units, of which two units were equipped with ESP's and the other three with VWS's. Elemental enrichment factors were relatively constant as a function of distance from the stack for Sc, Na, K, Cu, and the lanthanides. Enrichments for Mo, V, Ba, U, Ga, In, As, W, and Se increased from the stack to the plume. Subsequent plume samples indicated decreased EF's with distance from the point of release. The only elements displaying increased enrichments with increased distance from the stack were Br, Sb, Zn, and Co. The increased EF for Br was postulated to be because of mixing of plume aerosols with high background concentrations of Br, possibly because of automotive sources (Ondov *et al.*, 1977b).

In a further comparison of the stack fly ash from an ESP unit with that from a VWS unit, Ondov *et al.* (1979) reported that the mass median aerodynamic diameters (MMAD's) for the elements As, Ba, Sb, Se, U, V, and W in the ESP ash were approximately tenfold higher than in the VWS ash, which ranged from 0.47 to 0.59 μm. The authors concluded that despite an eleven-fold higher total particulate emission, the ESP unit is far more efficient at removing submicron particles than is the VWS unit. Thus, the scrubber unit tested appeared to be less effective at reducing potential inhalation hazards than the precipitator unit.

C. Surface Deposition Models

A number of investigators have presented mathematical models relating the concentrations of relatively volatile elements to geometric parameters associated with fly ash particles. Assuming a volatilization–condensation mechanism, Davison *et al.* (1974) proposed a simple mathematical model for elemental concentration as a function of particle size. Their model predicts that the elemental concentration of a volatile species will be inversely dependent on particle size. Kaakinen *et al.* (1975) presented a similar mathematical dependence based on the specific surface area (square meters per gram) of fly ash. If the specific surface area is proportional to the surface area:volume ratio and if particle sphericity is assumed, then elemental concentration is inversely proportional to particle size. Based on mass transfer arguments, Flagan and Friedlander (1976) indicated that concentration should be inversely dependent on particle size for Knudsen numbers >1 (i.e., for condensation when the particle size is greater than the mean free path of the depositing gas) and inversely dependent on the square of the particle size for Knudsen numbers <1. Application of this model fits existing data equally as well as the model of Davison *et al.* (1974). Smith *et al.* (1978) extended the Flagan and Friedlander (1976) and the Davison *et al.* (1974) models to include fine particles in which the thickness of the deposited surface layer approached the diameter of the total particle. This modification resulted in concentrations that asymptotically approached maxima at particle size <1 μm. The models were demonstrated to fit the concentration dependence on particle size of reaerosolized, ESP-collected fly ash. Biermann and Ondov (1978) have proposed a model with an inverse square dependence and an asymptotic maximum for concentration as a function of surface thickness. Their results indicated that the thickness of surface-deposited chemicals is inversely proportional to particle size and that total elemental composition is proportional to l/d^2, where l is the thickness of the surface layer and d the diameter of the particle. Analysis of 12-stage impactor data with increased resolution in the submicron region supported the mathematical model. Further studies are required, however, to extend the presently available data on concentration as a function of particle size thus allowing evaluation of the validity of the existing mathematical models.

D. Summary

In summary, most studies of the size dependence of elemental concentrations in coal fly ash support the hypothesis of Natusch *et al.* (1974); the more volatile elements (or chemical forms) are preferentially associated with fine particles. The fine particle mode (<1.0 μm) in the bimodal elemental distributions is generally considered to be because of coagulation of primary particles (Whitby, 1977). Bimodal size distributions may also result from the presence of multiple mineral forms, some of which may decompose or may be associated with a fine mineral grain size. It should be noted, however, that the bimodal distribution of the very volatile elements (Se, Br, and I) observed in impactor samples may be artifacts due to vapor condensation on the larger particles collected on the first impactor stages. Also, the bimodal distribution of metallurgical elements may be associated with entrainment of corrosion products in the flue gases. Similarly, small particle enhancement of relatively nonvolatile elements may be because of a combination of decomposition, chemical reaction, mineral grain size, or elemental association in the organic phase of coal.

V. MATRIX AND SURFACE COMPOSITION OF COAL FLY ASH

A. Matrix Composition

Elemental analyses of coal fly ash show that the major matrix elements are Al, Si, and Fe together with a few percent of Ca, K, Na, and Ti. Fly ashes derived from western United States sub-bituminous coals generally contain higher levels of calcium than do bituminous coals and lignites.

The actual compounds that constitute the fly ash matrix have been identified only for a comparatively small fraction of the mass. The techniques that have proved most useful for this purpose are x-ray powder diffraction and infrared spectroscopy (Natusch *et al.*, 1975). In addition, selected area electron diffraction has been employed in the identification of small crystals often found associated with the surface of fly ash.

X-Ray powder diffraction studies have demonstrated the presence of α-quartz (SiO_2), mullite ($3Al_2O_3 \cdot 2SiO_2$), hematite (Fe_2O_3), magnetite (Fe_3O_4), lime (CaO), and gypsum ($CaSO_4 \cdot 2H_2O$) in aged fly ash (Natusch *et al.*, 1975; Miguel, 1976). However, there is evidence to suggest that crystalline species, associated with aged fly ash may differ from those in the freshly collected material (Fisher *et al.*, 1976, 1978) because of either the presence or lack of moisture in storage atmospheres. In addition to these crystalline species, x-ray powder diffraction patterns indicate the presence of a substantial amount of material that is amorphous to x rays (Fig. 11). The composition of this material has not been established with certainty; however, it is widely accepted that it consists of an impure aluminosilicate glass and constitutes the bulk of the fly ash matrix (Natusch *et al.*, 1975; Walt and Thorne, 1965; Simons and Jeffery, 1960).

Infrared spectroscopic identification of inorganic compounds present in coal fly ash has largely been restricted to the tentative identification of residues of evaporated aqueous leachates (Jakobsen *et al.*, 1978). Several sulfate species have been identified; however, it is not clear whether these represent the actual compounds that existed prior to removal from the fly ash. In addition, studies have been made of glass melts derived from oxides of aluminum, iron, and silicon (Henry *et al.*, 1978). These have provided information that supports the contention that the matrix of coal fly ash is predominantly an aluminosilicate glass.

B.   Trace Elemental Distribution

Further insights into the factors that determine the distribution of elements in a bulk fly ash sample have been obtained from multiele-mental analyses of the 32 subsamples presented in Table IV. Specific concentrations of the elements Al, As, Ba, Ca, Co, Cr, Cs, Dy, Eu, Fe, Ga, Hf, K, La, Mg, Mn, Na, S, Sb, Sc, Si, Sm, Sr, Ta, Ti, Th, Y, and Zn were determined. In addition, x-ray powder diffraction patterns and BET surface areas (by nitrogen adsorption) were obtained (Natusch *et al.*, 1975).

As an aid to the interpretation of the extensive data sets obtained, multivariate statistical analyses, in the form of both common-factor analysis and hierarchical aggregative cluster analysis (Harmon, 1967; Blackith and Reyment, 1971) were employed. Common factor analysis makes it possible to determine the way in which each measured vari-able in the system is related to a set of n factors common to the system as a whole. The important causalities that give rise to the observed data can thus be inferred. By comparison, cluster analysis permits an objective assessment of the similarity between individual subsamples.

The results obtained indicated that the distributional pattern of trace elements in fly ash is controlled by five major factors. These factors have been interpreted to include particle size, particle composition, and the geochemical behavior of the elements.

Thus, specific distributional patterns are observed for the chalco-phile, lithophile, and siderophile elements as classified by Gold-schmidt's Geochemical Series (Bertine and Goldberg, 1971; Coles *et al.*, 1979). It would appear, therefore, that the size factor arises as a result of the volatilization and condensation of certain trace metals as de-scribed earlier (Davison *et al.*, 1974). The dependence on particle com-position possibly reflects the association of some elements (e.g., As and Mn) with certain types of mineral inclusions. The dependence on geochemical class of the elements, in all probability, reflects the differ-ent chemical characteristics of each of these classes under high tem-perature combustion conditions.

SEM-x-ray analysis has provided further insight into the complexity of the matrix composition of coal fly ash. Elemental analysis of mor-phologically similar fly ash particles from the NBS fly ash reference material indicated extreme matrix heterogeneity (Pawley and Fisher,

1977). Particles rich in K, Ti, Fe, S, or Ca were observed. Indeed, nearly all of the Ti in a field of 100 particles could be accounted for by a single Ti-rich particle. It is interesting to note the extreme matrix heterogeneity of individual particles in the NBS fly ash, a material that is well documented as being homogeneous by macroscopic analytical techniques.

C. Surface Composition

As pointed out in previous sections, the inverse dependence of trace elemental concentration on fly ash particle size is generally held to be due to condensation of metallic species onto particle surfaces from the vapor phase (Davison *et al.*, 1974). One would expect, therefore, to find certain volatilizable elements preferentially concentrated on particle surfaces. This has been observed (Linton *et al.*, 1976, 1977; Keyser *et al.*, 1978).

The techniques that have been employed, to date, in analyzing the surface regions of coal fly ash are electron spectrometry for chemical analysis (ESCA), Auger electron spectrometry (AES), and secondary ion mass spectrometry (SIMS). In addition, some surface analytical information is available using electron microprobe x-ray spectrometry. The operational characteristics of these techniques are summarized briefly as follows (Czandérna, 1975; Kane and Larrabee, 1974; Keyser *et al.*, 1978).

The electron microscope (EM) and microprobe (EP) bombard the sample with a focused beam of electrons to stimulate emission of x rays characteristic of the elements present. The technique is useful for analyses of individual micrometer-size particles and has a lateral and depth resolution of about 1 μm, determined by the x-ray emission volume. The electron probe microanalyzer is described in Chapter 48.

Surface analysis capabilities of EM and EP are poor since the depth resolution is very much greater than the thickness of the surface layer normally of interest. Indeed, information about elemental surface predominance can be obtained only by varying the energy of the electron beam (depth penetration) or by ion etching of the outer surface and by comparing elemental ratios for inner and outer surfaces.

The ESCA technique employs an x-ray source to eject core-level electrons from the sample. Energy analysis of the resulting photoelectrons provides chemical bonding information since the bonding energies of the core electron are sympathetic to changes in the electronic structure of the valence level. Elements present at levels greater than 1 at.% in the uppermost 20 Å are detected. Depth profiling is achieved by etching the surface with an ion beam between analyses. For details on ESCA (or XPS) see Volume I, Chapter 11.

The utility of ESCA for individual particle analysis is limited because of the difficulty of focusing x rays to a beam diameter smaller than 1 mm, although recent advances indicate that lateral resolutions of 10 μm are feasible. Normally, the sensitivity of ESCA is insufficient to enable observation of trace constituents unless considerable surface enrichment is encountered.

In AES the emission of Auger electrons is stimulated by bombarding the sample with a beam of electrons. The energy of the secondary Auger electrons is characteristic of the emitting element. Spectra are recorded in the first derivative mode to discriminate against a background of inelastically scattered electrons. Elemental detection limits lie in the range 0.1–1.0 at.% within the analytical volume (depth ~20 Å). Depth profiling is achieved by etching the sample surface with an ion beam (normally Ar^+) as in ESCA. Most AES spectrometers possess microprobe capabilities with incident beam diameters of 1–5 μm.

In SIMS the sample is bombarded with a stream of ions (most commonly, negative oxygen ions) and surface material is physically removed. About 1–10% of the sputtered material is in the form of secondary ions that are mass analyzed by a conventional mass spectrometer.

The ion microprobe represents a special configuration of SIMS in which the primary ion beam can be focused to a diameter of about 3–5 μm. Both individual particle analysis and elemental-mapping capabilities are thus available. Depth profiling constitutes an integral part of the process of secondary ion generation.

A major advantage of SIMS is its extremely high sensitivity, with elemental detection limits ranging from 10^{-2}–10^{-6} at.%, depending on the element and the primary ion used. Typically, it is possible to observe as little as 1 μg/g in the analytical volume, thereby enabling studies of species present at trace levels. Secondary ion mass spectrometry is, however, subject to several types of interferences and artifacts. In particular, spectral interferences from molecular- and multiple-charged ions make the high resolving power of a double-focusing mass spectrometer desirable. Also, volatilization losses and migration of sample ions under the influence of the primary ion beam can give rise to spurious depth profiles. Such effects are often difficult to identify in SIMS since removal of surface material is an integral part of the detection process.

Of the above techniques, AES and SIMS are generally most useful for surface analysis and the depth-profiling studies, owing to their sensitivity and good lateral and depth resolution. Electron spectrometry for chemical analysis, however, has the important advantage of providing information about the identity of molecular species present. With all the techniques, difficulties are encountered in establishing even semiquantitative depth scales, which are normally attempted by calibrating the rate of removal of surface material against that obtained for a standard having a surface layer of known thickness. The main problem, however, lies in matching the matrix composition of the standard to that of the material being studied, which, in the case of coal fly ash, is not well defined.

Surface analysis and depth profiling studies of both individual coal fly ash particles and groups of particles have established that a number of trace elements, including C, Cr, K, Mn, Na, Pb, S, Tl, V, and Zn, are substantially surface enriched, whereas the matrix and minor elements, Al, Ca, Fe, Mg, Si, and Ti, are not (Linton *et al.*, 1977). This observation clearly supports the hypothesis that the more volatile elements, or their compounds, are vaporized during combustion and then condense on the surfaces of coentrained fly ash particles at lower temperatures.

Depth profiling studies of fly ash have also demonstrated the utility of using instrumental techniques in conjunction with solvent leaching to remove soluble surface material. An example of this approach is presented in Fig. 12 for the elements Pb and Tl. This study demonstrated that extraction of fly ash with water or dimethyl sulfoxide removes the surface layer of both elements. Determination of the amounts of Pb and Tl in solution then enables estimation of the amounts present in the surface layer. Assuming a surface layer thickness of 300 Å, one obtains average concentrations of 2700 μg/g for Pb and 920 μg/g for Tl in the surface layer as compared to bulk particulate concentrations of 620 μg/g and 30 μg/g for these elements. Similar estimates for several other trace elements are presented in Table VII (Natusch, 1978a).

Solvent leaching can also provide some insight into the chemical forms of elements present. For example, although AES and SIMS indicate little surface enrichment of iron, aqueous leaching rapidly removes this element from the surface region, thereby indicating its presence in a readily soluble form. Similarly, comparison of the leaching and depth profiles of K, Fe, Na, and S suggests that these elements may be associated with each other in the surface layer, possibly as alkali–iron sulfates. Further support of the existence of simple and/or complex sulfates is provided by ESCA studies that show that the oxi-

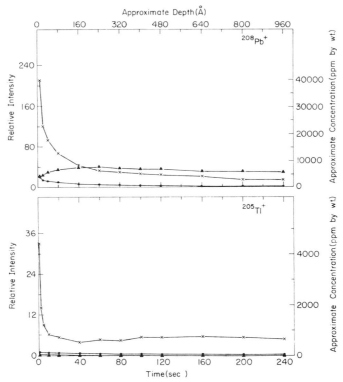

Fig. 12 Ion microprobe depth profiles for Pb and Tl for unextracted and extracted fly ash samples. X: unextracted, ▲: DMSO extracted, ●: H₂O extracted. (From Linton et al., 1977, p. 1514.)

dation states of Fe and S in the surface region are +3 and +6, respectively (Wallace, 1974).

Surface analytical results, such as those presented in Fig. 12 and Table VII demonstrate the considerable differences in composition that exist between the interior of fly ash particles and their external surface. Since it is the particle surface that is in contact with the external environment, determination of surface composition is of considerable importance, as previously discussed. Finally, it should be remarked that there are no coherent data that relate surface composition to particle size for fly ash. However, if the volatilization–condensation process is primarily responsible for surface enrichment of trace elements, then one would not expect surface concentrations to vary greatly with particle size. This is because the amount of vapor deposited is proportional to surface area, thereby resulting in a constant elemental con-

TABLE VII *Estimated Surface
Concentrations of Elements in Coal Fly Ash*[a]

Element	Bulk concentration ($\mu g/g$)	Estimated surface concentration in 300 Å layer ($\mu g/g$)
As	600	1500
Cd	24	700
Co	65	440
Cr	400	1400
Pb	620	2700
S	7100	252,000
V	380	760

[a] From Natusch (1978a).

centration per unit surface area. Of course, if other mechanisms are responsible for or contribute to surface enrichment (e.g., agglomeration of accumulation mode particles with coarse particles or thermal diffusion of trace species to the surface of molten particles), then some dependence on particle size would be anticipated.

D. Solubility and Leachability

A number of workers (Shannon and Fine, 1974; Theis and Wirth, 1977; James *et al.*, 1977; Dreesen *et al.*, 1977) have reported that the bulk solubility of coal fly ash in water is very low and rarely exceeds 2–3% by weight. The bulk solubility is clearly a property of both the glassy and crystalline matrix materials identified earlier, and one would expect elements that are either chemically or physically trapped within this matrix to exhibit low solubility. On the other hand, at least some of the material present in the surface skin is readily soluble in water (Fig. 12). Indeed, it is now quite well established (Linton *et al.*, 1977; Natusch, 1978a; Fisher *et al.*, 1978e) that most of the soluble fraction of fly ash is derived from this surface layer and is thus very rich in trace elements.

At the present time, there is considerable confusion involved in interpreting and understanding results obtained from different studies of fly ash solubility. Specifically, quite different results are obtained by workers using apparently similar laboratory leaching techniques and none are readily transferrable to field studies. It is appropriate, therefore, to consider the factors that control leachate composition under both laboratory and field conditions and to standardize one or more laboratory leaching techniques whose results bear some relationship to real field conditions.

Some insight into leaching behavior can be obtained by recognizing that soluble inorganic species M_iA_i associated with fly ash particles can dissociate into their component cations M_i and anions A_i in aqueous solution, and that both dissolution and deposition (e.g., precipitation) processes can occur. Furthermore, cations and anions present in solution can interact so as to set up multiple equilibria that may involve ion pairing, complexation, precipitation, or acid–base behavior (M_iA_j or M_jA_i). The result can be expressed, simplistically by the equations

$$
P \sum_i^n M_iA_i \underset{k_{-1}}{\overset{k_1}{\rightleftharpoons}} P + \sum_i^n M_i + \sum_i^n A_i
$$

$$
\begin{array}{cc}
+ & + \\[4pt]
\displaystyle\sum_j^m A_j & \displaystyle\sum_j^m M_j \\[6pt]
k_{-2} \updownarrow k_2 & k_{-3} \updownarrow k_3 \\[6pt]
\displaystyle\sum_{ij}^{mn} M_iA_j & \displaystyle\sum_{ij}^{mn} M_jA_i.
\end{array} \tag{10}
$$

Here P represents the parent particles and k_n, k_{-n} are the rate constants for forward and reverse reactions, respectively. It is apparent from Eq. (10) that, when leaching studies are conducted under batch conditions, such that the amount of fly ash and solvent are maintained constant, an equilibrium will be established between particulate and solution species. Consequently, only a fraction of the potentially soluble material will enter solution. On the other hand, if conditions are such that soluble material is continuously removed by provision of fresh solvent or by providing a large solution sink in the form of complexing ligands or acids, then all potentially soluble material will ultimately enter solution.

Matusiewicz and Natusch (1979) have conducted very extensive studies to demonstrate the validity of Eq. (10). They have established that the rate and extent of leaching depend upon the leaching method, the fly ash:solution ratio, temperature, pH, complexing agents, particle size, and fly ash origin exactly as predicted from Eq. (10) for both equilibrium and nonequilibrium conditions. When equilibrium is established between particulate and solution species, as in batch leaching, little dependence on particle size is observed since the amount of a given element is determined by its solution concentration (solubility) that is only weakly related to the amount of solid phase present. Exhaustive (nonequilibrium) leaching, however, removes all soluble material, the amount of which is directly related to particle size because of its presence in the fly ash surface layer.

It is apparent from the foregoing remarks that any leaching process that establishes the solid–solution equilibrium given in Eq. (10) will result in changes in solution composition with equilbrium position. It is hardly surprising, therefore, that widely differing results are obtained by workers who use different leaching conditions. Nevertheless, some significant generalizations can be made. First, it is readily apparent that much higher proportions of most trace elements are soluble than is the case for matrix elements (Table VIII). This is due, in part, to the predominance of trace elements in the particle surface layers in quite soluble forms (probably sulfates, oxides, and carbonates). Second, it is clear that, under batch leaching conditions, such as are most likely to occur in the field, the amount of each element entering solution is strongly dependent on the dilution (fly ash:water ratio) and the initial pH (Matusiewicz and Natusch, 1979; Dreesen *et al.*, 1977; Theis and Wirth, 1977).

TABLE VIII *Percentage of Elements Leached from a Typical Coal Fly Ash[a]*

Element	% Leached
Al	0.2
B	5
Ba	4
Ca	35
Cr	30
K	40
Mg	0.2
Mn	0.4
Mo	85
Na	10
P	6
Pb	100
Si	0.1
Sr	6
Zn	6

[a] From Matusiewicz and Natusch (1979).

E. Organic Constituents

To date, no exhaustive determination of organic species associated
with coal fly ash has been reported. Rather, emphasis has been placed
primarily on the determination of polycylic organic matter (POM) in
fly ash, due to the potential carcinogenicity of several compounds of
this type (Committee on Biological Effects of Atmospheric Pollutants,
1972). For the most part, the several studies of POM in fly ash have
indicated either extremely low or undetectable levels (Committee on
Biological Effects of Atmospheric Pollutants, 1972).

In a survey of fly ashes representing several coals and combustion
conditions, Aslund et al., (1978) found no individual species of POM
to be present at concentrations greater than 20 ng/g. A number of other
unidentified organic compounds were, however, observed at somewhat higher (10×) concentrations. It is important to note that all of
these studies have considered fly ash collected in bulk from power
plant control devices.

Only a few studies have been made of POM present in fly ash that
was actually emitted and collected from the atmosphere (Natusch,
1978b; Tomkins, 1978; Stahley, 1976). However, all have indicated concentrations that are very much greater than encountered in fly ash
collected within the plant. This apparent paradox has been explained
by Natusch and Tomkins (1977) who postulate that POM (and probably
other organic species) are present as gases at the temperatures encountered within a power plant but rapidly and quantitatively adsorb onto
surfaces of emitted fly ash particles as the temperature falls on leaving
the stack. Both laboratory (Miguel et al., 1979) and field (Miguel, 1976;
Natusch, 1978b) studies support this hypothesis.

The actual compounds that have been identified in emitted fly ash
are listed in Tables IX and X, which present the results of two separate
studies in which specific concentrations inside and outside the plant,
and volume concentrations in the plume, were determined. To our
knowledge, only one study has actually measured POM concentrations
as a function of particle size for emitted fly ash (Natusch, 1978b). The
results indicated little convincing dependence of concentration on aerodynamic particle size over the range <1.1 to >7.0 μm. However, the
fly ash in question was derived from a small plant that employed a
chain grate stoker, and the particles were found to be extremely irregular in outline. Furthermore, there was very little change in specific
surface area over the size range collected. We do not, therefore, consider these results to be conclusive.

In fact, if the temperature dependent adsorption mechanism pro-

TABLE IX *Measurement of Polycyclic Organic Matter Emitted from a Coal-Fired Power Plant Stack[a]*

Compound	Specific concentration (μg/g)	
	Inside stack	Outside stack
Fluorene	ND[b]	Trace
Phenanthrene	ND	9
Fluoranthene	ND	19
Pyrene	ND	12
Benzofluorene	ND	2
1-Methylpyrene	ND	1
Benzophenanthrene	ND	3
Benzo[a]pyrene	ND	5
Total fluorescence	3.61×10^{-3} units	3.68 units

[a] From Tomkins (1978).

[b] Not detectable.

posed by Natusch and Tomkins (1977) is correct, one would expect the specific concentration of organic species to vary in proportion to the surface area of the fly ash particles. There is some indirect evidence for this behavior (Chrisp *et al.*, 1978; Fisher *et al.*, 1979c), but further work is clearly required. It has been established, however, that adsorption of POM (pyrene) onto fly ash under laboratory conditions occurs, to significantly different extents, on different fly ashes and on magnetic and nonmagnetic fractions of a given fly ash (Miguel, 1976; Korfmacher *et al.*, 1979a).

Finally, it should be mentioned that POM associated with fly ash may undergo chemical transformation following adsorption and emis-

TABLE X *Emission Factors for Polycyclic Organic Matter from Coal Fired Furnaces in (pounds/ton of coal) $\times 10^4$ [a]*

Species	Pulverized firing	Chain grate stoker	Hand fired
Benzo[a]pyrene	0.2–0.52	0.3	3520
Pyrene	0.8–1.6	3.5	5260
Benzo[e]pyrene	0–2.3	1.1	880
Perylene	0–0.6	—	526
Fluoranthene	—	6.0	8800

[a] Committee on Biological Effects of Atmospheric Pollutants, 1972.

sion. In this regard, Korfmacher *et al.* (1979b) have shown that adsorption onto coal fly ash effectively stabilized most POM against photochemical decomposition, but actually promotes rapid (hours to days) nonphotochemical oxidation of polycyclic aromatic compounds possessing one benzylic carbon atom. Furthermore, Hughes and Natusch (1978) have shown that exposure of POM adsorbed on fly ash to typical plume concentrations of sulfur dioxide and nitrogen oxides results in very rapid formation of a variety of derivatives having sulfur or nitrogen containing substituents. It is possible, therefore, that the chemical nature of POM associated with coal fly ash emitted from a power plant is likely to change dramatically with time and distance from the plant.

ACKNOWLEDGMENTS

The authors are endebted to the untiring and enthusiastic support of Mr. Bruce A. Prentice, to the technical assistance and advice of Dr. David R. Taylor, and to the excellent clerical support of Mr. Charles E. Baty.

REFERENCES

Andren, A. W., and Klein, D. H. (1975). *Environ. Sci. Technol.* **9**, 856–858.
Aslund, A., Miller, M., Natusch, D. F. S., and Taylor, D. R. (1978). Unpublished results.
Bertine, K. K., and Goldberg, E. D. (1971). *Science* **173**, 233–235.
Bickelhaupt, R. E. (1974). *J. Air Pollut. Control Assoc.* **24**, 251–255.
Bickelhaupt, R. E. (1975). *J. Air Pollut. Control Assoc.* **25**, 148–152.
Biermann, A. H., and Ondov, J. M. (1978). "Application of Surface-Deposition Models to Size-Fractionated Coal Fly Ash," Prepr. UCRL-81361. Lawrence Livermore Lab., Livermore, California.
Billings, C. E., and Matson, W. R. (1972). *Science* **176**, 1232–1233.
Billings, C. E., Sacco, A. M., Matson, W. R., Griffin, R. M., Coniglio, W. R., and Harley, R. A. (1973). *J. Air Pollut. Control Assoc.* **23**, 773–777.
Blackith, R. E., and Reyment, R. A. (1971). "Multivariate Morphometrics." Academic Press, New York.
Butcher, S. S., and Charlson, R. J. (1972). "An Introduction to Air Chemistry." Academic Press, New York.
Campbell, J. A., Laul, J. C., Nielson, K. K., and Smith, R. D. (1978). *Anal. Chem.* **50**, 1032–1040.
Carter, J. L., and MacGregor, I. D. (1970). *Science* **167**, 661–663.
Cheng, R. J., Mohnen, V. A., Shen, T. T., Current, M., and Hudson, J. B. (1976). *J. Air Pollut. Control Assoc.* **26**, 727–824.
Chrisp, C. E., Fisher, G. L., and Lammert, J. E. (1978). *Science* **199**, 73–74.
Coles, D. G., Ragaini, R. C., and Ondov, J. M. (1978). *Environ. Sci. Technol.*, **12**, 442–446.
Coles, D. G., Ragaini, R. C., Ondov, J. M., Fisher, G. L., Silberman, D., and Prentice, B. A. (1979). *Environ. Sci. Technol.*, **13**, 455–459.
Committee on Biological Effects of Atmospheric Pollutants (1972). "Particulate Polycyclic Organic Matter." Natl. Acad. Sci., Washington, D.C.
Czandérna, A. W., ed. (1975). "Methods of Surface Analysis." Elsevier, Amsterdam.

Davison, R. L., Natusch, D. F. S., Wallace, J. R., and Evans, C. A., Jr. (1974). *Environ. Sci. Technol.* **8**, 1107–1113.

Diehl, R. C., Hattman, E. A., Schultz, H., and Haren, R. J. (1972). "Fate of Trace Mercury in the Combustion of Coal," Tech. Prog. Rep. No. 54. Bur. Mines, Managing Coal Wastes Pollut. Program, Pittsburgh Energy Res. Cent., Pittsburgh, Pennsylvania.

Dreesen, D. R., Gladney, E. S., Owens, J. W., Perkins, B. L., Wienke, C. L., and Wangen, L. E. (1977). *Environ. Sci. Technol.* **11**, 1017–1019.

Fisher, G. L., and Chrisp, C. E. (1978). *Symp. Appl. Short-Term Bioassays Fractionation Anal. Complex Environ. Mixtures* (Waters, M. D., Nesnow, S., Huisingh, J. L., Sandhu, S. S., and Claxton, L., eds.) *EPA 60019-78-027, pp. 441–462. Williamsburg, Va.*

Fisher, G. L., and Natusch, D. F. S. (1979). U.S. Dept. Energy, NTIS UCD-472-502.

Fisher, G. L., Chang, D. P. Y., and Brummer, M. (1976). *Science* **192**, 553–555.

Fisher, G. L., Prentice, B. A., Silberman, D., Ondov, J. M., Ragaini, R. C., Biermann, A. H., McFarland, A. R., and Pawley, J. B. (1977). *Jt. Chem. Inst. Can./Am. Chem. Soc. Meet., 2nd, Div. Fuel Chem. Symp., Properties Coal Ash, Montreal.*

Fisher, G. L., Prentice, B. A., Silberman, D., Ondov, J. M., Ragaini, R. C., Biermann, A. H., and McFarland, A. R. (1978). *Environ. Sci. Technol.* **12**, 447–451.

Fisher, G. L., Chrisp, C. E., and Hayes, T. L. (1979a). *Conf. Carbonaceous Part. Atmos., Berkeley, Calif.* in press.

Fisher, G. L., Hayes, T. L., Prentice, B. A., and Lai, C. (1979b). *In* "Radiobiology Laboratory Annual Report," UCD 472-125, in press, Univ. of California, Davis.

Fisher, G. L., Chrisp, C. E., and Raabe, O. G. (1979c). *Science* **204**, 879–881.

Fisher, G. L., Silberman, D., Prentice, B. A., Heft, R. E., and Ondov, J. M. (1979d). *Environ. Sci. Technol.* **13**, 689–693.

Flagan, R. C., and Friedlander, S. K. (1976). *Symp. Aerosol Sci. Technol., Meet. Am. Inst. Chem. Eng., 82nd, Atlantic City, N.J.*

Gay, N. C. (1976). *Science* **194**, 724–725.

Gibbon, D. L. (1978). Personal communication, Guilford Coll., Greensboro, North Carolina.

Gladney, E. S., Small, J. A., Gordon, G. E., and Zoller, W. H. (1976). *Atmos. Environ.* **10**, 1071–1077.

Gordon, G. E., and Zoller, W. H. (1973). *Proc. Annu. NSF Trace Contaminants Conf., 1st, Oak Ridge Natl. Lab., Oak Ridge, Tenn.* CONF-730802, pp. 314–325.

Gordon, G. E., Zoller, W. H., and Gladney, E. S. (1974). *In* "Trace Substances in Environmental Health—VII. A Symposium" (D. D. Hemphill, ed.), pp. 161–166. Univ. of Missouri, Columbia.

Harmon, H. H. (1967). "Modern Factor Analysis," 2nd Ed. Univ. of Chicago Press, Chicago, Illinois.

Henry, W. M., Mitchel, R. I., and Knapp, K. T. (1978). *Proc. Workshop Primary Sulfate Emiss. Combust. Sources, Southern Pines, North Carolina* EPA Rep. No. 600/9-78-020b, Vol. II, pp. 185–208.

Hughes, M. M., and Natusch, D. F. S. (1978). Unpublished results.

Jakobsen, R. J., Gendreau, R. M., Henry, W. M., and Knapp, K. T. (1978). *Proc. Workshop Primary Sulfate Emiss. Combust. Sources, Southern Pines, North Carolina* EPA Rep. No. 600/9-78-020b, Vol. I, pp. 253–274.

James, W. D., Janghorbani, M., and Baxter, T. (1977). *Anal. Chem.* **49**, 1994–1997.

Kaakinen, J. W., Jorden, R. M., Lawasani, M. H., and West, R. E. (1975). *Environ. Sci. Technol.* **9**, 862–69.

Kalb, W. G. (1975). *In* "Trace Elements in Fuel" (S. P. Babu, ed.), Advances in Chemistry Series, No. 141, pp. 154–174. Am. Chem. Soc., Washington, D.C.

Kane, P. F., and Larrabee, G. B., eds. (1974). "Characterization of Solid Surfaces." Plenum, New York.

Keyser, T. R., Natusch, D. F. S., Evans, C. A., and Linton, R. W. (1978). *Environ. Sci. Technol.* **12,** 768–773.

Kim, H., and Natusch, D. F. S. (1978). Unpublished results.

Klein, D. H., Andren, A. W., Carter, J. A., Emery, J. F., Feldman, C., Fulkerson, W., Lyon, W. S., Ogle, J. C., Talmi, Y., Van Hook, R. I., and Bolton, N. E. (1975). *Environ. Sci. Technol.* **9,** 973–979.

Koltrappa, P., and Light, M. E. (1972). *Rev. Sci. Instrum.* **43,** 1106–1112.

Korfmacher, W. A., Miguel, A. H., Mamantov, G., Wehry, E. L., and Natusch, D. F. S. (1979a). *Environ. Sci. Technol.,* in press.

Korfmacher, W. A., Natusch, D. F. S., Taylor, D. R., Mamantov, G., and Wehry, E. L. (1979b). *Science,* in press.

Linton, R. W., Loh, A., Natusch, D. F. S., Evans, C. A., Jr., and Williams, P. (1976). *Science* **191,** 852–854.

Linton, R. W., Williams, P., Evans, C. A., Jr., and Natusch, D. F. S. (1977). *Anal. Chem.* **49,** 1514–1521.

McCrone, W. C., and Delly, J. G. (1973). "The Particle Atlas," 2nd Ed. Ann Arbor Sci. Publ., Ann Arbor, Michigan.

McFarland, A. R., Bertch, R. W., Fisher, G. L., and Prentice, B. A. (1977). *Environ. Sci. Technol.* **11,** 781–784.

Matthews, B. J., and Kemp, R. F. (1971). "Development of Laser Instrumentation for Particle Measurement—Final Report," TRW Inc. Rep. No. 14103-6003-R0-00, Redondo Beach, California.

Matusiewicz, H., and Natusch, D. F. S. (1979). Submitted for publication.

Miguel, A. H. (1976). Ph.D. Thesis, Univ. of Illinois, Urbana.

Miguel, A. H., Schure, M. R., and Natusch, D. F. S. (1979). Submitted for publication.

Nagy, B., Scott, W. M., Modzeleski, V., Nagy, L. A., Drew, C. M., McEwan, W., Thomas, J. E., Hamilton, P. B., and Urey, H. C. (1970). *Nature (London)* **225,** 1028–1032.

Natusch, D. F. S. (1978a). *Proc. Workshop Primary Emiss. Combust. Sources, Southern Pines, North Carolina.* EPA Rep. No. 600/9-78-020b, Vol. II, pp. 149–164.

Natusch, D. F. S. (1978b). *Environ. Health Perspect.* **22,** 79–90.

Natusch, D. F. S. (1978c). Unpublished results.

Natusch, D. F. S., and Tomkins, B. (1977). In "Carcinogenesis—A Comprehensive Survey" (P. W. Jones and R. I. Freudenthal, eds.), Vol. 3, pp. 145–153. Raven, New York.

Natusch, D. F. S., and Wallace, J. R. (1974). *Science* **186,** 695–699.

Natusch, D. F. S., Wallace, J. R., and Evans, C. A., Jr. (1974). *Science* **183,** 202–204.

Natusch, D. F. S., Bauer, C. F., Matusiewicz, H., Evans, C. A., Jr., Baker, J., Loh, A., Linto, R. W., and Hopke, P. K. (1975). *Int. Conf. Heavy Met. Environ., Toronto* pp. 553–576.

Natusch, D. F. S., Bauer, C. F., and Loh, A. (1978). In "Air Pollution Control" (W. Strauss, ed.), Part III, pp. 217–315. Wiley, New York.

Newton, G. J., Raabe, O. G., and Mokler, B. V. (1977). *J. Aerosol Sci.* **8,** 339–347.

Olsen, K. W., and Skogerboe, R. K. (1975). *Environ. Sci. Technol.* **9,** 227–230.

Ondov, J. M., Ragaini, R. C., and Biermann, A. H. (1976). *Prepr. Pap. Natl. Meet., Div. Environ. Chem., Am. Chem. Soc., San Francisco, Calif.* 200–203.

Ondov, J. M., Ragaini, R. C., Heft, R. E., Fisher, G. L., Silberman, D., and Prentice, B. A. (1977a). *Mater. Res. Symp., 8th, Natl. Bur. Stand., Gaithersburg, Md.* pp. 565–572.

Ondov, J. M., Ragaini, R. C., Biermann, A. H., Choquette, C. E., Gordon, G. E., and

Zoller, W. H. (1977b). *Prepr. Pap. Natl. Meet., Div. Environ. Chem., Am. Chem. Soc., New Orleans, La.* 000-000.

Ondov, J. M., Ragaini, R. C., and Biermann, A. H. (1977c). "Elemental Emissions and Particle-Size Distributions of Minor and Trace Emissions at Two Western Coal-Fired Power Plants Equipped with Cold-Side Electrostatic Precipitators," Prepr. UCRL-80254. Lawrence Livermore Lab., Livermore, California.

Ondov, J. M., Ragaini, R. C., and Biermann, A. H. (1979) *Environ. Sci. Technol.* **13,** 598-607.

Padia, A. S., Sarofim, A. F., and Howard, J. B. (1976). *Combust. Inst. Cent. States Sect. Spring Meet.* (Available from: A. S. Padia, Halcon International, Inc., 2 Park Ave, N.Y.)

Parks, N. J., Raabe, O. G., Bradley, E. W., Teague, S., and Hopkins, B. (1977). *In* "Radiobiology Laboratory Annual Report," NTIS UCD 472-124, pp. 77-82. Univ. of California, Davis.

Pawley, J. B., and Fisher, G. L. (1977). *J. Microsc. (Oxford)* **110,** 87-101.

Raabe, O. G. (1976). *APCA J.* **26,** 856-860.

Raabe, O. G. (1978). *Environ. Sci. Technol.* **12,** 1162-1167.

Raask, E. (1966). *ASME Trans. J. Eng. Power* **88,** 40-44.

Raask, E. (1968). *J. Inst. Fuel* **43,** 339-44.

Ragaini, R. C., and Ondov, J. M. (1977). *J. Radioanal. Chem.* **37,** 679-691.

Ruch, R. R., Gluskoter, H. J., and Shimp, J. F. (1974). *Ill. State Geol. Surv., Rep.* No. 72.

Sarofim, A. F., Howard, J. B., and Padia, A. S. (1977). *Combust. Sci. Technol.* **16,** 187-204.

Shannon, D. G., and Fine, L. O. (1974). *Environ. Sci. Technol.* **8,** 1026-1028.

Simons, H. S., and Jeffery, J. W. (1960). *J. Appl. Chem.* **10,** 328-336.

Small, J. A. (1976). Ph.D. Thesis, Univ. of Maryland, College Park.

Smith, R. D., Campbell, J. A., and Nielson, K. K. (1979). *Environ. Sci. Technol.* **13,** 553-558.

Southern Research Institute (1975). "Survey of Information on Fine Particle Control," Rep. No. 259. Electr. Power Res. Inst., Palo Alto, California.

Stahley, S. (1976). Personal communication.

Task Group on Lung Dynamics. (1966). *Health Phys.* **12,** 173-207.

Theis, T. L., and Wirth, J. L. (1977). *Environ. Sci. Technol.* **11,** 1096-1100.

Tomkins, B. A. (1978). Ph.D. Thesis, Univ. of Illinois, Urbana.

Vandegraft, A. E., Shannon, L. J., and Gorman, P. G. (1973). *Chem. Eng.* **80,** 107-114.

Waits, G. A., McKay, D. S., and Gibbon, D. L. (1978). Personal communication, SN6, NASA-JSC, Houston, Texas.

Wallace, J. R. (1974). Ph.D. Thesis, Univ. of Illinois, Urbana.

Walt, J. D., and Thorne, D. J. (1965). *J. Appl. Chem.* **15,** 585-604.

Wangen, L. E., and Wienke, C. L. (1976). "A Review of Trace Element Studies Related To Coal Combustion in the Four Corners Area of New Mexico," Informal Rep. LA-6401-MS. Los Alamos Sci. Lab., Los Alamos, New Mexico.

Whitby, K. T. (1977). *Mater. Res. Symp., 8th, Natl. Bur. Stand., Gaitherburg, Md.* pp. 165-173.

White, H. J. (1963). "Industrial Electrostatic Precipitation." Addison-Wesley, Reading, Massachusetts.

Yeh, H. C., Phalen, R. F., and Raabe, O. G. (1976). *Environ. Health Perspect.* **15,** 147-156.

Zoller, W. H., Gladney, E. S., Gordon, G. E., and Bors, J. J. (1974). *In* "Trace Substances in Environmental Health—VIII. A Symposium" (D. D. Hemphill, ed.), pp. 167-172. Univ. of Missouri, Columbia.

Chapter 55

Problems of Oxygen Stoichiometry in Analyses of Coal and Related Materials

Alexis Volborth†

NUCLEAR RADIATION CENTER AND
DEPARTMENT OF GEOLOGY
WASHINGTON STATE UNIVERSITY
PULLMAN, WASHINGTON

I. INTRODUCTION

Classical coal analysis today still utilizes many methods developed nearly 100 years ago. Some of these cannot be considered sufficient any more, and do not provide the type of information required by modern technology. Also, while some of these methods must still be considered relatively accurate and adequate, e.g., hydrogen and carbon in the ultimate analysis, others are techniques that result in bulk data of indeterminate character signifying neither pure chemical nor pure physical properties of the coal analyzed, e.g., "moisture" in the proximate coal analysis. Furthermore, the uncertainties of the classical coal analysis are compounded by the use of direct "estimations" and reporting such quantities as "oxygen by difference" and "fixed carbon,"

† Former address: Chemistry Department, University of California, Irvine; and Chemistry and Geology Departments, North Dakota State University, Fargo.

both derived by the expedient method of subtraction of the sum of the determined quantities from 100.

These practices lead to artificially balanced summations while all the errors and biases of the results accumulate in the "by difference" figures and thus remain undetected and, what is worse, reports based on such appear to be misleadingly correct.

Coal chemists and engineers are well aware of the possible inaccuracies, errors, and confusion that result when these "estimated" quantities are used in research or to rank and classify the coal or coal products. The purpose of this chapter is to attempt to outline some of the problems that exist in coal analysis today and to try to point toward possible better solutions using instrumental techniques. A few more quantitative and specific approaches utilizing fast-neutron activation for oxygen determination, described in Chapter 47, are suggested in this chapter. The stoichiometry of oxygen in coal and coal related materials in general, and some oxidation and coalification processes, are discussed more fundamentally.

II.   OXYGEN AND ANALYSIS OF COAL

A.   Proximate Analysis and Upgrading of Coal

1.   *Moisture*

Proximate analysis of coal consists of determination, by ASTM standardized methods, of moisture, volatile matter, and ash, and calculating fixed carbon "by difference." These techniques are described in detail in the ASTM Standards (ASTM, 1974, 1975, 1976) and in U.S. Bureau of Mines (1967) Bulletin 638. See also Volume I, Chapter 6, Section III. An excellent critique on limitations (Rees, 1966), and an up-to-date discussion of problems and solutions of classical coal analysis (Given and Yarzab, 1975, see also Volume II, Chapter 20) emphasize the inadequacies, the difficulties in interpretation, and the misleading nature of reporting of coal analysis results in the context of what should be "modern" coal chemistry.

Volborth (1976c,d) and co-workers (Volborth *et al.*, 1977b,c) have pointed out the significance of adding oxygen determination to that of "moisture" before and after drying of coal. The increasing role of instrumental analysis of coal has been reviewed by Ignasiak *et al.* (1975). In this chapter only the information supplied by the oxygen determination in coal prior to and after determination of moisture in coal ash will be discussed.

The reason that the term "moisture" is misleading is because a chemist new to coal science will often equate this with water (H_2O), whereas, according to the ASTM approved procedures, it is actually a weight loss determined under strictly controlled drying conditions. Therefore, "moisture" is a quantity "by definition" determined during coal analysis. To make the matter more complicated, several standardized procedures are used. This is because of the nature of coal, its instability and varying handling and the changes of ambient conditions in laboratories. Rees (1966) distinguishes four fundamental categories of moisture in coal: (1) inherent or equilibrium moisture that a coal can hold when in equilibrium with the atmosphere; (2) water of hydration moisture held by the mineral matter in coal; (3) surface moisture in excess of the inherent moisture, and (4) decomposition moisture derived theoretically from the decomposing "coal molecule" at temperatures above 200°C. For more information on moisture in coal, see Volume I, Chapter 7.

In practice, the coal chemist most frequently determines the weight loss at 105°C by the so-called "oven method" of which there are two main variants, depending on the coal grain size. This routinely reported moisture value does not correspond to any of the previously mentioned fundamental categories. Besides, only if the water is specifically absorbed in a U-tube filled with a dessicant is it possible to determine the H_2O by weighing. Distillation and collection in a graduated tube usually results in two immiscible liquids, one of which is water, the other a mixture of hydrocarbons. The lighter hydrocarbons escape as gases unless specifically trapped. This process shows that in addition to water, the quantity called "moisture" contains varying losses other than H_2O depending on the coal, the temperature, and the time period. Heating in air may, in addition, cause oxidation and weight gain of the residue, depending on the type of coal, thus resulting in low data for the "weight loss."

These considerations apply to coal conversion when working at higher temperatures with coal as in coal gasification, liquefaction, solvent refining, combustion, carbonization, and in fluidized-bed utilization processes where the relative quantities of water (H_2O) versus evolved hydrocarbons and other gases are even more difficult to estimate unless one also monitors the oxygen content starting with the "feed stocks" or original material, including oxygen analyses of all the interim products up to the final ash materials. It is evident that if oxygen is monitored during all stages of moisture determination, any gross discrepancies between moisture calculated as H_2O and the equivalent oxygen loss as compared with oxygen retained in the solid product will signal to the analyst and the engineer that either significant volatile hydrocarbons are lost or, if more oxygen is found in the solids than expected, that oxidation has occurred and the "moisture" results thus must be low. It is for this purpose that we have mentioned industrial processes. The same considerations used by us in monitoring moisture should apply to a greater extent to industrial processes, even though these industrial interim steps do not constitute nor require at present moisture determination as such.

The complexities and difficulties of "moisture" determination and of the interpretation of results, as well as the existence of a multitude of different methods, have prompted us to attempt to simplify and further quantify the coal analysis by the use of FNAA determined oxygen data (see Chapter 47) before and after drying ("as received" and "dried" 105°C) coal. This approach can be applied irrespective of the method of drying. It consists of calculating the theoretical quantity of water (H_2O) lost during the drying of coal, based on the total loss of oxygen between the steps, and assuming that all of the loss was water. The equation for this calculation is

$$\% \text{ moisture calculated} = 100 \left(\frac{O_{ar} - O_d}{88.81 - O_{ar}} \right) \bigg/ 1 + \left(\frac{O_{ar} - O_d}{88.81 - O_{ar}} \right), \quad (1)$$

where O_{ar} is the percent oxygen determined by FNAA on the coal sample as received by the laboratory, and O_d is the percent oxygen determined on the dried sample (Miller and Volborth, 1976). In Table I six lignites from the Wyoming Wyodak Bed, kindly supplied to us by Gary B. Glass of the Wyoming Geological Survey, have been analyzed and recalculated in this manner. The moisture determination, repeated at our laboratories at the University of California at Irvine (UCI) and compared with the U.S. Bureau of Mines (USBM) report data, and oxygen are determined on wet samples "as received" as well as on samples "dried" at 105°C. This compilation shows that the assumption that all of the weight loss was water (H_2O) is probably correct in all of these lignites within about 1%. It also shows that by performing the two oxygen determinations on coals that do not contain considerable light hydrocarbons and do not markedly oxidize, one can estimate the "moisture" in terms of water, which is more informative for the coal chemist. In the last listed lignite K-46430, the UCI-determined "moisture" is proven to be more correct on this sample than that reported by the USBM [The fact the USBM report may include a printing error (21.9

TABLE I *Comparison of Moisture Calculated Based on FNA Analyses for Oxygen with Moisture by the Classical Methods in Six Wyoming Lignites*

Lignite, Wyoming Wyodak bed coal no.	% Moisture		% Oxygen		
	USBM Report	UCI (±0.1)	As received	Dried (105°C)	% Moisture[a] calculated
K-46565	29.9	31.1	43.64	22.99	31.37
K-46566	27.2	27.5	42.83	23.23	29.89
K-46216	22.6	21.1	36.38	21.48	22.13
K-46217	21.4	21.1	36.10	22.20	20.86
K-46218	21.1	20.9	35.90	22.02	20.79
K-46430	21.9	27.4	42.23	25.10	26.92

[a] See text and Eq. (1).

= 27.9?) does not alter this demonstration of our ability to detect serious discrepancies when they appear in coal analysis reports and this is not meant as a criticism of the U.S.B.M. Rather, the data further demonstrate the high accuracy of these U.S.B.M. coal analyses.] Extending this approach to coals dried at higher temperatures should produce larger discrepancies and enable a rough estimate of the percentage of nonoxygen containing gases, despite the fact that evolution of such gases as CO_2, CO, N, and SO_2, with concurrent oxidation complicates the calculation, requiring a study by gas chromatography or mass spectroscopy of the evolving gas compositions. That considerable discrepancies between the "moisture" based on oxygen determination and moisture based on the old methods exist is shown on some HVA and medium volatile bituminous coals kindly supplied to us by A. Davis of The Pennsylvania State University (Miller and Volborth, 1976, Table II). Further work to explain the nature and reasons for these differences is needed and is presently in progress.

2. Ash

The ASTM standards (ASTM, 1974, 1975, 1976), the U.S. Bureau of Mines (1967) Bull. 638, Shipley (1962), and Rees (1966) describe in detail the determination of high temperature ash (HTA) and the calculations of results (see also British Standards Institution, 1971) as well as the nature and origin of coal ash. Problems encountered in recalculation of ash as mineral matter are also specifically discussed by Parr (1932, pp. 49–50) and King *et al.* (1936) and very thoroughly treated by Given and Yarzab (1975) and in Volume II, Chapter 20, Section IV,B,1.

Direct determination of oxygen in coal ash was accomplished by

Block and Dams (1974), James *et al.* (1975, 1976), Hamrin *et al.* (1975, 1977), Volborth (1976c), and Volborth *et al.* (1977b). Referring to Hamrin *et al.* (1975), a warning not to equate determined oxygen by FNAA with "oxygen by difference" computed on dry "ash-containing" basis was sounded by Given (1976). This is because "oxygen by difference" is calculated by subtracting the sum of the ultimate coal analysis results plus the ash, which contains an indeterminate quantity of oxygen, from 100. However, Hamrin *et al.* (1975) did analyze low temperature ash (LTA) obtained by the method described by Gluskoter (1965) and did emphasize the importance of subtracting the determined LTA oxygen in order to obtain a better value for organic oxygen by the formula

$$O_{org} = O_{tot} - O_{inorg}. \tag{2}$$

Ideally, if the low temperature oxygen plasma ashing did result in an ash consisting of all the mineral matter in the coal in its original state, this would be right. Unfortunately, this is not strictly so. A few words concerning the determination of mineral matter in general are in order here.

Fraser and Belcher (1973) have proposed the use of direct determination of mineral matter by LTA as a regular routine. Kinson and Belcher (1975), by using high temperature radio-frequency decomposition of coal at 1950°C and a gravimetric determination of evolved carbon monoxide, have determined total oxygen and calculated organic oxygen in coal and coke using in their calculations the LTA ash. They propose three possible calculation methods for oxygen in this ash. The first two are (1) from a complete elemental and phase analysis of the mineral matter; or (2) by using a formula developed by Brown *et al.* (1965):

$$O_{inorg} = 0.89\,H_2O^i + 0.73\,CO_2{}^i + 0.5\,(MM_d - H_2O^i - CO_2{}^i + 1.88S^P), \tag{3}$$

where water of hydration of minerals in the dry sample $= H_2O^i$; the carbon dioxide of carbonates in the dry sample $= CO_2{}^i$; the mineral matter in the dry sample $= MM_d$; and the pyritic sulfur in the dry sample $= S^P$. This formula is adequate but less rigorous than the first method. The third method (3) is by approximation in low mineral matter coals:

$$O_{inorg} = 0.5MM_d. \tag{4}$$

Parr (1932) has proposed the calculation of the mineral matter from the HTA, taking the water of hydration of minerals and the pyritic sulfur into account:

$$\% \text{ Mineral Matter} = 1.08 \times \% \text{ HTA in coal} + 0.55\% \text{ S in coal}. \tag{5}$$

King *et al.* (1936) have developed a more rigorous formula also taking into account the CO_2, the SO_3, and the Cl in coal as well as the remaining SO_3 in the HTA:

$$\% \text{ MM} = 1.09 \times \% \text{ HTA} + 0.5 \times \% \text{ S}^P \text{ in coal} + 0.8 \, CO_2 \text{ in coal}$$
$$- 1.1 \times \% \, SO_3 \text{ in HTA} + \% \, SO_3 \text{ in coal} + 0.5 \times \% \text{ Cl in coal.} \quad (6)$$

Given and Yarzab (1975) give a "most recent" form of this formula with the 1.13 factor for water in ash based partially on Millot's (1958) work:

$$\text{MM} = 1.13 \text{ ash} + 0.5 \, S^P + 0.7 \, CO_2 + 2.8 \, (S_{SO_4} - S_{Ash}) + 0.5 \, Cl, \quad (7)$$

where CO_2 is the yield of gas when the sample is treated with acid; and S_{SO_4} and S_{Ash} represent sulfatic sulfur in original coal and ash, respectively.

To get the organic oxygen one could then apply an approximate factor such as 0.5, for example, as suggested by Kinson and Belcher (1975) and used in Eq. (4), to mineral matter and simply subtract.

The direct determination of mineral matter appears to be the most useful method because it is relatively simple to use when modern radio-frequency discharge ovens with controlled oxygen gas flow are available. Fraser and Belcher (1973), Given and Yarzab (1975), and Miller *et al.* (1978) point, nevertheless, toward several problems in this LTA determination: some pyrite may be oxidized and some organic sulfur may be fixed in the ash as sulfate, especially in sub-bituminous coals and lignites. Some SO_3 fumes may form—partially decomposing the carbonates, and considerable nitrogen and some unburned carbon may be present. The authors suggest pretreatment with dilute hydrochloric acid to minimize sulfur fixation and emphasize that different rank coals behave differently, which inhibits accurate interpretation and makes empirical corrections difficult.

Considering the complications in estimating the ash content and its composition in order to determine the organic oxygen of the "coal molecule," where it can be present in the form of carboxyl (—COOH), hydroxyl (—OH), methoxyl (—OCH$_3$), carbonyl (=C=O), ether (—C—O—C—), and phenolic (—OH) functional groups, one obviously would save much trouble by determining the total oxygen in coal ash, coal mineral matter, and coal. Such data alone would provide much specific information.

It so happens that the FNAA method for the direct determination of oxygen is virtually interference free and is very sensitive, rapid, and inexpensive. We have in this instrumental technique a tool that provides quantitative data for oxygen perhaps of higher quality and accuracy than can be obtained for any other element analyzed within the coal analysis family. Therefore, we have made attempts to base the interpretation and calculations of the coal analysis first on the oxygen determined directly. Let us consider some of these advantages.

First, the cumbersome calculations [Eqs. (3),(5),(6), and (7)] of oxygen in the mineral matter become unnecessary irrespective of the method used to ash. Comparison of different ashing methods by equating the total oxygen of the products will be more meaningful, giving, for example, such information as the relative degree of oxidation. If total oxygen in "as received" samples is compared with the dried sample's oxygen, the hygroscopic water content can be calculated without destroying the sample. The sum of all cations, including anions other than oxygen, is thus also obtained directly, permitting one to distinguish easily between ashes of basically different composition, e.g., pyrite rich versus kaolinite rich. The rate of oxidation of ash can be studied directly, permitting the selection of best timing for the reaction, and instead of using approximate empirical factors for inorganic oxygen in ash [Eq. (4)], one uses the true value, improving the organic oxygen estimate. Any unusual compositions can be easily detected, and summations, when other elements are also determined, become much more meaningful, permitting a double check of the validity of the sum of the other elements.

Relatively few data exist on direct determination of oxygen in HTA. Block and Dams (1974) have analyzed 20 Belgian coal ashes with a mean oxygen content of 46.15% and a range of about 5%. Volborth *et al.* (1977b) have analyzed seven USBM coal ashes with a mean of 45.54% and a range of about 5%.

Since the USBM and the ASTM methods calculate oxygen by difference by subtracting the total ash without regard to its oxygen content, Volborth *et al.* (1978) have proposed that if the determined or estimated oxygen content in the ash were added to the "classical" oxygen-by-difference value, a closer agreement may result in many coals between calculated and determined oxygen. This seems to be the case for many coals (see Table II). One sees from Table II that, in all cases, oxygen calculated by this method is closer to the determined value. Admittedly, this is a new manipulation of coal analysis, and it is not strictly valid because HTA is not mineral matter. In that sense it is wrong, but it permits a closer estimate of "true" oxygen in the total coal than the total bulk ash subtraction method. If used, this approach can indicate unusual ash compositions and, conversely, should give new values closer to "organic oxygen," especially in ash-rich coals. No suggestion nor pretense is made that this method of calculation should be univer-

TABLE II *Comparison of Oxygen by Difference with Oxygen Determined by FNA and Oxygen Calculated Assuming 46% Oxygen in Ash*

Coal No.	O_{diff} (wt.%) as reported USBM "as received"	Oxygen Determined FNA	O_{diff} calc. (ash 46% O)	Ash (wt.%)
K-46566	36.6	42.34	44.34	7.70
K-46218	32.3	35.90	37.82	5.50
K-46217	31.7	36.15	37.89	6.20
K-46216	33.2	36.85	38.13	4.90
K-46565	38.8	43.28	44.19	5.40

sally adopted. As in most other coal analysis calculations, this is only an approximation. It should be of value in initial stages of coal analysis when only ash, moisture, and volatiles are determined. FNA oxygen determination in ash and total coal then should be sufficient to estimate the organic oxygen and indicate if the coal has unusual characteristics. For example, if routine oxygen by difference is approximately equal to determined oxygen, the ash content must be low; or if the ash is low in oxygen, Fe_2O_3, or pyrite, or both are indicated. In low ash coals, an estimated value for oxygen in ash based on known mean values determined previously on similar coals should give close agreement as in Table II without the necessity of using the complex formulas recalculating the ash to mineral matter. When this latter correction is made, the agreement between the oxygen values should improve if the assumptions are correct or if the actual mineralogical data are used. Further, negative values for oxygen by difference, calculated by subtracting the determined oxygen in ash, will tell the chemist that his or her assumptions of the coal mineral composition must be changed. For example, considerable oxidized pyrite or other sulfides will have this effect. Thus, oxygen determined in wet coal as received, in dried coal, and in coal ash may give more information than if a complete proximate analysis were done.

The procedure above gives only a rough estimation. If the Parr (1932) formula is used, and the additional oxygen due to moisture is empirically calculated, a still closer agreement between the calculated and directly determined oxygen should be obtained. Further refinement should result when the King *et al.* (1936) mineral matter formula is used, and possibly a still better agreement if a properly prepared LTA is analyzed as Hamrin *et al.* (1975), James *et al.* (1975, 1976), and Volborth *et al.* (1977b, 1978) have done and suggested. Recently Schlyer *et al.* (1979), at Brookhaven National Laboratory, have used 10 MeV

He-3 ions to produce C-1. N-13, and F-18 isotopes, calculating the oxygen content in coal from the $^{18}F/^{11}C$ ratios. They also subtracted the estimated oxygen in the mineral matter to obtain approximate organic oxygen as suggested above by Volborth *et al.* (1978). This charged particle activation analysis (CPAA) appears to provide an independent nuclear activation method suitable to check the NAA method described in this volume, Chapter 47.

The total data so far collected are insufficient to confirm a general applicability of the procedure and Given's (1976) warning not to compare the oxygen data by difference directly with determined oxygen has to be taken seriously. The Mott and Spooner (1940) empirical formulas, which use "oxygen by difference" values to calculate the calorific values (BTU/lb) of coal on dry mineral free basis, have been used by Given and Yarzab (1975) to check the accuracy of dmmf analyses successfully. More accumulated FNA analyses for oxygen are needed to try to substitute "determined oxygen" for "oxygen by difference" in the above formula.

3. Volatile Matter

Volatile Matter is determined by gradually but rapidly heating air-dried coal up to 950°C (±20°C) (U.S. Bureau of Mines, 1967) in a closed platinum crucible in a special oven where the crucible can be gradually lowered to avoid too rapid losses, splatter, and, as much as possible, the phenomenon called "sparking" that may cause particulate losses. See Volume I, Chapter 6, Section III,E, and Fig. 2. Volatile matter is reported as loss of weight minus the moisture determined at 105°C. Volatile matter, therefore, does not represent any specific compound. It mostly is composed of volatile hydrocarbons and gases, such as CO and CO_2, escaping during the high temperature decomposition of coal. This value is entirely empirical and depends on the rate of heating, temperature, and timing. It results in a so-called "coke button" that is the residue that contains all the mineral matter, considerable carbon, etc. It is later used to determine caking or plastic properties of coal in the so-called agglomerating index test that is based on comparison charts depending on the state of this residue (U.S. Bureau of Mines, 1967, p. 35).

Another test common in evaluation of coal properties is the so-called free-swelling index of coal used to determine expansion properties of coal in coke ovens. It is performed in a special covered silica crucible over a gas burner, heating rapidly to about 800°C, and results also in coke buttons of various shapes and sizes calibrated in a chart (U.S. Bureau of Mines, 1967, p. 37). See Volume I, Chapter 6, Section XIII.

There are no correlation studies based on determined oxygen in coal versus retained oxygen in the residues of these important tests. We are presently initiating comparative studies of oxygen in these procedures and attempting to correlate the plastic properties of coal and coke with oxygen, total ash content, and the rank of coal. Preliminary tests indicate that additional industrially valuable information may be obtained by these studies. Correlation of the initial oxygen in the wet coal, in the dried coal, and in the residue of the volatile matter as well as in the same coal's HTA may help to classify coals better in the industrial-use sense than the present methods based on fixed carbon and volatile matter (U.S. Bureau of Mines, 1967, p. 59).

4. Fixed Carbon

Fixed carbon values versus volatile matter of the proximate coal analysis are used to classify coal by rank (U.S. Bureau of Mines, 1967, p. 59). Since both these quantities are purely empirical, the attempt to correlate them with oxygen in coal before and after the test in the residue may provide us with more informative industrial characteristics and help to classify and predict the behavior of coal and coke better. This work is also in progress, and we have indications that when determined oxygen is substituted for "oxygen by difference" in the well known Van Krevelen and Schuyer (1957) plot of H/C versus O/C (see Schopf and Long, 1966, p. 193) a better correlation and less scatter may result.

B. Ultimate Analysis and Oxygen

Direct oxygen determination belongs in the category of determinations called, collectively, the ultimate analysis family because other main constituents of coal are determined in this group quantitatively by selective and relatively accurate gravimetric methods. These elements are carbon, hydrogen, sulfur, nitrogen, chlorine, and carbon dioxide (U.S. Bureau of Mines, 1967). Oxygen is customarily determined only "by difference" in the ultimate analysis, meaning that all errors and bias accumulate in its reported value. The most important contribution of FNAA of oxygen in coal and coal products analysis is therefore to this group. Oxygen is the only major constituent of coal for which until recently there were no straightforward methods. Ignasiak *et al.* (1969) have developed a pyrolysis method measuring the volume of gases and determining carbon monoxide by gas chromatography. The most commonly used method was developed by Schutze (1939) and Unterzaucher (1940, 1952) and further studied and modified by Abernethy and Gibson (1966). This technique consists of demineralization of coal with hydrofluoric acid, pyrolysis in nitrogen atmosphere, and measurement by titration of iodine evolved from reaction with liberated carbon monoxide after passing the gases over a platinum catalyst at about 900°C. The complexity of this and similar methods is obvious and explains their infrequent use. The FNAA method developed by this author and others is described in Chapter 47.

When oxygen can be determined in the ultimate analysis, the informative content of all the other determinations increases and the summations become more meaningful. The preparation of true material balances (see Table II, Chapter 47) becomes possible, and several checks of accuracy can be made and the oxidative behavior of coal studied quantitatively (Volborth *et al.*, 1977a,b,c, 1978; Miller and Volborth, 1976). Also, once oxygen is accurately known in coal and the other major elements are determined with care, total sulfur can be estimated in many coals "by difference." This is complicated, however, by the presence of pyritic sulfur. Nevertheless, when the analyses do not balance out, we usually find that it is because of a large quantity of sulfur that can partially substitute for oxygen in the coal molecule (Volborth *et al.*, 1978). Large percentages of ash in coal also may cause problems. However, the knowledge of oxygen in coal and its ash does narrow the possible sources of error, and as data accumulate and correlations between true oxygen and other constituents and their ratios are established, it should become easier to interpret coal analyses with considerably less work and time. Coal upgrading products, and interim solid materials produced during industrial utilization and testing, that are subject to ultimate elemental analysis can all be better characterized when oxygen is also determined.

C. Oxygen Concentration and Rank of Coal

Several classifications for coal exist. The ASTM (1965) classification by rank is most frequently used in the United States (U.S. Bureau of Mines, 1967) and has been used by us. Other classification schemes are described by Francis (1961), Parks (1963), and Van Krevelen (1961). Classification schemes are described in detail in Volume I, Chapter 6, Section XII, and Tables I and II. With increase in coal rank, quite generally carbon content increases, oxygen decreases, volatile matter decreases, moisture decreases, hydrogen decreases, and the calorific value and density increase. Hickling (1931) has prepared a chart (see Francis, 1961, p. 376) showing linear correlation between carbon and oxygen content of coals. This chart demonstrates that coals from lignite and sub-bituminous to bituminous, and also anthracites, may be ranked by this continuous variation of only two main constituents. The main weakness of this approach, as Hickling pointed out, is the absence of hydrogen as one coordinate. Hydrogen content affects greatly the distillation, gasification, liquefaction, and pyrolytic properties of coals. This is true despite the fact that hydrogen weight percentage generally varies within the narrow range of 4–6% (see p. 47, Fig. 1, Rose, 1945).

Nevertheless, with the development of direct and rapid oxygen determination by FNAA, one is tempted to revive Hickling's approach, especially because his plot was based on "oxygen by difference." One should expect, therefore, less scatter and possibly a more refined relationship with hydrogen, especially when organic oxygen of the coal is plotted. This work is presently in progress at Washington State University. Meanwhile, a tentative plot of determined oxygen versus carbon of 33 coals (Volborth *et al.*, 1978) has shown that the correlation is linear and that, as expected, the lignites can be clearly distinguished from bituminous coals and the latter from anthracites. Comparison of two plots (Fig. 1) of the ratios of "oxygen determined" to carbon and "oxygen by difference" to carbon, show that there is more scatter in the first plot signifying in this instance that there is more information in this plot than in the "oxygen by difference" plot. Greatest deviations from the least squares line happen to occur in coals rich in sulfur, for example. More data are needed on coals of varying rank, and further correlation of organic oxygen and hydrogen on a dry-ash-free basis are required to establish more quantitative charts of coal characteristics based on these three most critical constituents. Especially in the anthracite region, where G. Hickling's (1927, plate VI) plot becomes nonlinear, more detailed study is indicated.

III. OXYGEN IN COAL RELATED MATERIALS

A. Oxygen in Coal

Recent literature reviews on this subject have been made by Volborth (1976a,b). Studies have shown that during the metamorphism of organic matter from peat to lignite and bituminous and brown coal to

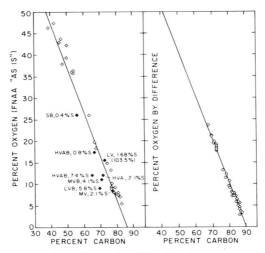

Fig. 1 O/C plot of 33 coals on "as received" basis using oxygen values determined by FNAA, and "by diff." The coals group by rank so that lignites are in the upper left corner, sub-bituminous coals in the middle, high volatile bituminous, and medium and low volatile bituminous in the right lower corner.

anthracite, condensation of aromatic nuclei seems to occur. This is accompanied by a loss of oxygen and hydrogen and an increase of carbon. In this process the functional groups such as carboxyl (—COOH), hydroxyl (—OH), methoxyl (—OCH₃), and carbonyl (=C=O) are progressively depleted in the coal but are not entirely lost, even in the high rank coals (Van Krevelen, 1961, pp. 160–170; Blom *et al.*, 1957).

Yohe (1958) and Dryden (1962) give detailed literature reviews of coal oxidation. Oxidation studies of coal have shed considerable light on the chemical structure of coal. Such work is of great value to the coal mining and utilization industry. Much *practical information* concerning the nature and the effects of atmospheric oxygen on coal exists.

For example, coal is susceptible to spontaneous combustion when stored in piles, especially when fines are intermixed with coarse material. The initial absorption of oxygen is relatively rapid on the fresh coal surface, and ignition temperatures can be very low. A high percentage of iron in coal may accelerate oxidation whereas a high percentage of mineral matter (ash) retards oxidation. Coal in storage piles can oxidize rapidly, and its properties as well as heating value will deteriorate with time. Low rank coals are especially affected by storage. High temperature, fine particle size, high percentage of bright constituents, and high oxygen concentration in the storage area all have an accelerating effect on coal degradation. Storage in covered piles under water or in "compacted layer piling," as well as addition of antioxidants, prevents fast weathering of coal. Very dry coals and very moist coals oxidize slowly, whereas coals with intermediate moisture content tend to oxidize faster. Time is an important factor in coal oxidation.

Oxidation generally impairs or even destroys the coking properties of coal, especially in marginal low rank coking coals. The extent of oxidation of coal can be measured because it affects the plastic and caking properties of coal as well as the ignition temperatures. Progressive oxidation of coal stored at power plants decreases the heating value of the fuel, thus increasing the costs. Coal stored for coking purposes, if not protected, may deteriorate to the point where coke of inferior quality is produced (Jackman *et al.*, 1957; Ignasiak *et al.*, 1970). Spontaneous ignition of coal piles and coal in ships' holds can cause major losses and accidents.

Controlled oxidation studies of coal permit us to learn how to utilize this fuel more economically; how to convert it to new useful products; how to prevent the undesired attack of oxidizing media; how to store it better; how to get the highest heating values; how to control its combustion; how to best fluidize it, gasify it, or produce synthetic oils from it by reacting it with oxygen, carbon monoxide, hydrogen, and water.

The coal surface reacts with oxygen the moment coal is mined. Coal absorbs oxygen at a rate that is first rapid, then falls off if the temperature remains constant.

In general, early work in this field in the 1940–1950s has established that (1) the quantity of oxygen functional groups is reduced with increasing rank and is higher in the younger coals; (2) hydroxyl groups account for one-fourth to one-eighth of the total oxygen in the coal, with lignites containing about 10% by weight of —OH groups; (3) in high rank coals of 80% carbon, only about 7% of functional groups remain and drop to near zero in coals with 90% carbon; (4) the carboxyl groups are present in brown coals and lignite but are practically absent in bituminous coals, with brown coal containing about 5–6% carboxyl (Blom *et al.*, 1957), and coal with 85% carbon none; (5) carbonyl groups are found generally in coals of all ranks, but their concentration is low in bituminous coal; brown coal contains about 3.5% carbonyl groups, coals with 83% or more of carbon average 0.5% carbonyl, associated mainly with quinoid structures; (6) methoxyl groups while forming a large part of the "fresh" lignin [7–17% according to Flaig's (1966) data] represent only an insignificant percentage in coal. This decreases from 0.8% in brown coal to about 0.4% in coal with 70% carbon and drops to about 0.2% or less methoxyl in a coal with 80% carbon. Not all oxygen in coal is detected by the determination of the functional groups. Like nitrogen (Francis, 1961), part of the oxygen can be built into the heterocyclic rings and thus may be liberated only by the pyrolysis of coal. Quantitative determination of functional groups in coal presents considerable difficulties, as discussed in Volume II, Chapter 20, Section VII,A.

Factors complicating the study of the structure and oxidation of coal are the presence of water of crystallization and of OH groups in the mineral matter associated with coal, and the oxidation of the iron sulfides pyrite and marcasite upon heating in air. These are sources for evolution of water from coal, and also for consumption of oxygen. It is very difficult to distinguish water evolved from mineral matter from the water evolved because of the pyrolytic reaction and destruction of the oxygen functional groups of the organic matter. Decomposition of coal by radio-frequency techniques often permits better retention of mineral matter, such as sulfide and carbonate minerals [for example, see Fraser and Belcher (1973)], and thus may lead to a better understanding of the composition of the organic coal material. Low temperature ashing techniques developed by Gluskoter (1965) permit decomposition of coal at temperatures between 150° and 200°C with residues composed primarily of calcite, pyrite, kaolinite, and undecomposed clay minerals. This technique combined with fast-neutron activation determination of oxygen shows promise for more accurate interpretation of the true organic oxygen content of the functional groups in coal (Hamrin *et al.*, 1975; Block and Dams, 1974) and has been previously discussed in more detail and in Chapter 47.

Oxidation reactions of coal with air, oxygen, hydrogen peroxide, nitric acid, potassium permanganate, sulfuric acid, and other reagents (Howard, 1945, 1947; Montgomery, *et al.*, 1956; Montgomery and Holly, 1957; Montgomery and Bozer, 1959) produce a mixture of water-soluble acids. Chromatographic and mass spectrometric studies of these acids, obtained from oxidation of bituminous coal, have detected mainly benzoic acid, phthalic acid and its isomers, as well as mellitic acid and trimellitic acid and its isomers. For detailed information, see Volume II, Chapter 21, Section IV.

The detection of phthalic acid in oxidized bituminous coal may be an indication of napthalene, anthracene, and phenanthrene type polycyclic aromatic nuclei in coal (I).

The presence of mellitic acids in oxidized coal products is significant because it indicates that the original material has probably contained larger aromatic clusters such as coronene (II).

Naphthalene — oxidation → Phthalic acid

I

Coronene

II

Since mellitic acids are obtained also from the oxidation of graphite (III), coal can be assumed to have aromatic structural units similar to

Graphite — KMnO$_4$ → Mellitic acid

III

the graphite structure. It was also found that the yield of mellitic acid increases with the rank of coal (Van Krevelen and Schuyer, 1957). The yield of aromatic acids in general increases when oxidizing lignin → peat → brown coal → lignite → bituminous coal → anthracite. This indicates increasing aromaticity and an increasing size of the aromatic clusters or nuclei. Based on these studies, models of the chemical structure of coal have been proposed. These consist of polyaromatic clusters bound together by heterocyclic rings containing oxygen, nitrogen, and sulfur, with occasional carbonyl groups, as shown in the model proposed by Fuchs and Van Krevelen (Van Krevelen and Schuyer, 1957).

Comparison of the ultimate analyses of subbituminous and low-to-high rank bituminous coals with the composition of hypericin, anthraquinone, anthrone, benzanthrone, and dibenzanthrone (see Tables III,

TABLE III *Composition of Some Products of Enzymatic Hydrolysis (rotting, humification) of Lignin Plotted in Approximate Sequence of Their Formation and Their Possible Polymerization Products*

| | | Weight percent | | |
		C	H	O
Coniferyl alcohol	$C_{10}H_{12}O_3$	66.65	6.71	26.64
Vanillin (aldehyde)	$C_8H_8O_3$	63.15	5.30	31.55
Vanillic acid	$C_8H_8O_4$	57.14	4.80	38.06
Dimethoxybenzoquinone	$C_8H_8O_4$	57.14	4.80	38.06
p-Hydroxybenzoic acid	$C_7H_6O_3$	60.86	4.38	34.75
p-Benzoquinone	$C_6H_4O_2$	66.67	3.73	29.60
Emodin 1,3,8-Trihydroxy-6-methylanthroquinone	$C_{15}H_{10}O_5$	66.67	3.73	29.60
Alizarin 1,2-Dihydroxy-anthraquinone	$C_{14}H_8O_4$	69.98	3.36	26.64
Humic acids from sub-bituminous coal (7 samples)	(Flaig, 1966)	69.2	4.5	26.3
Hypericin	$C_{30}H_{16}O_8$	71.43	3.20	25.38
Naphthoquinone	$C_{10}H_6O_2$	75.94	3.82	20.23
Anthraquinone	$C_{14}H_8O_2$	80.76	3.87	15.37
Anthrone	$C_{14}H_{10}O$	86.75	5.19	8.24
Benzanthrone	$C_{17}H_{10}O$	88.67	4.38	6.95
Dibenzanthrone (violanthrone)	$C_{34}H_{16}O_2$	89.46	3.53	7.01
Low to high rank bituminous coals (Table IV)		79–92	5.5–4.5	14–2

IV) indicates very similar composition, supporting Fuchs' and Van Krevelen's proposed model for the chemical structure of coal that contains polycyclic aromatic clusters interconnected by heterocyclic rings containing oxygen, sulfur, and nitrogen, as well as a few aliphatic chains, carbonyl, carboxyl, hydroxy, and methoxy functional groups.

Referring to the model coal structure proposed by Van Krevelen and Schuyer (1957), one could easily imagine that a relatively mild degree of oxidation would have a strong effect on the physical properties of the whole coal. As the heterocyclic structures are more affected by oxidation, additional aliphatic chains with carbonyl, carboxyl groups, and ether linkages can be formed, and the nature of the internal surface changes and swelling takes place. Recent studies using $^{18}O_2$ by Ignasiak et al. (1972) have shown that even mild weathering introduces sufficient crosslinks to have a profound influence on the physical properties of the coal. Ignasiak et al. (1974a) have further demonstrated that weathering influences the plasticity and dilatation properties of coking coal. Washowska et al. (1974) have also found the dilatation effect and dem-

TABLE IV *Composition of Some Carbohydrates, Lignin, Wood, Peat, and Coalification Products*

	C	H	O	N
		Weight percent		
Glucose $C_6H_{12}O_6$	40.00	6.72	53.29	
Sucrose $C_{12}H_{22}O_{11}$	42.10	6.48	51.42	
Cellobiose $C_{12}H_{22}O_{11}$	42.10	6.48	51.42	
Cellulose $(C_6H_{10}O_5)_n$	44.45	6.22	49.34	
Wood (Mason, 1962)	49.65	6.23	43.20	0.92
Peat[a] (Swain, 1970)	51.13 →	6.05 →	40.99 →	1.83 →
	58.48	5.64	33.54	2.34
Bjorkman Lignin (fresh straw)[b]	60.68	5.79	33.11	0.4
Lignitic soft coal, 2 samples[b]	63.2	4.5	32.3	1.0
Sub-bituminous coal, 10 samples[b]	73.2	5.3	21.5	0.8
Low rank bituminous coal[a]	79	5.45	14	1.55
Bituminous coal, 5 samples[b]	84.3	5.3	10.4	0.8
Medium rank bituminous coal[a]	85–88	5–5.2	7–5	1–1.75
High rank bituminous coal[a]	91.7	4.5	2.2	1.6
Semianthracite coal[a]	92.3	4.2	2.0	1.5
Anthracite[a]	93–97	3.7–0.6	1.9–1.8	1.4–0.6

[a] Variation in composition with increasing maturity. From Swain (1970, p. 339).
[b] From Flaig (1966).

onstrated that the ether linkages are also broken in the coal's internal structure by progressive oxidation at low temperatures (85°C) in air for three days, and Swann *et al.* (1974) have detected notable changes in the internal surface properties of brown coal after low temperature oxidation.

Tronov (1940a,b) has proposed an oxidation sequence of coal starting with formation of polyhydric phenols and proceeding through carbonyl compounds (quinones) to acid anhydrides and carboxylic acids that is still valid today, as shown in a modified form (IV).

Yohe and Blodgett (1947) concluded from studies of oxidation of various coals methylated by dimethyl sulfate that the content of phenolic structures in lower rank medium volatile bituminous coals decreased with the increase of coal rank. This was confirmed by infrared studies (Brown, 1955). Van Vucht *et al.* (1955) also found an increase in OH groups during the oxidation of coal. Atmospheric oxidation of Illinois coals at 100°C for about 100 days yielded essentially constant

Phenols   Quinone

Acid anhydride   Dicarboxylic acid

IV

concentration of the phenolic group (Yohe, *et al.*, 1955). These studies confirmed Tronov's postulation concerning the presence of phenolic structures in oxidizing coal. Dry oxidation studies of subbituminous coal by Jensen *et al.* (1966) have shown that phenolic structures appear well before appreciable amounts of the carboxylic acids are detected. Jensen *et al.* (1966) have also detected acid anhydrides by infrared spectroscopy, but after the appearance of carboxylic acid. This is explained by the fact that aliphatic structures in coal are the more probable loci for the initial carboxyl formation. Referring to the commercially used gas-phase oxidation of naphthalene to produce phthalic and maleic anhydrides, a modified form of the scheme by Jensen *et al.* (1966) is shown in V. Jensen and co-workers (1966) propose a scheme

V

for coal oxidation similar to Tronov's (VI). The scheme above is partially based on work by Moschopedis (1962) and Wood *et al.* (1961). Moschopedis has also shown that quinones and hydroxyquinones are present in coal and in humic acids derived from it.

The actual process of coal oxidation is much more complex. The recent detection of lactones on the surface of oxidizing graphite, char, and coal by Yang and Steinberg (1975) and Barton and Harrison (1975) further complicates the scheme presented above. If glycosides indeed seem to form on the surface in the so-called "coal–oxygen complex" in the early stages of oxidation of coals, further oxidation and breaking of the glycoside bond could lead directly to aliphatic small chain car-

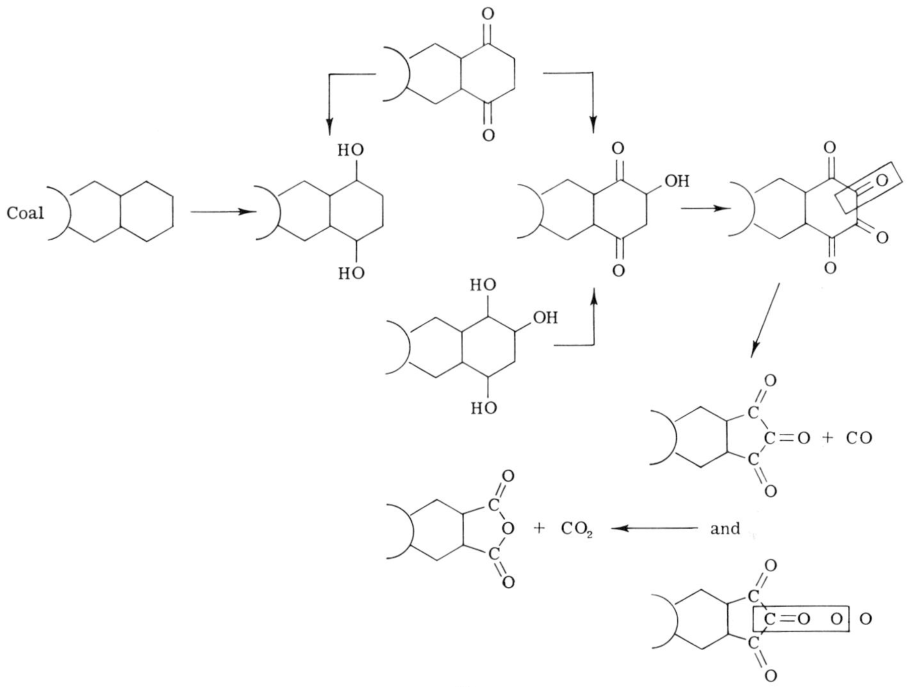

Coal

HO · · · HO · · · OH · · · O

HO · · · OH · · · HO

$C=O + CO$

$+ CO_2$ ← and

$C=O$ | O | O

VI

boxylic acids, bypassing the phenol and the hydroxyquinone stages. This oxidation path may further elucidate the early appearance of the carboxyl acids in coal oxidation before the acid anhydrides, which later could be the products of dehydration (see Jensen *et al.*, 1966, p. 639).

B. Oxygen Concentration in Organic Compounds Derived from Oxidation of Coal

Certain organic compounds are oxidized more readily than others. The ease with which an organic substance will oxidize depends on the functional groups present in the molecule. Oxidation of coal is affected by incorporation of oxygen functional groups, such as peroxide, hydroxyl, aldehyde, carbonyl, carboxyl, ether linkage, and methoxyl groups. Once oxygen has been combined with an organic molecule, further oxidation may occur more readily. Compounds with oxygen functional groups attached are more susceptible to further oxidation than their progenitors. For example, alcohol and acetaldehyde oxidize more readily than ethane and phenol, naphthol more easily than benzene and napthalene, and benzaldehyde more readily than toluene.

Saturated and purely aromatic hydrocarbons are more difficult to oxidize than their derivatives containing oxygen functional groups.

Some unsaturated structures and ethers are known to form peroxides
easily (Yohe, 1958, p. 17):

Group	Structure	Peroxide formed
Methyl	CH_3	$-OOCH_3$
Ether	$C-O-C-H$	$-C-O-C-OOH$
Olefin	$C:C-C-H$	$-C:C-C-OOH$
Alkylated aromatic	⟨benzene ring⟩$-\overset{\mid}{\underset{\mid}{C}}-H$	⟨benzene ring⟩$-\overset{\mid}{\underset{\mid}{C}}-OOH$

The formation of peroxide appears to be one of the initial steps of coal
oxidation. It can lead then to the formation of other oxygen functional
groups.

Molecular oxygen is an efficient breaker of chain reactions. The formation of the peroxy radical (R—O—O) by auto-oxidation of coal with
atmospheric oxygen may first attack, cleave, and weaken the aliphatic
domains of the coal miscelles, then gradually destroy the heterocyclic
and hydroaromatic regions, thus exposing the aromatic clusters to further degradation.

Kinetic studies of oxidation of coal are complicated considerably by
the fact that several different reactions seem to occur concurrently;
chemically different structural units and different functional groups are
involved, and the nature of coal is very variable. The measurement of
reaction rates on coal surfaces is therefore very difficult. Whether one
determines the oxygen adsorbed, the heat produced, or the concentration of compounds formed, one tends to measure the result of a series
of complex processes often unrelated and that are difficult or impossible
to separate. This explains why attempts at accurate and consistent
kinetic measurements have produced somewhat inconsistent results.
Nevertheless, the general agreement, considering these difficulties, is
remarkably good.

The weathering of coal increases the concentration of the oxygen
functional groups (Kucharenko, 1960), and some reduction of carbon
content results (Howard, 1947), technological quality of coal changes,
water content increases, and alkali-soluble products similar to the
humic acids form, as well as water-soluble lower molecular weight
acids. In the early stages of oxidation, oxygen is being added to the
coal bulk composition and water is lost, while during the later stages
of humification of coal, CO_2, CO, and H_2O evolve. The oxidation

proceeds along the periphery of the macromolecular clusters in coal and finally affects the C—C bonds, rupturing the aromatic rings.

The terms "coal–oxygen complex" and "unstable surface combination" have been used by a number of early workers in the field of coal oxidation and weathering (Yohe, 1958, p. 11). The types of interaction between solid coal and gaseous oxygen are dual in nature. *First*, the preferential adsorption of oxygen rather than that of nitrogen on the surface boundary of coal particles may produce an increased concentration of it on the surface. This attraction is not of a chemical nature, and the gas may be desorbed by increasing the temperature or evacuation, in vacuum. *Second*, a process of chemisorption may occur in connection with adsorption. This would involve initial oxidation of coal surfaces and the formation of hydroperoxides (—C—O—O—H) by the interaction of molecular oxygen with the —C—H groups. Such a process is well known to occur in the "drying" or "aging" of oil paints and the development of rancidity of butter. Such reacted oxygen cannot be removed by heating or evacuation as oxygen, but evolves as H_2O or CO_2. Both processes described above probably occur simultaneously. The chemisorbed oxygen of the hydroperoxides as well as the adsorbed oxygen can further react with the coal "molecules" forming more stable and permanent organic groups. Yohe and Harmon (1941) have studied the surface oxidizing power of Illinois coals and suggested that both processes exist. They emphasized that the rapid formation of the coal–oxygen complex in only a few minutes is indicative of the adsorption type reactions, while the fact that simple evacuation does not lead to full recovery of adsorbed oxygen implies that chemical combination also occurs. The rapidity of this reaction greatly complicates the weighing of dry or wet coal samples (see Chapter 47, Table I). Yohe (1950) also demonstrated, using the darkening effect on a photographic plate (Russell Effect), that peroxide compounds exist on coal surfaces. In earlier work Yohe and Wilt (1942a,b) found that the surface oxidizing power of several high-volatile bituminous coals reached a maximum at 10 to 25 days of exposure, then diminished slowly. They interpreted this phenomenon by suggesting that after the fast initial adsorption of oxygen the hydroperoxide phase is intermediate and leads to the formation of more stable oxidation products.

Oreshko (1949a,b,c) found in heating studies of Donets Basin coals in air and oxygen that in all cases there was first a gain in weight, then a drop, then another gain, and yet a higher weight gain before a sharp drop when combustion occurred. He was able to distinguish three stages of reaction depending on the temperature:

70°–80°C Formation of peroxide type complexes

80°–130°C Decomposition of peroxide complexes

130°–160°C and 180°–290°C Formation of "stable" coal–oxygen complexes destroyed only by ignition

These studies have an important bearing on selecting the right temperature for drying coal.

Jones and Townsend (1949) studied oxidation of British coals, and Chakravorty *et al.* (1950) studied British and Indian coals. These authors also found that the "peroxide" complex started to decompose at 80°C. Chakravorty *et al.* found that the decomposition of this complex approximates a first-order reaction and were able to determine that the activation energy is approximately that for hydroperoxide decomposition. Higushi and Shibuya (1954) have also detected peroxides in non-volatile air-oxidized lignite.

Yamasaki (1953) oxidized coal of different screen sizes and determined oxygen absorbed at temperatures under 100°C. He found a linear relationship between oxygen absorbed and temperature between 30° and 100°C. Below 30°C, values obtained for oxygen deviated greatly from his empirical equation. Yamasaki suggested that this phenomenon is due to Van der Waal's adsorption of oxygen on the surface of coal, which occurred predominantly under 30°C, whereas chemisorption occurred predominantly between 30° and 100°C.

In recent oxidation studies of coal, Albers *et al.* (1974) have shown that auto-oxidation of coal can proceed through the formation of hydroperoxides to ketones (VII).

$$\text{Ar}-\text{CH}_2-\text{R} \xrightarrow{\text{O}_2} \text{Ar}-\underset{\underset{\text{R}}{|}}{\overset{\overset{\text{O}-\text{O}-\text{H}}{|}}{\text{C}}}\text{H} \xrightarrow{-\text{H}_2\text{O}} \text{Ar}-\overset{\overset{\text{O}}{\|}}{\text{C}}-\text{R}$$

VII

Chakrabartty (1975), however, has recently raised serious questions concerning the evidence and probability of lignin decomposition through the formation of quinones, which he points out, if formed, would certainly react first with other reactive compounds available in the decomposing peat rather than get concentrated so that they could interact to form larger aggregates. He also emphasizes that before the cellulose can be degraded bacterially, it has to be transformed to noncyrstalline forms.

Chakrabartty presents a simplified scheme (VIII) for coal formation.†
He considers the present concept of an average coal "molecule" having polycondensed aromatic ring clusters as a predominant structural fea-

† See Fig. 1, Chapter 40.

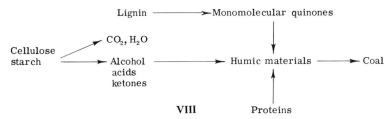

ture of coal unsatisfactory, because oxidation studies with hypohalite revealed reactivities and oxidative properties inconsistent with that widely accepted structure. See Volume II, Chapter 21, Section IV,C for results with NaOCl.

Oxidation of coal may also be represented by a scheme proposed by Oreshko (1953);

$$\text{Coal} + O_2 \rightarrow \text{coal-oxygen complex} \xrightarrow{O_2} \text{humic acid} + CO_2 + CO + H_2O$$

Oreshko emphasized the importance of the "coal–oxygen complex," the nature of which changes with progressive oxidation.

Among the acids formed in extracts from oxidized bituminous coal, mellitic, trimellitic, and phthalic acids as well as condensed aromatic acids dominate. The oxidation end-products are the oxalic, malonic, and acetic acids, indicating that the phenol ring is broken during the final stages of oxidation.

We have selected a representative array of aromatic and aliphatic acids and other compounds found in coal oxidation studies by Fester and Robinson (1966), Chakrabartty (1975), Flaig (1960, 1966), Swain (1963, 1970), Jensen *et al.* (1966), and Yohe (1958), and others, and listed them in a sequence representing progressive oxidation of coal (Table V). In Table V, we have also included gluconolactone and gluconic acid, because recent infrared reflection spectrometry and mass spectrometry studies have shown that lactones may form on the surface of oxidized graphite, char, and coal (Yang and Steinberg, 1975; Barton and Harrison, 1975). It may be significant that when listed according to increasing oxygen content, these compounds form distinct groups in the sequence of polyaromatic acids—carboxyl phenolic acids—polycarboxylic phenolic acids—polycarboxylic heterocyclic polyaromatic acids (humic acids)—carboxylic acids—glycosides.

When compounds found in coal oxidation studies are listed and their composition compared in the sequence of increasing oxygen content as in Table V, a few observations may be made that could be useful.

First, the increase in oxygen and the decrease in carbon seem to follow a pattern similar to that when different rank coals are grouped in a sequence of decreasing rank. While oxygen decreases some ten-

TABLE V *Organic Compounds Detected in Coal Oxidation Studies Listed According to Increasing Oxygen Content*

	% C	% H	% O	% N
Anthracene: $C_{14}H_{10}$	94.34	5.66	0.00	
Squalene: $C_{30}H_{30}$	87.73	12.27	0.00	
Anthracite, Mason (1962)	93.50	2.81	2.72	0.97
Friedeline: $C_{30}H_{50}O$	84.44	11.81	3.75	
Cholesterol: $C_{27}H_{45}(OH)$	83.87	11.99	4.14	
Phytol: $C_{20}H_{39}(OH)$	81.01	13.60	5.40	
Betuline: $C_{30}H_{50}O_2$	81.39	11.38	7.23	
Gillet's (1951) "fundamental coal formula" $C_{29}H_{22}O_2$	86.54	5.51	7.95	
Anthranol: $C_{14}H_9(OH)$	86.57	5.19	8.24	
Bituminous coal, Mason (1962)	84.24	5.55	8.69	1.52
Abietic acid: $C_{20}H_{30}O_2$	79.42	9.99	10.58	
Naphthol: $C_{10}H_7(OH)$	83.31	5.59	11.10	
Anthraquinone: $C_{15}H_{10}O_2$	80.76	3.87	15.37	
Phenol: $C_6H_5(OH)$	76.57	6.43	17.00	
Naphthoic acid: $C_{10}H_7(COOH)$	76.73	4.68	18.58	
Naphthoquinone: $C_{10}H_6O_2$	75.94	3.82	20.23	
Lignite, Mason (1962)	72.95	5.24	20.50	1.31
Cinnamic acid: $C_6H_5CH:CH(COOH)$	72.96	5.44	21.60	
Toluic acid: $CH_3C_6H_4(COOH)$	70.57	5.92	23.50	
Benzoic acid: $C_6H_5(COOH)$	68.84	4.95	26.20	
Vanillin: $CH_3OC_6H_3(OH)CHO$	63.15	5.30	31.55	
Anthraxylic Acid: $C_{20}H_8O_8$	63.84	2.14	34.02	
p-Hydroxybenzoic acid: $C_6H_4(OH)(COOH)$	60.86	4.38	34.75	
Purpurogallin: $C_{11}H_8O_5$	60.00	3.66	36.33	
Peat, Mason (1962)	55.44	6.28	36.56	1.72
Phthalic acid: $C_6H_4(COOH)_2$	57.83	3.64	38.52	
Syringic acid: $(CH_3O)_2C_6H_2(OH)(COOH)$	54.55	5.09	40.37	
Mevalonic acid: $C_5H_9(OH)_2(COOH)$ (also in lactone form)	48.64	8.16	43.20	
Wood, Mason (1962)	49.65	6.23	43.20	.92
Adipic acid: $(CH_2)_4(COOH)_2$	49.31	6.90	43.79	
Trimellitic acid: $C_6H_3(COOH)_3$	51.44	2.88	45.68	
Gallic acid: $C_6H_2(OH)_3(COOH)$	49.42	3.55	47.02	
Pyromellitic acid: $C_6H_2(COOH)_4$	47.26	2.38	50.36	
Acetic acid: CH_3COOH	40.00	6.71	53.29	
Gluconolactone: $C_6H_{10}O_6$	40.45	5.66	53.89	
Succinic acid: $(CH_2)_2(COOH)_2$	40.68	5.12	54.20	
Mellitic acid: $C_6(COOH)_6$	42.12	1.77	56.11	
Gluconic acid: $C_6H_{12}O_7$	36.74	6.17	57.10	
Malonic acid: $CH_2(COOH)_2$	34.62	3.87	61.50	
Formic acid: HCOOH	26.10	4.38	69.52	
Oxalic acid $(COOH)_2$	26.68	2.24	71.08	
Carbon dioxide	27.29	—	72.91	

to twenty-fold when comparing a high-rank anthracite with wood, the carbon concentration changes only by about a factor of two, and the bulk hydrogen concentration remains about the same while fluctuating within relatively narrow limits of about 3–5% total. The wide range of the oxygen content, between about 3–72 wt%, demonstrates the important role of this element in the oxidation processes of coal. This leads from pure hydrocarbons and anthracite to carbon dioxide. Determination of oxygen in these products should therefore provide more diagnostic information concerning the rank of coal or the degree of oxidation of the derivative products than the analysis for hydrogen or even carbon.

Second, even a fleeting comparison of composition of coals and their derivatives, when listed in this manner, groups together the saturated cyclic hydrocarbons, the steroids, the terpenes, and their derivatives according to their low oxygen and high hydrogen content. When the composition of these compounds is compared with the composition of coal, the abnormally high content of *hydrogen,* about double that of the typical coals, is obvious. Coals do not have nearly enough hydrogen should they be assumed to contain considerable quantities of such compounds or similar saturated structures.

Third, the comparison of compositions of several aromatic and polyaromatic as well as heterocyclic compounds of coal decomposition products with those of coals reveals remarkable similarities in oxygen, hydrogen, and carbon content.

Fourth, the arrangement of this list coincides with the general sequence of oxidation and decomposition products formed from coal, lignite, peat, and organic material.

To demonstrate these relationships better, concentration of oxygen, carbon, and hydrogen in products of coal oxidation are plotted in Fig. 2. Some typical organic compounds detected in these materials, and mostly listed in Table V, were selected for this purpose, including anthracite, bituminous coals, peat, and wood. For the latter substances approximate ranges of oxygen, carbon, and hydrogen concentrations are also indicated. The sequence selected is as much as possible the sequence of their formation or appearance during the processes of oxidation. It can be seen that if this sequence were selected on the basis of oxygen content, the picture would not have been much different.

C. Oxygen and Humic Acids

In the oxidation of coal, the coalification sequence represented in Tables III and IV is reversed (see Table V). The important intermediate

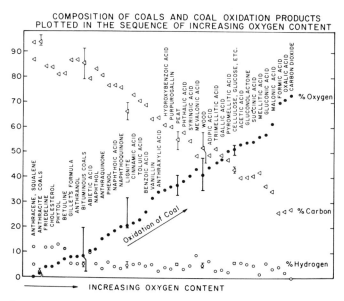

Fig. 2 Composition of coals and coal oxidation products plotted in the sequence of increasing oxygen content and approximately the sequence of formation when coal is oxidized.

products in both cases are the humic acids. In coalification these acids are decarboxylated and dehydroxylated, and further aromatization results. In the oxidation of coal the aromatic molecule is then reconstituted, cleaved stepwise into smaller units, and carboxyl and hydroxyl groups are added to the phenolic nuclei that are broken in the final stages of oxidation to form smaller aliphatic molecules rich in carboxyl oxygen. The sequence in which the compounds are listed in Tables III, IV, and V is not intended in the sense that the compounds are necessarily immediate precursors or are formed directly from the compounds listed above them. Such constraints as equilibria and the fact that the oxidation reactions proceed simultaneously in a variety of complex organic molecules and aromatic heterocyclic clusters affecting different active sites make a simple scheme very inadequate. However, we believe that the general path of coal weathering, degradation, and oxidation is well represented by plotting the products of these processes according to the increasing oxygen content of compounds identified by modern techniques.

Oxygen plays an important role in the first stages of the decomposition of organic substances derived from plant material. The decomposition and reformulation of such organic matter as cellulose and lignin leads, through humification and a sequence of metamorphic processes, to the formation of coal. Initially, oxidation reactions cause the formation of dark-colored humic acids; later under more anaerobic conditions, pressure, and higher temperatures, polymerization occurs as the sediment becomes buried. Under these conditions phenolic compounds are more stable and during the processes of decomposition phenolic substances are more resistant to microorganisms, and thus seem to accumulate.

A very complex series of reactions results in an organic rock we call coal. This process also consists of a sequence of bacterial decomposition of starches and proteins and fixation of nitrogen derived from peptides, amino acids, and ammonia. Breaking of the cellulose chains by enzymatic action (giving rise to the formation of aliphatic compounds later under reducing conditions), and decomposition of lignin through oxidation of the methoxyl groups, and formation of quinones that later polymerize, all lead to a higher degree of aromatization.

The decomposition of lignin is similar to that of starches and cellulose, but slower. Many primary lignin decomposition products have been identified. These are mostly phenolic compounds. Aromatic molecules with side chains of one or three carbon atoms are derived from the polymeric lignin. Flaig (1960, 1964, 1966), Swain (1963, 1970), Steelink (1966), Albers *et al.* (1974), and Martin (1975) in their studies of humification and humic acids have proposed the reactions shown in IX. Through cleavage or oxidation reactions the side chain is shortened. Also the compounds shown in X have been identified. The oxidation may proceed in the general sequence (Flaig, 1966; Steelink, 1966; Albers *et al.*, 1974; Martin, 1975) shown in XI. Equally, *cellulose, hemicellulose,* and other higher polymer *carbohydrates* can decompose through bacterial action to trioses and further to methyl glyoxal that may form quinones (XII) by condensation (Swain, 1970, p. 343). Quinones can thus be formed through condensation of short aliphatic chains derived from cellulose and sugars as well as by cleavage, hydroxylation, demethylation, decarboxylation, and demethoxylation of phenolic compounds with side chains. Starting with benzoquinone, naphthoquinone, anthraquinone, anthrone, benzanthrone, dibenzanthrone, and similar compounds may form (XIII) through condensation. We note that as the condensation proceeds in this direction, the oxygen content is decreased.

Examples of naturally occurring polycyclic hydroxyquinones that may be produced by polymerization of quinones, according to Steelink (1966), are (XIV) Emodin (1,3,8-Trihydroxy-6-methylanthraquinone and (XV) alizarin (1,2-dihydroxyanthraquinone).

R = —CHO
p-hydroxybenzaldehyde

R = —COOH
p-hydroxybenzoic acid

R = —CH=CH—COOH
p-hydroxycinnamic acid

Lignin

IX

Lignin →

R = —CHO
(vanillin)

R = —COOH
(vanillic acid)

R = —CH=CH—CHO
(coniferyl aldehyde)

R = —CH=CH—COOH
(ferulic acid)

R = —CHOH—CHOH—CH$_2$OH
(guajacyl-glycerol)

R = —CH$_2$—CO—COOH
(guajacyl-peruvic acid)

Dehydrodivanillin

X

According to Swain (1970), polymerization of hydroxyquinone may proceed as shown in XVI, where the dashed lines indicate the transfer of ring hydrogens to hydroxyl groups. If nitrogen is present, it can be incorporated by forming similar heterocyclic molecules (XVII).

Polymerization also may lead to polycyclic hydroxyquinones and

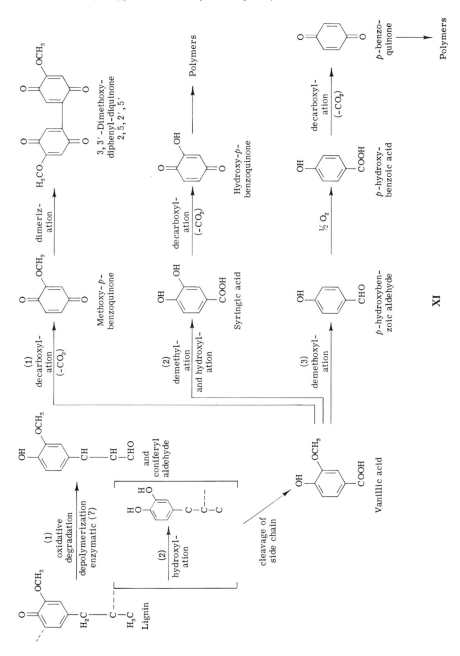

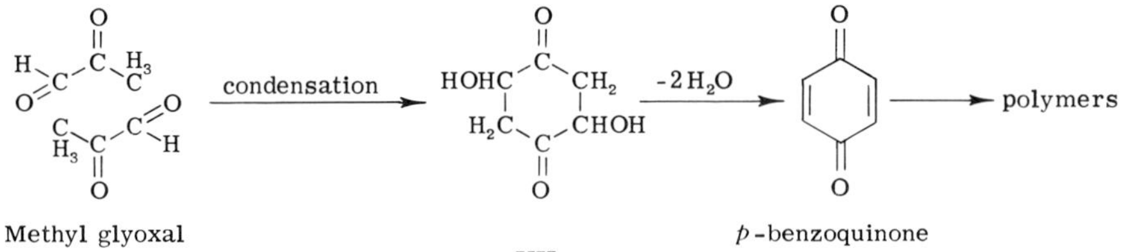

condensation
-2 H₂O
polymers
Methyl glyoxal
p-benzoquinone

XII

complex heterocyclic molecules. Humic acids are examples of such polymeric matrixes. The structure of these is not yet completely understood. Models of humic acids were worked out by Gillet (1956), such as for anthraxylic acid (XVIII), and (XIX) by Flaig (1960, 1964).

To emphasize the complexity of coal oxidation processes, Jensen *et al.* (1966) point out that coal breaks down on oxidation by two simul-

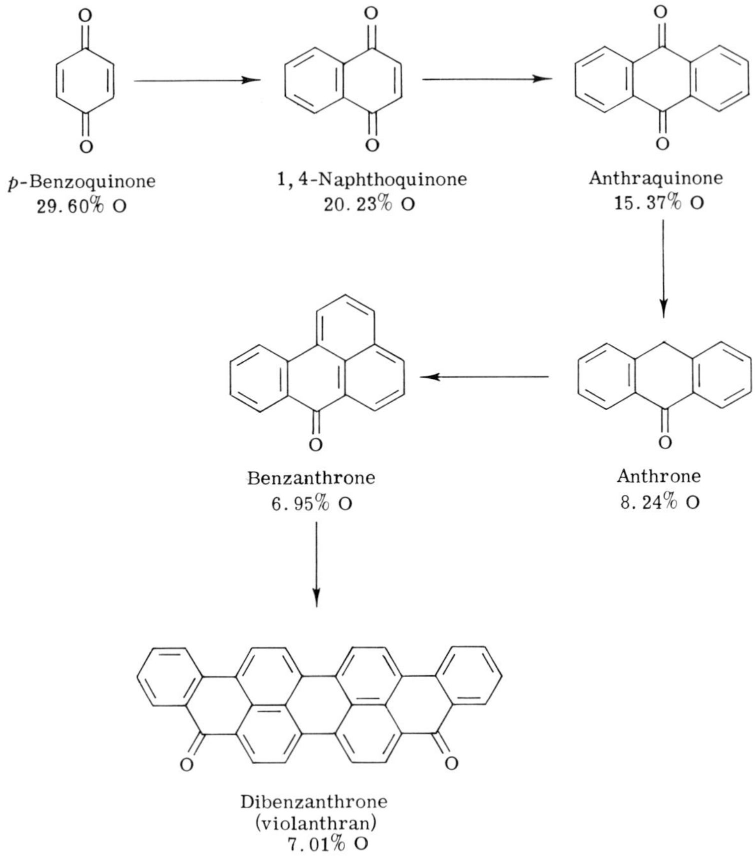

p-Benzoquinone
29.60% O
1,4-Naphthoquinone
20.23% O
Anthraquinone
15.37% O
Benzanthrone
6.95% O
Anthrone
8.24% O
Dibenzanthrone
(violanthran)
7.01% O

XIII

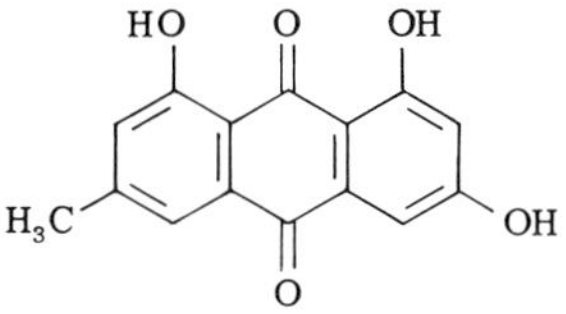

29.6% O

XIV

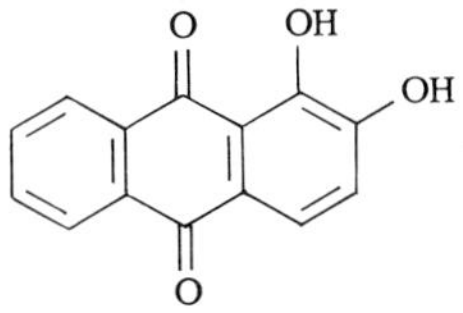

26.64% O

XV

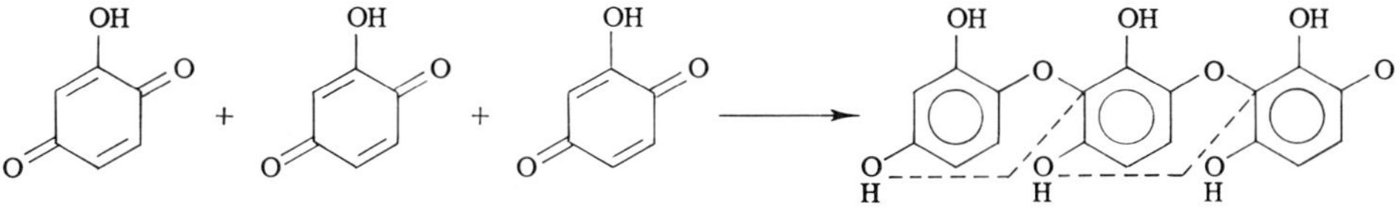

XVI

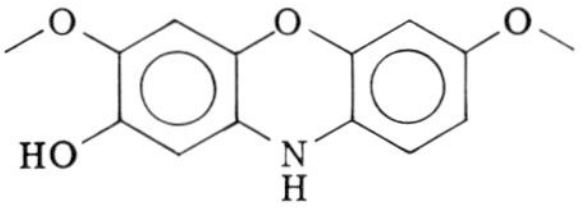

XVII

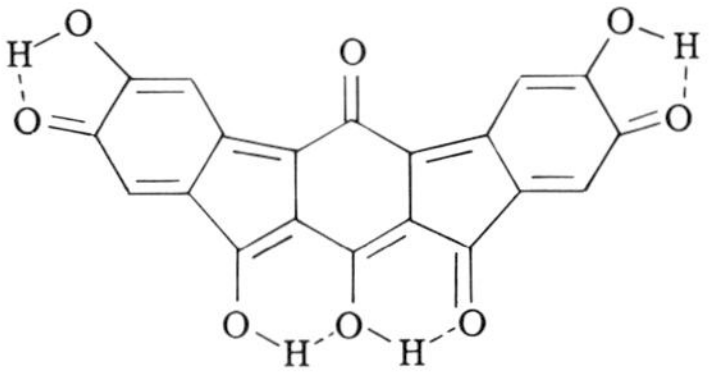

XVIII

taneously progressing types of reactions. One of these involves the physical skeletal breakdown and degradation of the coal infrastructure into progressively smaller fragments, which in itself results in increased alkali solubility whereas fresh coal is *alkali insoluble*. The other process is chemical in nature, causing increased acidity and the gradual evolution of "humic acids." Jensen *et al.* (1966) note that the "acidity" builds up more rapidly than the *alkali soluble* material and, therefore, suggests, that alkali solubility of coal is not a good measure of the acidity and should not be used to correspond to the accepted definition of "humic acid." To demonstrate this, Jensen *et al.* (1966) propose a scheme for the skeletal breakdown during oxidation of coal, as shown in a modified form by (XX).

These qualitative schemes also illustrate the wide spread in molecular weights of humic acids in the literature, indicating that 600 may be as valid as 10,000. Jensen *et al.* (1966) emphasize that attempts to give kinetic expression to directly measured humic acid concentrations based on alkali solubility are unjustified.

Table III gives a list of derivatives formed in oxidation studies of lignin and the compositions of the products of the polymerization of

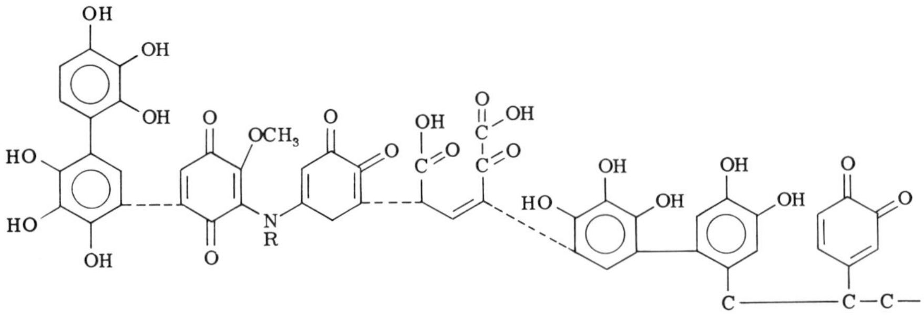

XIX

the quinoid compounds and humic acids compared with the composition of bituminous coals. The organic compounds are listed in the approximate sequence of their formation, as previously delineated. When compositions listed in this manner are compared, one notices the role of oxygen in this process. The oxygen content increases slightly during the oxidation of the depolymerized lignin derivative, coniferyl alcohol, then drops with progressive condensation and aromatization. Humic acid compositions indicate that these substances may be intermediate products of oxidation of lignin.

In general, although the content of oxygen decreases during the processes of coalification, the first stages of decomposition of plant material that lead to humification or formation of the "dark colored" humic substances are triggered by decay organisms, enzymes, and the oxidizing effect of atmospheric oxygen. Gradual oxidation of sugars derived from depolymerized cellulose and through enzymatic hydrolysis (rotting) of coniferyl alcohol (XXI) derived from lignin by similar

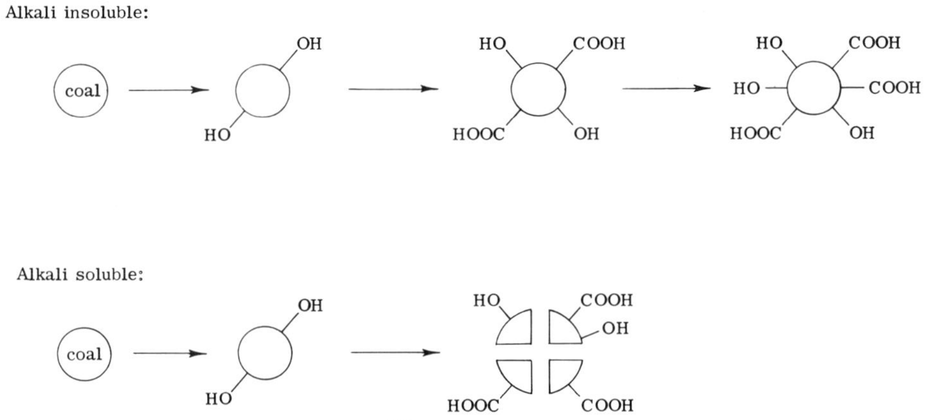

Alkali insoluble:

Alkali soluble:

XX

H_3CO ⟶ CH=CHCH$_2$OH

HO

XXI

processes, leads to the formation of humic substances; these also contain nitrogen compounds of plant, bacterial, and animal origin, which can form through interaction of carbohydrates with amino acids, resulting as shown by Maillard (1913) in melanoidins. These are dark-colored, amorphous nitrogen-containing substances, similar to humic acids.

Chitin, $C_8H_{13}NO_5$ (C 47.29%, H 6.45%, O 39.37%, N 6.89%) and glucosamine, $C_6H_{13}NO_5$ (C 40.22%, H 7.31%, O 44.65%, N 7.82%) may be precursors of melanoidins through the Maillard reaction. Melanoidins are important constituents of the humic acid fractions in soils, peat, and lignite, and probably essential building blocks within the complex humic acid molecules. Analyses of preparations of melanoidin and glucosamine show high carbon, oxygen, and nitrogen contents, similar to that of humic acids (see Table VI) (Manskaya and Drozdova, 1968).

Howard (1947) notes that the "regenerated" humic acids from oxidation of coal of higher rank have fewer active acidic groups, fewer ether links, and a higher degree of condensation of aromatic nuclei.

TABLE VI   *Composition of Humic Acids, Melanoidin, and Glucosamine*

	Weight percent			
	C	H	N	O
Humic acids:				
From peat	61.78	4.28	1.72	32.22[a]
From chernozem	57.32	4.25	4.04	34.34[a]
From average podzol soil	57.94	5.79	4.86	31.41[a]
From humified plant deposits	57.57	6.35	7.64	28.48[a]
Melanoidin from glucosamine	52.51	4.99	4.5	38.0[a]
Melanoidin from glucosamine and glycine	52.8	5.78	8.96	32.46[a]
Humic acid from straw	53.74	5.19	3.44	37.63[b]
Humic acid from chernozem	58.9	3.3	2.5	34.2[b]
Humic acid from lignitic soft coal, 2 samples	61.2	4.1	~0.6	34.8[b]
Humic acid from subbituminous coal, 7 samples	69.2	4.5	—	26.3[b]

[a] From Manskaya and Drozdova (1968, pp. 42, 76).
[b] Humic acids from Flaig (1966, p. 65).

Humic acids extracted from different types of coal have different optical density, depending on the degree of condensation of the aromatic clusters and the concentration of the oxygen groups, according to Kucharenko (1953, 1955).

By oxidizing coal in oxygen atmosphere at 200°C, Gillet (1958) obtained a humic acid with a molecular weight of 376 and proposed the formula $C_{20}H_8O_8$ for the dry material. Gillet and Pirghaye (1960) were able also to separate the anthraxylic acid of identical composition ($C_{20}H_8O_8$) from peat, coal, and lignite. Humic acids obtained from anthracite were labeled "regenerated" by Gillet (1958).

An interesting aspect of the first stages of the bacterial degradation of carbohydrate chains is the sudden drop in the oxygen content of the fresh organic matter from about 43% in wood and 41% in fresh peat to about 26–32% in the humic acid concentrates and to 32–22% in lignite soft coal and subbituminous coal (see Tables IV and VI).

Table IV shows the wide range of oxygen in plant matter that is mainly a mixture of cellulose (49% O) and lignin (33% O). The composition varies depending on the plant species. Lignin usually amounts to 10 to 35% of the total dry weight, and cellulose 20–55 wt.% (Flaig, 1968; Brauns, 1952; Brauns and Brauns, 1960). Sugars, derived from cellulose, such as glucose and cellobiose, have 51–53% O. The first steps of enzymatic decomposition (oxidation) of cellulose seem, therefore, to increase somewhat the oxygen content of the rotting plant matter if sugars are formed. The abrupt lowering of the oxygen content by some 25% relative to the original composition must be because of considerable evolution of water and carbon dioxide and is the result of metamorphic processes.

The enzymatic hydrolysis of lignin proceeds more slowly, but seems to involve a similar sequence, consisting of initial oxygen loss due to degradation of lignin (~33% O), to coniferyl alcohol (26.64% O), or to similar compounds (see Table III). Then an increase is noted due to oxidation (hydroxylation) to vanillin (31.55% O) and vanillic acid (38.06% O) and through decarboxylation to dimethoxybenzoquinone (38.06% O). Further oxidation (decarboxylation) reduces the oxygen content due to the evolution of carbon dioxide, thus p-hydroxybenzoic acid (34.75% O) is formed and p-benzoquinone (29.60% O). Aromatization of the quinones leads to a further drop in oxygen content, e.g., emodin (29.60% O) a hydroxyanthraquinone found in nature (Steelink, 1966), and alizarin (26.64% O) (see Table III).

It may be of interest to note that humic acids extracted by Flaig (1966) from subbituminous coals seem to have a nearly identical composition

(Table III) and little or no nitrogen. It also should be noted that the natural oxidative degradation of chitin (39.37% O) to glucosamine (44.65% O) to melanoidins (31–35% O) (Manskaya and Drozdova, 1968) seems to show a pattern of an initial increase, then an abrupt drop of the oxygen content by some 20% relative, followed by a more prolonged and gradual removal of oxygen from the system, similar to the degradation of cellulose and lignin. If these processes of enzymatic decomposition of organic matter do occur simultaneously, the bulk oxygen content variations in the decaying matter may serve as a good indicator of the respective stages of humification and coalification and thus of the rank of coal. This is shown in Table IV, which gives a compilation of analyses of lignite to anthracite coals taken from numerous sources. In Table IV comparison is made with compositions of wood, fresh lignin, and some carbohydrates to emphasize the whole spectrum of oxygen depletion during the coalification processes.

In summary, the humification process may be considered as the first step in coalification of plant matter. It starts by rapid decomposition of cellulose and by enzymatic degradation of lignin of the rotting plant substance to form C_3 or C_2 benzene compounds. These loose methoxyl groups and carboxyl groups and can form hydroquinones that may polymerize and combine, forming humic acids. Degradation may proceed also to aliphatic compounds that again serve as energy supply for the microorganisms. Most of the reactions seem to lead to benzoquinones that dimerize and polymerize further, causing an increase in aromatization with age (see Table III) and under more anaerobic conditions later during coalification. During the later stages of the process of humification when conditions become anaerobic, melanoidin and glucosamine compounds form and nitrogen fixation occurs when decomposing nitrogen compounds, such as protein-derived α-amino acids react with the dimerized and polymerized aromatics. This explains the presence of about 1 to 3.5% nitrogen in humic acid concentrates, lignin, lignite, subbituminous and bituminous coal (see Table IV). The fixation of nitrogen also results in further reduction of carbon in humic substance during the later stages of humification. Further coalification of buried humified strata of decomposed organic material causes reduction as the oxygen functional group content decreases, and CO and CO_2 gases and H_2O evolve and gradual dehydration occurs. This process leads to lignite, subbituminous coal, bituminous coal, and finally anthracite. Increasing pressure, temperature, and time are obvious critical factors in the formation of coal (Teichmuller and Teichmuller, 1966; Van Krevelen, 1961, 1963).

ACKNOWLEDGMENT

The author wishes to acknowledge the invaluable help received from Dr. G.E. Miller, Reactor Supervisor and Lecturer in Chemistry at the University of California at Irvine during the development of the methods and systems described in this chapter.

REFERENCES

Abernethy, R. F., and Gibson, F. H. (1966). *U.S. Bur. Mines, Rep. Invest.* No. 6753.

Albers, G., Lenart, L., and Oelert, H. H. (1974). *Fuel* **53,** 47-53.

ASTM (1965). "1965 Annual Book of ASTM Standards. Part 19: Gaseous Fuels; Coal and Coke," pp. 73-78. Am. Soc. Test. Mater., Philadelphia, Pennsylvania.

ASTM (1974). "1974 Annual Book of ASTM Standards. Part 26: Gaseous Fuels; Coal and Coke; Atmospheric Analysis." Am. Soc. Test. Mater., Philadelphia, Pennsylvania.

ASTM (1975). "1975 Annual Book of ASTM Standards. Part 26: Gaseous Fuels; Coal and Coke; Atmospheric Analysis." Am. Soc. Test. Mater., Philadelphia, Pennsylvania.

ASTM (1976). "1976 Annual Book of ASTM Standards. Part 26: Gaseous Fuels; Coal and Coke; Atmospheric Analysis." Am. Soc. Test. Mater., Philadelphia, Pennsylvania.

Barton, S. S., and Harrison, B. H. (1975). *Carbon* **13,** 283-288.

Block, C., and Dams, R. (1974). *Anal. Chim. Acta* **71,** 53-65.

Blom, L., Edelhausen, L., and Van Krevelen, D. W. (1957). *Fuel* **36**(2), 135.

Brauns, F. E. (1952). "The Chemistry of Lignin." Academic Press, New York.

Brauns, F. E., and Brauns, D. A. (1960). "The Chemistry of Lignin: Supplement Volume (Covering the Literature for the Years 1949-1958)." Academic Press, New York.

British Standards Institution (1971). British Standard 1016, Part 16, Reporting of Results, pp. 1-39.

Brown, J. K. (1955). *J. Chem. Soc.* (London) pp. 744-752.

Brown, N. A., Belcher, C. B., and Callcott, T. G. (1965). *J. Inst. Fuel* **38,** 198.

Chakrabartty, S. K. (1975). *proc. Workshop NSF, Univ. Tennessee, Knoxville* pp. 89-104.

Chakravorty, S. L., Long, R., and Ward, S. G. (1950). *Fuel* **29,** 68-69.

Dryden, I. G. C. (1963). *In* "Chemistry of Coal Utilization: Supplement Volume." (H. H. Lowry, ed.), pp. 232-295. Wiley, New York.

Fester, J. I., and Robinson, W. E. (1966). *Adv. Chem. Ser.* **55,** 22-30.

Flaig, W. (1960). *Suom. Kemistil.* **33,** 229.

Flaig, W. (1964). *Geochim. Cosmochim. Acta* **28,** 1523-1535.

Flaig, W. (1966). *Adv. Chem. Ser.* **55,** 58-68.

Flaig, W. (1968). *"Biochemical Factors in Coal Formation* In "Coal and Coal Bearing Strata." (D. Murchison and T. S. Westoll, eds.), pp. 197-232. Elsevier, Amsterdam.

Francis, W. (1961). "Coal." Arnold, London.

Fraser, F. W., and Belcher, C. B. (1973). *Fuel* **52,** 41-46.

Gillet, A. (1951). *Fuel* **30,** 181-187.

Gillet, A. (1956). *Brennst. Chem.* **37,** 395.

Gillet, A. (1958). *Brennst. Chem.* **38**(9/10), 151.

Gillet, A. C., and Pirghaye, A. (1960). *Sci. Proc. R. Dublin Soc., Ser. A* **1**(4), 133.

Given, P. H. (1976). *Fuel* **55,** 256.

Given, P. H., and Yarzab, R. F. (1975). "Problems and Solutions in the Use of Coal Analyses," Tech. Rep. FE-0390-a, pp. 1-40. Pennsylvania State Univ., University Park.

Gluskoter, H. J. (1965). *Fuel* **44,** 285-291.

Hamrin, C. E., Maa, P. S., Chyi, L. L., and Ehmann, W. D. (1975). *Fuel* **54,** 70-71.

Hamrin, C. E., Johannes, A. H., James, W. D., Sun, G. H., and Ehmann, W. D. (1979). *Fuel* **58,** 48-54.

Hickling, G. (1927). *Trans. Inst. Min. Engrs.* **72,** 261-276.

Hickling, H. G. A. (1931). *Proc. S.W. Inst. Eng.* **46,** 911.

Higushi, K., and Shibuya, Y. (1954). *Nenryo Kyokai-Shi* **33,** 366-373 (Chem. Abstr. **48,** 11029i).

Howard, H. C. (1945). *In* "Chemistry of Coal Utilization" (H. H. Lowry, ed.), Vol. 1, p. 346. Wiley, New York.

Howard, H. C. (1947). *Trans. Am. Inst. Min. Metall. Eng.* **177,** 523-524.

Ignasiak, B. S., Nandi, B. N., and Montgomery, D. S. (1969). *Anal. Chem.* **41,** 1676-1678.

Ignasiak, B. S., Nandi, B. N., and Montgomery, D. S. (1970). *Fuel* **49,** 214-221.

Ignasiak, B. S., Clugston, D. M., and Montgomery, D. S. (1972). *Fuel* **51,** 76-80.

Ignasiak, B. D., Szladov, A. J., and Montgomery, D. S. (1974a). *Fuel* **53**(1), 12-15.

Ignasiak, B. S., Szladov, A. J., and Berkowitz, N. (1974b). *Fuel* **53,** 229-231.

Ignasiak, B. S., Ignasiak, T. M., and Berkowitz, N. (1975). *Rev. Anal. Chem.* **2**(3).

Jackman, H. W., Eissler, R. L., and Reed, F. H. (1957). *Ill. State Geol. Surv., Circ.* **227,** 1-22.

James, W. D., Chyi, L. L., and Ehmann, W. D. (1975). *Trans. Am. Nucl. Soc.* **21,** Suppl. No. 3, 30.

James, W. D., Ehmann, W. D., Hamrin, C. E., and Chyi, L. L. (1976). *J. Radioanal. Chem.* **32,** 195-205.

Jensen, E. J., Melnyk, N., Wood, J. C., and Berkowitz, N. (1966). *Adv. Chem. Ser.* **55,** 621-642.

Jones, R. E., and Townsend, D. T. (1949). *J. Soc. Chem. Ind., London* **68,** 197-201.

King, J. G., Maries, M. B., and Crossley, H. E. J. (1936). *J. Soc. Chem. Ind., London* **55,** 277.

Kinson, K., and Belcher, C. B. (1975). *Fuel* **54,** 205-209.

Kucharenko, T. A. (1953). *Dokl. Akad. Nauk SSSR* **89**(1), 133.

Kucharenko, T. A. (1955). *Tr. Inst. Goryuch. Iskop., Moscow* **5,** 11.

Kucharenko, T. A. (1960). "Chemistry and Genesis of Mineralized Coals." Gosgortech-izdat, Moscow.

Maillard, L. C. (1913). *C. R. Acad. Sci., Paris* **156,** 1159.

Manskaya, S. F., and Drozdova, T. V. (1968). "Geochemistry of Organic Substances" Pergamon, Oxford.

Martin, F. (1975). *Fuel* **54,** 236-240.

Mason, B. (1962). "Principles of Geochemistry," p. 310. Wiley, New York.

Miller, G. E., and Volborth, A. (1976). *Sci. Ind. Appl. Small Accel., 4th I. E. E. E.* pp. 288-292.

Miller, R. N., Yarzab, R. F., and Given, P. H. (1979). *Fuel* **58,** 4.

Millot, J. O'N. (1958). *Fuel* **37,** 71.

Montgomery, R. S., and Bozar, K. B. (1959). *Fuel* **38**(3), 400.

Montgomery, R. S., and Holly, E. D. (1957). *Fuel* **36**(4), 493.

Montgomery, R. S., Holly, E. D., and Gohlke, R. S. (1956). *Fuel* **35**(1), 60.

Moschopedis, S. E. (1962). *Fuel* **41,** 425.

Mott, R. A., and Spooner, C. E. (1940). *Fuel* **19,** 266.

Oreshko, V. F. (1949a). *Izv. Akad. Nauk SSSR, Otd. Tekh. Nauk* pp. 249-257 [*Chem. Abstr.* **44,** 2200i (1950)].

Oreshko, V. F. (1949b). *Izv. Akad. Nauk SSSR, Otd. Tekh. Nauk* pp. 748-759 [*Chem. Abstr.* **44,** 2201f (1950)].

582   *Alexis Volborth*

Oreshko, V. F. (1949c). *Izv. Akad. Nauk SSSR, Otd. Tekh. Nauk* pp. 1642-1648 [*Chem. Abstr.* **44,** 2202e (1950)].
Oreshko, V. F. (1953). *Coll. Chim. Genesis Tverd. Goryuch. Iskop.* Akad. Nauk SSSR, p. 162.
Parks, B. C. (1963). *In* "Chemistry of Coal Utilization: Supplementary Volume" (H. H. Lowry, ed.), pp. 1-34. Wiley, New York.
Parr, S. W. (1932). "The Analysis of Fuel, Gas, Water and Lubricants." McGraw-Hill, New York.
Rees, D. W. (1966). *Ill. State Geol. Surv., Rep. Invest.* **220,** 1-55.
Rose, H. J. (1945) "Chemistry of Coal Utilization" (H. H. Lowry, ed.) Vol. 1, pp. 25-85.
Schlyer, D. J., Ruth, T. J., and Wolf, A. P. (1979). *Fuel* **58,** 208.
Schopf, J. M., and Long, W. E. (1966). *Adv. Chem. Ser.* **55,** 156-195.
Schutze, M. (1939). *Z. Anal. Chem.* **118,** 245.
Shipley, D. E. (1962). *Br. Coal Util. Res. Assoc., Mon. Bull.* **26,** 3.
Steelink, C. (1966). *Adv. Chem. Ser.* **55,** 80-90.
Swain, F. M. (1963). *In* "Organic Geochemistry," Earth Science Series, Monogr. No. 16, pp. 87-147. Macmillan, New York.
Swain, F. M., (1970). "Non-Marine Organic Geochemistry," Earth Science Series, Cambridge Univ. Press, London and New York.
Swann, P. D., Allardice, D. J., and Evans, D. G. (1974). *Fuel* **53,** 85-87.
Teichmuller, M., and Teichmuller, R. (1966). *Adv. Chem. Ser.* **55,** 133-155.
Tronov, B. V. (1940a). *J. Appl. Chem. (USSR)* **13,** 1053-58 [*Chem. Abstr.* **35,** col. 1966 (1941)].
Tronov, B. V. (1940b). *Izv. Tomsk. Ind. Inst.* **60**(3), 11-36 [*Chem. Abstr.* **37,** col. 1029 (1943)].
Unterzaucher, J. (1940). *Ber. Deut. Chem. Ges. B* **73,** 391-404.
Unterzaucher, J. (1952). *Analyst* **77,** 584.
U.S. Bureau of Mines (1967). *Bull.* No. 638, 1-85.
Van Krevelen, D. W. (1961). "Coal." Elsevier, Amsterdam
Van Krevelen, D. W. (1963). *In* "Organic Geochemistry," Earth Science Series, Monogr. No. 16, pp. 183-247. Macmillan, New York.
Van Krevelen, D. W., and Schuyer, J. (1957). "Coal Science." Elsevier, Amsterdam.
Van Vucht, H. A., Rietveld, B. J., and Van Krevelen, D. W. (1955). *Fuel* **34,** 50-59.
Volborth, A. (1976a). "Fossil Energy," DOE COO-2898-2. U.S. Gov. Print. Off., Washington, D.C.
Volborth, A. (1976b). "Fossil Energy," DOE COO-2898-3. U.S. Gov. Print. Off., Washington, D.C.
Volborth, A. (1976c). "Fossil Energy," DOE COO-2898-4. U.S. Gov. Print. Off., Washington, D.C.
Volborth, A. (1976d). "Fossil Energy," DOE COO-2898-7. U.S. Gov. Print. Off., Washington, D.C.
Volborth, A., Miller, G. E., Garner, C. K., and Jerabek, P. A. (1977a). "Fossil Energy," DOE COO-2898-10. U.S. Gov. Print. Off., Washington, D.C.
Volborth, A., Miller, G. E., Garner, C. K., and Jerabek, P. A. (1977b). *Fuel* **56,** 204-208.
Volborth, A., Miller, G. E., Garner, C. K., and Jerabek, P. A. (1977c). *Fuel* **56,** 209-215.
Volborth, A., Miller, G. E., Garner, C. K., and Jerabek, P. A. (1978). *Fuel* **57,** 49-55.
Washowska, H. M., Nandi, B. N., and Montgomery, D. S. (1974). *Fuel* **53,** 212-219.
Wood, J. C., Moschopedis, S. E., and den Hertog, W. (1961). *Fuel* **40,** 491.
Yamasaki, T. (1953). *Nippon Kogyo Kaishi* **69,** 253-258 [*Chem. Abstr.* **48,** 11756g (1954)].
Yang, R. T., and Steinberg, M. (1975). *Carbon* **13,** 411-416.
Yohe, G. R. (1950). *Fuel* **29,** 163-165.

Yohe, G. R. (1958). *Ill. State Geol. Surv., Rep. Invest.* **207,** 1–51.

Yohe, G. R., and Blodgett, E. O. (1947). *J. Am. Chem. Soc.* **69,** 2644–2648.

Yohe, G. R., and Harman, C. A. (1941). *J. Am. Chem. Soc.* **63,** 555–556.

Yohe, G. R., and Wilt, M. M. (1942a). *J. Am. Chem. Soc.* **64,** 1809–1811.

Yohe, G. R., and Wilt, M. M. (1942b). *Ill. State Geol. Surv., Circ.* **201,** 1–6.

Yohe, G. R., Wilt, M. M., Kaufmann, H. F., and Blodgett, E. O. (1955). *Trans. Ill. State Acad. Sci.* **47,** 77–80.

Chapter 56

Sulfur Groups in Coal and Their Determinations

A. Attar

DEPARTMENT OF CHEMICAL ENGINEERING
UNIVERSITY OF HOUSTON
HOUSTON, TEXAS

I. INTRODUCTION

Technical decision making is based on the results of analysis and on their interpretation. Therefore, the reliability of a decision depends to a great extent on the accuracy and the precision of the analysis, on the

585

selection of the sample, and on the interpretations of the analytical results.

Analysis of coal and coal-derived materials (CDM) for their sulfur content and its forms is done for the following objectives:

(1) to determine the quality of a coal lot or reserve;
(2) to optimize processing of coal;
(3) to evaluate coals as blends components.

In all these cases, a representative sample has to be selected, handled, stored, and analyzed. Then, the results have to be interpreted in a meaningful way that permits some generalizations.

Following a brief qualitative discussion of problems associated with the selection and preparation of a sample, common and novel methods of sulfur analysis are discussed. Since the details of many methods were already presented in Volume I, Chapter 9, the discussion here will concentrate on novel methods of analysis, on the determination of individual sulfur functional groups, and on errors associated with various methods of analysis. Particular attention is devoted to sources of errors that may lead to a wrong technical decision.

II. SAMPLE PREPARATION FOR SULFUR ANALYSIS

Sampling of coal or coal-derived materials for sulfur analysis is done everyday in many laboratories. The objective of this section is to discuss some of the potential sources of error in sulfur analysis due to inadequate sampling and handling techniques.

The presentation is informal and dwells on the major issues only. Recommendations based on the practice in different laboratories are given at the end of this section.

Errors in sulfur analysis are due predominantly to the following:

(1) the selection of a representative sample,
(2) the segregation of the sulfur-containing components during preparation of the sample,
(3) the oxidation of pyritic sulfur by air,
(4) the use of inadequate analytical techniques,
(5) misinterpretation of the results of an analysis.

Accurate quantitative analysis of sulfur can be accomplished only if the form of the sulfur is known. Therefore, the physical forms and chemical compounds in which the sulfur appears are discussed first. A more comprehensive review has been published recently by Attar (1978).

A. Physical Form and Appearance of Sulfur in Coal

In Volume I, Chapter 9 three classes of sulfur compounds were distinguished: pyritic sulfur (FeS_2); sulfatic sulfur, predominantly gypsum, ($CaSO_4$); and organic sulfur, the sulfur bound to the organic matrix. Forms of sulfur are also discussed in Volume II, Chapter 20, Section IV,C,2. Other inorganic sulfides were also found in coal, notably FeAsS, ZnS, and PbS (Gluskoter, 1975). A clear distinction should be made between the inorganic sulfur compounds and the organic sulfur compounds. The inorganic sulfur is always present in the form of small concentrated lumps of material, often crystalline. FeS_2 crystals were found in the following forms:

(1) isolated crystals,
(2) strings of isolated crystals, and
(3) small crystallites attached to other minerals.

Washing of crushed coal removes large FeS_2 crystals (larger than about 200 μm). However, smaller particles remain attached to the coal. Isolated FeS_2 crystals were found embedded in the organic matrix (Whelan, 1954). Table I summarizes the difference in the physical forms of the organic and the pyritic sulfur.

Figure 1 (Attar and Messenger, 1978) shows the interior of a "capsule" of isolated iron pyrite crystals found in a Kentucky No. 9 coal.

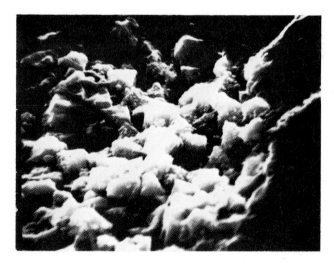

Fig. 1 A "capsule" of iron pyrite crystallites from a Kentucky No. 9 coal. (From Attar and Messenger, 1978.)

TABLE I *Physical Forms and Appearance of Organic and Inorganic Sulfur in Coal*

Category	Organic	Inorganic
Major compounds	Thiophenes	FeS_2 (pyrite), FeS_2 (marcasite), $CaSO_4$, $FeSO_4$. Exposure of coal to air converts FeS_2 to $FeSO_4$. Predominant specie: iron pyrite.
	Aromatic and aliphatic Sulfides ϕ-S-ϕ R-S-R	
	Thiols and thiophenols RSH, ϕSH	
Relation to the organic structure	Uniformly distributed in the organic matrix	Present in the form of inorganic patches, often crystalline, embedded in or on the surface of the organic matrix
Accessibility to chemical reagents	Reagents have to diffuse through the organic matrix to reach the organic sulfur	The surface inorganic sulfur is accessible. Reagents have to diffuse in the organic matrix to reach embedded sulfur. Crushing makes embedded sulfur accessible.
Density	Density of organic matrix about 1.3 g/cm³	Pyrite, 5.0 g/cm³, marcasite, 4.87 g/cm³
Particle size in washed coal		<200 μm. Most in the range of 5–65 μm
Desulfurization potential	By chemical reaction only.	By chemical reaction or by physical separation.

The size of each crystallite† is approximately 5–10 μm, and the size of the "capsule" is 150–300 μm. Strings of 3 to 10 such capsules were found in Kentucky No. 9 and to a lesser extent in Illinois No. 6 coals. Figure 2 shows one such string of capsules found in a sample of Kentucky No. 9 coal.

Small crystals with ill-defined shapes were found on silicates and aluminosilicates, and on calcium-containing minerals. The size of these crystals was of the order of 1–15 μm.

Calcium and iron sulfate are the two most important sulfates in coal. In fresh coal they always constitute a small fraction of the total sulfur, and their concentration rarely exceeds 0.3 wt% (Gluskoter *et al.*, 1977). Exposure of coal to air increases the concentration of sulfates since

† See Chapter 42, Fig. 3.

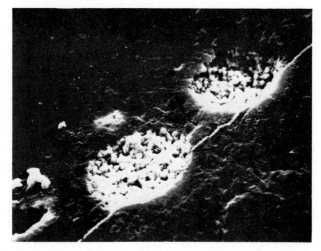

Fig. 2 A string of "capsules" of iron pyrite crystallites from a Kentucky No. 9 coal. (From Attar and Messenger, 1978.)

some of the pyrite is oxidized to iron sulfates. See more details in a subsequent section.

The organic sulfur is distributed in the organic matrix, and the origin of at least part of it is believed to be the sulfur that was in the original plant material. However, since many high sulfur coals contain large concentrations of organic sulfur or organic and inorganic sulfur, one must conclude that some sulfur was trapped by the organic matrix during the coalification process. One possible mechanism of "trapping" of sulfur could have involved an aerobic bioreduction of sulfates to H_2S and exchange of oxygen functional groups by sulfur groups, e.g.,

$$SO_4^{2-} + (\text{Bacteria}) - H_{10} \rightarrow H_2S + 4H_2O + (\text{Bacteria}), \tag{1}$$

$$ROH + H_2S \rightarrow RSH + H_2O. \tag{2}$$

Whenever minerals were present that could form stable inorganic sulfides, they competed with the organic groups for the H_2S. For example, siderite ($FeCO_3$) can be found only in low-sulfur (total) coals. The author of this chapter believes that since the rate of reaction of siderite with H_2S is so much larger than that of the organic functionalities, trapping of H_2S by the organic matrix could have occurred only after all the siderite had been converted to stable FeS_2 and could react with no more H_2S. Detailed reviews of the forms of the organic sulfur-containing functionalities in coal were published recently by Attar and Corcoran (1977) and by Attar (1978). More details are also presented in the following section.

B. Selection of a Representative Coal Sample† and Its Preparation for Sulfur Analysis

The philosophical concept of "representative" sample needs clarification. Consider a lot of 10,000 tons coal, and suppose that you are asked to prepare a "representative" sample for determination of the sulfur in the lot. "Total sulfur" analysis may require 10–1000 mg coal, while determination of the classes of sulfur requires 1–5 g. Thus, you are asked to select approximately $1/10^{10}$ of the lot to represent all of the lot.

Since coal is a heterogeneous material, the chances of success are very limited. However, had the material been homogeneous, selection of the sample would have been simple. Therefore, intuitively we should be looking for a sampling process that is as close as possible to selecting a sample from a homogeneous material.

All of the sampling processes of heterogeneous materials involve selecting large samples that contain many of the heterogeneous particles, reducing the particle size of *all* of the portion and selecting a smaller sample of smaller particles. The assumption is that if a sufficiently large number of particles are included in each lot, then each component of the mixture will be properly represented. Needless to say, analysis of several samples taken from different portions also increases the accuracy of the representation of the lot.

Preparation of a representative coal sample for analysis involves three types of operations: (1) collection of a sample, (2) pulverizing and crushing of the coal, and (3) separation of large samples into smaller samples. In addition, the samples have to be stored and handled in a way that does not change them. ASTM suggests rigid rules for the preparation of a coal sample in D2234, D271, and D2013 (ASTM, 1973). However, the procedure is lengthy, costly, and involves many steps. The exact procedure is often not followed, and different laboratories frequently adopt various shortcuts. Keller *et al.* (1968) discussed some of the arguments behind the design of a proper sampling procedure of coal for ash analysis. The general rule is that a greater number of larger portions have to be combined in order to obtain a representative sample of coal with larger particle size or heterogeneity.

† See also Volume II, Chapter 20, Section IV,A.

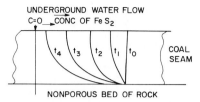

Fig. 3 Geochemical leaching of iron pyrite from the upper part of a coal seam by aerated acidic water. $t = 0 < t_1 < t_2 < t_3 < t_4$. t is the leaching time, t_0 is the time when geochemical leaching began.

Preparation of samples for sulfur analysis requires great care. Since the density of iron pyrite is about 5 g/cm³, and that of the mineral matter is about 2.7 g/cm³ they may segregate during the preparation since the density of the organic material is about 1.31 g/cm³. Preparation of a "uniform" mixture of particles with a large difference in their density is very difficult and can be accomplished only if the mixture contains very small particles, which size depends on the magnitude of the difference in the densities and on the surface properties of the solids. Whenever samples are prepared for sulfur analysis, we suggest selecting 1.5 times the number of portions that Keller *et al.* (1968) recommends for ash analysis.

The sulfur content of coal from the same mine may be different. Evidence exists (Gluskoter and Simon, 1968) that show that the sulfur content of coal varies with the depth of the seam. Geochemical processes are known that can enrich or deplete the sulfur content of the upper or the lower parts of a coal seam (Gluskoter, 1978). The most important processes are associated with the flow of water above or below the coal seam and with the acidity, sulfate, and oxygen content of this water. Figure 3 shows an example of a depletion process that preferentially reduces the pyrite content of the upper layer of a coal seam.

III. RECOMMENDATIONS RELEVANT TO SAMPLE PREPARATION

A. Recommendations Relevant to the Collection of Samples

Gluskoter *et al.* (1977) used the method recommended by Holmes (1911) to collect samples "representative" of Illinois coals. Three face channel samples were hand picked at three different locations in each mine. Each sample was collected from the total width of the seam, however, mineral bands thicker than 1 cm were rejected. The three samples were mixed, crushed, and used as a representative of the coal of that mine.

Deurbrouck (1972) and co-workers collected 600 lb samples from channel faces that were carefully selected and cleaned (Volume I, Chapter 15, Sections II,A,B). Each sample was air dried and crushed to top size $1\text{-}\frac{1}{2}$ in. Smaller portions were then selected, crushed, and the process was repeated until an appropriate sample was obtained.

Either method can be used; however, in both cases, the selection of the sample requires judgment from the person who does the work with regard to where to dig the coal and which parts to reject.

Manual collection of samples has the advantage, and potential pitfall, that the collector can appropriately reject portions of the coal that are grossly nonrepresentative.

Periodic automatic sampling, crushing, and separation of samples can allow selection of samples that represent the average sulfur content more accurately. However, care should be exercised to avoid systematic errors.

B.   Recommendations Relevant to the Preparation of Coal Samples for Sulfur Analysis

The main source of error in the preparation step that could affect the results of the analysis for sulfur is density segregation. The density of iron pyrite is about 5.0 g/cm^3, mineral matter about 2.7 g/cm^3, and coal about 1.3 g/cm^3. Since iron pyrite is inorganic material, most of which is present on the surface of the organic material, gravity forces will tend to concentrate the dense pyrite at the bottom of the coal containers. Complete gravity separation is impossible, however, because part of the pyrite is embedded in the organic matrix. Grinding the coal to fine increases the fraction of pyrite present as isolated crystals but also helps to form a "homogeneous-like" mixture since the relative magnitude of the surface forces to the gravity forces is larger for smaller particles. *All grinding operations should be conducted* when the *coal is cold and preferably under inert environment, e.g.,* N_2.

We found that the segregation error becomes negligible when samples of Illinois No. 6 coal were grounded to −170 mesh. The analysis of samples of Pittsburgh seam −65 + 100 mesh coal had a normalized standard deviation,

$$\sqrt{\frac{\Sigma\sigma^2}{n}}\Bigg/S_{tot}$$

smaller than 5% when −210 mesh iron pyrite was added to the coal. It is recommended to grind coal and char samples below 210 mesh when test samples of the order of 20 to 60 mg are used for total sulfur analysis, e.g., by the rapid combustion method, and to run duplicate or triplicate samples. Analysis of 5 g samples can yield more accurate results; however, the data may be erratic if segregation has occurred prior to the selection of the 5 g. It is not recommended to use instruments that require samples smaller than 10 mg. Such equipment is common in many organic chemistry laboratories but can yield very serious errors in coal analysis. It is recommended to pulverize the lot of coal (e.g., 500 g) to −170 mesh prior to selecting the 5 g samples. We found that segregation is more important in some coals than in others and that crushing to −60 mesh or even to −100 is not always sufficient.

C. Recommendations Relevant to Storing Coal Samples for Sulfur Analysis

The class distribution of sulfur in coal samples can change during drying and storage. The single most important reaction that effects the sulfur in coal samples is oxidation. There is ample evidence that suggest that iron pyrite slowly oxidizes to iron sulfate,† probably according to

$$FeS_2 + 3O_2 \rightarrow FeSO_4 + SO_2. \tag{3}$$

We believe that moist air oxidizes pyrite faster than dry air. For example, Illinois No. 6 coal, $-\frac{3}{8}$ in. with 4.05 wt% total sulfur, 1.56 wt% FeS_2, and 0.38 wt% sulfate was stored in a steel drum that was opened occasionally. In a period of one year we found that the pyritic sulfur dropped to 0.23 wt% and the sulfate sulfur increased to 1.04 wt%. We established that 1.00 wt% of the sulfate sulfur was due to iron sulfate, which was formed from iron pyrite. This was deduced from the concentration of the Fe^{2+} ions as well as that of the SO_4^{2-} ions in the 2:3 HCl extract of the coal (see ASTM D2492 for details). Iron pyrite is insoluble in dilute HCl, but iron sulfate is soluble. Therefore, the concentrations of iron ions in the HCl extract of coal is a measure of the concentrations of $FeSO_4$ in the sample, provided that it did not contain FeS and/or $FeCO_3$. $FeOH(SO_4)$ and $Fe_2(SO_4)_3$ are also soluble in HCl.

Large chunks of coal were stored in polyethylene bags for a few months without a noticeable change.

Powdered coal and coal samples treated with HCl, NaOH, $ZnCl_2$,

† The various iron sulfate minerals found in coals are discussed in Chapter 50, Section III,E.

K_2CO_3, Na_2CO_3, H_2, H_2O_2, and many other reagents were oxidized by air relatively fast. We usually store such samples in air-tight plastic or glass bottles under N_2 or He atmosphere. Storage under water can also be used; however, water dissolves sulfates and leaches a few other components.

Extreme care should be exercised with samples that are retrieved from a reactor to be analyzed later. Rapid washing of the samples in oxygen-free solvents, e.g., boiled water or alcohol as the case may require, and drying the samples in a vacuum oven help preserve them. It is recommended to predry containers for samples of coal taken from a reactor. We dry containers and coal samples 4 hr at 60°C in a vacuum oven.

It is recommended to dry coal samples right before their processing or testing since dry coal from which the surface moisture was removed seems to oxidize faster than wet coal. Water vapors, however, accelerate the oxidation of FeS_2.

IV. SULFUR IN COAL DERIVED MATERIALS (CDM)

Processing of coal changes the chemical and physical structure of coal and often produces new structures that were not present in the original coal. Therefore, application of methods that were developed for coal analysis cannot be used without a careful examination of the types of compounds present in the products.

The identity and the quantity of CDM will therefore depend on the structure of the original coal and on the processing conditions applied and their intensities. Mild processing of coal does not completely destroy the original structure of the coal; however, intensive processing of coal produces simple molecules that bear little relationship to the structure of the original coal. The types of sulfur compounds that can be found in the products of coal processing are summarized in Table II.

The analysis of sulfur compounds in gases and liquids is documented in the literature; Karchmer (1970, 1971, 1972) edited an excellent set of books titled "The Analytical Chemistry of Sulfur and its Compounds." Another excellent compilation of analytical methods for sulfur and its compounds was recently published by Ashworth (1972, 1976, 1977). Analysis of sulfur-containing gases is documented very intensively in the analytical chemistry and air pollution literature (e.g., West and Gaeke, 1956; Jacobs, 1967; Attari *et al.*, 1970; Federal Regulations, 1971; Garber and Wilson, 1972; Winkler and Syty, 1976; Chamberland and

TABLE II *Major Sulfur Compounds in Products of Coal Processing*

	Processing in reducing conditions		Processing in oxidizing conditions	
	Intense	Mild	Intense	Mild
Typical process	Hydrogen donor and other liquefaction processes	Flash hydrogenation, low temperature pyrolysis, etc.	Combustion	Air drying and storage. Miscel. oxidation at temperature, $<180°C$
In solid residue	FeS, CaS Thiophenes Aromatic sulfides	As under intense (HD) plus: aliphatic sulfides and FeS_2	$Fe_2(SO_4)_3$ $CaSO_4$	$FeSO_4$, $CaSO_4$ $Fe_2(SO_4)_3$, sulfones
In liquids	Thiophenes Aromatic sulfides	Thiophenes, aliphatic, and aromatic sulfides and thiols		Sulfones, sulfoxides, and sulfonic acids
In gases	H_2S	H_2S, CS_2; CH_3SH, Thiophene, $(CH_3)_2S$	SO_2, SO_3	SO_2, COS

Ganthier, 1977). Only limited reliable information can be derived on sulfur in solid coal and its solid products. Therefore, most of our effort will concentrate on the analysis of sulfur in solids.

There is an urgent need to determine the physical forms and the structure of the sulfur-containing components in the solid products of coal hydrogenation and in the products of mild oxidation of coal. The following points summarize the current consensus on the forms of sulfur in the residue of coal liquefaction and other processes conducted in a reducing environment:

(1)   most of the inorganic sulfur is present in the form of iron pyrite and partially reduced pyrite, e.g. *iron sulfide* or pyrrhotite (Jenkins, 1977),

$$FeS_2 + H_2 \rightarrow FeS + H_2S \tag{4}$$

(2)   a limited portion of the sulfur could be present in association with calcium, e.g., CaS and $CaSO_4$. The source of CaS is reactions between the basic minerals and H_2S, e.g., Attar and Dupuis (1978b),

$$CaO + H_2S \rightarrow CaS + H_2O. \tag{5}$$

Two theories were proposed with regard to the form of the organic sulfur in the solid residues of coal pyrolysis and liquefaction. Wibaut (1919) suggested that solid complexes are formed between the organic matrix and elemental sulfur. However, elemental sulfur can be extracted with many solvents, e.g., CS_2 and benzene. No elemental sulfur was detected in extracts of pyrolyzed coal. It is consistent with the thermodynamics and kinetics of the reactions of sulfur (Attar, 1978) to assume that all the sulfur in the solid products of coal pyrolysis is bound in thiophenic structures.

Intense oxidation of coal converts 95% or more of the sulfur to sulfur oxides, although some of it is trapped in the ash as sulfates. The pyritic sulfur is oxidized rather rapidly to iron sulfates, e.g.,

$$2FeS_2 + 7O_2 \rightarrow Fe_2(SO_4)_3 + SO_2. \tag{6}$$

Mild oxidation of coal can oxidize some of the organic sulfur to the corresponding sulfoxides, sulfones, or sulfonic acids; however, no positive proof has been presented to date for their presence. Mild oxidation is not believed to effect the thiophenic sulfur. The literature on the rate of formation of sulfoxides and sulfones under different conditions was recently reviewed by Attar and Corcoran (1978).

V. ANALYSIS OF SULFUR IN COAL AND SOLID
COAL-DERIVED MATERIALS

It is convenient to distinguish among three types of data on the sulfur in solids:

(1)   total sulfur,
(2)   class distribution of the sulfur among pyrite, sulfate, and organic, and
(3)   detailed quantitative information on the sulfur in each form.

A. Total Sulfur Analysis

Total sulfur analysis usually consists of two steps:

(1)   complete destruction of the solid structure and the formation of a simple sulfur compound, i.e., H_2S or SO_2 and
(2)   a finishing technique in which the amount of the simple sulfur compound formed is determined.

A detailed survey of the methods used to determine total sulfur is presented in Volume I, Chapter 9; therefore, the discussion here will concentrate only on sources of errors and inconsistencies in the analysis. Two types of destruction techniques are used:

(1)   intense oxidation that converts the sulfur to a mixture of SO_2 and SO_3 or SO_3^{2-} and SO_4^{2-} and

(2)   intense hydrogenation that converts the sulfur to H_2S.

A variety of finishing techniques are available for H_2S and SO_x as discussed in detail by Ashworth (1972) and in Volume I, Chapter 9.

1.   Oxidative Destruction of the Coal Matrix

The most commonly used destruction techniques involve combustion of the coal, for example, the Eschka method and the oxygen bomb method D3177 (ASTM, 1973) that are the accepted standard methods for total sulfur analysis in coal. Both methods can be used for the analysis of sulfur in coal-derived materials; however, it is recommended not to use Eschka when the sample contains more than 6 wt% sulfur and not to use the bomb method when the sulfur content exceeds 4 wt%. Other methods that are used on a routine basis are the high temperature combustion method (Mott and Wilkinson, 1956) and the induction furnace method of Leco Corporation (1974). The latter is a rapid and convenient method; however, great caution and frequent calibrations are needed in order to avoid errors. The latter method is not accepted by ASTM as a method for determining the total sulfur in coal. The peroxide bomb method (Selvig and Fieldner, 1927), which had been a standard method for years, was phased out by ASTM, although many keep using this method.

Several companies (e.g., Leco Corporation, 1974) are offering new equipment for total sulfur analysis that uses fast combustion and determination of the sulfur oxides by infrared analysis. These methods are rapid and reasonably accurate when the operation procedure is followed carefully and when frequent calibration is made. As in the Leco method, the combustion temperature influences the results of the analysis. In addition, halogens, such as chlorine, interfere with the determination.

The most common finishing techniques are:

(1)   Iodometric titration (e.g., Leco Corporation, 1974):

$$SO_2 + I_2 + 2H_2O \rightarrow H_2SO_4 + 2HI. \tag{7}$$

(2)   Oxidation to sulfate and determination of the sulfate (ASTM D3177):

$$SO_2 + H_2O_2 \rightarrow H_2SO_4, \tag{8}$$

$$SO_2 + Br_2 + 2H_2O \rightarrow H_2SO_4 + 2HBr. \tag{9}$$

Gravimetric determination of the sulfate (ASTM D3177):

$$H_2SO_4 + BaCl_2 \rightarrow BaSO_4 \downarrow + 2HCl. \tag{10}$$

Titration of the sulfuric acid (Mott and Wilkinson, 1956) and a variation on Leco's method:

$$H_2SO_4 + 2NaOH \rightarrow Na_2SO_4 + 2H_2O. \tag{11}$$

(3) Infrared analysis of SO_2 (Leco Corporation, 1974).

(4) Coulometric or amperometric titration [e.g., Wallace et al. (1970) and Fisher Scientific Company (1978), respectively].

Care should be exercised in the choice of a finishing method. For example, iodometric titration is not applicable to coals with a high chlorine content since some of the bound chlorine is oxidized during combustion to elemental chlorine. Elemental chlorine oxidizes iodide solutions to elemental iodine:

$$2I^- + Cl_2 \rightarrow I_2 + 2Cl^-. \tag{12}$$

Such a reaction is the reverse of that of SO_2, which consumes I_2:

$$I_2 + SO_2 + H_2O \rightarrow 2I^- + SO_3{}^{2-} + 2H^+. \tag{13}$$

The results of such tests will therefore be biased downward, i.e., indicate a smaller amount of sulfur.

Chlorine also interferes with acidimetric finishing but does not interfere with the gravimetric determination of sulfur as barium sulfate, $BaSO_4$.

2. Reductive Destruction of the Coal Matrix

Coal can be gasified in a stream of hydrogen at 1200°–1350°C. The sulfur is released predominantly as H_2S that can be determined by chromatography, amperometry, coulometry, gravimetry, or one of many other techniques. Complete reduction of all the sulfur to H_2S is difficult to achieve. Some of the coal sulfur is retained in the solid residue as polynuclear aromatic thiophenic sulfur or as CaS. Large errors can be expected, therefore, when low ranked coals are analyzed that contain a large fraction of CaO in their ash.

Although many convenient finishing techniques are available for H_2S, reduction has never been an important method for determining the total sulfur in coal.

3. Comparison of Destructive Techniques and Recommendations

Destruction of the coal matrix by oxidation (combustion) is clearly advantageous to destruction by hydrogenation. Out of the many oxidative methods available, the Eschka method (ASTM D3177) is the simplest and the most popular method. However, if many samples are to be analyzed, one of the instrumental methods can very well pay off, e.g., Leco's rapid combustion instruments with titration or infrared determination of SO_x, or Fisher Scientific's combustion apparatus with coulometric titration of SO_2.

B. Class Distribution of Sulfur in Coal

A detailed description of the philosophy of the class distribution of sulfur in coals is given in Volume I, Chapter 9. Shimp *et al.* (1975) compared the available alternative methods for the determination of the classes of sulfur.

The objective of this presentation is to analyze sources of errors associated with the standard methods and to discuss possible solutions. However, a very brief description of the philosophy of the accepted technique (e.g., D3177) will be given.

1. Class Distribution—Logics of the Analysis

The three classes of sulfur that are determined in coals on a routine basis are the pyritic (S_P), the sulfatic (S_S), and the organic sulfur (S_O). However, since no generally accepted method is available that could yield the organic sulfur, it is common to estimate its value by subtracting S_P and S_S from the total sulfur S_T. Since the total sulfur is the sum of the three classes of sulfur

$$S_T = S_P + S_S + S_O, \qquad (14)$$

S_O can be calculated by difference

$$S_O = S_T - S_P - S_S. \qquad (15)$$

Obviously, the uncertainty about the value of S_O is large since the error, ΔS_O, is the cummulative error in the determination of S_T, S_P, and S_S

$$\Delta S_O \leq |\Delta S_T| + |\Delta S_P| + |\Delta S_S|. \qquad (16)$$

The values of S_S and ΔS_S are usually small. However, large errors can be realized in the determination of S_T or S_P. The determination of S_P often introduces the largest error, therefore it will be discussed in detail.

2. The Determination of Pyritic Sulfur, S_P

Determination of pyritic sulfur can be done using any property of the iron pyrite that distinguishes it from other components of the coal (i.e., any selective property). Three approaches were used: the nitric acid extraction method [ASTM D3173, D2492 (British Standards Institution, 1975; International Organization for Standardization, 1960), the ashing method (Young and Zawadzki, 1967), and the x-ray method (e.g., Pollack, 1971; Kuhn et al., 1973; Schehl and Friedel, 1973; Solomon and Manzione, 1977).

   a. Nitric Acid Extraction of Iron Pyrite.† The simplest version of nitric acid extraction was proposed by Powell (1920). The fundamental presumption behind this method is that all the iron in coal that is soluble in dilute nitric acid (HNO_3), is in the form of iron pyrite FeS_2. Therefore, the number of moles of IRON that dissolves in HNO_3 is equivalent to half the number of moles of pyritic sulfur in the sample.

   Two clear sources of errors can be identified in the use of the nitric acid method:

   (1)   incomplete extraction of all the iron pyrite and
   (2)   the presence of soluble iron salts other than FeS_2.

Incomplete Extraction of all the Iron Pyrite.   Incomplete extraction of all the iron pyrite by nitric acid is due predominantly to the iron pyrite that is embedded in the organic matrix and for which contact with the oxidizing solution is limited by diffusion of the acid through the organic matrix. Therefore, it can be expected that a more complete extraction will be achieved when smaller coal particles are extracted.

   Data are available in the literature that show that in some cases the amount of pyrite extracted is independent of the coal particle size (e.g., data quoted by Shimp et al., 1975), but in other cases more pyrite is extracted from smaller coal particles (e.g., data cited in James and Severn, 1967; Edwards et al., 1964; Brown et al., 1964).

   The fraction of iron pyrite that is embedded in the organic matrix is different in different coals, and therefore it can be anticipated that a larger fraction of the pyrite will be retained in the coal when a larger fraction of the pyrite is embedded in the organic matrix. The depth of penetration of nitric acid into the organic material can be estimated as follows:

   Consider spherical coal particles of radius R and assume that they contain S'_{pi} g embedded FeS_2/unit volume and S_{ps} g of surface FeS_2/g coal and that the density of the coal is ρ. If the HNO_3 penetrates to a depth of δ, then the only embedded pyrite that will be extracted and determined, S_{pd} g FeS_2/g coal is that which is in the shell of thickness

---

† For ASTM D2492 see Volume I, Chapter 6, Section VII.

δ on the surface of the coal. Therefore,

$$S_{pd} = S_{ps} + N\tfrac{4}{3}\pi[R^3 - (R - \delta)^3]S'_{pi}, \tag{17}$$

where N is the number of coal spheres per gram of coal. Therefore,

$$N = 3/\rho 4\pi R^3 \tag{18}$$

and

$$S_{pd} = S_{ps} + \frac{[R^3 - (R - \delta)^3]}{\rho R^3} S'_{pi}$$

$$= S_{ps} + \left[1 - \left(1 - \frac{\delta}{R}\right)^3\right] \frac{S'_{pi}}{\rho}. \tag{19}$$

If large particles are used where $\delta/R \ll 1$, then to a first-order approximation:

$$S_{pd} \cong S_{ps} + (3\delta/R)(S'_{pi}/\rho). \tag{20}$$

Equation (20) shows that if the iron pyrite extracted (i.e., determined by the ASTM method) is plotted versus $1/R$, where R is the coal particle size, then the intercept with $1/R = 0$ will give an indication of the "surface" FeS_2, and the slope of the curve will be $3\delta S'_{pi}/\rho$. If we redefine the embedded FeS_2 per unit mass of coal, i.e.,

$$S_{pi} = \text{g embedded } FeS_2/\text{g coal} = S'_{pi}/\rho, \tag{21}$$

then

$$S_{pd} = S_{ps} + (3\delta/R)S_{pi} = S_{ps} + (3\delta/R)S_{pi}, \tag{22}$$

and the slope of the plot of S_{pd} versus $1/R$ will be $3\delta S_{pi}$.

Figure 4 shows the data of James and Severn (1967) on the concentration of pyritic sulfur determined in different size fractions of the same coals plotted versus $1/D$, where D is the top sized opening of the screens used to classify the samples. The data on the slope and intersections of the curves with $1/D = 0$ are summarized in Table III. The sample numbers used by James and Severn (1967) are used in Table III and in Fig. 4. The true concentration of pyrite is

$$S_{pt} = S_{ps} + S_{pi}. \tag{23}$$

Therefore,

$$S_{pd} = S_{pt} - S_{pi} + 6\delta S_{pi}/D. \tag{24}$$

If it is assumed that when the smallest particles are tested, $S_{pd} \approx S_{pt}$, i.e., $6\delta/D \sim 1$, then the value of δ can be estimated. The values of δ

TABLE III *Effect of Particle Size on the Determination of Pyritic Sulfur in Coal[a]*

Mesh		Top size (μm)				Calculated values Wt% S			
Coal rank code	$10^3/D$ (μm^{-1})	210 4.76	124 8.60	76 13.16	53 18.87	S_{pt}	S_{pi}	S_{ps}	δ
102		0.89	1.16	1.39	1.42	1.42	0.832	0.588	12.3
402		0.63	0.70	0.89	1.07	1.07	0.688	0.388	8.8
602		0.49	0.52	0.52	0.56	0.56	0.091	0.469	9.2
802		0.32	0.32	0.45	0.44	0.44	0.165	0.275	9.4
902		0.88	0.88	1.04	1.31	0.31	0.648	0.663	10.3

[a] Data of James and Severn (1967).

that were determined from the data of James and Severn (1967) are also listed in Table III. For all the coals tested $\delta \approx 10.2 \pm 2$ μm, which suggests that complete extraction of embedded pyritic sulfur will be achieved if the coal will be crushed to -20 μm. In reality, it is not necessary to crush coal to such fine particles because the coal particles are not spherical and because the size of typical embedded pyrite particles is 5–15 μm. Therefore, if the coal is crushed to -325 mesh (-43 μm), most of the embedded material will be reached by the acid. Edwards *et al.* (1964) examined 40 coals of different rank using microscopy and report penetration that ranges from 2–3 μm for high-ranked coals to 14–18 μm in low-ranked coals. Several low-ranked coals were "completely penetrated" according to their tests in which -72 mesh particles were used. The agreement between the calculated and the measured values is good.

It is recommended that one crush coal samples with unknown geochemical origin to -325 mesh and then analyze the product. However, if most of the pyritic sulfur is present as surface pyrite, -200 mesh can also be used. If many coal samples with similar geochemical origin are to be analyzed, it is suggested that one establish the largest particle size that gives the "same" value for the pyritic sulfur as -325 mesh particles and to use the large particles on a routine basis. However, the agreement with the results obtained with -325 mesh particles has to be occasionally reexamined. In general, high-ranked coals have to be ground to a finer particle size than low-ranked coals in order to assure full extraction of the pyrite. The calculated values of δ are given in the last column of Table III; they are the average values obtained by testing of approximately 1 g coal.

Stable Iron Compounds other than Iron Pyrite.   The most important

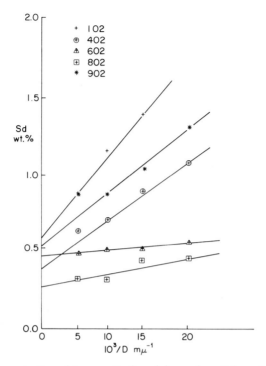

Fig. 4 The concentration of iron pyrite found for coal particles with different sizes. (Data from James and Seven, 1967.)

iron compounds that are found in coals are FeS_2, FeS, $Fe_2O_3 \times H_2O$, $FeCO_3$, and $CuFeS_2$. Iron silicates are also present in coals but they are insoluble in HNO_3. The solubility of the hydrated iron oxide in various acids depends on the acid, on the temperature of the acid, and on the history of the oxide. Weathering of coal can produce additional amounts of iron oxides and sulfates from FeS_2. If the coal is first extracted with boiling 2N HCl and then *the same* coal sample is extracted with HNO_3, one will find that the HCl extracted all or most of the FeS, Fe_2O_3, $FeCO_3$, and the iron and calcium sulfates. The FeS_2, $CuFeS_2$, the part of the Fe_2O_3 that was not extracted by HCl, plus some of the embedded iron compounds, will be extracted by HNO_3.

The implications to the determination of the pyritic sulfur are as follows:

(1) If HCl extraction is not used prior to the HNO_3 extraction, all the $FeCO_3$, FeS, Fe_2O_3, etc., will dissolve in the HNO_3 and will be

counted subsequently as FeS_2. The value so obtained for the pyritic sulfur will be too high and the value for the organic sulfur too low.

(2)   Even when HCl extraction preceeds the HNO_3 extraction, low organic sulfur will be estimated if not all the Fe_2O_3 was extracted by the HCl.

(3)   The FeS content of a coal can be estimated from the amount of Fe^{2+} in the HCl extract, provided that the $FeCO_3$ and Fe_2O_3 content of the sample is small.

The presence of $FeCO_3$ (siderite) has been established in many Australian and Russian coals. $FeCO_3$ has been found in low-ranked, low sulfur coals where the geochemical conversion of the $FeCO_3$ to FeS_2 was incomplete. To detect and determine $FeCO_3$, extract the coal sample with a warm dilute solution of acetic acid, which dissolves $FeCO_3$ but does not attack FeS or FeS_2. If $FeSO_4$ is also present, it will dissolve in the acetic acid. However, its quantity can be estimated from the concentration of SO_4^{2-} in the solution.

b. Determination of Pyritic Sulfur by Analysis of the Ash.   Young and Zawadzki (1967) ashed coal and used the iron content of the ash as a measure of the pyritic iron in the coal. Since in pyrite each iron atom is tied to two sulfur atoms, the iron content of the ash can be used to calculate the concentration of pyritic sulfur in the original coal sample.

Estimation of the pyritic sulfur by this method is simple and convenient, however, several precautions should be taken to avoid errors.

(1)   Ashing the coal at too high a temperature can fuse the iron and the silicates and form iron silicates.

(2)   Iron compounds other than iron pyrite will be counted as iron pyrite. Therefore, in the presence of $FeCO_3$, values that are too large will be obtained for pyritic sulfur and values that are too low for organic sulfur.

(3)   The method of finishing used to determine the iron in the ash has to be selected so that the only iron compounds that will be determined will be these produced by the oxidation of FeS_2, i.e., Fe_2O_3 and iron sulfates. Iron silicates may present a problem when some of the modern instrumental methods are used. Young and Zawadzki (1967) obtained very consistent results by their procedure compared with the nitric acid method. They ashed the coal for 1 hr at 500°C and 2 hr at 750°C. They extracted the ash for 60 min in boiling concentrated hydrochloric acid under reflux and determined the iron volumetrically by titration.

c. Determination of Pyritic Sulfur by X-Ray Analysis.   Two methods are included under the umbrella of x-ray methods:

(1)   x-ray diffraction (e.g., Pollack, 1971), and
(2)   x-ray fluorescence (e.g., Kuhn *et al.*, 1973; Solomon and Manzione, 1977).

The x-ray diffraction method uses the fact that the energies of the Bragg lines of crystalline FeS_2 (cubic) are unique to a crystal and that the intensity of each line is a function of the amount of crystalline FeS_2 in the coal sample. The five most important crystalline spacings of cubic FeS_2 are 2.709, 2.423, 2.212, 1.916, and 1.633 Å (Volume II, Chapter 26, Table II). Carbon seems to interfere with most of the lines; however, the lines at 1.633 Å may be used to determine the pyrite.

The most important advantages of the x-ray diffraction methods are:

(1)   specificity—the signals obtained are unique to cubic FeS_2,
(2)   rapidity—analysis can be accomplished in 10–30 min, and
(3)   the method is instrumental and can therefore be easily automated.

The major disadvantages are:

(1)   lack of sensitivity,
(2)   lack of accuracy—x-ray diffraction will not recognize small or ill-defined crystals of FeS_2.

X-Ray fluorescence (XRF) has been the subject of several recent studies, since developments of instrumentation made routine testing possible. The coal sample is irradiated with a beam of short wavelength x-rays. As a result, each element emits a series of x rays with characteristic wavelengths (energies). Therefore, monitoring the intensity of the emitted beam at a given characteristic wavelength (energy) gives a measure of the amount of the corresponding element in the sample.

Typically, the sample is sputtered with a conducting metal, e.g., gold, and then the fluorescence is examined at the characteristic wavelength of the desired element. Iron fluoresces at 6.394 eV and sulfur at 2.307 eV. When the objective is to measure only the iron, gold can be used to sputter the sample. However, the M line of gold (energy 2.137 eV) interferes with the determination of the sulfur. We used silver coating whenever it was desired to determine both the iron and the sulfur. Solomon and Manzione (1977) examined many coal samples using x-ray fluorescence and used gold-palladium to sputter their samples. Solomon and Menzione (1977) determined the ratio of iron to sulfur in many samples and used the data to estimate the inorganic and organic sulfur in the coal.

The x-ray fluorescence method is an advanced instrumental method that can be used on a routine basis. However, great care should be used in interpreting the data. A few potential sources of error are:

(1)   iron compounds other than FeS_2,
(2)   inorganic sulfur compounds like $Fe_2(SO_4)_3$, $FeSO_4$, and $CaSO_4$.
(3)   sampling and sample preparation.
(4)   sample and data statistics—a sufficient number of pictures have to be processed before conclusions can be made. More pictures have to be processed when low-sulfur coals are examined (see, e.g., Solomon and Manzione, 1977).

Hurley and White (1974) used XRF to estimate the total sulfur and suggested some correlations for the organic and pyritic sulfur (Volume I, Chapter 9, Section V,D,3).

3.   Determination of Organic Sulfur

Direct determination of organic sulfur is difficult because it is present in many different organic functional groups and is dispersed throughout the organic matrix. A recent review of the organic functional groups in coal was published by Attar and Corcoran (1977).

The only methods that were used to determine the total organic sulfur were indirect methods. The inorganic sulfur was either determined independently or first removed from the sample. The total sulfur in excess of the pyritic sulfur was assumed to be organic.

The inorganic sulfur can be removed by extraction with HNO_3, as described previously, and can be estimated from the iron content of the solution. However, all the potential sources of error that can effect the determinations of iron pyrite or its removal can effect the accuracy of the value assigned to the organic sulfur. Moreover, the accuracy of this value also depends on the accuracy of the determination of total sulfur and sulfate sulfur.

Methods for the direct analysis of organic sulfur could potentially be developed and be based on:

(1)   the characteristics of the sulfur–carbon bond,
(2)   the difference in the rate of reactions of organic and inorganic sulfur, and
(3)   reactions with specific reagents.

Infrared and Raman spectra of coal are specific for the C—S bond but cannot be used to determine organic sulfur because they lack the sensitivity needed, because of the dark color of coal, or because of the luminescence of coal.

ESCA does not seem to be an adequate method for functional groups analysis, and ESCA, as well as IR and Raman, give information on the composition of the surface layer only. For ESCA (XPS) of sulfur functional groups see Volume I, Chapter 11, Table I.

Recently, Sutherland (1975) described a method for the direct determination of organic sulfur based on microprobe analysis (XRF). He used a calibration factor based on iron pyrite and presented results that are consistent with the ASTM method. However, additional development is needed before this method can be used on a routine basis.

Paris (1977) described a novel approach to the determination of organic sulfur: The coal sample is oxidized at such mild and selective conditions that only the organic matrix is combusted. The organic sulfur is oxidized to volatile oxides that are collected and subsequently analyzed. Pure iron pyrite added to the coal samples did not release sulfur oxides upon oxidation. The results that were obtained by this method were in fair agreement with the results obtained by the ASTM method.

Selective oxidation of the organic matrix has a good potential of becoming a method for the direct determination of organic sulfur. However, more research is needed in order to have a reliable method.

VI. DISTRIBUTION OF THE ORGANIC SULFUR FUNCTIONAL GROUPS

Three approaches have been used in the examination of the organic sulfur compounds in coal: (1) direct examination of the coal; (2) decomposition of the coal, or "depolymerizing" the organic matrix, followed by examination of the fragments; and (3) chemical incorporation of sulfur compounds into an organic matrix ("simulated coal") and examination of the properties of the products. The most straightforward method is the first; however, very little information on sulfur compounds has been derived by direct examination of coal samples. Some information has been obtained from examination of the products of destruction of coal, but the nature of the products depends on the method of destruction used. Very little relevant knowledge resulted from the work on simulated coal. The basic difficulty seemed to be the lack of a method that would unambiguously determine if the simulated coal indeed was analogous.

Direct spectroscopic examination of coal was carried out using infrared, Raman, and ultraviolet techniques. The spectroscopic approach has been reviewed by Speight (1971) and Volume II, Chapter 22. The main useful method for analysis of functional group distribution is by infrared, but coal cannot be totally solubilized so that examination in solution or suspension is of little use. Pellets and mulls of finely divided coal have been studied, but no data relevant to the functional groups of sulfur has been mentioned. The infrared bands of sulfur are not very strong and are rarely well resolved. Because Raman spectrometry is relatively more sensitive to sulfur-containing groups, it has some potential for sulfur group analysis, even though the dark color of coal samples, degradation of the coal under the laser beam, and chemiluminescence may be severe problems. Although Raman spectra of coal have been obtained (e.g., Friedel and Carlson, 1972), no specific work on sulfur compounds was mentioned. An investigation by infrared of the mineral matter in coal, notably FeS_2, was made by Estep *et al.* (1968). The reaction of sulfur functional groups with selective reagents was used with moderate success in solid coal analysis. Methylation by methyl iodide was used by Postovski and Harlampovich (1936) to estimate the amount of sulfidic sulfur in accord with the reaction

$$\begin{array}{c} R \\ \diagdown \\ S \\ \diagup \\ R \end{array} + CH_3I \longrightarrow \begin{array}{c} R \\ \diagdown \\ SCH_3^+I^- \\ \diagup \\ R \end{array} \qquad (25)$$

I

Determination of the amount of I^- left in the sample is a measure of the concentration of organic sulfides. Mercaptans and thiophenols release HI when reacted with CH_3I:

$$RSH + CH_3I \rightarrow RSCH_3 + HI. \qquad (26)$$

The HI released thus can be used to estimate the amount of —SH groups in the sample.

More recently similar ideas were used by Bogdanova and Boranski (1961) and Prilezhaeva *et al.* (1963) to evaluate the organic sulfur functional groups distribution in coal. The data suggest that, in bituminous coals, the organic sulfides constitute some 5–20% of the organic sulfur while the rest is assumed to be thiophenic. About 70% of the sulfidic sulfur is in an unstable form. Thiophenic, condensed thiophenic, and aryl sulfides were determined by difference because they do not react with MeI (Prilezhaeva *et al.*, 1963).

Nondestructive extraction of coal, followed by examination of the sulfur compounds, showed that the ratio of the total C to the organic sulfur does not change as a result of the extraction (Van Krevelen, 1961). That does not necessarily imply that the sulfur-functional group distribution is identical in the solid and extract. However, because sulfur is a minority element, it is unlikely that it will induce selective extraction. Attempts to determine the sulfur-functional group distribution in some bituminous coal extracts have been reported by Minra and Yanagi (1963). Tetrahydrofuran, dimethylformamide, and benzene were usd. The following distribution was found for Mieke coal: thiol (—SH), 3–9%; disulfide (—S—S), 6–13%; aliphatic sulfide (R—S—R), 28–37%; thiophenic and aryl sulfide (Ar—S—), 7–19%; and undetermined ~30%. We believe that all of the undetermined S is in a condensed thiophenic structure. Moreover, it is unlikely that unstable groups like thiolic or disulfidic will survive the coalification process. Therefore, more work should be done on the analysis of sulfur functional groups. Note should be made that thiophenic sulfur compounds are the major constituents of the sulfur compounds in shale oil (Veretennekova and Petrov, 1964; Paulson, 1975).

Depolymerization of coal was accomplished by hydrogenation, pyrolysis, oxidation, and reactive solvents. Only the most stable sulfur compounds will survive catalytic hydrogenation at high temperatures and pressures. Hydrodesulfurization of coal at a hydrogen pressure of 2000–4000 psi and a temperature of 450°C, followed by examination of the product by mass spectroscopy, allowed the identification of 14 different organosulfur compounds. Thirteen contained thiophene or condensed thiophene rings (Akhtar *et al.*, 1974; see also Volume I, Chapter 17). The thiophenic compounds that were not hydrodesulfurized accounted for at least 20–40% of the total organic sulfur; however, no data were given that would allow a quantitative estimate.

Another decomposing technique that has been used extensively is coking. Coked coal releases hydrogen sulfide as well as other volatile sulfur compounds. Most of the conclusions of the classic work of Powell (1920, 1923) are still valid. The important ones are (1) iron pyrite decomposes when heated and releases half of its sulfur (some of the pyrite sulfur reacts with the coal and forms H_2S and other compounds as well, e.g., CS_2); (2) some of the pyritic sulfur reacts with the coal and forms very stable compounds that do not readily decompose; (3) sulfate sulfur remains in the coal, probably as a calcium sulfide; and (4) the H_2S that is released may react with the hot coke to form CS_2.

Interaction between the pyritic sulfur and the organic matrix during carbonization has been discussed broadly in the literature (Georgiodis and Gaillord, 1954; Howard, 1963; Cerni-Simic, 1962; Peet et al., 1969; El-Koddah and Ezz, 1973; Given and Jones, 1966; Maa et al., 1975; Attar et al., 1976). Several conclusions can be drawn, while acknowledging that contradictions exist.

(1)   Carbonization at temperatures up to 1400°C will not desulfurize the coal completely. Sulfur will be retained both in organic and inorganic forms. Coking at 1600°C desulfurizes the coal (95%), but substantial loss of material occurs (El-Koddah and Ezz, 1973).

(2)   Approximately 66% of the inorganic sulfur and 73% of the organic sulfur are retained in the coke. About 23% of the inorganic and 26% of the organic sulfur goes with the gas (Eaton et al., 1948). Slightly different numbers are quoted by Fuchs (1951). He found that about 50% of the total sulfur is retained in the coal; 3% goes to the tar; and 45% goes out in the gas. The actual distribution depends strongly on the rank of the coal.

(3)   Larger ash content increases the amount of sulfur that is retained in the coke.

(4)   A larger amount of sulfur goes to gas and tar when the volatile percentage is larger.

(5)   The overall desulfurization that is achievable reaches a plateau at 800°C. Further increase of the temperature does not substantially contribute to the desulfurization.

In addition to H_2S and CS_2, thiophene and its derivatives are the major components of the tar oils and gases that evolve (Muder, 1963). Some sulfides and disulfides can be found in the light oil; mercaptans were recognized in the gases (Peet et al., 1969). We believe that the mercaptans and disulfides are reaction products of the pyrite or the H_2S with the organic matrix. Moreover, it is very likely that the disulfide is a secondary product of mercaptans oxidation, because the —S—S— bond is too weak to withstand the coalification and/or coking processes. The organic sulfur that is retained in the coke is probably in the form of condensed thiophenic rings or aryl sulfides. No other C—S bond, namely, a bond without resonating electrons, can stand the high coking temperatures. Because dehydrocyclization and aromatization can occur as a result of the carbonization process, the distribution of sulfur functional groups in the tars does not necessarily represent the distribution in the original coal. The feasibility of dehydrocyclization and aromatization support the well-accepted assumption that most of the organic sulfur in coal, and probably all the sulfur in coke, is in the form of thiophenic structures or aryl sulfides with a high resonance energy.

Other techniques have been used in an attempt to learn about the sulfur-containing compounds in coal, notably, oxidation by nitric acid, chlorine in aqueous and nonaqueous media, air in basic media, and potassium permanganate; no significant ideas relevant to the chemical nature of the sulfur, however, have been obtained.† A more promising approach is via chemical depolymerization with a solvent and a strong Lewis acid, (e.g., BF_3 or p-toluenesulfonic acid (Darlage *et al.*, 1974). Such depolymerization is mild so that only aliphatic bridges between aromatic lamellae are broken. Smaller molecules are produced that can potentially be separated and characterized more easily. Depolymerization of some Russian bituminous coals, using the same method, led Rodionova and Barauskii (1970) to the following conclusions: (1) the ratios of sulfides and disulfides in coal do not vary much (variation in the total organic sulfur is due to the major sulfur constituents, the thiophenic compounds); and (2) thiolic groups do not exist in coal in a detectable amount. However, depolymerization is still new, and it has a large potential for determination of functional groups.

Many investigators have studied simulated coals and cokes. Basically, sulfur was incorporated into an organic matrix that was then coked at a high temperature and pressure. Not much was learned from these studies on the actual existence of sulfur in coal. Some inference on the relative stability of sulfur-containing groups in coking may be made, however. The data of Robertson and Steedman (1966), Aitken *et al.* (1968), and Scott and Steedman (1972), as well as those of Blayden and Patrick (1970), support the following conclusions: (1) thiol groups are unstable and tend to be eliminated as H_2S; (2) thiophenic structures are very stable and tend to condense with the organic matrix as the temperature is increased; (3) complete desulfurization of thiophenic sulfur compounds by coking is impossible (larger sulfur-containing structures are formed on carbonization). These conclusions support the information that was derived by coking. They do not, however, add new insights on coal. The major problem in the work on simulated coal seems to stem from the lack of a method to determine unambiguously in what ways the simulated coal does indeed simulate coal.

Attar and Dupuis (1978a) have recently proposed a method for the determination of the organic sulfur functional groups in coal. The method seems to offer capabilities that are not available in any other way. Therefore, the rest of the discussion will concentrate on this method.

---

† Several thiophenic sulfur compounds have been identified in coal oxidation products (Volume II, Chapter 21, Section IV).

VII. DETERMINATION OF THE DISTRIBUTION OF ORGANIC SULFUR FUNCTIONAL GROUPS USING KINETOGRAMS†

A. Principle of the Method and Scientific Ground

The classical definition of an organic functional group is a group of atoms that are bound in a specific structure and that reacts in a specific manner. All the molecules that include a given functional group will react in a similar manner when the reaction involves the common functional group. Traditionally, the concept of a "functional group" has been associated with a given molecular structure. Each group was detected and determined by its specific reactions with a given reagent. However, under certain conditions, the *rate* of reaction of a given group can be used to determine its identity and quantity.

Consider the reaction of a solid matrix that contain the functional group F_iX with a given reagent, RH_2.

$$F_iX + RH_2 \rightleftarrows [F_iXRH_2]^{\neq} \rightarrow F_iR + H_2X, \tag{27}$$

where $[F_iXRH_2]^{\neq}$ is an intermediate activated complex that can decompose to the products $F_iR + H_2X$ or to the original reagents. Let $[F_iXRH_2]^{\neq}$ as well as F_iX be part of the solid. Then, the rate constant of the reaction with the ith groups, k_i, will be

$$k_i = \frac{k'T}{h} \frac{Q^{\neq}}{Q_{F_iX}Q_{RH_2}} e^{-E_i/RT}, \tag{28}$$

where k' is the Boltzmann constant, h Plank's constant, Q denotes the partition function, and E_i is a constant that is very close to the activation energy of the reaction of RH_2 with the functional group F_iX. Since it is assumed that both F_iX and $[F_iXRH_2]^{\neq}$ are a part of the solid matrix, they have no translational degrees of freedom and to a first-order approximation, it can be assumed that

$$Q^{\neq} = Q_{F_iX}. \tag{29}$$

Therefore,

$$k_i = \frac{k'T}{hQ_{RH_2}} e^{-E_i/RT} = k_0 e^{-E_i/RT} \tag{30}$$

The last equation shows that to a first-order approximation, the rate of reaction of a functional group that is a part of a solid, with a given reagent,

† See Attar and Dupuis (1978a).

depends only on the activation energy of the reaction. The frequency constant of the reaction, k_0, is therefore,

$$k_0 = k'T/hQ_{RH_2},\tag{31}$$

and the activation energy is E_i. The conclusion is that *the activation energy of the reaction of a given functional group with a given reagent is a characteristic of the group and can be used to identify the group.*

In the following discussion it will be shown how this principle has been translated into a method for the quantitative and qualitative determination of sulfur functional groups.

1. Chemistry of the Reduction of Sulfur Compounds

All sulfur compounds can be reduced to hydrogen sulfide. However, different groups are reduced at different rates. For example,

$$RSH \xrightarrow{\text{[H]}} RH + H_2S\tag{32}$$

$$\underset{\text{II}}{\boxed{}_S} \longrightarrow C_4H_{10} + H_2S\tag{33}$$

The previous discussion and stoichiometry lead to the following conclusions:

(1)   Every atom of sulfur when hydrogenated produces one molecule of H_2S. Therefore, the number of moles of H_2S formed during the reduction of a given group is equivalent to the number of moles of the sulfur group reduced (provided that the sulfur group contained only one sulfur atom).

(2)   When the reducing conditions are mild, only the easily reduceable groups will release H_2S. When high reduction potential is applied, e.g., severe reduction conditions are used, it is possible to reduce *all* the sulfur groups. If the severity of the reduction conditions in a cell that contains a mixture of sulfur groups is gradually increased, the easily reduceable groups will release H_2S first, and the groups that are more difficult to reduce, will release their H_2S later.

Reduction of FeS_2 can proceed as follows:

$$FeS_2 \xrightarrow{\text{[H]}} FeS + H_2S,\tag{34}$$

and, subsequently,

$$FeS \xrightarrow{\text{[H]}} Fe + H_2S.\tag{35}$$

However, while the organic sulfur is dispersed in the organic phase, most of the pyritic sulfur is present as small lumps of FeS_2 crystallites. Therefore, reduction of a crystal of FeS_2 may result in the formation of a layer of FeS or Fe on the surface of the crystal which would limit the rate of mass transport of reducing agent to the core of the FeS_2.

Reduction of FeS_2 to FeS proceeds rather rapidly, even in a mild reducing environment. However, reduction of FeS to Fe proceeds at a very slow rate.

If the rate of the chemical reaction controls, then conversion of the FeS_2 to FeS will be complete before conversion of the FeS to Fe could begin. When hydrogen is used to reduce pyrite, the rate of the chemical reaction controls. However, when a very strong reducing agent is used, the rate of either reaction step or both can become diffusion controlled. In such a case, a layer of Fe can form on the surface of the FeS, block the diffusion, and effectively stop the reduction process altogether.

Table IV shows that reduction of pure iron pyrite with a very strong reducing mixture never proceeded beyond 1–2% of the material for any of the particle sizes tested.

2. Kinetics of the Reduction of Sulfur Groups in a Temperature-Programmed Cell

Reduction of the organic sulfur in coal under the conditions described here is a first-order reaction for the reducing agent and sulfur compound. When the reducing agent is in a large excess, one obtains

$$d[H_2S]_i/dt = -d[F_iS]/dt = k_0 \, e^{-E_i/RT}[F_iS]. \qquad (36)$$

Since the quantity of each group is finite in a fixed sample and is

TABLE IV   *The Effect of Pyrite Particle Size on the Recovery of Sulfur*

Particle size		% Sulfur
Mesh	μm	recovered[a]
−60+100	149–250	0.9
−100+120	125–149	1.9
−120+170	88–125	1.6
−170+200	77–88	1.5
−200+270	53–74	1.5
−270+325	44–53	0.9
−325	<44	2.1

[a] The estimated error is ±0.4%.

initially equal to the total amount of this group in the sample, $[F_iS]_0$, one obtains the following for an *isothermal* reaction:

$$\ln \frac{[F_iS]}{[F_iS]_0} = -k_0 \, e^{-E_i/RT} t = -kt. \tag{37}$$

When the temperature is increased, $k = k_0 \, e^{-E_i/RT}$ increases and the available pool of $[F_iS]$ decreases faster than described in Eq. (37). When the temperature is increased at the rate α,

$$T = T_0 + \alpha t \tag{38}$$

and

$$dt = \alpha^{-1} dT. \tag{39}$$

Therefore,

$$-\frac{d[F_iS]}{dT} = \frac{d[H_2S]_i}{dT} = \frac{k_0 \, e^{-(E_i/RT)}}{\alpha} [F_iS]. \tag{40}$$

Change of the variables yields

$$-\frac{d[F_iS]}{[F_iS]} = \frac{k_0 \, e^{-(E_i/RT)}}{\alpha} \qquad dT = -\frac{k_0 \, E_i \, e^{-X}}{R\alpha \, X^2} dX, \tag{41}$$

where

$$X = E_i/RT \tag{42}$$

and

$$dX = -(E_i/RT^2)dT = -(R/E_i)X^2 dT. \tag{43}$$

Therefore,

$$\ln \frac{[F_iS]}{[F_iS]_0} = \frac{k_0 \, E_i}{R\alpha} \int_{X=X_0}^{X} \frac{e^{-X}}{X^2} dX = \left(\frac{e^{-X}}{X} - \int_{X_0}^{X} \frac{e^{-X}}{X} \frac{\alpha k_0 \, E_i}{R} \right). \tag{44}$$

And if $X_0 \gg 1$ then,

$$\frac{d[H_2S]_i}{dT} = \frac{k_{0i}[F_iS]_0}{\alpha} \exp \left[-\frac{E_i}{RT} - \frac{k_{0i}R}{\alpha E_i} T^2 \exp\left(-\frac{E_i}{RT}\right) \right]. \tag{45}$$

The last equation gives the dependence on the temperature of the rate of evolution of H_2S from the ith functional group. The exponent has two terms, one decreases as T increases and the other increases. Therefore, the curve shows a maximum at T_{mi} given by

$$e^{-E_i/RT_{mi}} = E_i \alpha/RT_{mi}^2 k_{oi} \tag{46}$$

or

$$e^{-Q_{mi}} = Q_{mi}^2 B_i,$$ (47)

where

$$Q_{mi} = E_i/RT_{mi}$$ (48)

and

$$B_i = \alpha R/k_{oi}E_i.$$ (49)

Since to a first-order approximation k_{oi} is the same for the reduction of *all* the sulfur groups in the solid, the value of T_{mi} is determined essentially only by the value of the activation energy needed to reduce the ith groups, E_i. Therefore, the location of the maximum T_{mi} can be used to identify the group reduced, provided that α is maintained constant. The actual value of α can be used to control the resolution, $\Delta_{ij}T_m$, which gives the temperature difference between the maxima of two peaks with a close activation energy. The following equation can be used to choose α if the approximate values of E_i and k_o are known:

$$\frac{dT_{mi}}{d\alpha} = 1/k_o \left(\frac{2}{k_{oi}T_{mi}} + e^{-E_i/RT_{mi}} \right) = T_{mi} \Big/ \left(2 + \frac{E_i}{RT_{mi}} \right).$$ (50)

For example, if a test using the rate of heating, α, produced two poorly resolved peaks around T_m and if the activation energy is estimated to be E_i, then a change $\Delta\alpha$ in the rate of heating will result in a change ΔT_m in the resolution:

$$\frac{\Delta T_m}{\Delta\alpha} = T_m \Big/ \left(2 + \frac{E_i}{RT_m} \right).$$ (51)

B. Experimental System

1. Components of the Experimental System and Their Relationships

The experimental system consists of six parts: (1) a reduction cell, (2) a gas feed and monitoring system, (3) a hydrogen sulfide detector, (4) an XY recorder, (5) an integrator, and (6) a temperature programmer. The connections among the various units are described in Fig. 5.

A sample of coal is placed in the reduction cell with a solvent, reducing agent, and a catalyst. The cell is swept with a constant flow of gas which carries the H_2S that evolves from the cell into the detector. The detector produces a signal that is proportional to the concentration

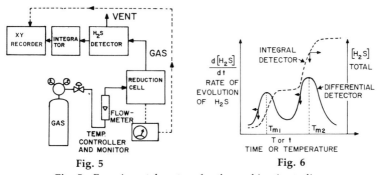

Fig. 5   Fig. 6

Fig. 5 Experimental system for thermokinetic studies.

Fig. 6 Analysis of a mixture containing two groups using differential and integral detectors.

of H_2S in the gas. The temperature of the cell is programmed up and monitored. As shown in Fig. 6; the H_2S signal from the detector is integrated and plotted versus the cell temperature. The location of the maximum of each peak of the signal of $d[H_2S]/dT$ versus T is used to identify the functional group reduced, and the area of each peak is proportional to the amount of the ith sulfur group in the sample. Figure 7 shows the kinetogram for the reduction of thianthrene (I).

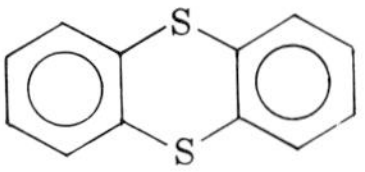

III

The reduction of the first sulfur proceeds as if it were an *aryl* thiol and the second as if it were an aryl sulfide. The corresponding process could be:

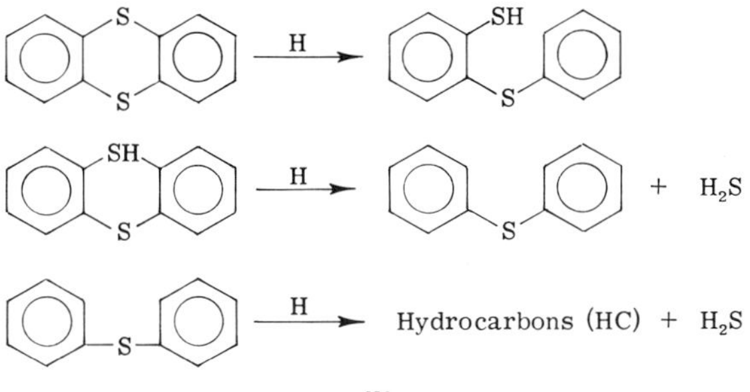

$$\text{(structure)} \xrightarrow{\text{H}} \text{(structure)} \qquad \text{fast} \qquad (52)$$

$$\text{(structure)} \xrightarrow{\text{H}} \text{(structure)} + H_2S \quad \text{first peak} \qquad (53)$$

$$\text{(structure)} \xrightarrow{\text{H}} \text{Hydrocarbons (HC)} + H_2S \quad \text{second peak} \qquad (54)$$

IV

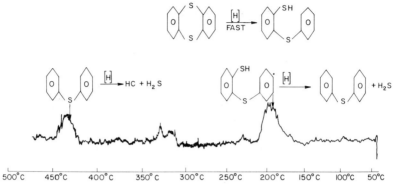

Fig. 7   Kinetogram of Thianthrene.

where the peaks are numbered according to increasing temperature from right to left in Fig. 7.

2.   Experimental Conditions

Most of the results listed below were derived using the following mixture:

(1)   Coal 20–120 mg, according to the sulfur content. When $S_{tot} \cong$ 4% 20 mg, and when $S_{tot} = 0.5\%$ 120 mg.

(2)   Solvent: Pyrogallol 0.250 g
                Resorcinol 0.125 g
                Tetralin 0.200 g

(3)   Catalyst: 1–2 mg sulfidized CoMo. (Harshaw 0402)

(4)   Rate of flow of nitrogen: 200 cm³/min

(5)   Rate of temperature programming: 10°C/min

Polymers were synthesized with well-characterized sulfur functional groups in them. The reduction of the polymers under the previously mentioned conditions gave the values of the peak temperature for each group. Poor cell design may reduce the apparent resolution, however, the following T_{mi} values may serve as approximate values for the peak temperatures: Aliphatic thiols, 160°–180°C; disulfides, 200°–220°C; aromatic thiols and thiolates, 220°–250°C; aliphatic sulfides, 240°–280°C. The latter are poorly resolved from the iron pyrite peak that usually contributes about 1.5% of the ASTM pyritic sulfur. Alicyclic sulfides are at 290°–330°C, aryl sulfides 450°–470°C, and thiophenes 500°–550°C. About 95–99% of the organic sulfur in the model polymers has been accounted for using our thermokinetic method.

TABLE V   *Sulfur Class Distribution in Five Coals*

Coal	Denote	Total S (wt%)	Pyritic S (wt%)	Sulfatic S (wt%)	Organic S (wt%)
Illinois (No. 6), Montrey	A	4.5	1.23	0.06	3.2
Kentucky (No. 9+14)	B	6.6	5.05	0.135	1.43
Martinka, Lower Kittaning, W.Va.	C	2.20	1.48	0.12	0.60
Westland, Pittsburgh Seam, Pa.	D	2.60	1.05	0.07	1.48
Texas lignite, Wilcox Co.	E	1.20	0.4	—	0.80

A warning may be in place here: *all work* with the *hot mixture* of solvent and reducing agent should be carried out *in a hood*. The fumes from the reduction cell are carcinogenic and may cause blood poisoning. Pyrogallol should be stored under dry pure nitrogen, and a new bottle should be opened and used every 2–3 months. Attar and Dupuis (1978a, 1979) described the method and presented results using this method of analysis. Some of the implications of their work are presented in the following section.

C. Implications of the Method for Understanding the Functional Structure of Sulfur in Coal

Interesting information can be obtained by applying the analysis method to coal samples and to treated coal samples. Table V shows the classes of sulfur in five coal samples, and Table VI shows the distributions of the organic sulfur groups in the same coals. The main

TABLE VI   *Distribution of Organic Sulfur Groups in Five Coals*[a]

Coal	% Organic S accounted for	Thiolic	Thiophenolic	Aliphatic sulfide	Aryl sulfide	Thiophenes[r]
A	11	7	15	18	2	58
B	46.5	18	6	17	4	55
C	81	10	25	25	8.5	21.5
D	97.5	30	30	25.5	—	14.5
E	99.5	6.5	21	17	24	31.5

[a] Percent of organic sulfur.

[b] Corrected for "unaccounted for" sulfur; C and E are calculated based on total sulfur content.

conclusions from tests on raw coals are:

(1) The majority of the organic sulfur in high-ranked coals, i.e., LVB, is thiophenic while in low-ranked coals, i.e., lignites, most of the organic sulfur is thiolic or sulfidic.

(2) 18–25% of the organic sulfur is in the form of aliphatic sulfides in all coals.

Current theories on coal structure presume that during coalification the organic structure of coal condenses and becomes more aromatic. The data on the organic sulfur suggested that during coalification, the sulfur groups also condense and become more aromatic. Therefore, it is plausible to assume that during coalification, the organic sulfur transforms according to

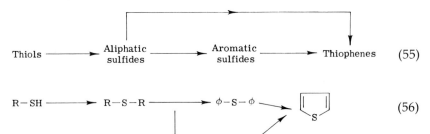

$$\text{Thiols} \longrightarrow \begin{array}{c}\text{Aliphatic}\\ \text{sulfides}\end{array} \longrightarrow \begin{array}{c}\text{Aromatic}\\ \text{sulfides}\end{array} \longrightarrow \text{Thiophenes} \qquad (55)$$

$$R{-}SH \longrightarrow R{-}S{-}R \longrightarrow \phi{-}S{-}\phi \qquad (56)$$

Since the aliphatic sulfides are an intermediate in the sequence, it is not surprising that their abundance reaches a pseudo steady state or a virtually constant value relative to the other forms of organic sulfur.

Figure 8 shows the kinetogram of a sample of raw Illinois No. 6 and the kinetogram of the same material after extraction with HCl. HCl extraction removes sulfates and dissolves the calcium and magnesium salts. The kinetograms show that the second peak essentially disappears as a result of the HCl extraction. Therefore, it is believed that some of the thiolic sulfur is bound in coal as calcium thiolates. Treatment of coal with CH_3I reduces the rate of diffusion in the coal and produces aryl sulfides from the aryl thiols. The concentration of aliphatic thiols remains essentially the same, which indicates that under the conditions used, CH_3I did not methylate the aliphatic thiols.

D.   Implications of the Method for Coal Reserve Evaluation

Most of the desulfurization processes under consideration remove the pyritic sulfur but have very inconsistent effect on the organic sulfur.

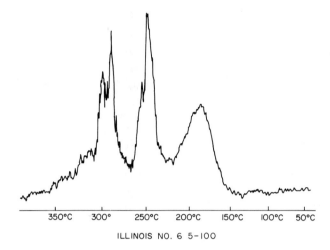

ILLINOIS NO. 6 5-100

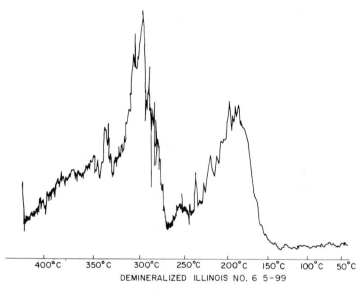

DEMINERALIZED ILLINOIS NO. 6 5-99

Fig. 8 Kinetogram of an Illinois No. 6 coal before and after demineralization with HCl.

The reason is that although the total organic sulfur in different coals could be the same, its distribution among different organic sulfur functional groups might be different. Since the rate of reaction of each group is different, it should not be surprising that desulfurization of the organic sulfur, as measured by the decrease in the *total* organic sulfur of different coals, will be inconsistent.

We have limited data that are consistent with the thesis that *all* desulfurization processes, excluding intensive liquefaction, can remove a fraction of the organic sulfur that does not exceed the thiolic sulfur fraction. The implication of these observations is that coals with a high content of organic sulfur can be desulfurized to the extent equivalent to the content of thiolic sulfur in them. Therefore, coals with a large content of organic sulfur can be divided into two groups: these in which most of the organic sulfur is thiolic, and can therefore be easily desulfurized, and those in which most of the organic sulfur is not thiolic and cannot be easily desulfurized. The thermokinetic method gives an easy test that can differentiate between these two types of coals.

E. Implications of the Method for Process Development and Evaluation

Different desulfurization processes use different reagents and apply them under different conditions. Since the rates of reaction of individual sulfur groups are different, one can expect that different processes will have different effects on the individual sulfur groups. Most desulfurization processes will remove some or all the thiolic sulfur. Therefore, the difference among different processes will be due to:

(1) the degree to which they can remove some of the aliphatic sulfides and

(2) the degree of resulfurization that can occur. "Resulfurization" is used to denote all the processes in which sulfur can react back and therefore be retained with the products instead of in a separate by-product stream.

If coal samples are taken during desulfurization, e.g., at different reaction times, and the distribution of sulfur groups in them is determined, then the results can be used to define the optimal processing conditions in which minimum resulfurization and maximum desulfurization of the sulfides occurs.

REFERENCES

Aitken, J., Heeps, T., and Steedman, W. (1968). *Fuel* **47**, 353–357.
Akhtar, S., Sharkery, A. G., Jr., Shultz, J. L., and Yavorsky, P. U. (1974). *Am. Chem. Soc., 167th Natl. Meet., Los Angeles, Calif.*
ASTM (1973). "1973 Annual Book of ASTM Standards. Part 19: 380." Philadelphia, Pennsylvania.
Ashworth, M. R. F. (1972). "The Determination of Sulphur-Containing Groups. Vol. I: Sulfoxides Sulfones."

Ashworth, M. R. F. (1976). "The Determination of Sulphur-Containing Groups. Vol. II: Thiols."

Ashworth, M. R. F. (1977). "The Determination of Sulphur-Containing Groups. Vol. III: Sulfides and Disulfides."

Attar, A. (1978). *Fuel* **57**(4), 201-212.

Attar, A., Corcoran, W. H., and Gibson, G. S., (1976). *Prepr. Div. Fuel Chem., Am. Chem. Soc.*, **21**(7), 106-117.

Attar, A., and Corcoran, W. H. (1977). *Ind. Eng. Chem., Prod. Res. Dev.* **16**(2), 168-170.

Attar, A., and Corcoran, W. H. (1978). *Ind. Eng. Chem., Prod. Res. Dev.* **17**(2), 102-109.

Attar, A., and Dupuis, F. (1978a). *Prepr. Div. Fuel Chem., Am. Chem. Soc.* **23**(2), 44-50.

Attar, A., and Dupuis, F. (1978b). *Prepr. Div. Fuel Chem., Am. Chem. Soc.* **23**(1), 214-227.

Attar, A., and Dupuis, F. (1979). *Prepr. Div. Fuel Chem., Am. Chem. Soc.*, **24**(1), 166-177.

Attar, A., and Messenger, L. (1978). Unpublished data.

Attari, A., Igielski, T. P., and Jaselskis, B. (1970). *Anal. Chem.* **42**, 1282-1285.

Blayden, H. E., and Partick, J. W. (1970). *Fuel* **49**, 257-270.

Bogdanova, V. A., and Boranski, A. O. (1961). *Kratk. Soobshch. Nauchno-Issled. Rabotakh* 2a, sb 68 [*Chem. Abstr.* **63**, 1774 le (1965)].

British Standards Institution (1975). British Standard 1016, Part 2.

Brown, H. R., Burns, M. S., Durie, R. A., and Swaine, D. J. (1964). *Fuel* **43**, 409-413.

Cerni-Simic, S. (1962). *Fuel* **41**, 141-151.

Chamberland, A. M., and Ganthier, J. M. (1977). *Atmos. Environ.* **11**, 257-261.

Darlage, L. J., Weidner, J. P., and Block, S. S. (1974). *Fuel* **53**, 53-59.

Deurbrouck, A. W. (1972). "Sulfur Reduction Potential of the Coals of the United States," R.I. 7633. U.S. Dep. Inter., Washington, D.C.

Eaton, S. E., Hyde, R. W., and Old, B. S. (1948). *Am. Inst. Min., Metall. Eng., Iron Steel Div., Met. Technol.* **15**(7), 343-363 (*Tech. Bull.* No. 2453).

Edwards, A. H., Jones, M. S., Durie, R. A., and Swaine, D. J. (1964). *Fuel* **43**, 55-62.

El-Koddah, N., and Ezz, S. Y. (1973). *Fuel* **52**, 128-129.

Estep, P. A., Kovach, J. J., and Karr, C., Jr. (1968). *Anal. Chem.* **40**(2), 358-363.

Federal Regulations (1971). **36**, 22384, (Nov. 25),

Fisher Scientific Company (1978). Sulfur Reporter, Bull. #470.

Friedel, R. A., and Carlson, G. L. (1972). *Fuel* **51**, 194-198.

Fuchs, W. (1951). *Brennsst.-Chem.* **32**, 274-276.

Garber, R. W., and Wilson, C. E. (1972). *Anal. Chem.* **44**, 1357-1360.

Georgiodis, C., and Gaillord, G. (1954). *C. R. Acad. Sci.* **238**, 355.

Given, P. H., and Jones, J. R. (1966). *Fuel* **45**, 151-158.

Gluskoter, H. J. (1975). *Prepr. Div. Fuel Chem., Am. Chem. Soc.* **20**(2), 94-98.

Gluskoter, H. J. (1978). *Short Course Sulfur Fossil Fuels, Houston, Tex.* To be published.

Gluskoter, H. J., and Simon, J. A. (1968). *Ill. State Geol. Surv., Circ.* **432**.

Gluskoter, H. J., Ruch, R. R., Miller, W. G., Cahill, R. A., Dreher, G. B., and Kuhn, J. K. (1977). *Ill. State Geol. Surv., Circ.* **499**.

Holmes, J. A. (1911). *U.S. Bur. Mines, Tech. Pap.* No. 1.

Howard, H. C. (1963). *In* "Chemistry of Coal Utilization: Supplementary Volume" (H. H. Lowry, ed.), pp. 340-394. Wiley, New York.

Hurley, R. G., and White, E. W. (1974). *Anal. Chem.* **46**, 2234-2237.

International Organization for Standardization (1960). R157.

Jacobs, M. B. (1967). "The Analytical Toxicology of Industrial Inorganic Poisons." Wiley (Interscience), New York.

James, R. G., and Severn, M. I. (1967). *Fuel* **46**, 476-478.

Jenkins, R. J. (1977). *EPRI Conf. Coal Liquefact. Res., Palo Alto, Calif.* pp. 252-275.

Karchmer, J. H., ed. (1970). "The Analytical Chemistry of Sulfur and Its Compounds," Vol. I. 00, 00

Karchmer, J. H., ed. (1971). "The Analytical Chemistry of Sulfur and Its Compounds," Vol. III. 00, 00. (Contains only NMR data.)

Karchmer, J. H., ed. (1972). "The Analytical Chemistry of Sulfur and Its Compounds," Vol. II. 00, 00

Keller, G. E., Aresco, S. J., and Visman, J. (1968). *In* "Coal Preparations" (J. W. Leonard and D. R. Mitchell, eds.), chapter 2. Am. Inst. Min., Met. Pet. Eng., New York.

Kuhn, J. K., Kohlenberger, L. B., and Shimp, N. F. (1973). *Ill. State Geol. Surv., Environ. Geol. Note* No. 66, p. 11.

Leco Corporation (1974). "Instruction for Analysis of Sulfur in Hydrocarbons by the High Frequency Combustion Titration Procedure." Leco Corp., St. Joseph, Michigan.

Maa, P. S., Lewis, C. R., and Hamrin, C. E., Jr. (1975). *Fuel* **54**, 62-69.

Minra, Y., and Yanagi, Y. (1963). *Nenryo Kyokai-Shi* **42**, 21-24 [*Chem. Abstr.* **61**, 10501 (1964)].

Mott, R. A., and Wilkinson, H. C. (1956). *Fuel* **35**, 6-18.

Muder, R. E. (1963). *In* "Chemistry of Coal Utilization: Supplementary Volume" (H. H. Lowry, ed.), pp. 629-675. Wiley, New York.

Paris, B. (1977). *Prepr. Div. Fuel Chem., Am. Chem. Soc.* **22**(5), 1-9.

Paulson, R. F. (1975). *Prepr. Div. Fuel Chem., Am. Chem. Soc.* **20**(2), 183-197.

Peet, N. J., Simeon, S. R., and Stott, J. B. (1969). *Fuel* **48**, 259-265.

Pollack, S. S. (1971). *Fuel* **50**, 453-454.

Postovski, J. J., and Harlampovich, A. B. (1936). *Fuel* **15**, 229-232.

Powell, A. R. (1920). *Ind. Eng. Chem.* **12**, 887-890, 1069-1087.

Powell, A. R. (1923). *J. Am. Chem. Soc.* **45**, 1-15.

Prilezhaeva, E. N., Fedorovskaya, N. P., Miesserova, L. V., Domanina, O., and Khaskina, I. M. (1963). *Tr. Inst. Goryuch. Iskop., Moscow* **21**, 202 [*Chem. Abstr.* **60**, 6217h (1964)].

Robertson, H. W., and Steedman, W. (1966). *Fuel* **45**, 375-379.

Rodionova, L. E., and Barauskii, A. O. (1970). *Izv. Nauchno-Issled. Inst. Nefte- Uglekhim. Sint. Irkutsk. Univ.* **12**, 93 [*Chem. Abstr.* **75**, 5122 (1971)].

Schehl, R. R., and Friedel, R. A. (1973). *U.S. Bur. Mines, Tech. Prog. Rep.* No. 71.

Scott, C. L., and Steedman, W. (1972). *Fuel* **51**, 10-13.

Selvig, W. A., and Fieldner, A. C. (1927). *Ind. Eng. Chem.* **29**, 729-733.

Shimp, N. F., Helfinstine, R. J., and Kuhn, J. K. (1975). *Prepr. Div. Fuel Chem., Am. Chem. Soc.* **20**(2), 99-107.

Solomon, P. R., and Manzione, A. V. (1977). *Fuel* **56**, 393-396.

Speight, J. G. (1971). *Appl. Spectrosc. Rev.* **5**(2), 211-264.

Sutherland, J. K. (1975). *Fuel* **54**, 132.

Van Krevelen, D. W. (1961). "Coal," Elsevier, Amsterdam.

Veretennekova, I. V., and Petrov, A. A. (1964). *In* "Chemistry of Organic Sulfur Compounds in Petroleum and Petroleum Products" (R. O. Obolensev, ed.), Vol. 6, p. 133. (Israel Program Sci. Transl., Jerusalem, 1967.)

Wallace, L. D., Kohlenberger, D. W., Jones, R. J., Moore, R. T., Riddle, M. E., and McNulty, J. A. (1970). *Anal. Chem.* **42**(3), 387-394.

West, P. W., and Gaeke, G. C. (1956). *Anal. Chem.* **28**, 1816-1819.

Whelan, P. F. (1954). *J. Inst. Fuel* **27**, 455-464.

Wibaut, J. P. (1919). *Recl. Trav. Chim. Pays-Bas Belg.* **38**, 159-162.

Winkler, H. E., and Syty, A. (1976). *Environ. Sci. Technol.* **10**(9), 913-916.

Young, R. K., and Zawadzki, E. A. (1967). *Fuel* **46**, 151-152.

Index

625